Industrial Robotics Fundamentals

Theory and Applications

Third Edition

by

Larry T. Ross

Department Chair—Department of Applied Engineering and Technology
Eastern Kentucky University
Richmond, KY

Stephen W. Fardo

Foundation Professor Emeritus—Department of Applied Engineering and Technology
Eastern Kentucky University
Richmond, KY

Michael F. Walach

Assistant Professor—Career and Technical Education
Eastern Kentucky University
Richmond, KY

Publisher
The Goodheart-Willcox Company, Inc.
Tinley Park, IL
www.g-w.com

Manufactured in the United States of America.

Library of Congress Catalog Card Number 2016048375

ISBN 978-1-63126-941-7

2 3 4 5 6 7 8 9 – 18 – 22 21 20 19 18

The Goodheart-Willcox Company, Inc. Brand Disclaimer: Brand names, company names, and illustrations for products and services included in this text are provided for educational purposes only and do not represent or imply endorsement or recommendation by the author or the publisher.

The Goodheart-Willcox Company, Inc. Safety Notice: The reader is expressly advised to carefully read, understand, and apply all safety precautions and warnings described in this book or that might also be indicated in undertaking the activities and exercises described herein to minimize risk of personal injury or injury to others. Common sense and good judgment should also be exercised and applied to help avoid all potential hazards. The reader should always refer to the appropriate manufacturer's technical information, directions, and recommendations; then proceed with care to follow specific equipment operating instructions. The reader should understand these notices and cautions are not exhaustive.

The publisher makes no warranty or representation whatsoever, either expressed or implied, including but not limited to equipment, procedures, and applications described or referred to herein, their quality, performance, merchantability, or fitness for a particular purpose. The publisher assumes no responsibility for any changes, errors, or omissions in this book. The publisher specifically disclaims any liability whatsoever, including any direct, indirect, incidental, consequential, special, or exemplary damages resulting, in whole or in part, from the reader's use or reliance upon the information, instructions, procedures, warnings, cautions, applications, or other matter contained in this book. The publisher assumes no responsibility for the activities of the reader.

The Goodheart-Willcox Company, Inc. Internet Disclaimer: The Internet resources and listings in this Goodheart-Willcox Publisher product are provided solely as a convenience to you. These resources and listings were reviewed at the time of publication to provide you with accurate, safe, and appropriate information. Goodheart-Willcox Publisher has no control over the referenced websites and, due to the dynamic nature of the Internet, is not responsible or liable for the content, products, or performance of links to other websites or resources. Goodheart-Willcox Publisher makes no representation, either expressed or implied, regarding the content of these websites, and such references do not constitute an endorsement or recommendation of the information or content presented. It is your responsibility to take all protective measures to guard against inappropriate content, viruses, or other destructive elements.

Cover image: megastocker/Shutterstock.com; **Back cover image:** SvedOliver/Shutterstock.com; **Careers in Robotics Photo:** megastocker/Shutterstock.com; **Unit Opening Photos:** Baloncici/Shutterstock.com (yellow production robot); CHAINFOTO24/Shutterstock.com (blue robot); **Chapter Opening Photos:** PP77LSK/Shutterstock.com (white grappler); Mopic/Shutterstock.com (orange robot); **Chapter 1 Opening Photo:** Zbynek Jirousek/Shutterstock.com; **Chapters 2, 4, 5 Opening Photos:** asharkyu/Shutterstock.com; **Chapters 3, 7, 9, 14 Opening Photos:** Vladimir Nenezic/Shutterstock.com; **Chapter 6 Opening Photo:** Maksim Dubinsky/Shutterstock.com; **Chapter 8 Opening Photo:** Baloncici/Shutterstock.com; **Chapter 10 Opening Photo:** Photorelease Stuttgart/Shutterstock.com; **Chapter 11 Opening Photo:** MOLPIX/Shutterstock.com; **Chapter 12 Opening Photo:** SvedOliver/Shutterstock.com; **Chapter 13 Opening Photo:** Comaniciu Dan/Shutterstock.com; **Chapter 15 Opening Photo:** science photo/Shutterstock.com

Library of Congress Cataloging-in-Publication Data
Names: Ross, Larry T. | Fardo, Stephen W. | Walach, Michael F.
Title: Industrial robotics fundamentals : theory and applications / by Larry T. Ross, Stephen W. Fardo, Michael F. Walach.
Description: Third edition. | Tinley Park, IL : The Goodheart-Willcox Company, Inc., [2018] | Series: Robotics theory and industrial applications | Includes index.
Identifiers: LCCN 2016048375 | ISBN 9781631269417
Subjects: LCSH: Robotics. | Robots, Industrial.
Classification: LCC TJ211 .R684 2018 | DDC 629.8/92--dc23 LC record available at http://lccn.loc.gov/2016048375

Preface

Industrial Robotics Fundamentals: Theory and Applications is an introductory text that explores many aspects of robotics in a basic and easy-to-understand manner. The key concepts are discussed using a "big picture" or systems approach that greatly enhances student learning. Many applications and operational aspects of equipment and robotic systems are discussed.

We continually considered the needs of both students and instructors while preparing this comprehensive text. Several additions have been made to this edition to make this text even more comprehensive, including:

- **Updated content.** The content of the text has been revised and reorganized, with new topics added to reflect changes in the field of robotics technology. The text has been continually updated through quality checks to ensure technical accuracy.
- **Chapter summary and analysis.** Each chapter ends with a summary of the chapter content. Following the Review Questions and Learning Extensions, a series of Advanced Analysis questions have been added to help students think critically about chapter content.
- **Activities.** Hands-on activities have been added to the end of many chapters to give students the opportunity to apply chapter content to real-world situations.

The development of the computer created what some experts have called the "Second Industrial Revolution." Many consider robots to be the prime movers of this revolution. It is more important than ever to have more than a basic knowledge of robots and robotics technology.

This text is a comprehensive approach to learning the technical aspects of robotics. It is divided into four units, covering broad areas of robotic principles, power supplies and movement systems, sensing and end-of-arm tooling, and control systems.

Unit I is devoted to the basic principles of robotic technology. Chapter 1 discusses the development and use of industrial robotics in the area of automation. Chapter 2 prepares a solid foundation for understanding the characteristics and fundamentals of robotics, including basics components and operation. Chapter 3 provides a general safety overview for use in any laboratory setting. Chapter 4 provides an overview of programming languages and techniques used to program industrial robots. Chapter 5 explores the many applications for industrial robots.

Unit II addresses robotic power supplies and movement systems. Chapter 6 provides an overview of the electromechanical systems used with robots. Chapter 7 presents fluid power systems.

Unit III presents robotic sensing systems and end-of-arm tooling. Chapter 8 discusses the various sensors commonly used by robots to gain information about the external environment. Chapter 9 provides information about various end effectors and tools used to move workpieces from one location to another within a robot's work envelope.

Unit IV covers robotic control systems, basic maintenance, and implementation planning. Chapter 10 presents the basics of digital electronics, which includes information on microcomputers and microprocessors. Chapter 11 explains how the robot controller communicates with peripheral equipment found in robotic workcells, including vision systems. Chapter 12 provides an overview of maintenance procedures. Chapter 13 discusses some of the major factors to consider when using robotic systems in an industrial environment. Chapter 14 presents the use of robots outside the factory, artificial intelligence and expert systems, and suggested coursework and training related to the field of robotics. Chapter 15 helps give examples of how to build smaller robots.

About the Authors

Tim Ross is a Professor and Chair of Applied Engineering and Technology at Eastern Kentucky University, where he has taught for 18 years. He earned his doctorate in Vocational and Industrial Education from the University of Kentucky, a master of science in Industrial Professional Technology (Manufacturing Concentration) from Indiana State University, and a bachelor of science in Industrial Technology (Computer Integrated Manufacturing) from Western Kentucky University. In addition to his time at Eastern Kentucky, Dr. Ross has also had seven years of teaching experience in secondary and postsecondary systems. Tim currently teaches courses with a focus on automation, industrial applications, and career and technical education. He is also involved in conducting industry training in automation and fluid power. Dr. Ross has participated in the Association for Technology, Management, and Applied Engineering (ATMAE); International Technology and Engineering Educator Association (ITEEA); and Kentucky Association for Career and Technical Education.

Stephen W. (Steve) Fardo is Foundation Professor Emeritus of Applied Engineering and Technology at Eastern Kentucky University. He earned his doctorate in Technical Education Leadership (EPDA Fellow) from the University of Kentucky, a master of science in Industrial Technology, and a bachelor of science in Industrial Education (Electronics) from Eastern Kentucky University. Dr. Fardo currently serves as a consultant/advisor for technical teachers and school systems through the Department of Applied Engineering and Technology at Eastern Kentucky and the Office of Career and Technical Education in the state of Kentucky. He previously taught and advised for 35 years at Eastern Kentucky in the Electronics and Career and Technical Education programs, teaching over 25 different courses. Dr. Fardo also served with the US Army Corps of Engineers (Reserve) for 28 years in various Training and Evaluation Command positions and retired as a Lieutenant Colonel. Dr. Fardo has participated in the International Technology and Engineering Education Association (ITEEA); Association for Technology, Management, and Applied Engineering (ATMAE); Mississippi Valley Technical Teacher Education Conference; and Kentucky Association for Career and Technical Education. He coauthored (with Dale R. Patrick) over 25 textbook and laboratory manual titles for several publishers during the evolution of postsecondary electronics programs in the 1970s, 1980s, and 1990s. He has been honored to receive several recognitions, including EKU Foundation Professorship, Kentucky Career and Technical Education Service Award, and Pendleton County High School (Kentucky) Wall of Fame.

Michael F. Walach is an Assistant Professor of Career and Technical Education at Eastern Kentucky University. He earned his doctorate in education from the University of Rhode Island, a master of arts in teaching from Sacred Heart University, a bachelor of science in technology education from Rhode Island College, and an associate of science in electronics from the Community College of Rhode Island. Dr. Walach currently teaches courses in the Department of Applied Engineering and Technology at Eastern Kentucky University and previously taught for 14 years in the public school system as an Engineering and Technology teacher. Dr. Walach also worked in the industrial x-ray industry for four years as a field service technician and applications engineer. He is the president of the Kentucky Technology and Engineering Education Association and Kentucky state representative for the International Technology and Engineering Educator Association (ITEEA). Dr. Walach presents annually at national and state conferences on engineering and technology topics, and has published work in the Journal of Technology Education. Dr. Walach was the 2013 ITEEA Rhode Island technology teacher of the year and the 2013 ITEEA distinguished program award recipient.

Reviewers

The author and publisher wish to thank the following industry and teaching professionals for their valuable input into the development of ***Industrial Robotics Fundamentals: Theory and Applications***.

William Chapman
Arcadia High School
Arcadia, CA

Frank Claude
Dunwoody College
Minneapolis, MN

James Mansfield
Parkland College
Champaign, IL

Stephen B. Marsh
Cabrillo High School
Long Beach, CA

Elizabeth Simpson
Greco Middle School
Temple Terrace, FL

Mike Wilson
McLane MS Robotics Engineering
Academy
Brandon, FL

Michael Zhang
Fanshawe College
London, Ontario

Acknowledgments

The author and publisher would like to thank the following companies, organizations, and individuals for their contribution of resource material, images, or other support in the development of ***Industrial Robotics Fundamentals: Theory and Applications***.

ABB Graco; Motoman

ABB Robotics

Accuray Incorporated

Adept Technology, Inc.

American Honda Motor Co., Inc.

Anaheim Automation, Inc.

Carnegie Mellon University

Cisco-Eagle

Daifuku Co., Ltd.

De-STA-Co

Edison Robot

Educational Robot Company

FANUC Robotics

florin operea

Grabit, Inc.

Infrared Cameras, Inc.

International Federation of Robotics

iRobot Corporation

Knightscope, Inc.

Mack Corporation

ME Labs, Inc.

MobileRobots, Inc.

Motoman

NEC Corporation

Northrop Grumman Corporation

Omron Adept Technologies, Inc.

Pacific Robotics, Inc.

PHD, Inc.

Pololu Robotics

PRI-Precision Robots, Inc.

Reis Machines, Inc.

Schunk

SCHUNK GmbH & Co. KG

SCHUNK Intec Inc.

Schunk-USA

Segway, Inc.

Servo-Robot Inc.

Sparkfun Electronics, Inc.

Sterling Detroit Company

Superior Electric Co.

Techno, Inc.

Tektronix, Inc.

University of Pennsylvania GRASP Lab, Vijay Kumar, Daniel Mellinger, and Alex Kushleyev

VEX Robotics, Inc.

Yamaha

G-W Integrated Learning Solution

Together, We Build Careers

At Goodheart-Willcox, we take our mission seriously. Since 1921, G-W has been serving the career and technical education (CTE) community. Our employee-owners are driven to deliver exceptional learning solutions to CTE students to help prepare them for careers. Our authors and subject matter experts have years of experience in the classroom and industry. We combine their wisdom with our expertise to create content and tools to help students achieve success. Our products start with theory and applied content based upon a strong foundation of accepted standards and curriculum. To that base, we add student-focused learning features and tools designed to help students make connections between knowledge and skills. G-W recognizes the crucial role instructors play in preparing students for careers. We support educators' efforts by providing time-saving tools that help them plan, present, assess, and engage students with traditional and digital activities and assets. We provide an entire program of learning in a variety of print, digital, and online formats, including economic bundles, allowing educators to select the right mix for their classroom.

Student-Focused Curated Content

Goodheart-Willcox believes that student-focused content should be built from standards and/or accepted curriculum coverage. ***Industrial Robotics Fundamentals*** uses a systems approach. It is composed of four descriptive units of chapters. We call on industry experts and teachers from across the country to review and comment on our content, presentation, and pedagogy. Finally, in our refinement of curated content, our editors are immersed in content checking, securing and sometimes creating figures that convey key information, and revising language and pedagogy.

Features of the Textbook

Features are student-focused learning tools designed to help you get the most out of your studies. This visual guide highlights the features designed for the textbook.

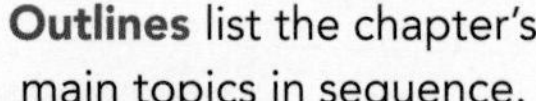

Outlines list the chapter's main topics in sequence.

Objectives clearly identify the knowledge and skills to be obtained when the chapter is completed.

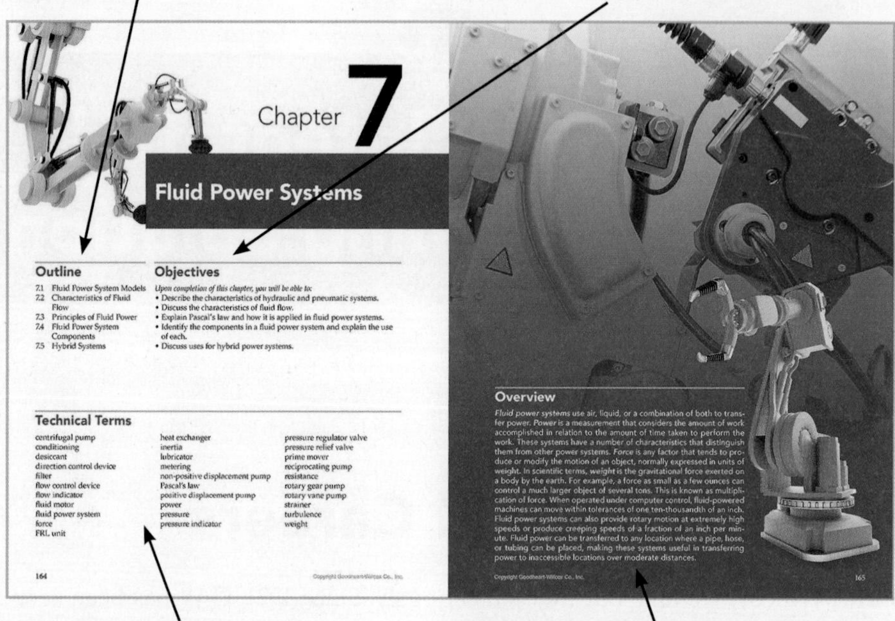

Chapter 7

Fluid Power Systems

Outline

7.1 Fluid Power System Models
7.2 Characteristics of Fluid Flow
7.3 Principles of Fluid Power
7.4 Fluid Power System Components
7.5 Hybrid Systems

Objectives

Upon completion of this chapter, you will be able to:
- Describe the characteristics of hydraulic and pneumatic systems.
- Discuss the characteristics of fluid flow.
- Explain Pascal's law and how it is applied in fluid power systems.
- Identify the components in a fluid power system and explain the use of each.
- Discuss uses for hybrid power systems.

Technical Terms

centrifugal pump	heat exchanger	pressure regulator valve
conditioning	inertia	pressure relief valve
desiccant	lubricator	prime mover
direction control device	metering	reciprocating pump
filter	non-positive displacement pump	resistance
flow control device	Pascal's law	rotary gear pump
flow indicator	positive displacement pump	rotary vane pump
fluid motor	power	strainer
fluid power system	pressure	turbulence
force	pressure indicator	weight
FRL unit		

164 Copyright Goodheart-Willcox Co., Inc.

Overview

Fluid power systems use air, liquid, or a combination of both to transfer power. *Power* is a measurement that considers the amount of work accomplished in relation to the amount of time taken to perform the work. These systems have a number of characteristics that distinguish them from other power systems. *Force* is any factor that tends to produce or modify the motion of an object, normally expressed in units of weight. In scientific terms, *weight* is the gravitational force exerted on a body by the earth. For example, a force as small as a few ounces can control a much larger object of several tons. This is known as multiplication of force. When operated under computer control, fluid-powered machines can move within tolerances of one ten-thousandth of an inch. Fluid power systems can also provide rotary motion at extremely high speeds or produce creeping speeds of a fraction of an inch per minute. Fluid power can be transferred to any location where a pipe, hose, or tubing can be placed, making these systems useful in transferring power to inaccessible locations over moderate distances.

Copyright Goodheart-Willcox Co., Inc. 165

Technical Terms list the key terms to be learned in the chapter.

Overview provides an introduction and preview of the chapter content.

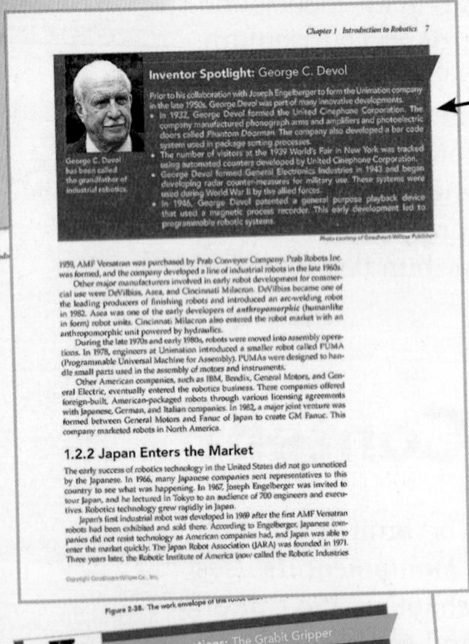

Chapter 1 Introduction to Robotics 7

Inventor Spotlight: George C. Devol

Prior to his collaboration with Joseph Engelberger to form the Unimation company in the late 1950s, George Devol was part of many innovative developments.
- In 1932, George Devol formed the United Cinephone Corporation. The company manufactured phonograph arms and amplifiers and photoelectric doors called *Phantom Doorman*. The company also developed a bar code system used in package sorting processes.
- The number of visitors at the 1939 World's Fair in New York was tracked using automated counters developed by United Cinephone Corporation.
- George Devol formed General Electronics Industries in 1943 and began developing radar counter-measures for military use. These systems were used during World War II by the allied forces.
- In 1946, George Devol patented a general purpose playback device that used a magnetic process recorder. This early development led to programmable robotic systems.

George C. Devol has been called the grandfather of industrial robotics.

Photo courtesy of Goodheart-Willcox Publisher

1959, AMF Versatran was purchased by Prab Conveyor Company. Prab Robots Inc. was formed, and the company developed a line of industrial robots in the late 1960s.

Other major manufacturers involved in early robot development for commercial use were DeVilbiss, Asea, and Cincinnati Milacron. DeVilbiss became one of the leading producers of finishing robots and introduced an arc-welding robot in 1982. Asea was one of the early developers of ***anthropomorphic*** (humanlike in form) robot units. Cincinnati Milacron also entered the robot market with an anthropomorphic unit powered by hydraulics.

During the late 1970s and early 1980s, robots were moved into assembly operations. In 1978, engineers at Unimation introduced a smaller robot called PUMA (Programmable Universal Machine for Assembly). PUMAs were designed to handle small parts used in the assembly of motors and instruments.

Other American companies, such as IBM, Bendix, General Motors, and General Electric, eventually entered the robotics business. These companies offered foreign-built, American-packaged robots through various licensing agreements with Japanese, German, and Italian companies. In 1982, a major joint venture was formed between General Motors and Fanuc of Japan to create GM Fanuc. This company marketed robots in North America.

1.2.2 Japan Enters the Market

The early success of robotics technology in the United States did not go unnoticed by the Japanese. In 1966, many Japanese companies sent representatives to this country to see what was happening. In 1967, Joseph Engelberger was invited to tour Japan, and he lectured in Tokyo to an audience of 700 engineers and executives. Robotics technology grew rapidly in Japan.

Japan's first industrial robot was developed in 1969 after the first AMF Versatran robots had been exhibited and sold there. According to Engelberger, Japanese companies did not resist technology as American companies had, and Japan was able to enter the market quickly. The Japan Robot Association (JARA) was founded in 1971. Three years later, the Robotic Institute of America (now called the Robotic Industries

Copyright Goodheart-Willcox Co., Inc.

Figure 2-38. The work envelope of this robot

Robotics Innovations: The Grabit Gripper

The Grabit company has developed a robotic gripper that uses static electricity to grip and hold objects. Until now, a typical robot gripper would use pressure to grasp objects, and often a set pressure would be used to grasp all objects. A slight improvement to a gripper that uses a set pressure would be a gripper with sensor feedback that adjusts pressure for different objects. The Grabit gripper is different, however, because it uses static electricity to grasp objects. The Grabit gripper can lift delicate and heavy objects alike without damaging them. Innovations such as this keep robot technology on the cutting edge. Robots will have to work in closer proximity to humans in the future, so they will need to be strong and gentle.

Photo courtesy of Grabit, Inc.

Copyright Goodheart-Willcox Co., Inc.

Inventor Spotlights introduce you to key figures in the development of robotics.

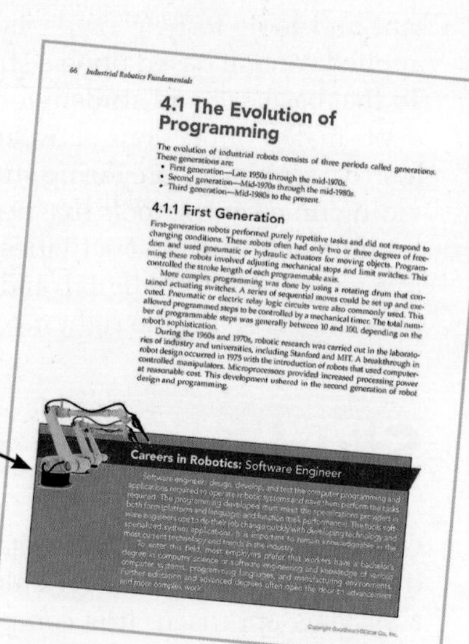

66 *Industrial Robotics Fundamentals*

4.1 The Evolution of Programming

The evolution of industrial robots consists of three periods called generations. These generations are:
- First generation—Late 1950s through the mid-1970s.
- Second generation—Mid-1970s through the mid-1980s.
- Third generation—Mid-1980s to the present.

4.1.1 First Generation

First-generation robots performed purely repetitive tasks and did not respond to changing conditions. These robots often had only two or three degrees of freedom and used pneumatic or hydraulic actuators for moving objects. Programming these robots involved adjusting mechanical stops and limit switches. This controlled the stroke length of each programmable axis.

More complex programming was done by using a rotating drum that contained actuating switches. A series of sequential moves could be set up and executed. Pneumatic or electric relay logic circuits were also commonly used. This allowed programmed steps to be controlled by a mechanical timer. The total number of programmable steps was generally between 10 and 100, depending on the robot's sophistication.

During the 1960s and 1970s, robotic research was carried out in the laboratories of industry and universities, including Stanford and MIT. A breakthrough in robot design occurred in 1975 with the introduction of robots that used computer-controlled manipulators. Microprocessors provided increased processing power at reasonable cost. This development ushered in the second generation of robot design and programming.

Careers in Robotics: Software Engineer

Copyright Goodheart-Willcox Co., Inc.

Careers in Robotics features explain different types of careers in robotics, including the education needed and the tasks related to that career.

Robotics Innovations inform you of current and upcoming robotics technology.

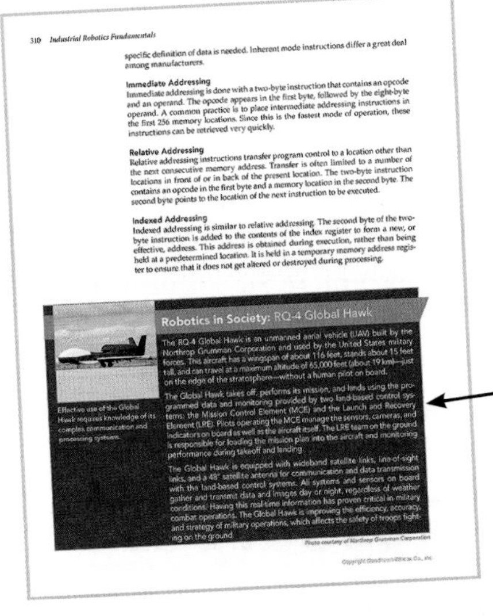

Illustrations have been designed to clearly and simply communicate the specific topic.

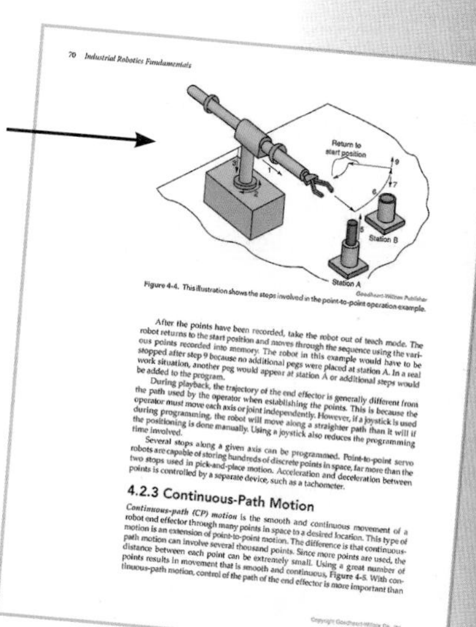

Robotics in Society features show the different ways robotics technology has impacted our world.

Summary feature provides an additional review tool for you and reinforces key learning objectives.

Learning Extensions allow you to research and to apply what you have learned in each chapter to real-world questions.

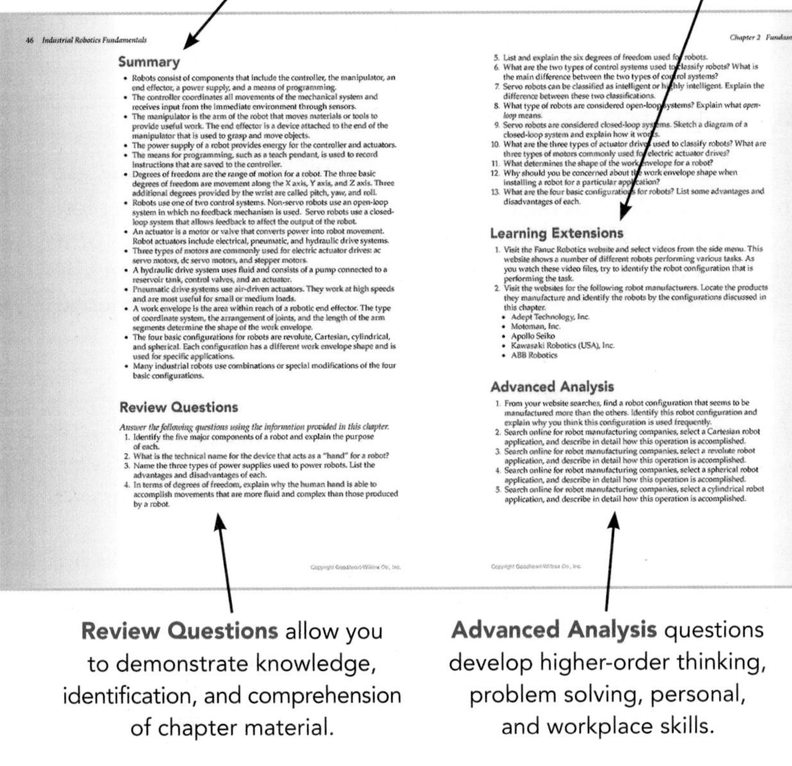

Review Questions allow you to demonstrate knowledge, identification, and comprehension of chapter material.

Advanced Analysis questions develop higher-order thinking, problem solving, personal, and workplace skills.

Activities at the end of select chapters extend your learning and help you analyze and apply knowledge.

Activity 6-3
Control Systems Overview
Objectives
Procedure
SPST switch
Battery
Incandescent lamp
Ammeter
Electrical circuit.

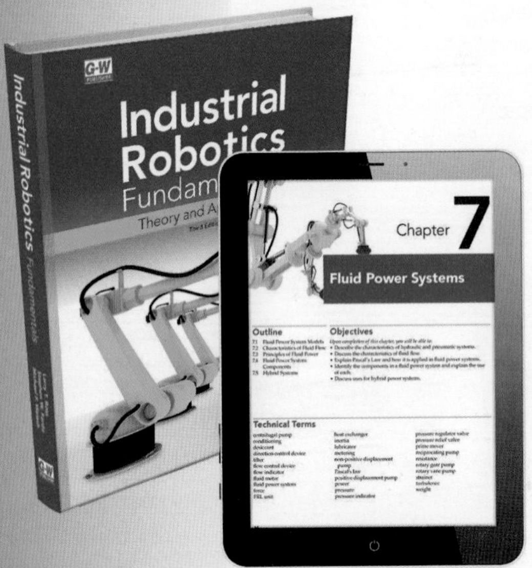

Student Resources

Textbook

The *Industrial Robotics Fundamentals* textbook provides an exciting, full-color, and highly illustrated learning resource. The textbook is available in print or online versions.

Instructor Resources

Instructor resources provide information and tools to support teaching, grading, and planning; class presentations; and assessment.

Instructor's Presentations for PowerPoint®

Help teach and visually reinforce key concepts with prepared lectures. These presentations are designed to allow for customization to meet daily teaching needs. They include objectives, outlines, and images from the textbook.

ExamView® Assessment Suite

Quickly and easily prepare, print, and administer tests with the ExamView® Assessment Suite. With hundreds of questions in the test bank corresponding to each chapter, you can choose which questions to include in each test, create multiple versions of a single test, and automatically generate answer keys. Existing questions may be modified and new questions may be added. You can prepare pretests, formative assessments, and summative assessments easily with the ExamView® Assessment Suite.

Instructor's Resource CD

One resource provides instructors with time-saving preparation tools, such as answer keys; lesson plans; and other teaching aids.

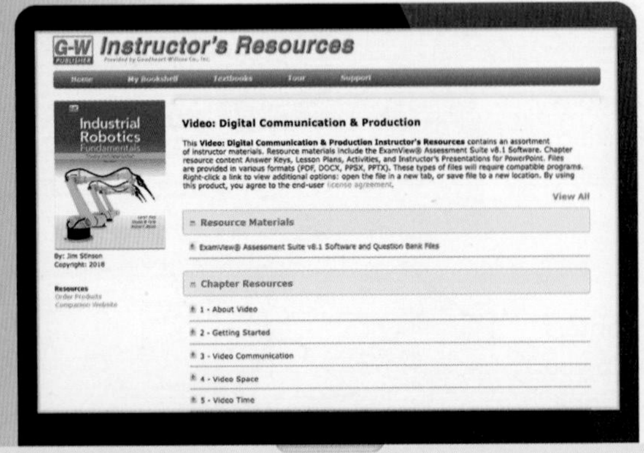

Online Instructor Resources

Online Instructor Resources provide all the support needed to make preparation and classroom instruction easier than ever. Available in one accessible location, support materials include Answer Keys, Lesson Plans, Instructor Presentations for PowerPoint®, ExamView® Assessment Suite, and more! Online Instructor Resources are available as a subscription and can be accessed at school or at home.

Brief Contents

Contents

Feature Contents

Inventor Spotlight

Robotics Innovations

Careers in Robotics

Robotics in Society

Activities

Unit I

Principles of Robotics

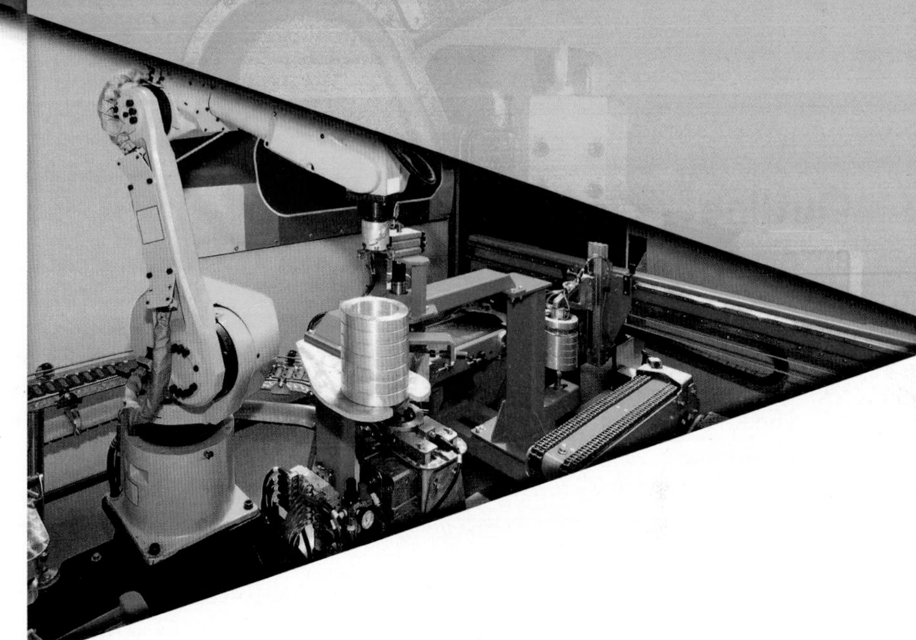

The use of robots in our society continues to increase. Since the first industrial robot was installed at a US automotive plant in 1961, robotics technology has become an integral factor in most types of manufacturing. Robots are widely used for applications that require extreme precision, repetitive and tedious tasks, and work that is considered unpleasant or dangerous for humans. Robots are also vital components of flexible manufacturing systems, which allow robotic configurations to be quickly changed to meet production requirements.

Chapter 1

Introduction to Robotics

Outline

1.1 Early Robots
1.2 The Evolution of Robots
1.3 Classifications of Industrial Robots
1.4 Types of Automation
1.5 The Role of Robots in the Workforce

Objectives

Upon completion of this chapter, you will be able to:

- Describe early robots and their role in literature.
- Cite important developments in the evolution of robots.
- List and explain the classifications of industrial robots.
- Define two types of automation.
- Discuss the role of robots in the workforce.

Technical Terms

anthropomorphic
artificial intelligence (AI)
automaton
fixed-sequence robot
flexible automation
hard automation
industrial robot
intelligent robot
manipulator
manual manipulator
numerically controlled (NC) robot
playback robot
reprogrammable
robot
robotics
Unimate
variable-sequence robot

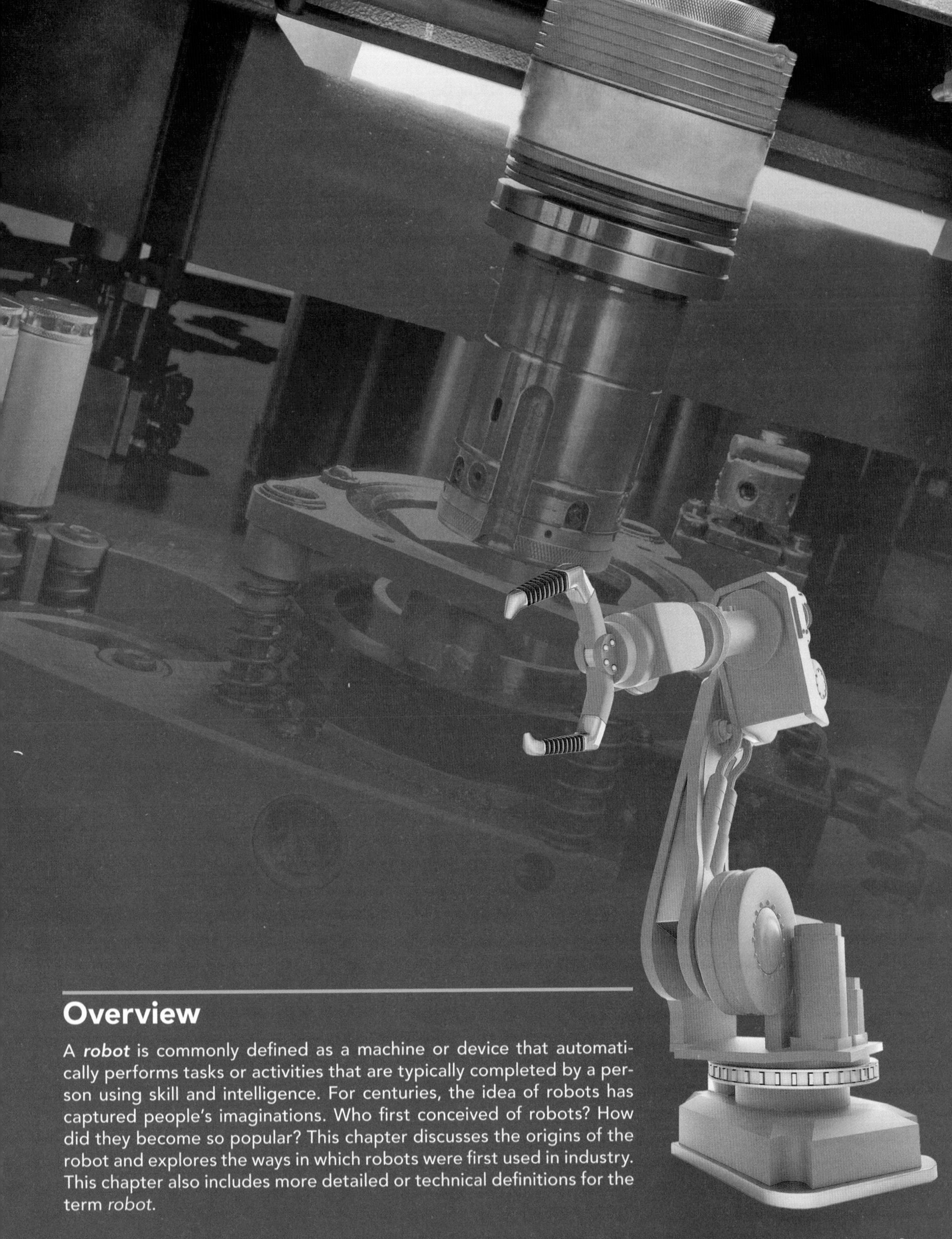

Overview

A *robot* is commonly defined as a machine or device that automatically performs tasks or activities that are typically completed by a person using skill and intelligence. For centuries, the idea of robots has captured people's imaginations. Who first conceived of robots? How did they become so popular? This chapter discusses the origins of the robot and explores the ways in which robots were first used in industry. This chapter also includes more detailed or technical definitions for the term *robot*.

1.1 Early Robots

The term ***automaton*** was originally used for what is now considered to be an early robot. An automaton is an object made by a person that moves automatically. The first useful automatons were clockwork devices introduced during the Middle Ages to keep track of time automatically. As these clocks became more complex, systems of gears and pulleys enabled the workings and figures attached to the clocks to move in lifelike ways. Many advanced automatons were built to entertain royalty and nobility. The English scientist Roger Bacon (ca. 1220–1292) had ideas for other automaton, such as a flying machine, a diving bell, a mechanical chariot, and a mechanical bird.

As people became more knowledgeable about physiology, some believed that even the human body was merely a complex automaton separate from the mind and soul. The seventeenth-century French philosopher René Descartes (1596–1650) strongly supported this viewpoint, and people had visions of building truly lifelike automatons. However, since the machines would be mindless and soulless, would they wreak havoc on humans if they were to go out of control? The story is told that Descartes built a female automaton and took it on a sea voyage. The captain of the ship, thinking it was the work of the devil, promptly threw it overboard.

1.1.1 Robots in Literature

Robots have long played important roles in the movies, in books, and on television. Characters such as R2-D2 and C-3PO of the *Star Wars* movies show robots in a favorable, comic light. Some real robots have a nonthreatening appearance with familiar, even humanlike, characteristics and qualities, **Figure 1-1**. However, some books and movies in the past have portrayed robots as being a threat to people

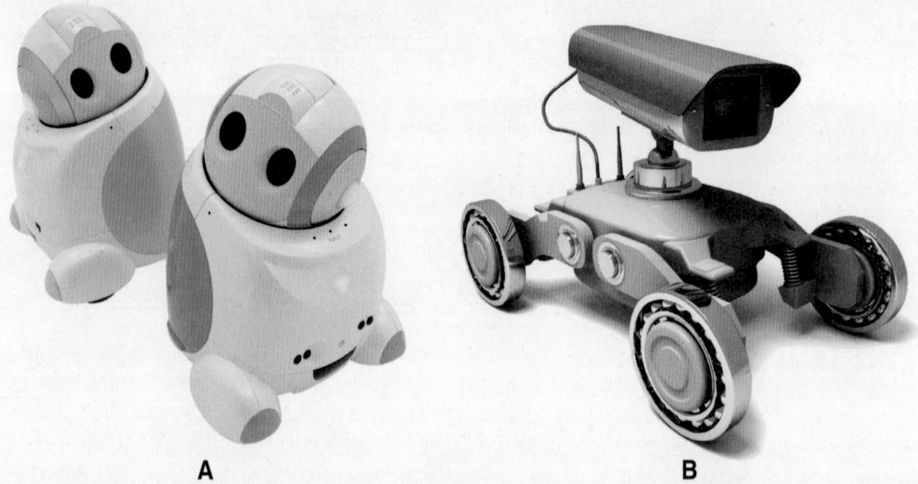

Courtesy of NEC Corporation. Unauthorized use not permitted.

Figure 1-1. These robots serve different purposes and vary in "humanlike" characteristics. A—This robot is used for educational activities and interacting with small children. B—This service robot may be used for video surveillance and security, and has the familiar characteristics of a family pet.

rather than as being helpful. Exploring the origins of the robot will help in understanding why some people have regarded robots as a threat in the past.

Mary Shelley's 1818 novel *Frankenstein* did little to ease people's fears of technology. Dr. Frankenstein's monster, an artificial man, was unhappy when his creator neglected him, so he took a horrible revenge and terrorized his creator.

In 1921, the Czech dramatist Karel Capek (1890–1938) wrote a play titled *R.U.R.* (*Rossum's Universal Robots*). In the play, Rossum develops a formula for making mechanical robots whose only function is to work. The robots are extremely strong and dedicated to their tasks. Some are even quite intelligent, but they lack emotions. A young woman in the story objects to the inhumane treatment given the robots and persuades the physiologist at the plant to change the formula for making the robots so that they become more human. As a result of this change, some robots develop interests other than work. The altered robots organize the original robots and turn them against humanity. Eventually, the robots annihilate everyone except the company's founder.

The word *robot* is derived from the Czech word *robota,* which means forced labor. Capek's play was so popular that the word *robot* began to be used instead of *automaton* in virtually every language. Today, robots are commonly considered to be any manufactured structure that performs functions normally done by human beings.

Not all writers have portrayed robots in a negative manner. In 1939, when Isaac Asimov was only nineteen, he began to write science fiction in which robots were machines that could be built with safety measures that would prevent them from harming people. Asimov's story *Runaround,* published in the March 1942 issue of *Astounding* magazine, included three fundamental laws of robotics:

1. A robot may not injure a human being or, through inaction, allow a human being to come to harm.
2. A robot must obey the orders given it by human beings, except where such orders would conflict with the first law.
3. A robot must protect its own existence as long as such protection does not conflict with the first and second laws.

Much later, he added the Zeroth Law: A robot may not injure humanity or, through inaction, allow humanity to come to harm. At the time, Asimov did not have a clear idea of how robots would be built. For instance, he described them as having *positronic,* rather than electronic, brains. Although the original three laws of robotics are part of a work of fiction, they have been taken seriously by some. His term ***robotics***, which means the engineering and use of robots, has become an established part of our language. Asimov's portrayal of robots as useful machines is now the more generally accepted one, and Capek's portrayal has faded into the past.

1.1.2 The Advent of Computers

By the end of World War II, people began building computers. To many people, it seemed that computers were thinking machines. Scientists combined the self-guided control systems developed during the war with computers. The idea of building humanlike machines came back into popularity. Both scientists and laypeople began to consider the consequences of merging computers with another new development, artificial intelligence. ***Artificial intelligence*** is the ability of a computer program to make decisions based on known information. Could a computer be installed into a structure resembling a human body and become a robotic human? Many people asked questions such as this and wondered what the future would hold for robotics.

1.2 The Evolution of Industrial Robots

The invention of the first ***industrial robot***, a multifunction manipulator programmed to perform various tasks, was now close at hand. A ***manipulator*** is the arm of the robot that moves materials, parts, tools, or special devices through various motions to provide useful work. After World War II, North American industry benefited from state-of-the-art equipment and technologies, while several other industrialized countries lay in shambles. By 1950, industry in the United States was at its peak. The demand for goods at home and abroad was overwhelming, and the need for increased production through automated systems became apparent.

1.2.1 Early Industrial Robots

George C. Devol Jr. patented the first industrial robot in 1954. However, he faced many problems in financing his invention and selling it to corporations in the United States. In 1956, Devol met a young engineer named Joseph Engelberger. Engelberger was impressed with Devol's idea and tried to persuade his employer, Aircraft Products, to help develop an industrial robot. However, it was not until Aircraft Products was acquired by Consolidated Diesel Electric Corporation that the necessary capital was made available.

The subsidiary, Unimation Inc. was formed in the late 1950s, with Engelberger as president, to develop Devol's invention. In 1961, the first Unimation robot, called the ***Unimate***, was sold to General Motors, **Figure 1-2**. Like many other early robots used in industry, the Unimate was not called a robot due to the many negative connotations of that term. The first Unimate was used in a die-casting operation. The robot was first guided through the desired sequence of steps, which were recorded. Using these recorded instructions, the robot automatically performed the required task.

AMF Versatran built the manipulators used on atomic energy projects and became interested in robots around the same time Unimation was being formed. In

Goodheart-Willcox Publisher

Figure 1-2. The Unimate was the first industrial robot produced by Unimation Inc. in 1961.

George C. Devol has been called the grandfather of industrial robotics.

Inventor Spotlight: George C. Devol

Prior to his collaboration with Joseph Engelberger to form the Unimation company in the late 1950s, George Devol was part of many innovative developments.

- In 1932, George Devol formed the United Cinephone Corporation. The company manufactured phonograph arms and amplifiers and photoelectric doors called *Phantom Doorman*. The company also developed a bar code system used in package sorting processes.
- The number of visitors at the 1939 World's Fair in New York was tracked using automated counters developed by United Cinephone Corporation.
- George Devol formed General Electronics Industries in 1943 and began developing radar counter-measures for military use. These systems were used during World War II by the allied forces.
- In 1946, George Devol patented a general purpose playback device that used a magnetic process recorder. This early development led to programmable robotic systems.

Photo courtesy of Goodheart-Willcox Publisher

1959, AMF Versatran was purchased by Prab Conveyor Company. Prab Robots Inc. was formed, and the company developed a line of industrial robots in the late 1960s.

Other major manufacturers involved in early robot development for commercial use were DeVilbiss, Asea, and Cincinnati Milacron. DeVilbiss became one of the leading producers of finishing robots and introduced an arc-welding robot in 1982. Asea was one of the early developers of ***anthropomorphic*** (humanlike in form) robot units. Cincinnati Milacron also entered the robot market with an anthropomorphic unit powered by hydraulics.

During the late 1970s and early 1980s, robots were moved into assembly operations. In 1978, engineers at Unimation introduced a smaller robot called PUMA (Programmable Universal Machine for Assembly). PUMAs were designed to handle small parts used in the assembly of motors and instruments.

Other American companies, such as IBM, Bendix, General Motors, and General Electric, eventually entered the robotics business. These companies offered foreign-built, American-packaged robots through various licensing agreements with Japanese, German, and Italian companies. In 1982, a major joint venture was formed between General Motors and Fanuc of Japan to create GM Fanuc. This company marketed robots in North America.

1.2.2 Japan Enters the Market

The early success of robotics technology in the United States did not go unnoticed by the Japanese. In 1966, many Japanese companies sent representatives to this country to see what was happening. In 1967, Joseph Engelberger was invited to tour Japan, and he lectured in Tokyo to an audience of 700 engineers and executives. Robotics technology grew rapidly in Japan.

Japan's first industrial robot was developed in 1969 after the first AMF Versatran robots had been exhibited and sold there. According to Engelberger, Japanese companies did not resist technology as American companies had, and Japan was able to enter the market quickly. The Japan Robot Association (JARA) was founded in 1971. Three years later, the Robotic Institute of America (now called the Robotic Industries

Association or RIA) was formed. Another three years later, the British Automation and Robotics Association (BARA) was formed. In 1978, the SCARA (selective compliance assembly robot arm) was developed at Yamanashi University in Japan and marketed by IBM and Sankyo. Japan became the world's largest user of robots.

1.3 Classifications of Industrial Robots

The most widely accepted definition for the term *robot* in the United States has been published by the Robot Institute of America (RIA): *A reprogrammable, multifunctional manipulator designed to move material, parts, tools, or specialized devices through various programmed motions for the performance of a variety of tasks.*

This definition contains several important points:

- The robot is a machine.
- The robot is ***reprogrammable***, meaning that it can be given new instructions that allow it to meet changed requirements and perform new tasks.
- The robot has a multifunction manipulator, which means it may be used in different ways even within the same program.
- The robot is flexible and can perform a variety of operations to meet special needs.

The robots pictured in **Figure 1-3** fulfill the requirements of the RIA definition. The robotic system consists of a motor-driven, multifunction manipulator, an electronic memory system containing the program that controls manipulator movement, and a microcomputer for reprogramming the robot for new tasks. The progression of three generations of manufacturing robots is illustrated in **Figure 1-4**. It lists several applications used today and possible applications for the future. Many of these applications will be discussed in Chapter 4. Because a robot can be reprogrammed, it can function in many different jobs. In the future, the majority of these jobs will be outside the manufacturing area.

As robots evolved, Japan categorized certain types of automated machinery as industrial robots. These machines would not be considered robots in the United States. This raised questions as to how many true robots were used by Japanese industry compared to those used in the United States. The Japanese used a wide range of classifications, from simple arms to what they called *intelligent robots*. These classifications include:

- ***Manual manipulator.*** A manipulator worked by a human operator.
- ***Fixed-sequence robot.*** A manipulator that repetitively performs successive steps of a given operation according to a predetermined order, condition, and position. Its instructions cannot be easily changed.
- ***Variable-sequence robot.*** A manipulator that repetitively performs successive steps of a given operation according to a predetermined order, condition, and position. The order of movement it follows can be easily changed.
- ***Playback robot.*** A manipulator that can reproduce operations originally executed under human control. An operator initially feeds in the instructions relating to a sequence of movements, conditions, and positions. These instructions are stored in memory.
- ***Numerically controlled (NC) robot.*** A manipulator that can perform the sequence of movements, conditions, and positions of a given task, which are communicated by means of numerical data.

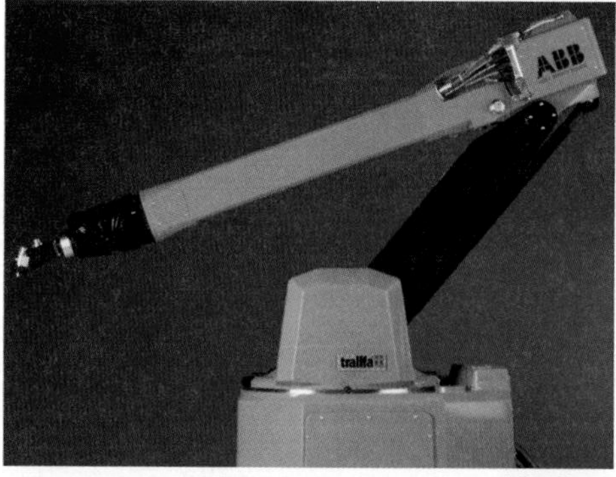

ABB Graco

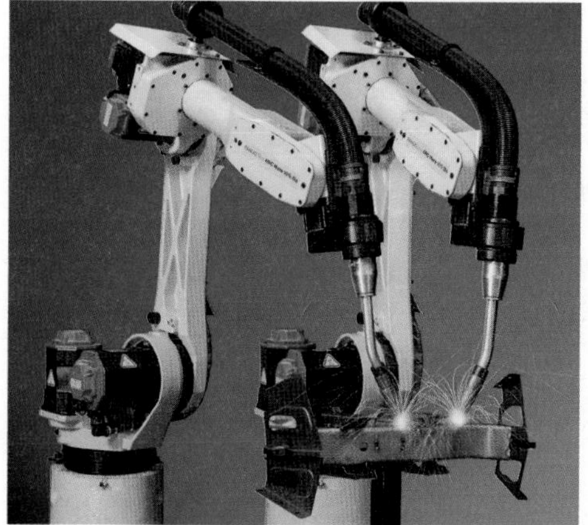

Fanuc Robotics

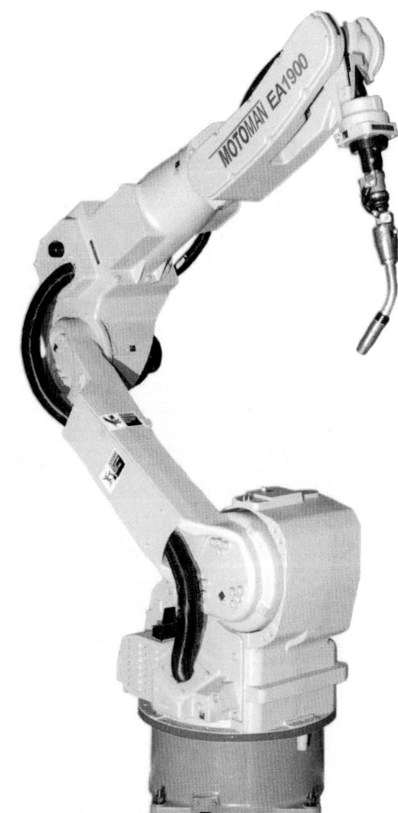

Motoman

Figure 1-3. These industrial robots meet the RIA definition.

- ***Intelligent robot.*** A robot that can, by itself, detect changes in the work environment by means of sensors (visual and/or tactile). Using decision-making capabilities, the robot can then proceed with the appropriate operations.

1.4 Types of Automation

Industries use various types of automation to manufacture parts and products. Automation is the use of control systems to operate machinery or perform processes with little human involvement. Two common classifications are hard automation and flexible automation.

1.4.1 Hard Automation

Hard automation refers to machinery that has been specifically designed and built to perform particular tasks within an assembly line. This approach works well when a very large number of items are to be produced over a long period of time. It is very reliable, and the machine precision is generally high. After an item the machine was designed to make is no longer needed, the machine is often discarded due to the high cost of retooling. Therefore, this kind of automation can be very costly. With the increasing demand for new products and new models of

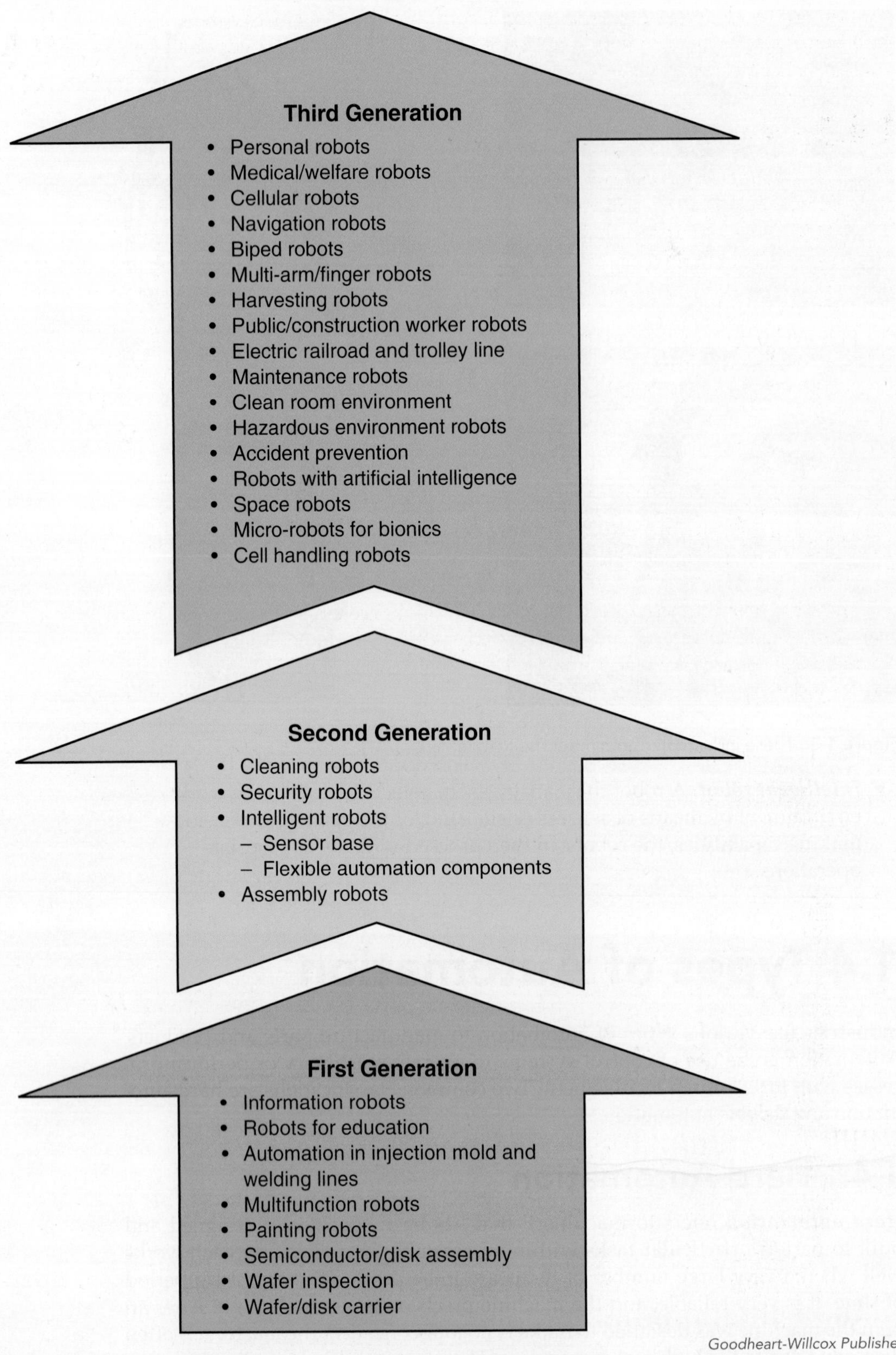

Goodheart-Willcox Publisher

Figure 1-4. Three generations of industrial robots show increasing ability to accomplish more difficult tasks.

existing products, product life spans are becoming shorter. To justify the use of a hard automation system, the product built must be produced at a very high volume and have the potential of being in use for a long period of time.

1.4.2 Flexible Automation

Flexible automation refers to machinery that is capable of performing a variety of tasks. Robots belong in this category. As new products or new models are needed, flexible machines can be reprogrammed to make the parts required. This flexibility can save money because equipment does not have to be discarded or rebuilt. In addition, it takes much less time to reprogram an existing machine than to build and install a new one. **Figure 1-5** shows a work cell for flexible automation.

Robots are the most flexible type of automation available today. They can be programmed easily, quickly, and economically. They can also be moved from one location in a plant to another.

1.5 The Role of Robots in the Workforce

Our thinking has come a long way since Karel Capek wrote his play in 1920. However, some negative ideas about robots are still being considered. Robots are more productive at some tasks than humans. They can perform work less expensively, and they do some jobs that were done by people in the past. For these reasons,

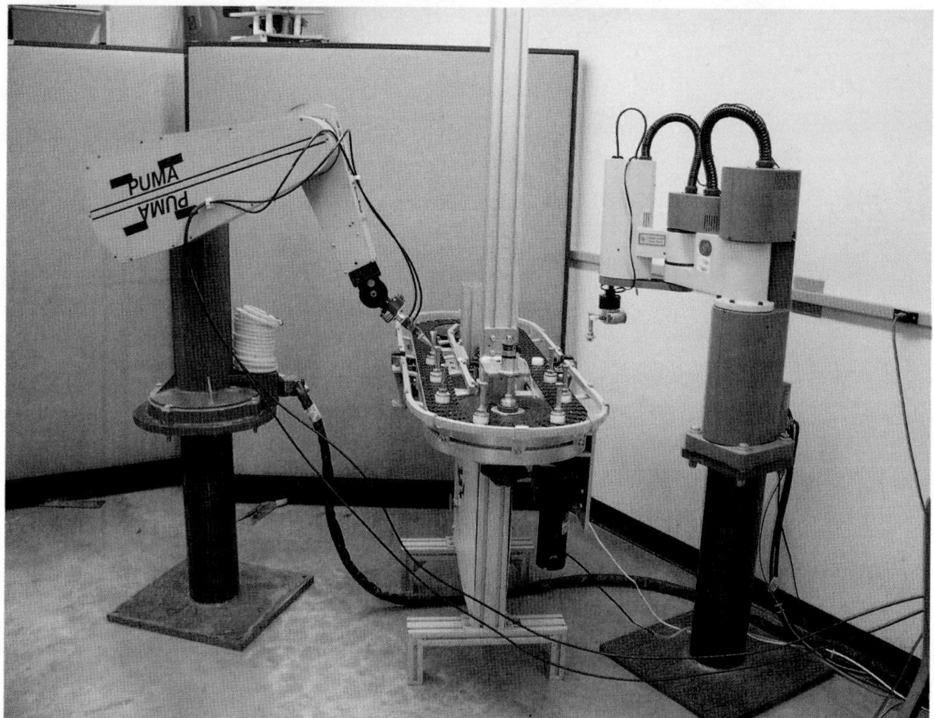

Goodheart-Willcox Publisher

Figure 1-5. A flexible manufacturing work cell can be programmed for more than one task.

workers and members of labor unions often see robots as a threat, thinking that robots may displace human workers.

Other people see robots as an aid to workers. Robots perform work that relieves men and women from repetitive operations that are monotonous. They also do jobs that are unpleasant or dangerous for people, **Figure 1-6**.

The addition of robots in the workforce does not necessarily mean that jobs will be lost. Technicians and other types of skilled laborers are needed to install, program, and maintain the robots. Perhaps it is best to keep the words of Joseph Engelberger in mind: "Nobody needs a robot. There isn't anything that a robot can do that a willing human being can't do better." Company managers must consider, however, whether a person or a robot can do a particular job more quickly, accurately, or safely.

iRobot Corporation

Figure 1-6. The SUGV 300 (small unmanned ground vehicle) is a remote-controlled robot used in military operations. This robot can be sent into hazardous or inhospitable areas for inspection or tactical maneuvers.

Summary

- A robot is a machine or device that automatically performs tasks or activities that are typically completed by a person using skill and intelligence.
- The term *automaton* was used for early robots. The first useful automatons were clockwork devices introduced during the Middle Ages.
- Robots have long played important roles in movies, books, and television shows. Some books and movies have portrayed robots as being a threat to people, and others have portrayed them as being helpful.
- Issac Asimov's fictional story *Runaround* included three fundamental laws of robotics. Asimov's portrayal of robots as useful machines is now generally accepted.
- By the end of World War II, people began building computers. People also began to consider the consequences of merging computers with artificial intelligence.
- An industrial robot is a multifunction manipulator programmed to perform various tasks. George C. Devol Jr. patented the first industrial robot in 1954. During the late 1970s and early 1980s, robots were moved into assembly operations.
- After the early success of robotics technology in the United States, robotics technology grew rapidly in Japan.
- The Robot Institute of America's definition for the term *robot* points out that a robot is a machine, is reprogrammable, has a multifunction manipulator, and can perform a variety of operations.
- Japanese classifications for robots include manual manipulator, fixed-sequence robot, variable-sequence robot, playback robot, and numerically controlled (NC) robot.
- Industries use automation to manufacture products. Hard automation refers to machinery that has been designed to perform particular tasks within an assembly line. Flexible automation refers to machinery that is capable of performing a variety of tasks.
- Robots can be more productive and work less expensively than people for some tasks. Some people see robots as a threat, thinking that robots may displace human workers. Other people see robots as an aid to workers because they can do monotonous or dangerous jobs.

Review Questions

Answer the following questions using the information provided in this chapter.

1. What is a common definition of the term *robot*?
2. What is an automaton? What is one example of an early automaton?
3. Who was the dramatist that introduced the word *robot* in his writing? What is the meaning of the Czech word *robota*?
4. How did Isaac Asimov's writings influence the way people think about robots?
5. In what year was the first industrial robot patented?
6. How were George Devol and Joseph Engelberger involved in the early development of industrial robots?
7. What are four companies that were involved in early robot development for commercial use?

8. Some robots are considered *anthropomorphic*. What does this term mean?
9. What is the definition of the term *robot* that was published by the Robot Institute of America? Discuss the key factors in that definition.
10. List and define the six Japanese classifications of robots.
11. What is hard automation and in what situation is the use of hard automation justified?
12. What is flexible automation and what are some advantages of using flexible automation?
13. What are some negative aspects and some positive aspects of having robots as part of the workforce?

Learning Extensions

1. Visit websites for Robotics Online and the Society of Manufacturing Engineers Robotics International. Describe two ways that robotics impacts our society.
2. Locate the websites for at least three professional and student associations related to robotics.
3. Create a time line that charts important dates, people, developments, and writings in the evolution of robots.

Advanced Analysis

1. Prepare a brief paper describing the early evolution of robots.
2. Select a current nontechnical book or article that includes reference to robots and describe the way in which robots are presented in a descriptive essay.
3. Prepare a book report on a book that has references to robots.
4. Advancements in computer technology have impacted the types and applications of robots. Find and discuss one example of this.
5. List and describe two applications/uses of robots in industry.
6. Some people believe that the use of automation and robots will replace human workers. Others believe that if automation and robots are not incorporated into manufacturing, we will lose a far greater number of jobs. Explain how both views can be defended.

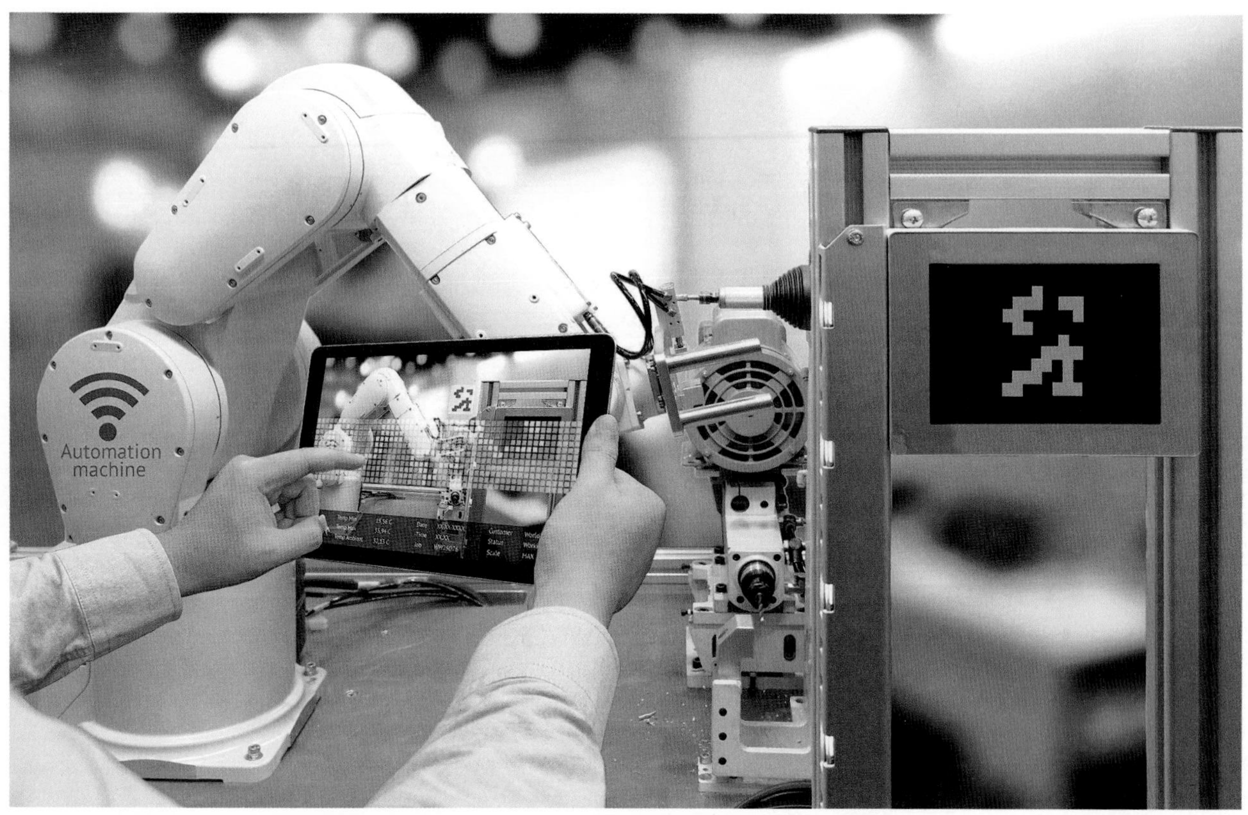

Zapp2Photo/Shutterstock.com

Robots have evolved over time, and today they can be programmed remotely.

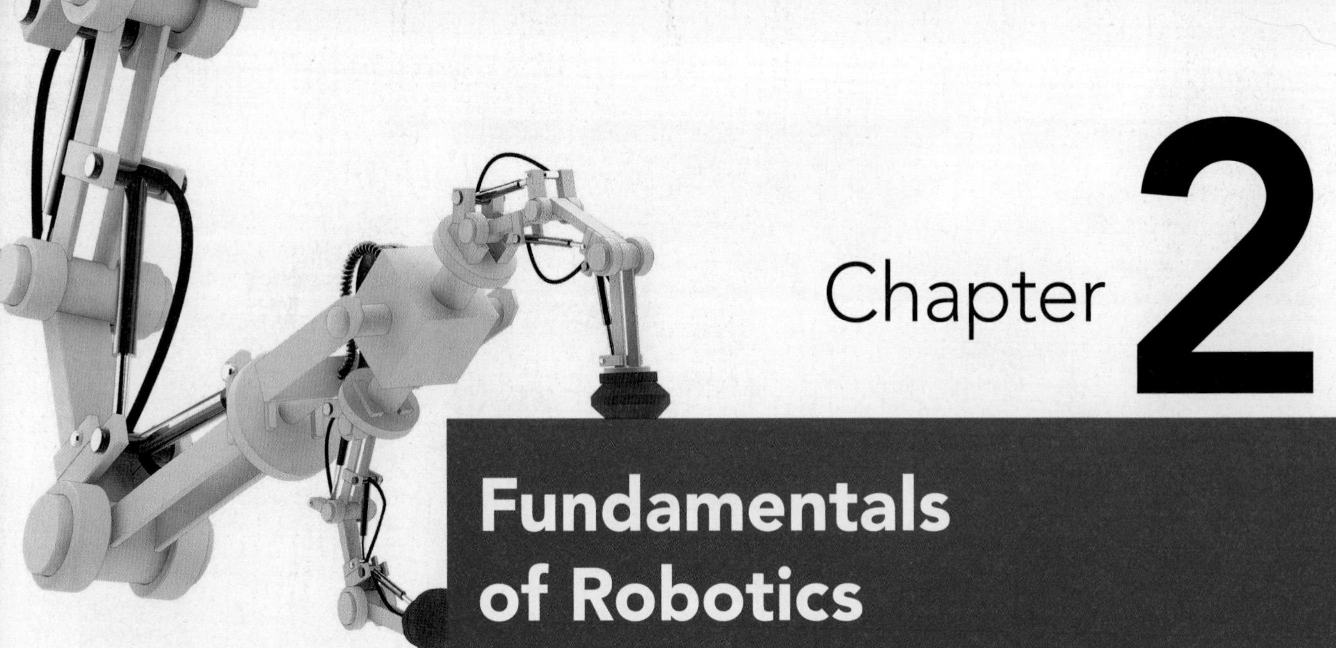

Chapter 2

Fundamentals of Robotics

Outline

Objectives

Upon completion of this chapter, you will be able to:

- Identify the five major parts of a robot.
- Explain degrees of freedom as applied to robots.
- Classify robots according to type of control systems and type of actuator drives.
- Discuss the four basic configurations for robots.

Technical Terms

actuator
Cartesian configuration
closed-loop system
controller
cylindrical configuration
degrees of freedom
direct-drive motor
end effector
error signal
hierarchical control
hydraulic drive
linear actuator
non-servo robot
open-loop system
pitch
pneumatic drive
power supply
program
radial traverse
revolute configuration
roll
rotary actuator
rotational traverse
SCARA configuration
servo amplifier
servo robot
spherical configuration
tachometer
teach pendant
teach pendant programming
trajectory
vertical traverse
work envelope
yaw

Overview

Even the most complex robotic system can be broken down into a few basic components. Studying these components provides an overview of how a robot works. Basic information about these components is covered in this chapter, and more details are provided in later chapters. Freedom of motion and the resulting shape of the robot's work area are also addressed in this chapter.

2.1 Parts of a Robot

Robots come in many shapes and sizes. The industrial robot illustrated in **Figure 2-1** resembles an inverted human arm mounted on a base. Robots consist of a number of components, **Figure 2-2**, that work together: the controller, the manipulator, an end effector, a power supply, and a means for programming. The relationship among these five components is illustrated in **Figure 2-3**. The functions of these components are explained in the following sections of this chapter.

2.1.1 Controller

The ***controller*** is a microprocessor linked to input/output and monitoring devices. It coordinates all movements of the mechanical system, **Figure 2-4**. It also receives input from the immediate environment through various monitoring devices, which are also called sensors. The commands issued by the controller activate the motion control mechanism, which consists of actuators and other devices. An ***actuator*** is a motor or valve that converts power into robot movement. This movement is initiated by a series of instructions, called a ***program***, which is stored in the controller's memory.

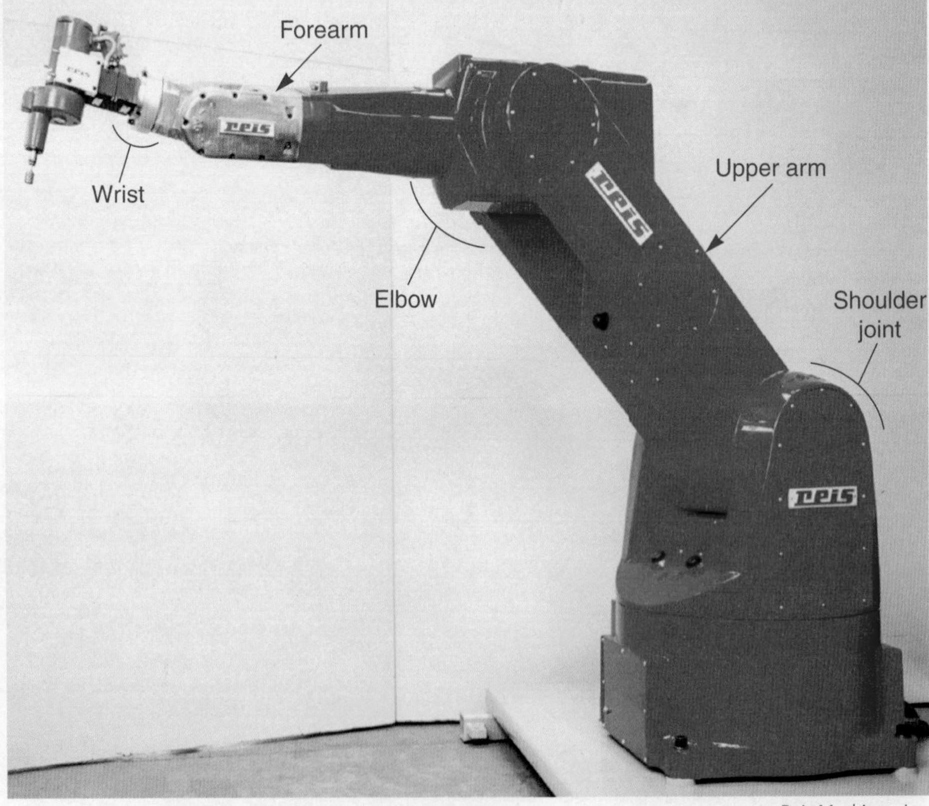

Reis Machines, Inc.

Figure 2-1. This robot has been designed expressly for use in precise path-oriented tasks, such as deburring, milling, sanding, gluing, bonding, cutting, and assembly.

ABB Robotics

Figure 2-2. This robot illustrates the systems of a typical industrial robot. This electric robot can be used in a variety of industrial applications.

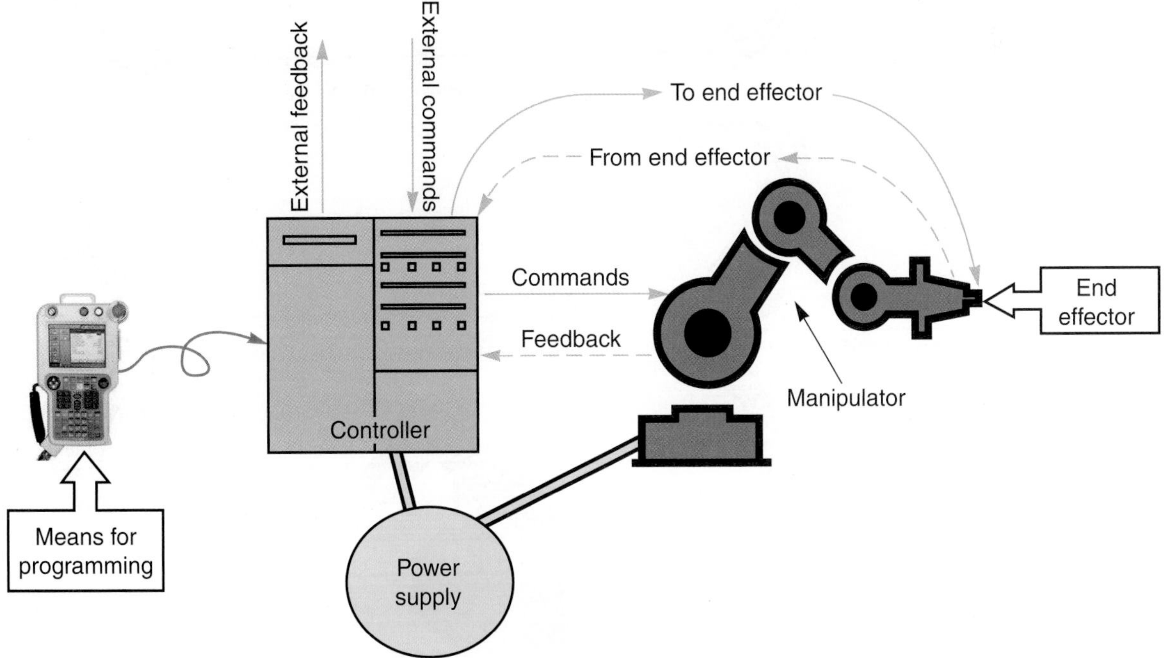

Motoman; Goodheart-Willcox Publisher

Figure 2-3. The relationships among five major systems that make up an industrial robot are shown in this diagram.

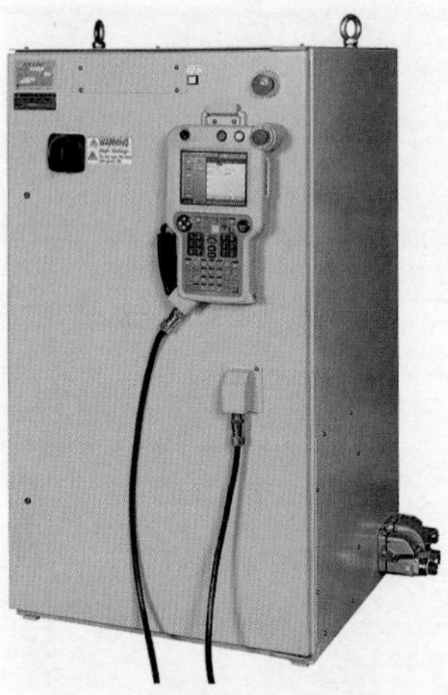

Motoman

Figure 2-4. A controller/power supply with a teach pendant.

The controller has three levels of hierarchical control. ***Hierarchical control*** is a system that organizes commands and instructions from the software to the parts and devices of the robot into levels or layers. Each level sends control signals to the level below and feedback signals to the level above. The levels become more elemental as they progress toward the actuator. Each level is dependent on the level above it for instructions, **Figure 2-5**. The three levels are:

- **Level I—Actuator Control.** This is the most elementary level at which separate movements of the robot along various planes, such as the X, Y, and Z axes, are controlled. These movements will be explained in detail later in this chapter.
- **Level II—Path Control.** The path control (intermediate) level coordinates the separate movements along the planes determined in Level I into the desired ***trajectory*** or path of movement.
- **Level III—Main Control.** The primary function of this highest control level is to interpret the written instructions from the human programmer regarding the tasks required. The instructions are then combined with various environmental signals and translated by the controller into the more elementary instructions that Level II can understand.

2.1.2 Manipulator

The manipulator is the arm of the robot (see **Figures 2-2** and **2-3**) that moves materials, parts, tools, or special devices through various motions to provide useful work.

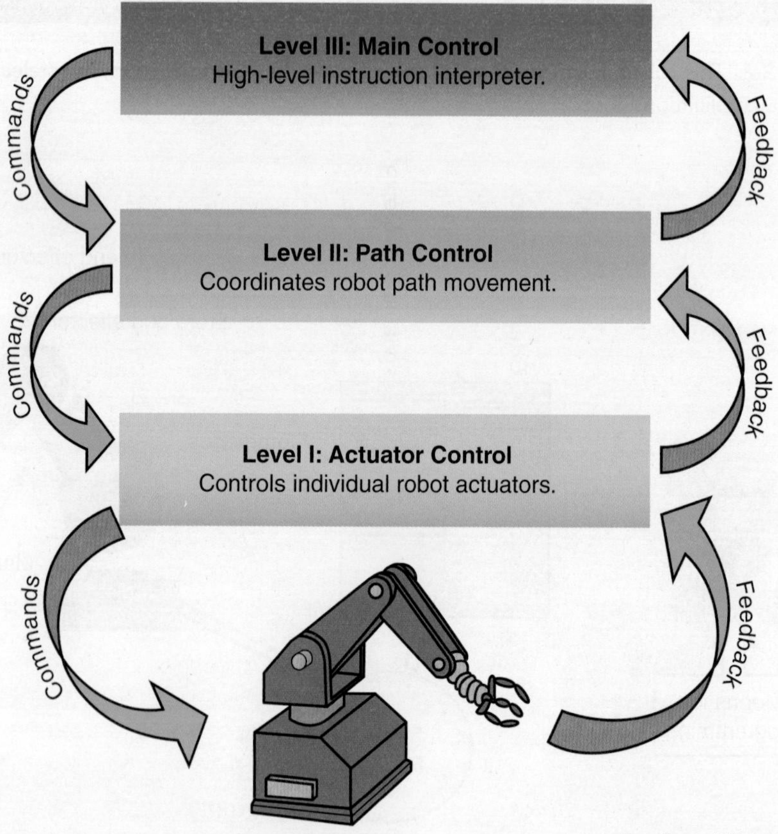

Goodheart-Willcox Publisher

Figure 2-5. The three basic levels of hierarchical control.

A manipulator can be identified by method of control, power source, the type of actuator used, and other factors. These factors help identify the best type of robot for the task at hand. For example, you would not use an electric robot in an environment where combustible fumes exist and a spark could cause an explosion.

The manipulator is made up of a series of segments and joints much like those found in the human arm. Joints connect two segments and allow them to move relative to one another. The joints provide either linear (straight line) or rotary (circular) movement, **Figure 2-6**.

A robot uses actuators to move its arm along programmed paths and then to hold its joints rigid once the correct position is reached. A ***linear actuator*** is a motor or valve that provides motion along a straight line. These actuators extend or retract the robot arm and any attached load (tools or objects), **Figure 2-7**. A ***rotary actuator*** is a motor or valve that provides rotation, moving a load in an arc or circle. Rotary motion can be converted into linear motion using a device called a lead screw or using other mechanical means. Actuators are also used at the end of the arm to move or perform work on a part or item (often called the *workpiece*) and provide other kinds of motion within the work area.

A ***tachometer*** is a device used to measure the speed of rotation of an object. In the case of robotic systems, a tachometer is used to monitor acceleration and deceleration of manipulator movements.

2.1.3 End Effector

The ***end effector*** is a device attached to the end of the manipulator (robot arm) for the purpose of grasping, lifting, transporting, maneuvering, or performing operations on a workpiece. The end effector is also called the end-of-arm tooling. The

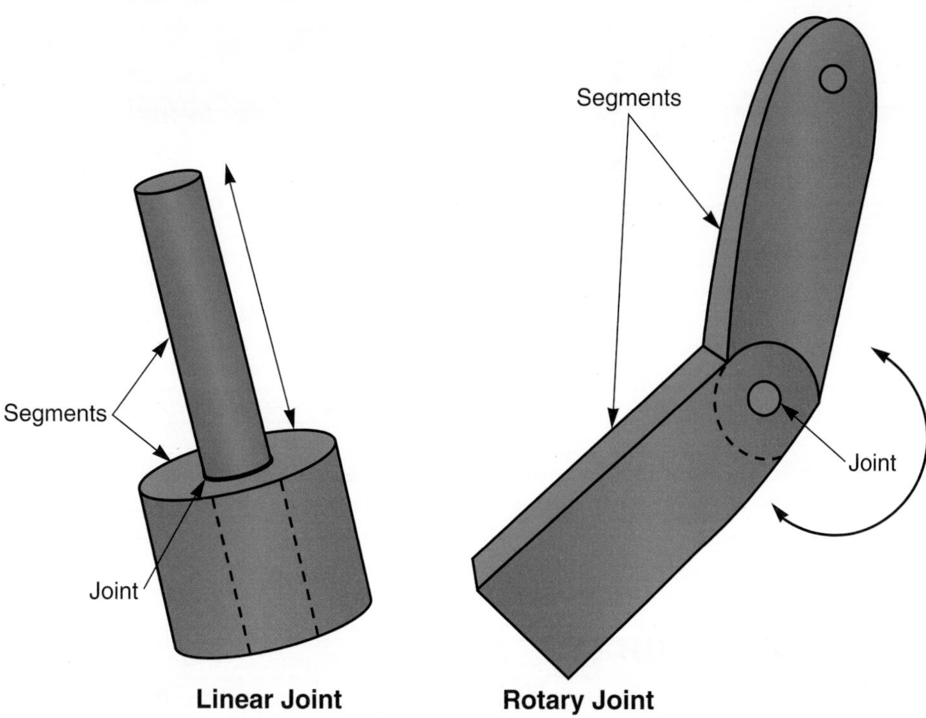

Goodheart-Willcox Publisher

Figure 2-6. Both linear and rotary joints are commonly found in robots.

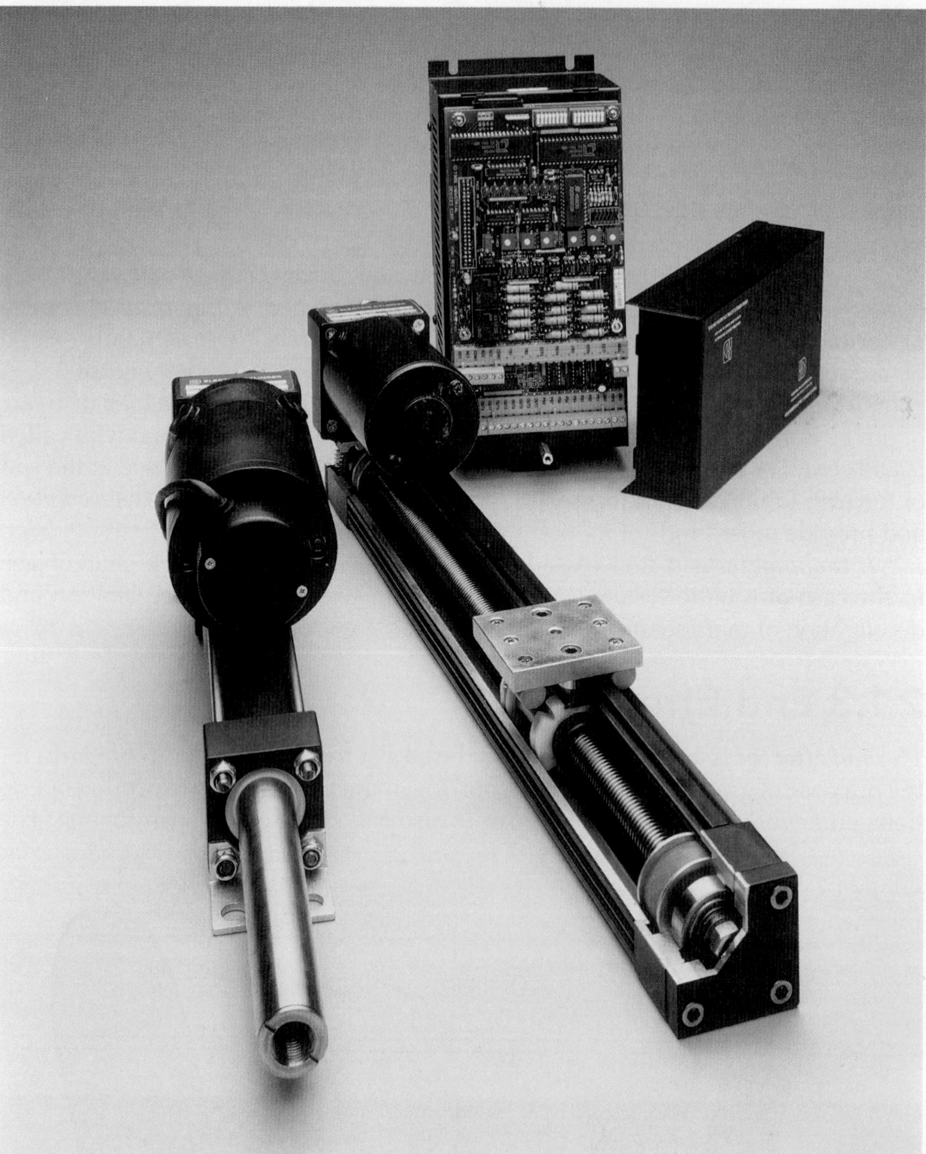

PHD, Inc.

Figure 2-7. Actuators can be powered by electric motors, pneumatic (air) cylinders, or hydraulic (oil) cylinders. Linear actuators provide straight-line movement. Rotational movement around an axis is provided by the angular (rotary) actuator.

end effector is one of the most important components of a robot system. The robot's performance is a direct result of how well the end effector meets the task requirements. The area within reach of the robot end effector is called its ***work envelope***. The work envelope is discussed in more detail later in this chapter.

2.1.4 Power Supply

The ***power supply*** is a mechanism that provides the energy to drive the robot's controller and actuators. It may convert alternating current (ac) to the direct current (dc) required by the robot's internal circuits, or it may be a pump or compressor providing hydraulic (fluid) or pneumatic (compressed air) power. The three basic types of power supplies are electrical, hydraulic, and pneumatic.

The most common power source available where industrial robots are used is electricity. The second most common power source is pneumatic, and the least common source is hydraulic power. These primary sources of energy must be converted into the form and amount required by the type of robot being used. The electronic part of the control unit and any electric actuators require electrical power. A robot with pneumatic actuators requires compressed air, which is usually supplied by a compressor driven by an electric motor. A robot containing hydraulic actuators requires the conversion of electrical power into hydraulic power through the use of an electric, motor-driven hydraulic pump.

2.1.5 Means for Programming

The means for programming is used to record instructions that are saved to the controller. A robot may be programmed using any of several different methods. The ***teach pendant***, also called a teach box or handheld programmer, **Figure 2-8**, is a device used to demonstrate the movements required for a robot to perform a useful task. ***Teach pendant programming*** is a method of creating instructions in which the operator moves the robot through the various positions involved in an operation.

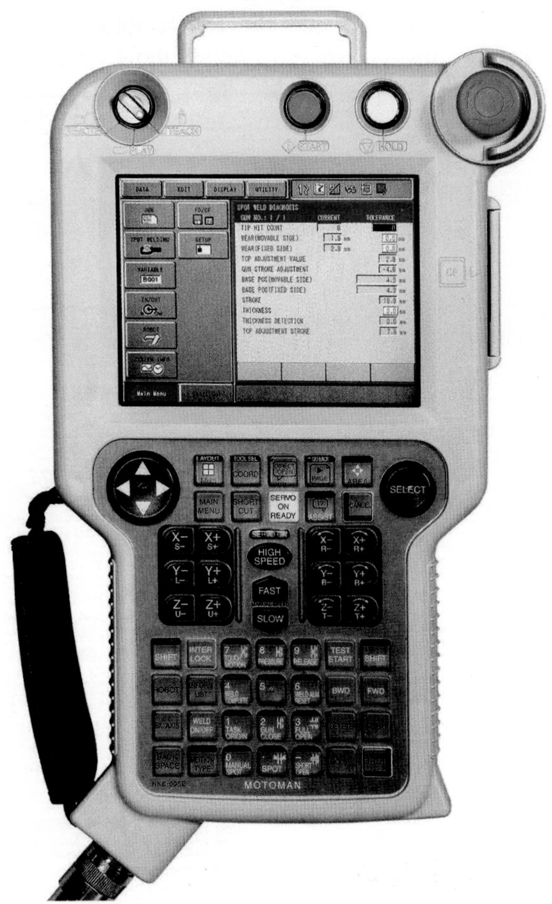

Motoman

Figure 2-8. This teach pendant is connected to a controller and is used to teach a robot how to complete a task.

Each desired point is recorded by pushing buttons on the teach pendant. The recorded points are used to generate a point-to-point path the robot will follow during operation.

2.2 Degrees of Freedom

Although robots have a certain amount of dexterity (skill in performing tasks with the hands), a robot's dexterity does not compare to that of a human. The movements of the human hand are controlled by 35 muscles. Fifteen of these muscles are located in the forearm. The arrangement of muscles in the hand provides great strength to the fingers and thumb for grasping objects. Each finger can act alone or together with the thumb. This enables the hand to do many intricate and delicate tasks. In addition, the human hand has 27 bones. **Figure 2-9** shows the bones found in the hand and wrist. This bone, joint, and muscle arrangement gives the hand its dexterity.

Degrees of freedom (DOF) is a term used to describe a range of motion (as for a robotic arm and end effector) in three-dimensional space—specifically, the ability to move forward and backward, up and down, and to the left and to the right. For each degree of freedom, a joint is required. A robot requires six degrees of freedom to be completely versatile. Its movements are clumsier than those of a human hand, which has 22 degrees of freedom.

The number of degrees of freedom defines the robot's configuration. For example, many simple applications require movement along three axes: X, Y, and Z. See **Figure 2-10**. These tasks require three joints, or three degrees of freedom. The three degrees of freedom in the robot arm are the rotational traverse, the radial

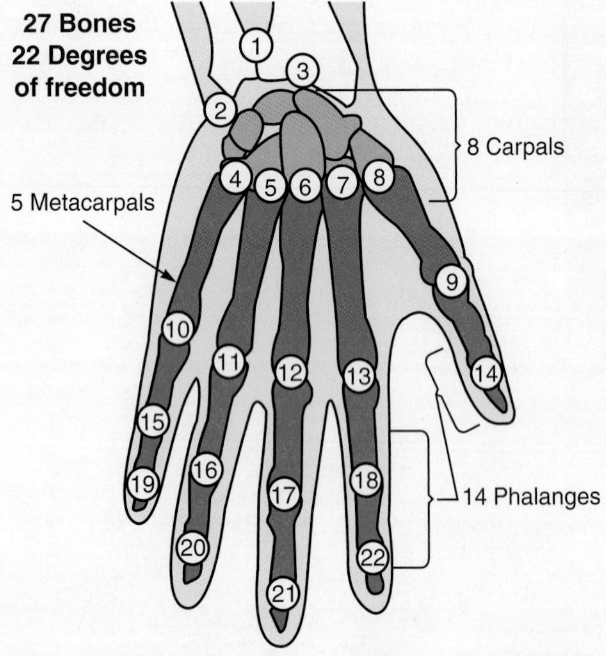

Goodheart-Willcox Publisher

Figure 2-9. The arrangement of bones and joints found in the human hand provides dexterity. Each joint represents a degree of freedom. There are 22 joints, and thus, 22 degrees of freedom in the human hand.

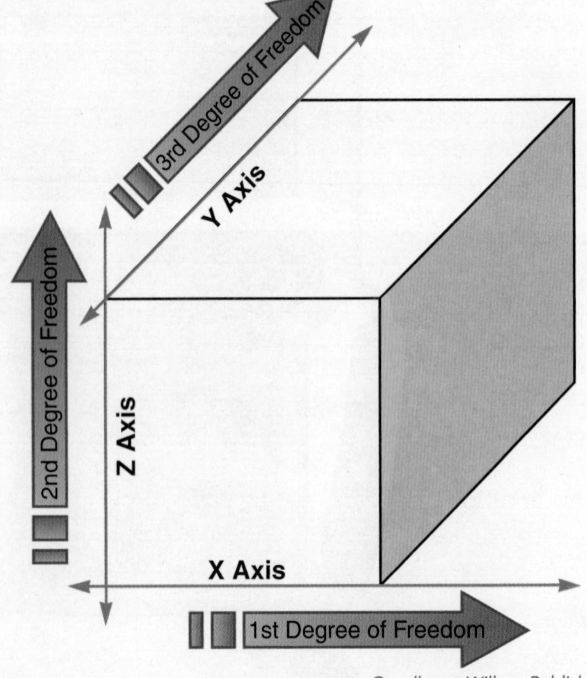

Goodheart-Willcox Publisher

Figure 2-10. The three basic degrees of freedom are associated with movement along the X, Y, and Z axes of the Cartesian coordinate system.

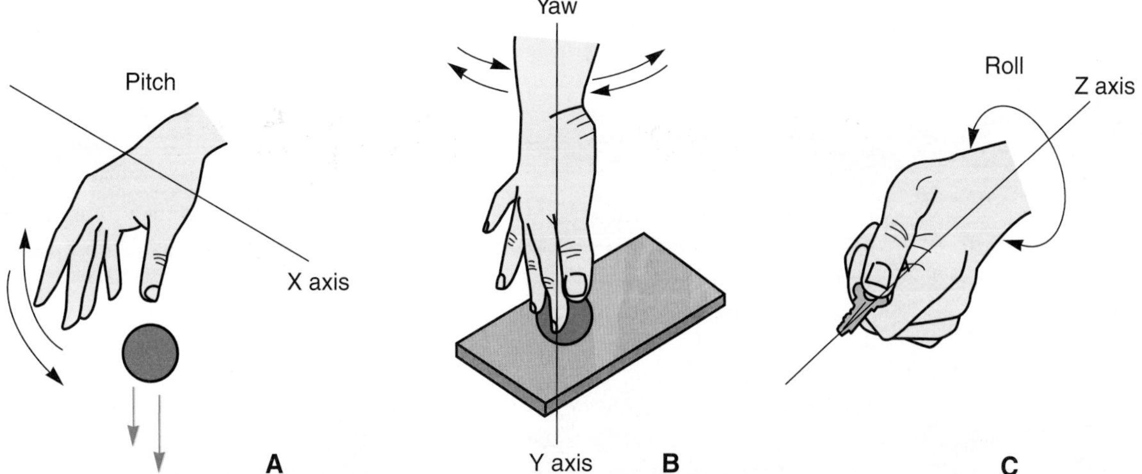

Mack Corporation

Figure 2-11. Three additional degrees of freedom—pitch, yaw, and roll—are associated with the wrist.

traverse, and the vertical traverse. The ***rotational traverse*** is movement on a vertical axis. This degree of freedom allows the side-to-side movement of the robot arm on the base. The ***radial traverse*** is the extension and retraction of the robot arm, which creates in-and-out motion relative to the base. The ***vertical traverse*** degree of freedom allows the up-and-down motion of the robot arm.

For applications that require more freedom, additional degrees can be obtained from the wrist, which gives the end effector its flexibility. The three degrees of freedom in the wrist have aeronautical names: pitch, yaw, and roll. See **Figure 2-11**. ***Pitch***, or bend, is the up-and-down movement of the robot wrist. ***Yaw*** is the side-to-side movement of the robot wrist, and ***roll***, or swivel, is the rotating movement of the robot wrist.

The degrees of freedom a robot uses affect its ability to locate and orient its end effector at any point in its work envelope, **Figure 2-12**. Although six degrees of freedom are required for maximum flexibility, most applications require only three to five. When more degrees of freedom are required, motions and controller design for the robot become more complex. Some industrial robots have seven or eight degrees of freedom. These additional degrees are achieved by mounting the robot on a track or moving base, as shown in **Figure 2-13**. The track-mounted robot shown in **Figure 2-14** has seven degrees of freedom. This addition also increases the reach of the robot.

Although the freedom of motion for a robot is limited in comparison with that of a human, the range of movement in each of the joints is considerably greater. For example, the human hand has a bending range of only about 165°. However, the end effector for a robot has a bending range of about 180°. The illustrations in **Figure 2-15** show the six major degrees of freedom by comparing those of a robot to a person using a spray gun.

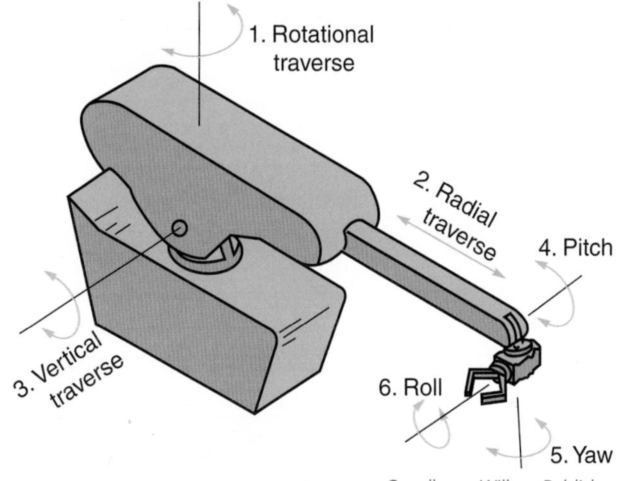

Goodheart-Willcox Publisher

Figure 2-12. Six degrees of freedom provide maximum flexibility for an industrial robot.

Schunk

Figure 2-13. Using a gantry robot creates a large work envelope (A) because the manipulator arm is mounted on tracks (B).

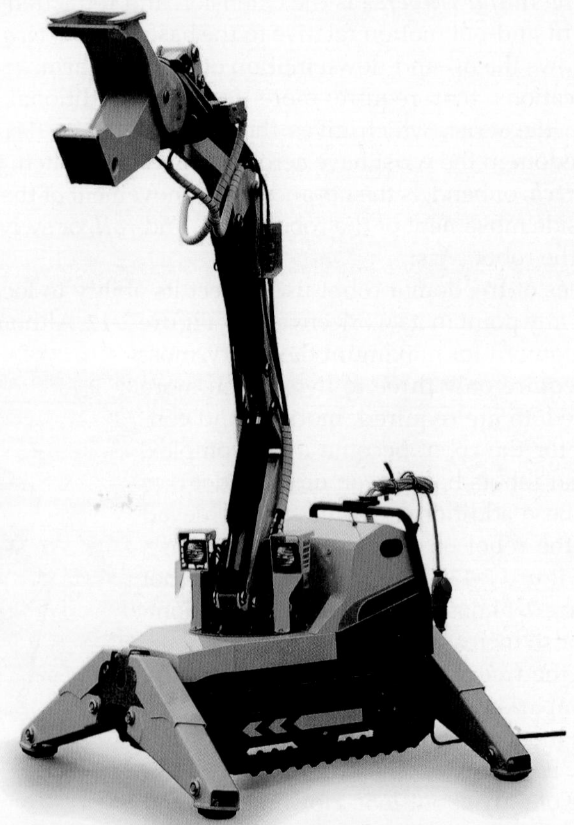

Goodheart-Willcox Publisher

Figure 2-14. Mounting this robot on tracks gives the system seven degrees of freedom—six from the configuration of the robot and one additional degree from the track mount.

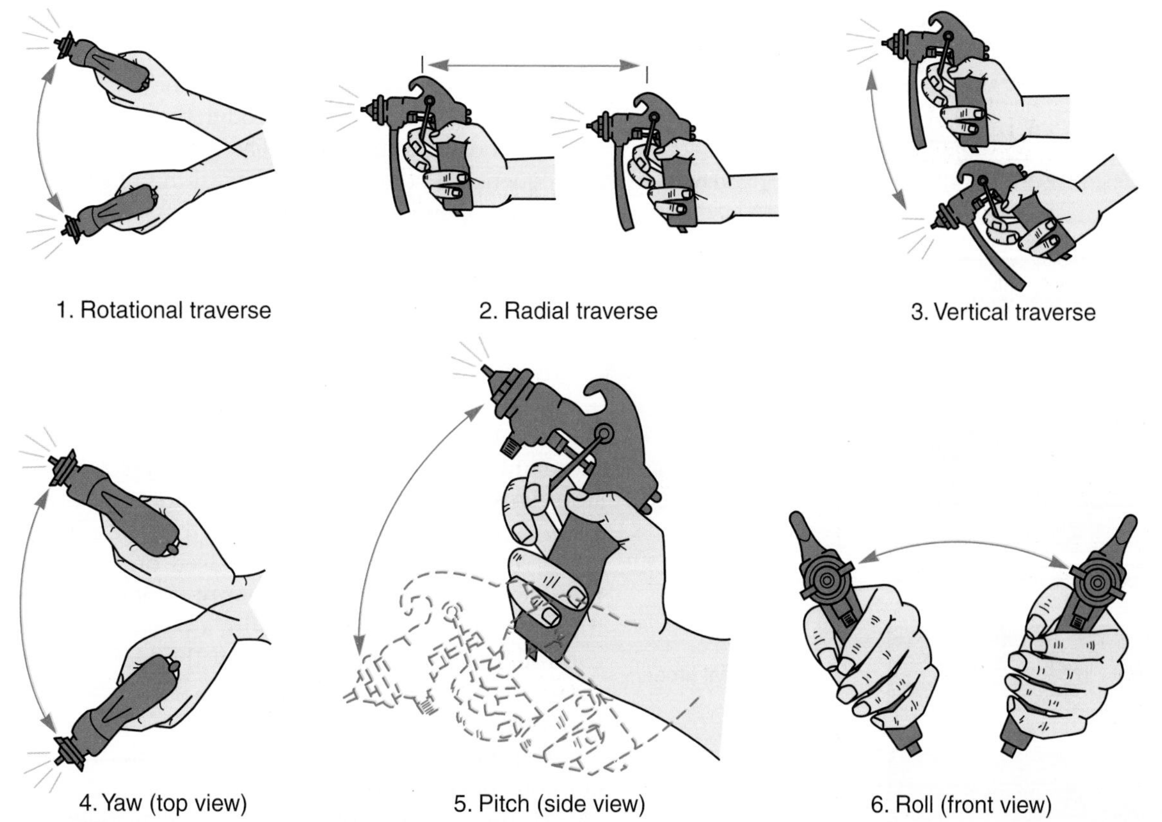

Goodheart-Willcox Publisher

Figure 2-15. The six degrees of freedom are demonstrated by a person using a spray gun. Illustrations 1, 2, and 3 are arm movements. Illustrations 4, 5, and 6 are wrist movements.

2.3 Classifying Robots

Robots can be classified in various ways according to their components, configuration, and use. Two common methods of classifying robots are by the type of control system used and the type of actuator drive used.

2.3.1 Types of Control Systems

Robots may use one of two control systems: non-servo and servo. The earliest type of robot was non-servo, which is considered a non-intelligent robot. Servo robots are classified as either intelligent or highly intelligent. The primary difference between an intelligent and highly intelligent robot is the level at which sensors provide information about the surrounding environment.

Non-Servo Robots

An ***open-loop system*** is a type of control method in which no feedback mechanism is used to compare programmed positions to actual positions. A robot that uses this type of system is called a ***non-servo robot***. Non-servo robots are the simplest robots and are often referred to as limited sequence, pick-and-place, or fixed-stop robots. A good example of an open-loop system is the operating cycle of a washing machine, **Figure 2-16**. At the beginning of the operation, the dirty clothes and

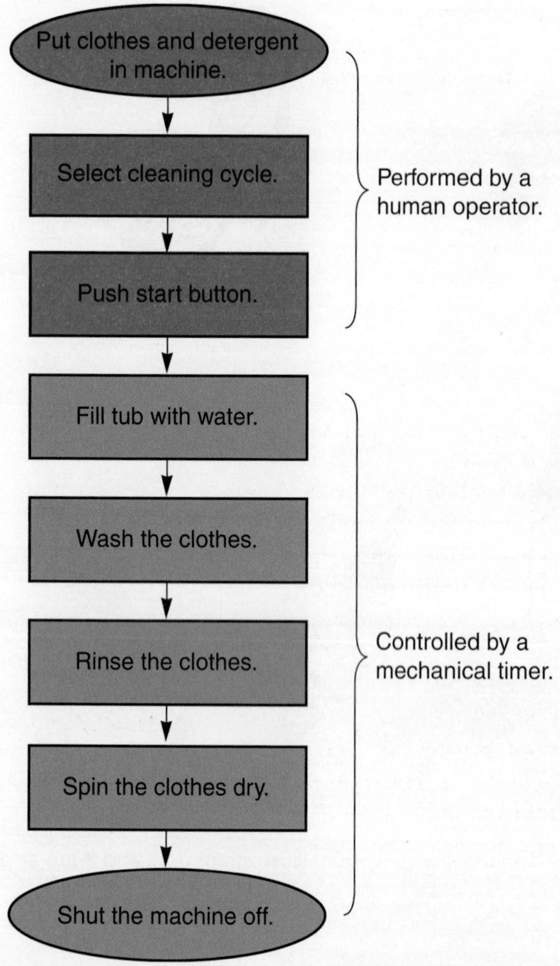

Goodheart-Willcox Publisher

Figure 2-16. This block diagram depicts the sequence of steps performed by a washing machine. Notice that no feedback is used. In such an open-loop control system, the condition of the clothes during the washing operation is not monitored and used to alter the process.

the detergent are placed in the washing machine tub. The cycle selector is set for the proper cleaning cycle, and the machine is activated by the start button. The machine fills with water and begins to go through the various washing, rinsing, and spinning cycles. The machine stops after the set sequence is completed. The washing machine is considered an open-loop system for two reasons:

- The clothes are not examined by sensors during the washing cycle to see if they are clean.
- The completion time for the cycle is not automatically adjusted to compensate for the amount of dirt remaining in the clothes. The completion time is fixed as determined by the cycle selected.

Non-servo robots also have limited movement, and these limitations are usually in the form of a mechanical stop. This form of robot is excellent for repetitive tasks, such as material transfer. You may question if the non-servo robots qualify as robots based on the definition provided by the Robot Institute of America. However, if these robots are equipped with a programmable logic controller (PLC), they easily meet the requirement of a reprogrammable device. This allows them to be classified as robots.

The diagram in **Figure 2-17** represents a pneumatic (air-controlled), non-servo robot.

1. At the beginning of the cycle, the controller sends a signal to the control valve of the manipulator.
2. As the valve opens, air passes into the air cylinder and causes the rod in the cylinder to move. As long as the valve remains open, this rod continues to move until it is restrained by the end stop.
3. After the rod reaches the limit of its travel, a limit switch tells the controller to close the control valve.
4. The controller sends the control valve a signal to close.
5. The controller then moves to the next step in the program and initiates the necessary signals. If the

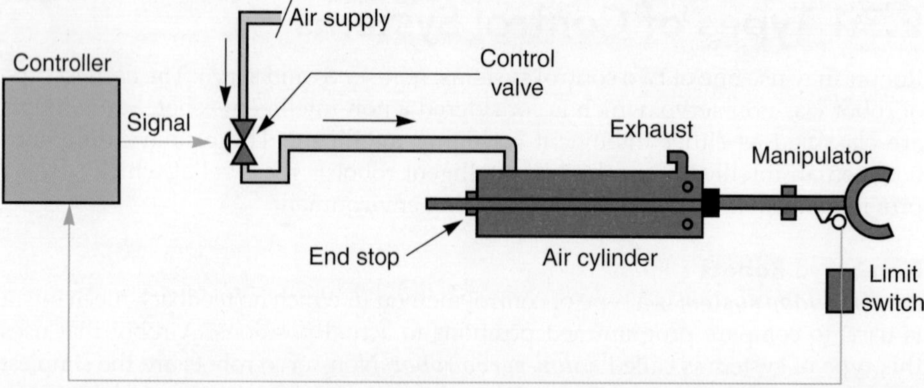

Goodheart-Willcox Publisher

Figure 2-17. In a non-servo system, movement is regulated by devices such as a limit switch, which signals the controller when it is activated.

signals go to the end effector, for example, they might cause the gripper to close in order to grasp an object.

The process is repeated until all the steps in the program have been completed.

Characteristics of non-servo robots include:

- Relatively inexpensive compared to servo robots.
- Simple to understand and operate.
- Precise and reliable.
- Simple to maintain.
- Capable of fairly high speeds of operation.
- Small in size.
- Limited to relatively simple programs.

Servo Robots

A ***closed-loop system*** is a type of control method that allows feedback to affect the output of the robot. A robot that uses this type of system is called a ***servo robot.*** In a closed-loop system, the feedback signal is sent to the servo amplifier. A ***servo amplifier*** is a device that translates signals from the controller into motor voltage and current signals. Servo amplifiers are used in motion control systems where precise control of position or velocity is necessary. In a sense, a servomechanism is a type of control system that detects and corrects for errors. **Figure 2-18** shows a block diagram of a servo robot system.

The principle of servo control can be compared to many tasks performed by human beings. One example is cutting a circle from a piece of stock (wood) on a power band saw, shown in **Figure 2-19**. The machine operator studies the position of the stock to be cut in relation to the cutting edge of the blade. The operator's eye transmits a signal to the brain. The brain compares the actual position to the desired position. The brain then sends a signal to the arms to move the stock beneath the cutting edge of the blade. The eyes are used as a feedback sensing

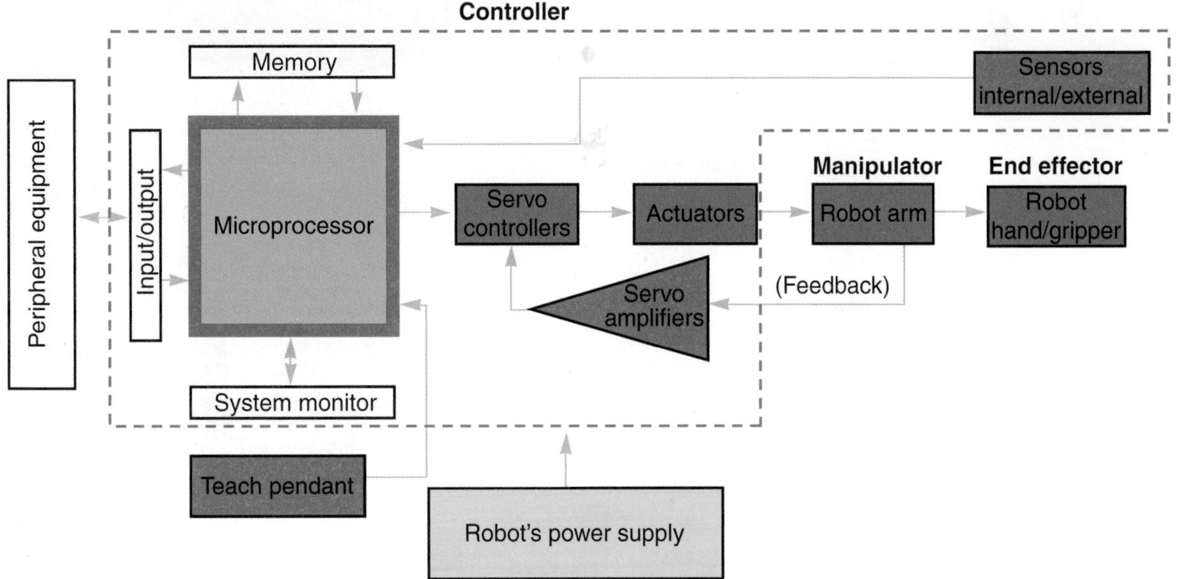

Goodheart-Willcox Publisher

Figure 2-18. A servo robot system, such as the one depicted in this block diagram, might be classified as intelligent or highly intelligent, depending on the level of sensory data it can interpret. Components in the dashed area are part of the control system.

device, and the brain compares desired locations with actual locations. The brain sends signals to the arms to make necessary adjustments. This process is repeated as the operator follows the scribed line during the sawing operation.

The diagram in **Figure 2-20** details one of the axes used in a hydraulic robot and helps to explain its operation.

1. When the cycle begins, the controller searches the program for the desired locations along each axis.
2. Using the feedback signals, the controller determines the actual locations on the various axes of the manipulator.
3. The desired locations and actual locations are compared.

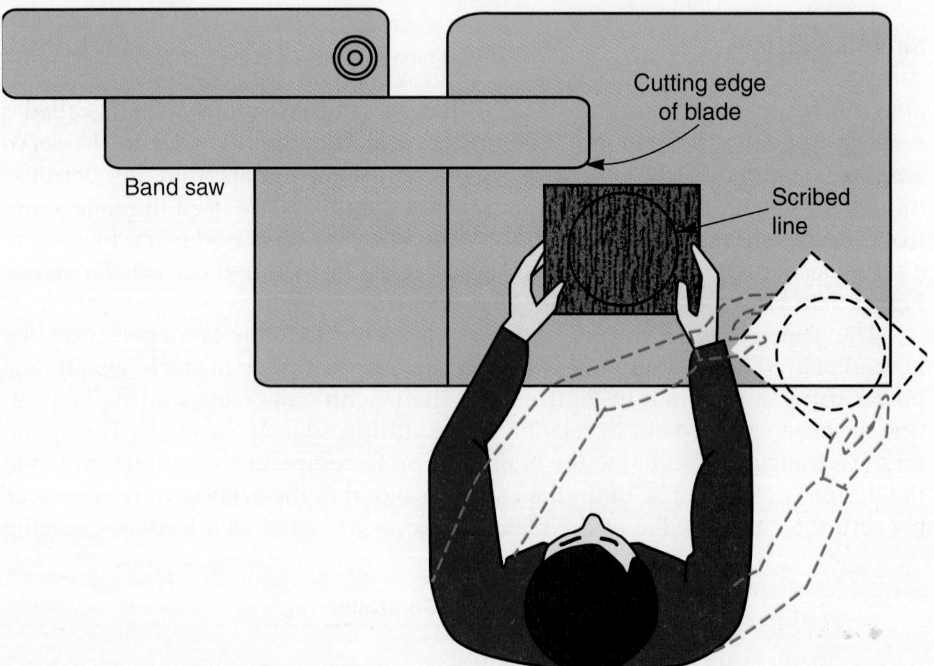

Goodheart-Willcox Publisher

Figure 2-19. Human beings make use of the servomechanism principle for many tasks, such as cutting a circle on a band saw.

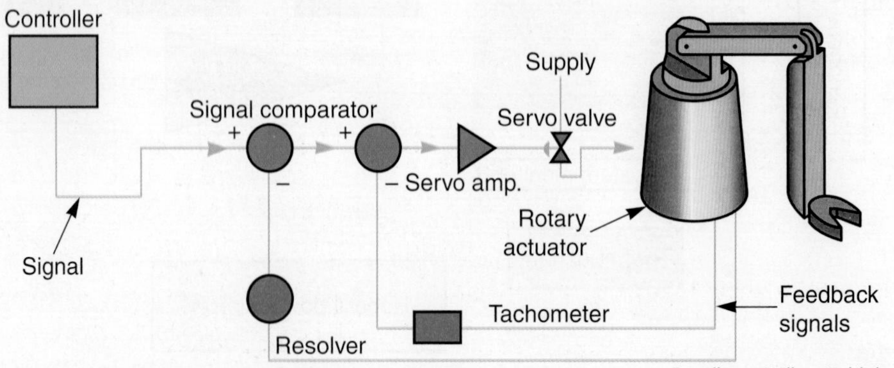

Goodheart-Willcox Publisher

Figure 2-20. Feedback signals from sensors allow the system to make corrections whenever the actual speed or position of the robot does not agree with the values contained in the program.

4. When these locations do not match, an error signal is generated and sent back to the servo amplifier. An ***error signal*** is a message generated when the programmed actions of a robot and the actual actions do not match. The greater the error, the higher the intensity of the signal.
5. These error signals are increased by the servo amplifier and applied to the control valve on the appropriate axis.
6. The valve opens in proportion to the intensity of the signal received. The opened valve admits fluid to the proper actuator to move the various segments of the manipulator.
7. New signals are generated as the manipulator moves.
8. The servo control valves close when there are no more error signals, shutting off the flow of fluid.
9. The manipulator comes to rest at the desired position.
10. The controller then addresses the next instruction in the program, which may be to move to another location or operate some peripheral equipment.

The process is repeated until all steps of the program are completed.

Characteristics of servo robots include:

- Relatively expensive to purchase, operate, and maintain.
- Sophisticated, closed-loop controller.
- Wide range of capabilities.
- Ability to transfer objects from one point to another as well as along a controlled, continuous path.
- Ability to respond to very sophisticated programming.
- A manipulator that can be programmed to avoid obstructions within the work envelope.

2.3.2 Types of Actuator Drives

One common method of classifying robots is by the type of drive required by the actuators. Three types of actuators include:

- Electric actuators that use electric drives.
- Hydraulic actuators, **Figure 2-21**, that use hydraulic drives.
- Pneumatic actuators that use pneumatic drives.

Electric Drive

Three types of motors are commonly used for electric actuator drives: ac servo motors, dc servo motors, and stepper motors. Both ac and dc servo motors have built-in methods for controlling exact position. Many newer robots use servo motors rather than hydraulic or pneumatic ones. Small and medium-size robots commonly use dc servo motors. Because of their high torque capabilities, ac servo motors are found in heavy-duty robots, **Figure 2-22**. A stepper motor is an incrementally controlled dc motor. Stepper motors are rarely used in commercial industrial robots, but they are commonly found in educational robots, **Figure 2-23**.

FANUC Robotics

Figure 2-21. A large hydraulic actuator provides up-and-down motion to the arm of this industrial robot.

Motoman

Figure 2-22. This heavy-duty industrial robot uses two ac servo motors in the operation of the robot arm.

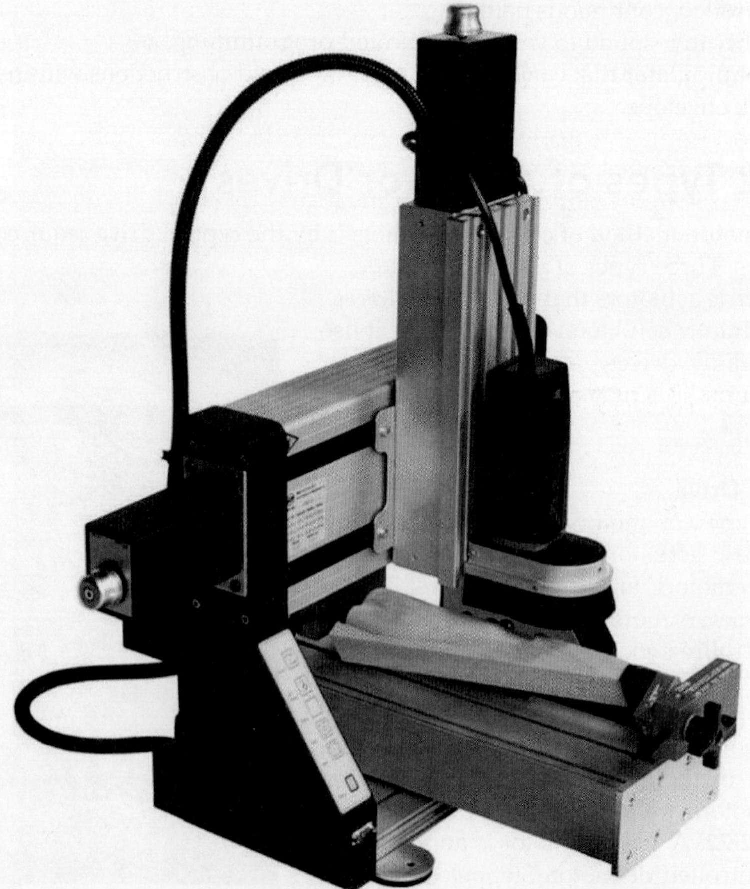

Techno, Inc.

Figure 2-23. DC stepper motors are used on this tabletop educational robot.

Conventional, electric-drive motors are quiet, simple, and can be used in clean-air environments. Robots that use electric actuator drives require less floor space, and their energy source is readily available. However, the conventionally geared drive causes problems of backlash, friction, compliance, and wear. These problems cause inaccuracy, poor dynamic response, need for regular maintenance, poor torque control capability, and limited maximum speed on longer moves. Loads that are heavy enough to stall (stop) the motor can cause damage. Conventional electric-drive motors also have poor output power compared to their weight. This means that a larger, heavier motor must be mounted on the robot arm when a large amount of torque is needed.

The rotary motion of most electric actuator drives must be geared down (reduced) to provide the speed or torque required by the manipulator. However, manufacturers are beginning to offer robots that use ***direct-drive motors***, which eliminate some of these problems. These high-torque motors drive the arm directly, without the need for reducer gears. The prototype of a direct-drive arm was developed by scientists at Carnegie-Mellon University in 1981.

The basic construction of a direct-drive motor is shown in **Figure 2-24**. Coupling the motor with the arm segment to be manipulated eliminates backlash, reduces friction, and increases the mechanical stiffness of the drive mechanism. Compare the design of a robot arm using a direct-drive motor in **Figure 2-25** to one with a conventional electric-drive motor (**Figure 2-22**). Using direct-drive motors in robots results in a more streamlined design. Maintenance requirements are also reduced. Robots that use direct-drive motors operate at higher speeds, with greater flexibility, and greater accuracy than those that use conventional electric-drive motors.

Applications currently being performed by robots with direct-drive motors involve mechanical assembly, electronic assembly, and material handling. These robots will increasingly meet the demands of advanced, high-speed, precision applications.

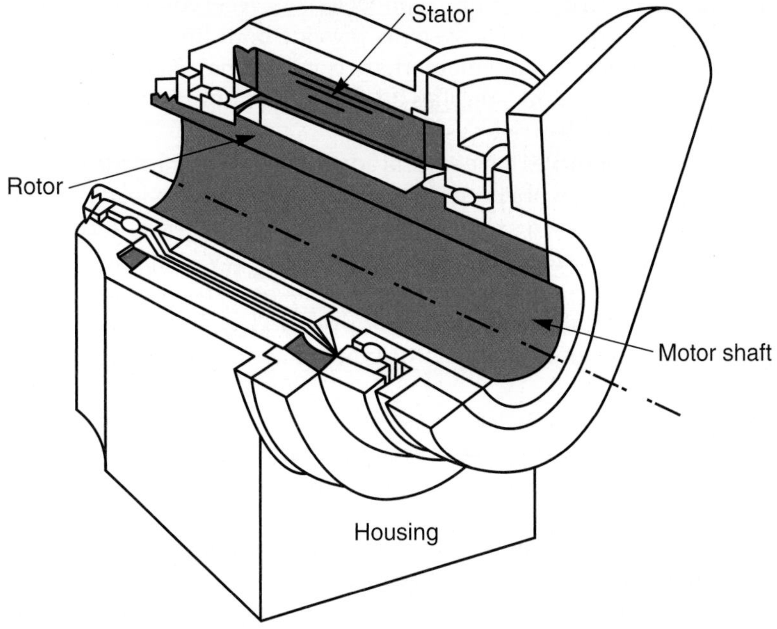

Goodheart-Willcox Publisher

Figure 2-24. This drawing shows the basic construction of a direct-drive motor.

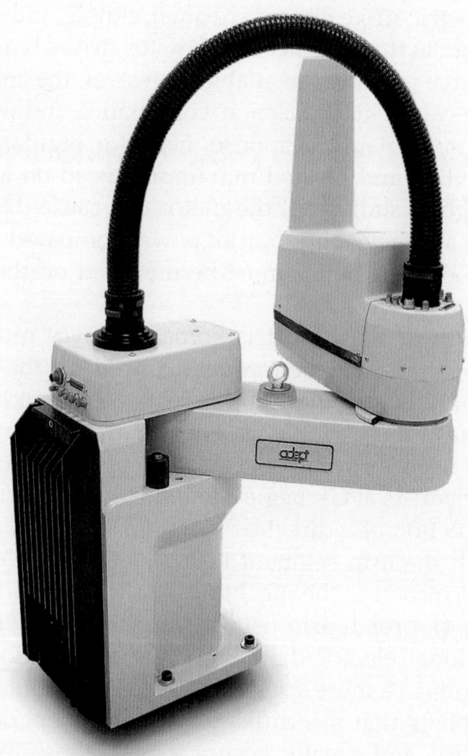

Adept Technology, Inc.

Figure 2-25. Note the simplified mechanical design of this direct-drive robot as compared to the conventional electric-drive robot shown in Figure 2-22.

Hydraulic Drive

Many earlier robots were driven by hydraulic actuator drives. A ***hydraulic drive*** system uses fluid and consists of a pump connected to a reservoir tank, control valves, and an actuator. Hydraulic drive systems provide both linear and rotary motion using a much simpler arrangement than conventional electric-drive systems, **Figure 2-26**. The storage tank supplies a large amount of instant power, which is not available from electric-drive systems.

Hydraulic actuator drives have several advantages. They provide precise motion control over a wide range of speeds. They can handle heavy loads on the end of the manipulator, can be used around highly explosive materials, and are not easily damaged when quickly stopped while carrying a heavy load. However, they are expensive to purchase and maintain and are not energy efficient. Hydraulic actuator drives are also noisier than electric actuator drives. They are not recommended for clean-room environments due to the possibility of hydraulic fluid leaks.

Pneumatic Drive

Pneumatic drive systems are power systems that use air-driven actuators. Since air is a fluid substance, many of the same principles that apply to hydraulic systems are applicable to pneumatic systems. Pneumatic and hydraulic motors and cylinders are very similar. Since most industrial plants have a compressed air system running throughout assembly areas, compressed air is an economical and readily available energy source. This makes the installation of robots that use pneumatic actuator drives easier and less costly than that of hydraulic robots. For lightweight

ABB Robotics

Figure 2-26. Large robots that use hydraulic drive systems perform a demonstration at a manufacturing trade show.

pick-and-place applications that require both speed and accuracy, a pneumatic robot is potentially a good choice.

Pneumatic actuator drives work at high speeds and are most useful for small or medium loads. They are economical to operate and maintain. They can be used in explosive atmospheres (in which combustible dust or gases are mixed with air). However, additional components are required to achieve the precise placement and positioning possible with a hydraulic system. These components are discussed in later chapters. It is also difficult to keep the air as clean and dry as the control system requires. Robots that use pneumatic actuator drives are noisy and vibrate as the air cylinders and motors stop.

2.4 Robot Configurations

Robots come in many sizes and shapes. The type of coordinate system used by the manipulator also varies. The type of coordinate system, the arrangement of joints, and the length of the manipulator segments all help determine the shape of the work envelope. To identify the maximum work area, a point on the robot wrist is used rather than the tip of the gripper or the end of the tool bit. Therefore, the work envelope is slightly larger when the tip of the tool is considered.

Work envelopes vary from one manufacturer to another, depending on the exact design of the manipulator. Combining different configurations in a single robot can result in another set of possible work envelopes. Before choosing a particular robot configuration, the application must be studied carefully to determine the precise work envelope requirements.

Some work envelopes have a regular shape, such as a circle, square, or triangle. Other work envelopes have an irregular shape. One method of classifying a robot is by its configuration and the resulting work envelope. Some robots may be equipped for more than one configuration. The four basic configurations for robots are revolute, Cartesian, cylindrical, and spherical. Each configuration is used for specific applications.

2.4.1 Revolute Configuration (Articulated)

The ***revolute configuration*** is a layout that is irregularly shaped and requires a jointed-arm robot. Also called the jointed-arm configuration, this layout is the most common. These revolute configuration robots are often referred to as anthropomorphic because their movements closely resemble those of the human body. Rigid segments resemble the human forearm and upper arm. Various joints mimic the action of the wrist, elbow, and shoulder. A joint called the *sweep* represents the waist. A revolute coordinate robot performs in an irregularly shaped work envelope. There are two basic revolute configurations: vertically articulated and horizontally articulated.

The vertically articulated configuration, shown in **Figure 2-27**, has five revolute (rotary) joints. A vertically articulated robot is depicted in **Figure 2-28**. The jointed-arm, vertically articulated robot is useful for painting applications because of the long reach this configuration allows.

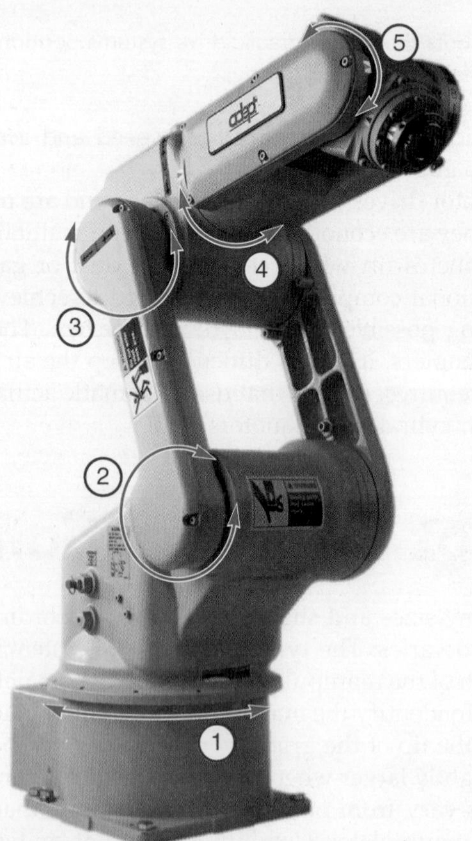

Adept Technology, Inc.

Figure 2-27. These five revolute (rotary) joints are associated with the basic manipulator movements of a vertically articulated robot.

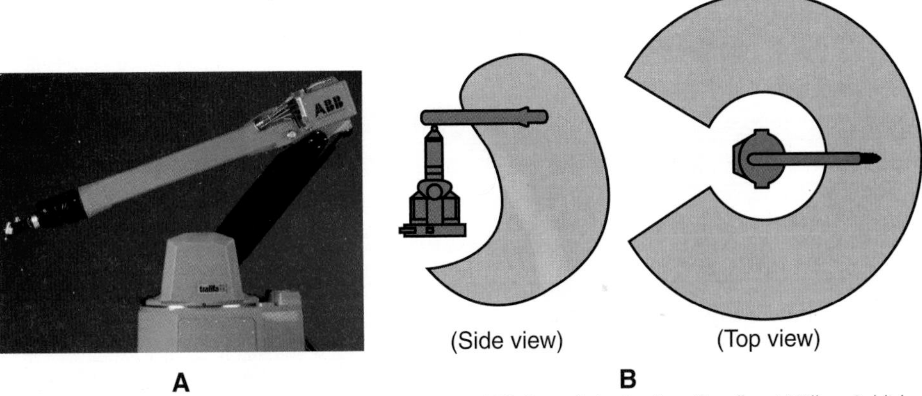

ABB Graco Robotics, Inc.; Goodheart-Willcox Publisher

Figure 2-28. A—This painting robot is vertically articulated. B—The shaded areas represent a top view of the work envelope for this robot.

The horizontally articulated configuration generally has one vertical (linear) and two revolute joints. It was designed by Professor Makino of Yamanashi University, Japan, and it is also called the ***SCARA configuration***. SCARA is an acronym for *selective compliance assembly robot arm.* The primary objective was a configuration that would be fairly yielding in horizontal motions and rather rigid in vertical motions. The basic SCARA configuration, **Figure 2-29**, is an adaptation of the cylindrical configuration, which is described later in this chapter. The SCARA robot shown in **Figure 2-30** is designed for clean-room applications, such as wafer and disk handling in the electronics industry.

SCARA robots are ideally suited for operations in which the vertical motion requirements are small compared to the horizontal motion requirements. Such an application would be assembly work where parts are picked up from a parts holder and moved along a nearly horizontal path to the unit being assembled.

The revolute configuration has several advantages. It is the most versatile configuration, and it provides a larger work envelope than the Cartesian, cylindrical, or spherical configurations. It also offers a more flexible reach than the other configurations, making it ideally suited to welding and spray painting operations. However, there are also disadvantages to the revolute configuration. It requires a very sophisticated controller, and programming is more complex than for the other three configurations. The accuracy, load-carrying capacity, dynamics, and ability of the robot to repeat a movement accurately varies depending on the placement of the workpiece within the work envelope. This configuration becomes less stable as the arm approaches its maximum reach. Industrial applications of revolute configurations are discussed in more detail in Chapter 4.

Typical applications of revolute configurations include the following:

- Automatic assembly.
- Parts and material handling.
- Multiple-point light machining operations.
- In-process inspection.
- Palletizing.
- Machine loading and unloading.
- Machine vision.
- Material cutting.
- Material removal.
- Thermal coating.

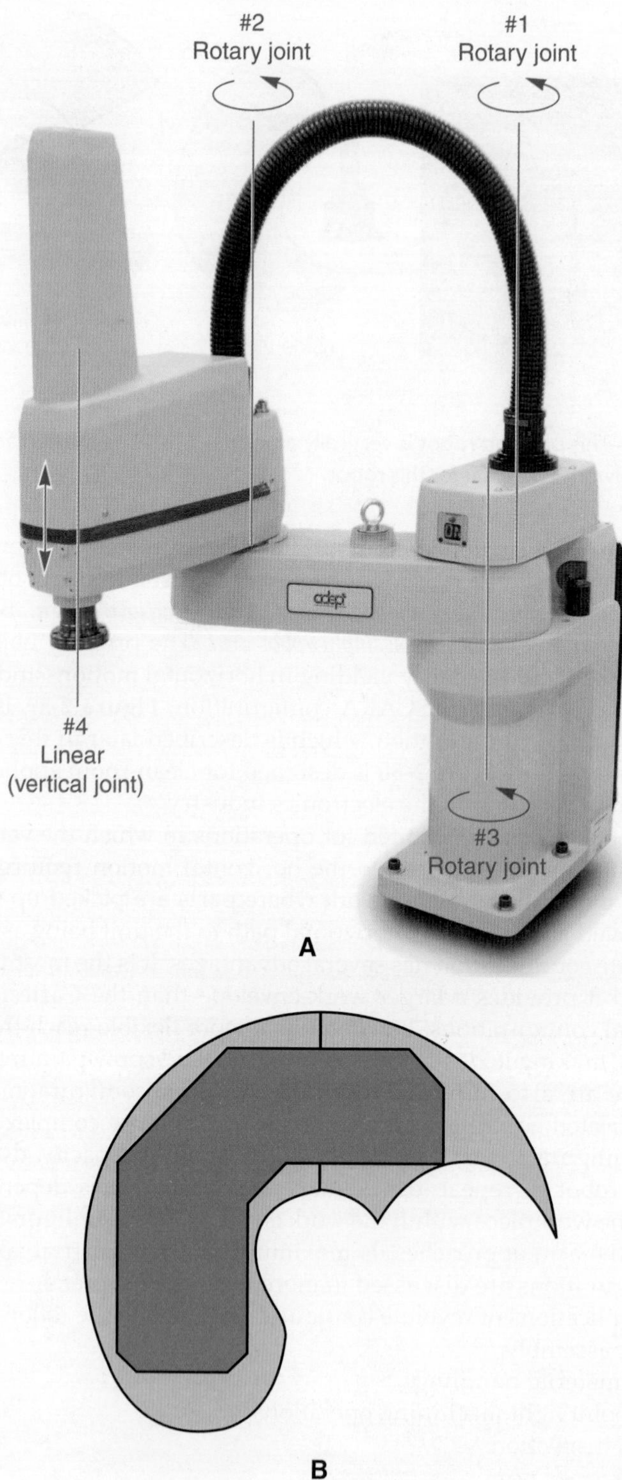

Adept Technology, Inc.

Figure 2-29. A—This is an example of a basic SCARA robot configuration. Note the three rotary joints and the single vertical joint used in this horizontally articulated configuration. B—This is a top view of the work envelope of a typical SCARA horizontally articulated robot configuration. This work envelope is sometimes referred to as the folded book configuration.

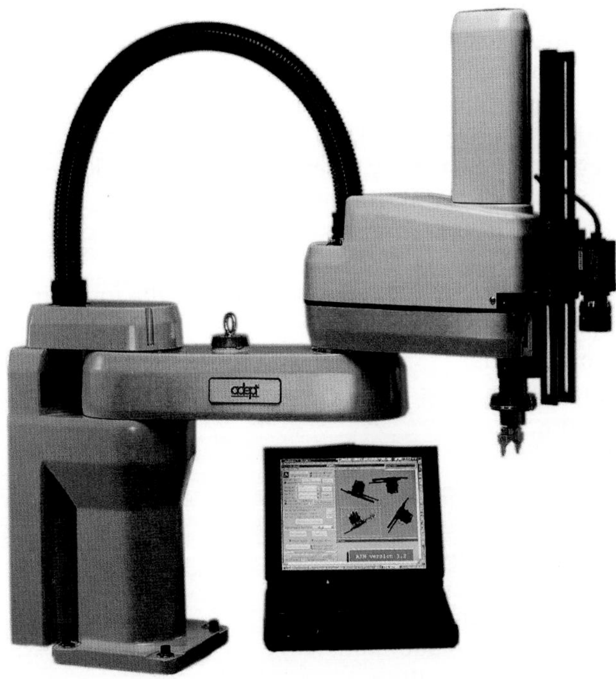

Adept Technology, Inc.

Figure 2-30. This SCARA robot is specifically designed for clean-room applications.

- Paint and adhesive application.
- Welding.
- Die casting.

2.4.2 Cartesian Configuration

The ***Cartesian configuration*** is a type of layout that allows the arm movement of a robot along three intersecting perpendicular straight lines that are referred to as the X, Y, and Z axes (**Figure 2-31**). Because movement can start and stop simultaneously along all three axes, motion of the tool tip is smoother. This allows the robot to move directly to its designated point instead of following trajectories parallel

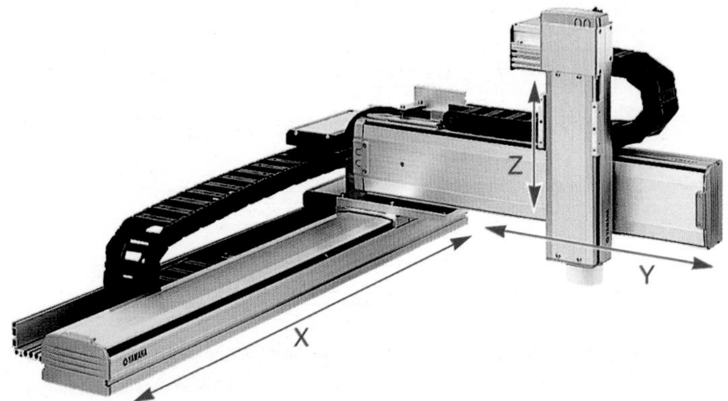

Yamaha

Figure 2-31. A robot with a Cartesian configuration moves along X, Y, and Z axes.

to each axis, **Figure 2-32**. The rectangular work envelope of a typical Cartesian configuration is illustrated in **Figure 2-33**. (Refer to **Figure 2-13** for an example of a Cartesian robot mounted on a gantry or overhead framework.)

One advantage of a Cartesian configuration is that the linear movement allows for simpler controls, **Figure 2-34**. They also have a high degree of mechanical rigidity, accuracy, and repeatability. They can carry heavy loads, and this weight-lifting capacity does not vary at different locations within the work envelope. A disadvantage of the Cartesian configuration is that the robot is generally limited in movement to a small, rectangular work space.

Typical applications for Cartesian robots include the following:

- Assembly.
- Machining operations.
- Adhesive application.
- Surface finishing.
- Inspection.
- Waterjet cutting.
- Welding.
- Nuclear material handling.
- Robotic X-ray and neutron radiography.
- Automated lathe loading and operation.
- Remotely operated decontamination.
- Advanced munitions handling.

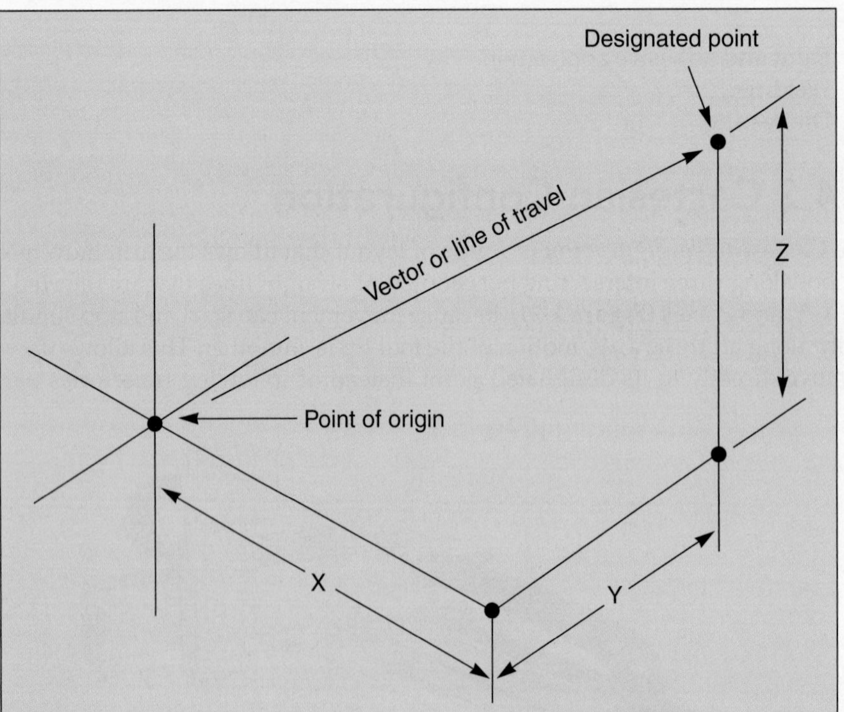

Goodheart-Willcox Publisher

Figure 2-32. With a Cartesian configuration, the robot can move directly to a designated point rather than moving in lines parallel to each axis. In this example, movement is along the vector connecting the point of origin and the designated point rather than first along the X axis, then Y, and then Z.

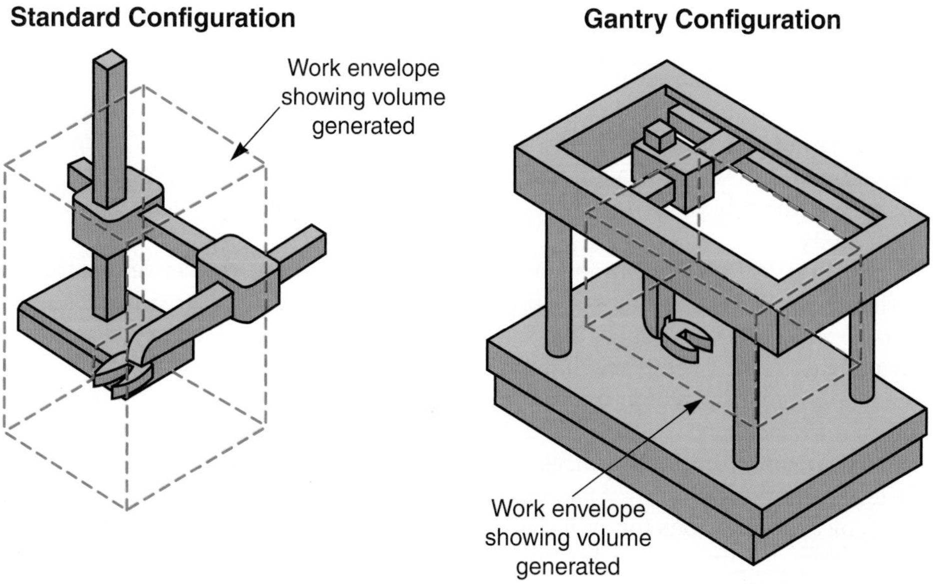

Standard Configuration

Work envelope
showing volume
generated

Gantry Configuration

Work envelope
showing volume
generated

Goodheart-Willcox Publisher

Figure 2-33. In either the standard or gantry construction, a Cartesian configuration robot creates a rectangular work envelope.

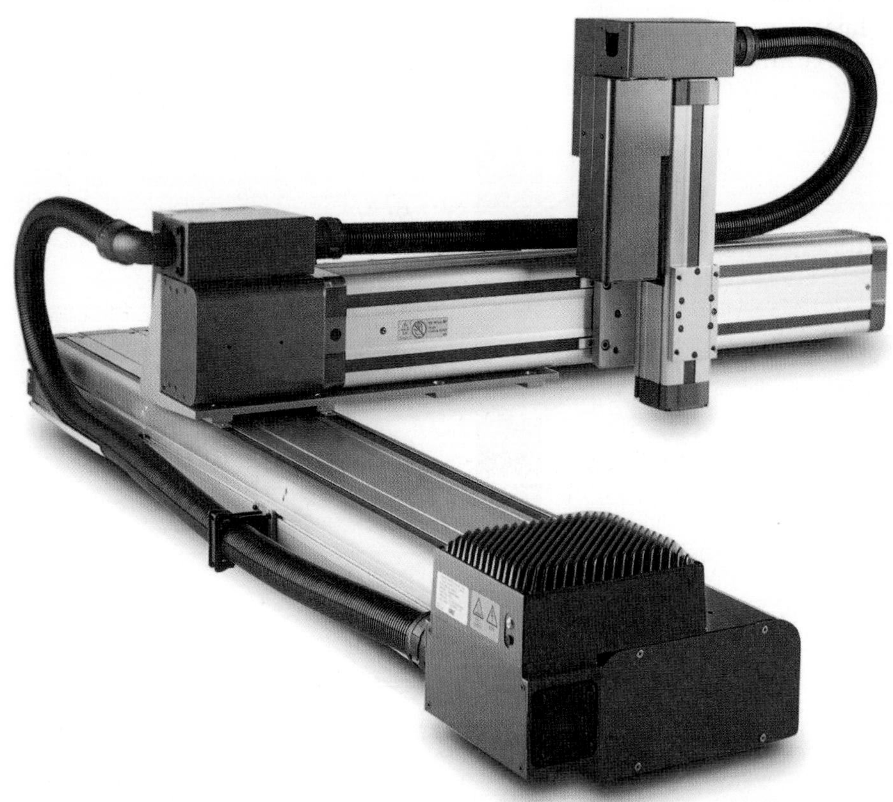

Adept Technology, Inc.

Figure 2-34. This robot has a Cartesian configuration and is used for high-precision jobs.

2.4.3 Cylindrical Configuration

A ***cylindrical configuration*** is a type of layout that consists of two slides placed at a 90° angle and mounted on a rotary axis, **Figure 2-35**. Reach is accomplished as the arm of the robot moves in and out. For vertical movement, the carriage moves up and down on a stationary post, or the post can move up and down in the base of the robot. Movement along the three axes traces points on a cylinder, **Figure 2-36**.

A cylindrical configuration generally results in a larger work envelope than a Cartesian configuration. Robots with a cylindrical configuration are ideally suited for pick-and-place operations. However, cylindrical configurations have some disadvantages. Their overall mechanical rigidity is reduced because robots with a rotary axis must overcome the inertia of the object when rotating. Their repeatability and accuracy is also reduced in the direction of rotary movement. The cylindrical configuration requires a more sophisticated control system than the Cartesian configuration.

Typical applications for cylindrical configurations include the following:

- Machine loading and unloading.
- Investment casting.
- Conveyor pallet transfers.
- Foundry and forging applications.
- General material handling and special payload handling and manipulation.
- Meat packing.
- Coating applications.
- Assembly.
- Injection molding.
- Die casting.

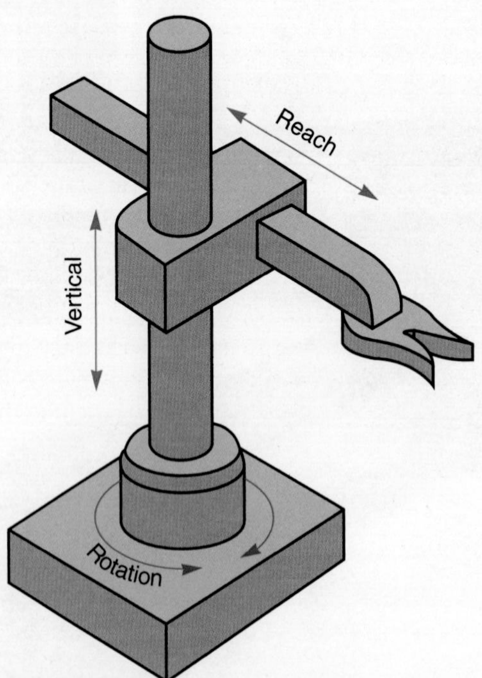

Goodheart-Willcox Publisher

Figure 2-35. The basic configuration for a cylindrical robot includes two slides placed at a 90° angle and mounted on a rotary axis.

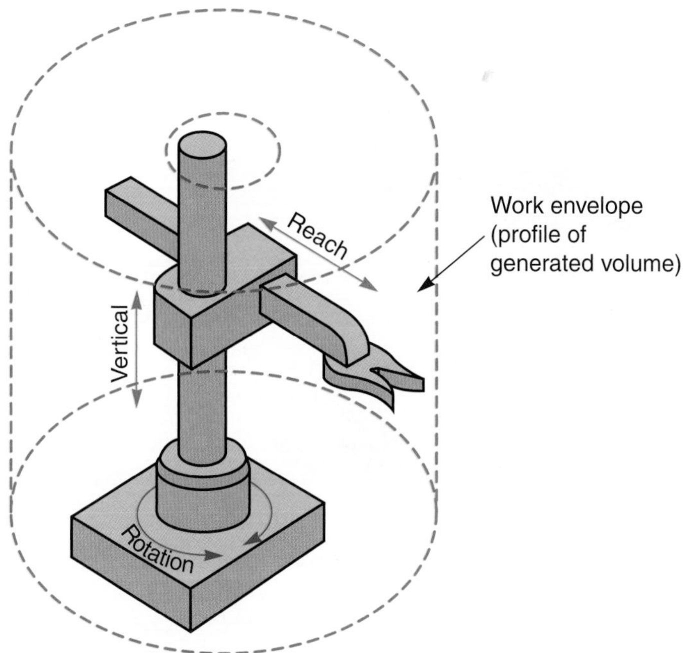

Goodheart-Willcox Publisher

Figure 2-36. Motion along the three axes traces points on a cylinder to form the work envelope.

2.4.4 Spherical Configuration (Polar)

The ***spherical configuration***, sometimes referred to as the polar configuration, is a layout that allows action similar to that of the turret on a military tank. A pivot point gives the robot its vertical movement, and a telescoping boom extends and retracts to provide reach, **Figure 2-37**. Rotary movement occurs around an axis perpendicular to the base. **Figure 2-38** illustrates the work envelope profile of a typical spherical configuration robot.

The spherical configuration generally provides a larger work envelope than the Cartesian or cylindrical configurations. The design is simple and provides good weight-lifting capabilities. This configuration is suited to applications where a small amount of vertical movement is adequate, such as loading and unloading a punch press. Its disadvantages include reduced mechanical rigidity and the need for a more sophisticated control system than either the Cartesian or cylindrical configurations. The same problems occur with inertia and accuracy in this configuration as they do in the cylindrical configuration. Vertical movement is limited as well.

Typical applications of spherical configurations include the following:

- Die casting.
- Injection molding.
- Forging.
- Machine tool loading.
- Heat treating.
- Glass handling.
- Parts cleaning.
- Dip coating.
- Press loading.
- Material transfer.
- Stacking and unstacking.

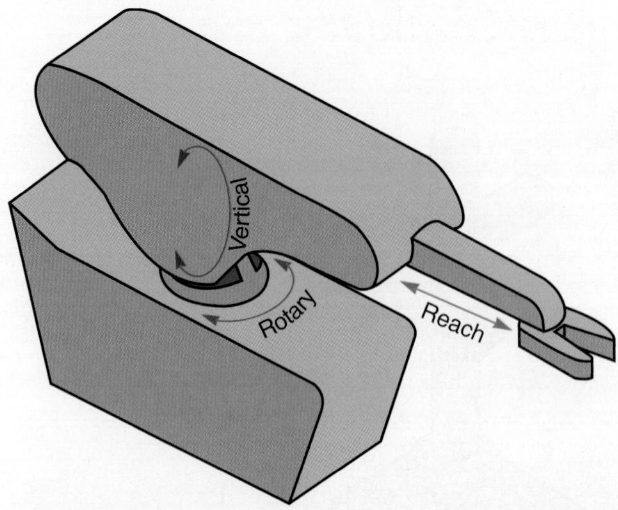

Goodheart-Willcox Publisher

Figure 2-37. A pivot point enables the spherical configuration robot to move vertically. It also can rotate around a vertical axis.

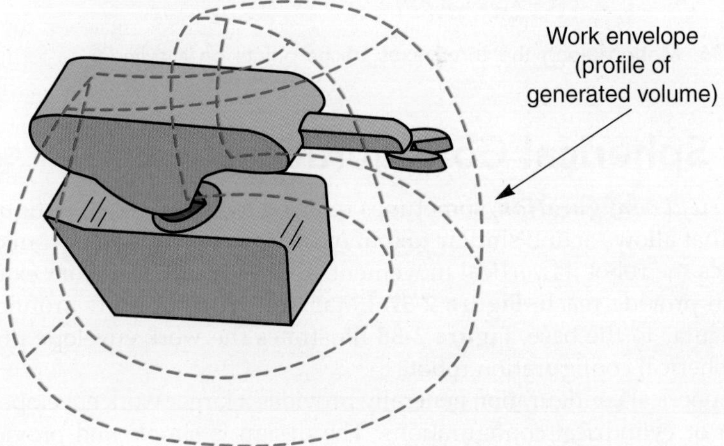

Goodheart-Willcox Publisher

Figure 2-38. The work envelope of this robot takes the shape of a sphere.

Robotics Innovations: The Grabit Gripper

The Grabit company has developed a robotic gripper that uses static electricity to grip and hold objects. Until now, a typical robot gripper would use pressure to grasp objects, and often a set pressure would be used to grasp all objects. A slight improvement to a gripper that uses a set pressure would be a gripper with sensor feedback that adjusts pressure for different objects. The Grabit gripper is different, however, because it uses static electricity to grasp objects. The Grabit gripper can lift delicate and heavy objects alike without damaging them. Innovations such as this keep robot technology on the cutting edge. Robots will have to work in closer proximity to humans in the future, so they will need to be strong and gentle.

Photo courtesy of Grabit, Inc.

2.4.5 Special Configurations

Many industrial robots use combinations or special modifications of the four basic configurations. The robot pictured in **Figure 2-39A** uses an articulated configuration, but the base does not rotate horizontally. It is designed to bend over backward in order to grasp objects behind it. This feature makes it possible to install these robots very close to other equipment. This minimizes space requirements while maintaining a large, effective work envelope, **Figure 2-39B**. These robots are used in applications such as spot welding and material handling.

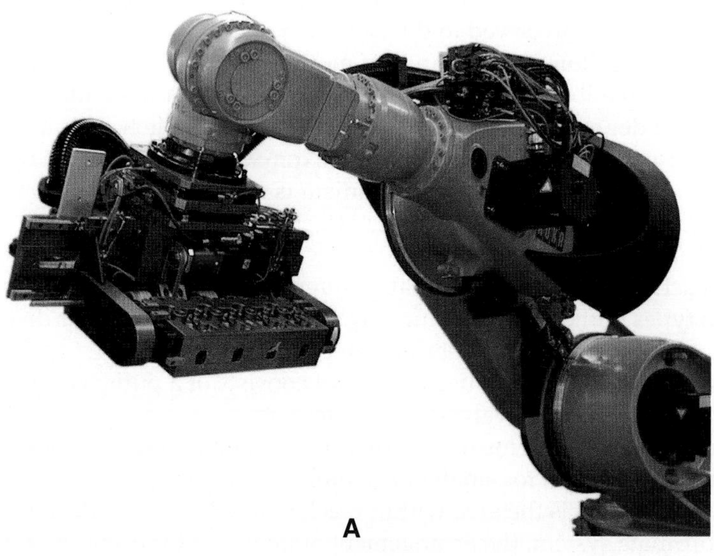

A

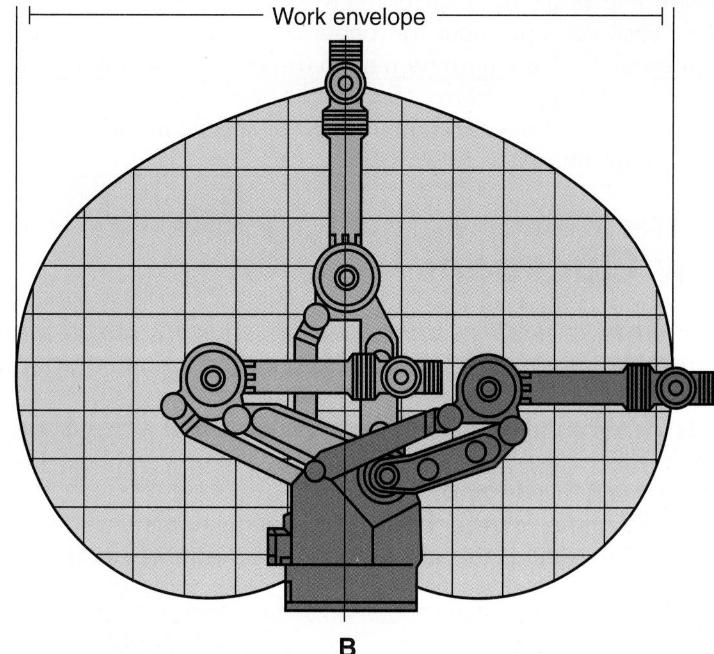

B

Schunk; Goodheart-Willcox Publisher

Figure 2-39. A—This heavy-duty robot literally bends over backward. B—The work envelope for this robot is large.

Summary

- Robots consist of components that include the controller, the manipulator, an end effector, a power supply, and a means of programming.
- The controller coordinates all movements of the mechanical system and receives input from the immediate environment through sensors.
- The manipulator is the arm of the robot that moves materials or tools to provide useful work. The end effector is a device attached to the end of the manipulator that is used to grasp and move objects.
- The power supply of a robot provides energy for the controller and actuators.
- The means for programming, such as a teach pendant, is used to record instructions that are saved to the controller.
- Degrees of freedom are the range of motion for a robot. The three basic degrees of freedom are movement along the X axis, Y axis, and Z axis. Three additional degrees provided by the wrist are called pitch, yaw, and roll.
- Robots use one of two control systems. Non-servo robots use an open-loop system in which no feedback mechanism is used. Servo robots use a closed-loop system that allows feedback to affect the output of the robot.
- An actuator is a motor or valve that converts power into robot movement. Robot actuators include electrical, pneumatic, and hydraulic drive systems.
- Three types of motors are commonly used for electric actuator drives: ac servo motors, dc servo motors, and stepper motors.
- A hydraulic drive system uses fluid and consists of a pump connected to a reservoir tank, control valves, and an actuator.
- Pneumatic drive systems use air-driven actuators. They work at high speeds and are most useful for small or medium loads.
- A work envelope is the area within reach of a robotic end effector. The type of coordinate system, the arrangement of joints, and the length of the arm segments determine the shape of the work envelope.
- The four basic configurations for robots are revolute, Cartesian, cylindrical, and spherical. Each configuration has a different work envelope shape and is used for specific applications.
- Many industrial robots use combinations or special modifications of the four basic configurations.

Review Questions

Answer the following questions using the information provided in this chapter.

1. Identify the five major components of a robot and explain the purpose of each.
2. What is the technical name for the device that acts as a "hand" for a robot?
3. Name the three types of power supplies used to power robots. List the advantages and disadvantages of each.
4. In terms of degrees of freedom, explain why the human hand is able to accomplish movements that are more fluid and complex than those produced by a robot.

5. List and explain the six degrees of freedom used for robots.
6. What are the two types of control systems used to classify robots? What is the main difference between the two types of control systems?
7. Servo robots can be classified as intelligent or highly intelligent. Explain the difference between these two classifications.
8. What type of robots are considered open-loop systems? Explain what *open-loop* means.
9. Servo robots are considered closed-loop systems. Sketch a diagram of a closed-loop system and explain how it works.
10. What are the three types of actuator drives used to classify robots? What are three types of motors commonly used for electric actuator drives?
11. What determines the shape of the work envelope for a robot?
12. Why should you be concerned about the work envelope shape when installing a robot for a particular application?
13. What are the four basic configurations for robots? List some advantages and disadvantages of each.

Learning Extensions

1. Visit the Fanuc Robotics website and select videos from the side menu. This website shows a number of different robots performing various tasks. As you watch these video files, try to identify the robot configuration that is performing the task.
2. Visit the websites for the following robot manufacturers. Locate the products they manufacture and identify the robots by the configurations discussed in this chapter.
   - Adept Technology, Inc.
   - Motoman, Inc.
   - Apollo Seiko
   - Kawasaki Robotics (USA), Inc.
   - ABB Robotics

Advanced Analysis

1. From your website searches, find a robot configuration that seems to be manufactured more than the others. Identify this robot configuration and explain why you think this configuration is used frequently.
2. Search online for robot manufacturing companies, select a Cartesian robot application, and describe in detail how this operation is accomplished.
3. Search online for robot manufacturing companies, select a revolute robot application, and describe in detail how this operation is accomplished.
4. Search online for robot manufacturing companies, select a spherical robot application, and describe in detail how this operation is accomplished.
5. Search online for robot manufacturing companies, select a cylindrical robot application, and describe in detail how this operation is accomplished.

Chapter 3

Safety

Outline

3.1 General Safety
3.2 Safety with Tools and Equipment
3.3 Robotic Safety Considerations

Objectives

Upon completion of this chapter, you will be able to:

- Discuss general safety practices for the laboratory and workshop.
- Discuss safety practices related to tools and equipment.
- Describe guidelines, barriers, sensors, and overload protection for robotics safety.

Technical Terms

circuit breaker
detents
electrical shock
fire extinguisher
first aid kit
fuse
grounding
hand tool
light curtain
lockout/tagout procedure
mechanical fuse
National Electrical Code (NEC)
Occupational Safety and Health Administration (OSHA)
personal protective equipment (PPE)
power tool
preloaded spring
pressure sensitive safety mat

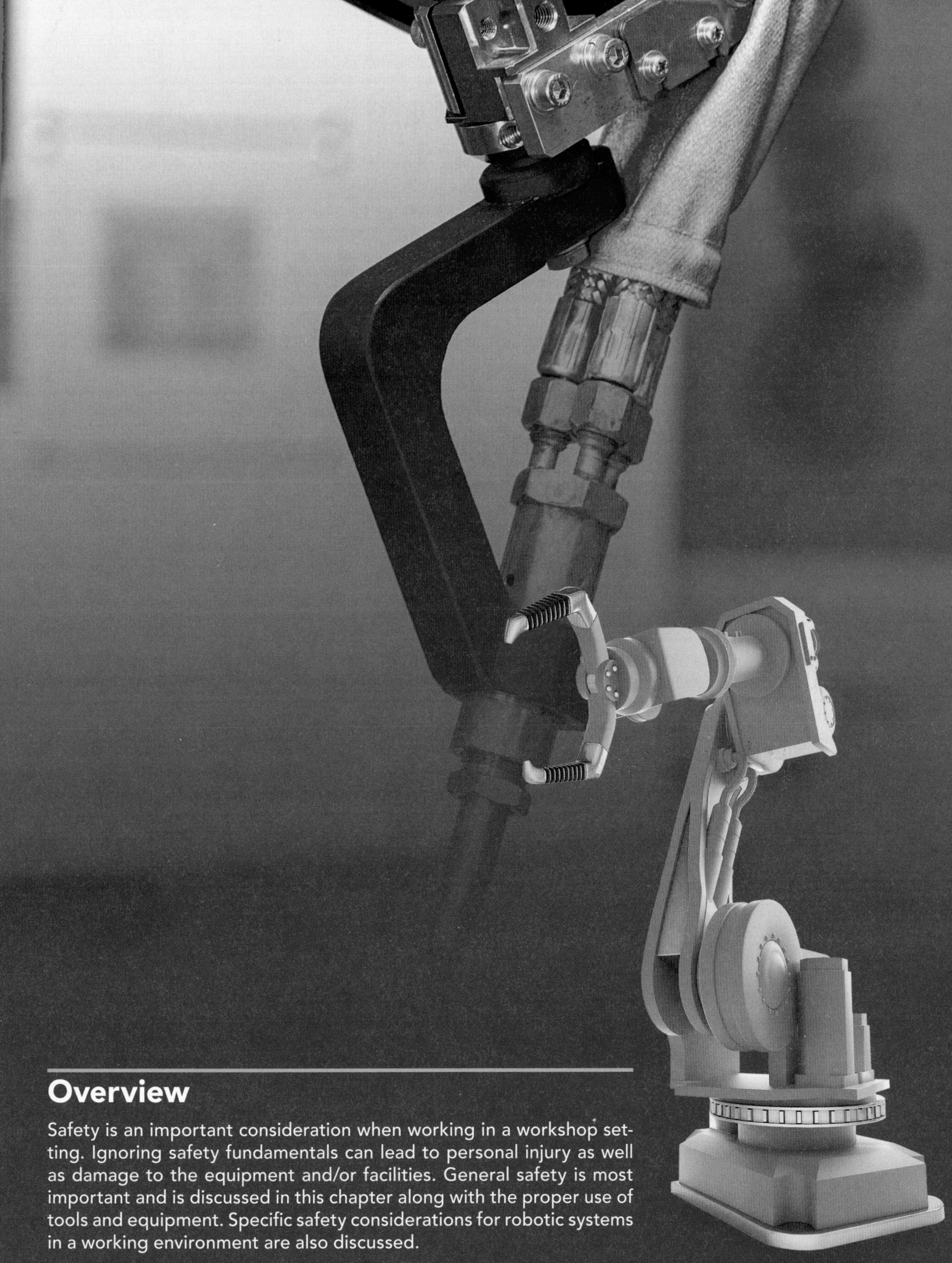

Overview

Safety is an important consideration when working in a workshop setting. Ignoring safety fundamentals can lead to personal injury as well as damage to the equipment and/or facilities. General safety is most important and is discussed in this chapter along with the proper use of tools and equipment. Specific safety considerations for robotic systems in a working environment are also discussed.

3.1 General Safety

General safety awareness is very important. While many safety hazards are easy to see or recognize, others may not be as readily apparent. For this reason, safety should be based on an understanding of basic rules and principles. Working safely is important in order to protect yourself and those working around you. The equipment in a lab or work area should be set up and placed to ensure the safety of operators and bystanders. Approved safety equipment should always be used rather than equipment that has been improvised for the sake of economy. All equipment should be made as safe to operate as is reasonably possible. This is especially true for devices and circuits that are built in a lab for experimentation. Many injuries are the result of taking shortcuts or performing work with too much haste. Following correct work procedures will help you avoid such accidents or injuries. Do not operate equipment or systems unless you have been fully trained in their use.

Workers in many jobs need to wear personal protective equipment (PPE). ***Personal protective equipment (PPE)*** is clothing or devices worn to shield the worker's body and help prevent injuries. Boots with steel toes, safety glasses or goggles, earplugs, gloves, respirators, and hard hats are examples of PPE. Always wear or use the PPE that is recommended for the job you are doing.

3.1.1 Laboratory Safety and Practices

Accidents and injuries in a laboratory can be related to electric current or fires. An ***electrical shock*** is an injury that occurs when a person comes into contact with electric current. The severity of the injury can vary from a mild burn to internal damage or cardiac arrest. The severity of the injury is related to the type of current and how long the person is in contact with the power source. Work surfaces in a laboratory should be covered with a material that does not conduct electrical current to avoid potential problems. An exception to this would be the use of electrostatic mats and wristbands to prevent static discharge to sensitive microelectronic circuits. The floor of a laboratory should also be made of or covered with nonconducting materials. Concrete floors should be covered with rubber tile or vinyl.

Fire extinguishers are safety devices that contain substances that can be used to put out flames. Typically, the substance is in a pressurized container and can be sprayed on the flames. Extinguishers that contain different substances are used to put out different types of fires, **Figure 3-1.** Fire extinguishers of the proper type should be placed in easily accessible locations. Extinguishers should be handled with caution. Workers should be trained to use and maintain the extinguishers properly.

Equipment in the laboratory should be plainly marked with hazard warning or safety labels, **Figure 3-2**. Workers should be trained to know and follow hazard warning symbols. Failure to do so may result in accidents or injuries.

Storage areas for supplies and equipment should be well maintained and properly organized. Good organization of equipment, tools, and supplies helps keep these items in good condition. Tools and small equipment should be maintained and stored in a tool panel, tool box, or marked storage area. Tools that have insulated handles should be used when working around electrical equipment that has the power turned on. Tools and equipment plugged into electrical outlets should be wired with grounded cords and plugs. ***Grounding*** is providing a path that allows current to flow from an electrical system or device to the ground in the

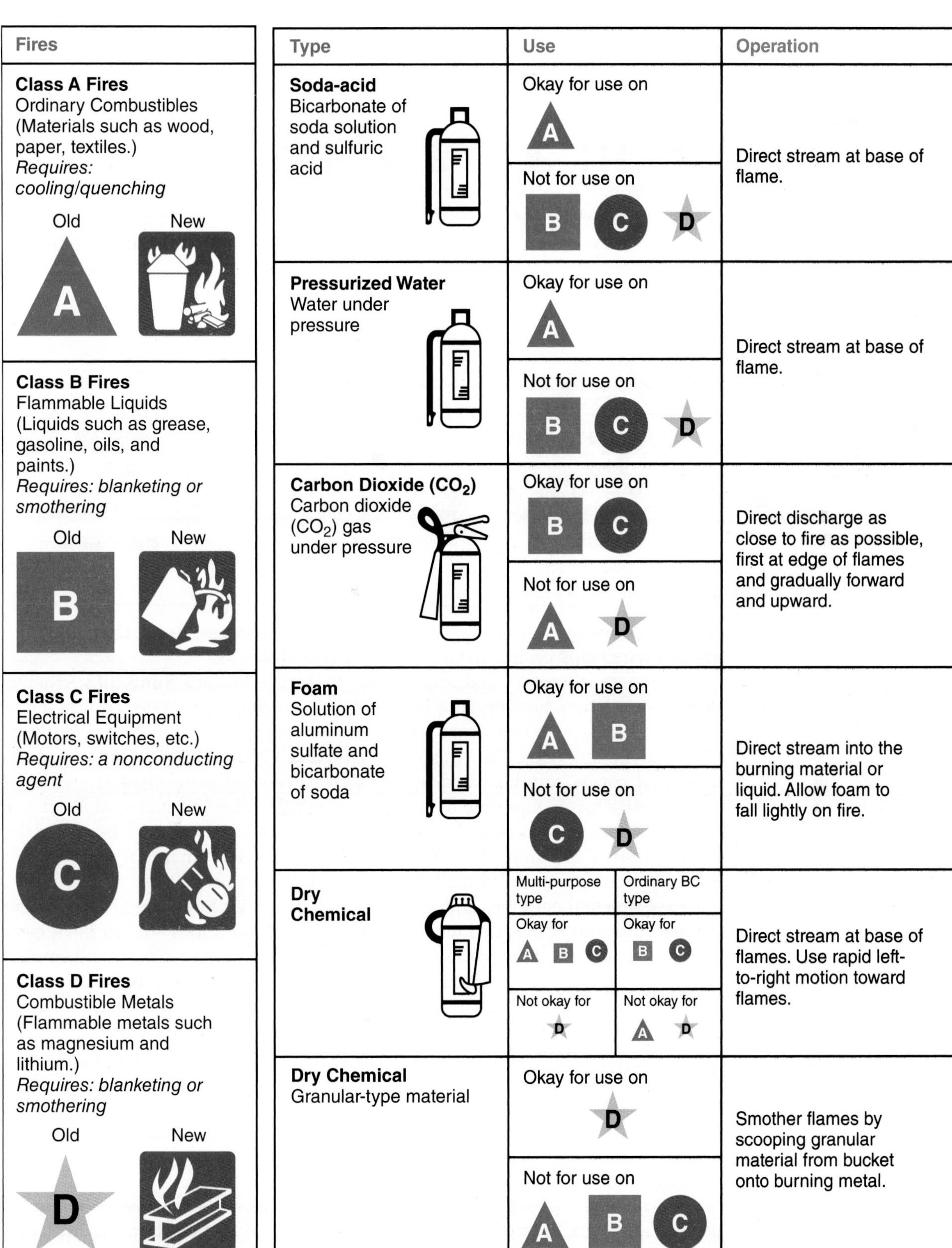

Fires
Class A Fires Ordinary Combustibles (Materials such as wood, paper, textiles.) *Requires: cooling/quenching* Old: A New
Class B Fires Flammable Liquids (Liquids such as grease, gasoline, oils, and paints.) *Requires: blanketing or smothering* Old: B New
Class C Fires Electrical Equipment (Motors, switches, etc.) *Requires: a nonconducting agent* Old: C New
Class D Fires Combustible Metals (Flammable metals such as magnesium and lithium.) *Requires: blanketing or smothering* Old: D New

Type	Use	Operation
Soda-acid Bicarbonate of soda solution and sulfuric acid	Okay for use on: A Not for use on: B C D	Direct stream at base of flame.
Pressurized Water Water under pressure	Okay for use on: A Not for use on: B C D	Direct stream at base of flame.
Carbon Dioxide (CO_2) Carbon dioxide (CO_2) gas under pressure	Okay for use on: B C Not for use on: A D	Direct discharge as close to fire as possible, first at edge of flames and gradually forward and upward.
Foam Solution of aluminum sulfate and bicarbonate of soda	Okay for use on: A B Not for use on: C D	Direct stream into the burning material or liquid. Allow foam to fall lightly on fire.
Dry Chemical	Multi-purpose type: Okay for A B C; Not okay for D Ordinary BC type: Okay for B C; Not okay for A D	Direct stream at base of flames. Use rapid left-to-right motion toward flames.
Dry Chemical Granular-type material	Okay for use on: D Not for use on: A B C	Smother flames by scooping granular material from bucket onto burning metal.

Goodheart-Willcox Publisher

Figure 3-1. This chart shows different types of fire extinguishers.

Safety goggles
vectorisland/Shutterstock.com

Heat
Ecelop/Shutterstock.com

Flames
Ecelop/Shutterstock.com

Corrosive chemical
Ecelop/Shutterstock.com

Animal safety
andromina/Shutterstock.com

Poison
Ecelop/Shutterstock.com

Scissors
Max Griboedov/Shutterstock.com

Biohazard
Ecelop/Shutterstock.com

Figure 3-2. Be aware of all safety symbols in the laboratory.

event of a short circuit or power surge. In electrical devices or tools, the equipment ground wire helps protect the user from electrical shocks.

Adequate laboratory space is needed to reduce the possibility of accidents. Proper ventilation, heating, and lighting also contribute to a safe working environment. Wiring in a laboratory should conform to specifications of the National Electrical Code (NEC). The ***National Electrical Code (NEC)*** is a set of regulations that covers safety related to all electrical wiring in residential and commercial structures. Equipment wiring is controlled by the National Electrical Manufacturers Association (NEMA) guidelines. Products should be tested by Underwriters Laboratory (UL). This company works with manufacturers, retailers, and governments on a broad range of safety issues.

Experimental activities should always be done with low voltage and current whenever possible. Instructions for performing lab activities should have clear directions and be written with safety in mind. All laboratory work should emphasize safety. Experimental circuits should be checked carefully before they are connected to a power source. Disconnect electrical equipment or experimental circuits from the source of power before working on these items.

A ***first aid kit*** consists of supplies and equipment used to treat a person in the event of an accident or injury. A kit that contains standard supplies, such as bandages, latex gloves, antiseptic ointment, and eye drops, can be purchased. A kit may be created or a standard kit may be supplemented to contain items needed to treat injuries likely to occur in a particular type of work setting. A good first aid kit should be in every laboratory, **Figure 3-3**.

The phone number of emergency medical services, as well as instruction in the use of 911 in case of emergencies, should be part of the safety training for workers in a laboratory or workshop. Accidents should be reported immediately to the instructor or supervisor of the laboratory facility. Procedures should be posted for treatment of minor cuts and bruises. The ***Occupational Safety and Health Administration (OSHA)*** is an agency of the US federal government. Its purpose is to ensure safe and healthful conditions for workers by setting and enforcing standards related to safety. OSHA typically requires records of accidents to be maintained.

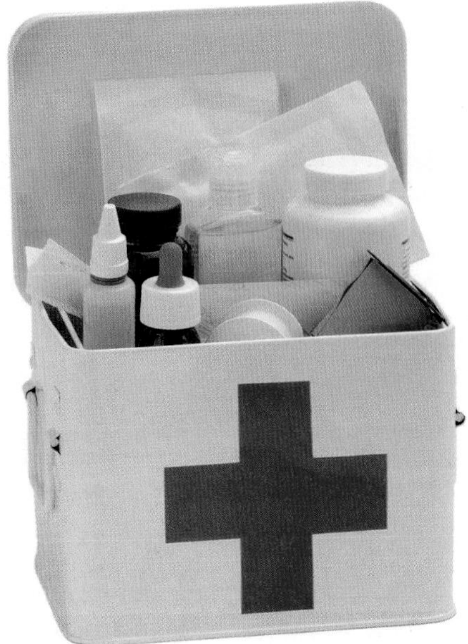

Pat_Hastings/Shutterstock.com

Figure 3-3. A good first aid kit should be available in every laboratory.

3.1.2 Circuit Breakers and Fuses

Circuit breakers and fuses are important safety devices. An electrical current travels along a path or channel made of conductive material, such as copper wire. A ***circuit breaker*** is a component of an electrical system that can be closed or opened. When closed, the circuit breaker allows the current to flow uninterrupted along the channel. When opened, the circuit breaker interrupts the channel and stops the flow of the current. The circuit breaker is designed to open when a short circuit occurs or a power surge reaches it. This stops the flow of the current and provides overload protection. When the power surge is no longer present, the circuit can be closed, allowing power to flow through it again. A circuit breaker is shown in **Figure 3-4**.

A ***fuse*** is another device that can be used to provide overload protection in an electrical system. When a power surge reaches a fuse, an element in the fuse melts, interrupting the flow of current. The fuse is no longer usable and is commonly called a blown fuse. The fuse must be replaced with a new one before power flow can be restored. When a fuse blows (or opens), this means that something is wrong in the circuit or equipment. Causes of blown fuses could be a short circuit (caused by two wires touching), too much electrical load on one circuit, or worn or damaged insulation that allows bare wires to touch grounded metal objects.

After correcting a problem, a new fuse of the proper size should be installed. Power should be turned off when replacing a fuse. Never use a makeshift device in place of a new fuse of the correct size. Fuses are used to cut off the power and prevent overheated wires and the potential to cause a fire. Some types of fuses are shown in **Figure 3-5**.

Circuit breakers are more commonly used for safety than fuses. Always remember to use caution when working with circuits or equipment in the laboratory. Safe practices should be followed in the laboratory and in the home. Detailed safety information that can be studied is available from the National Safety Council and other organizations.

a_v_d/Shutterstock.com

Figure 3-4. Circuit breakers are more commonly used safety devices.

dcwcreations/Shutterstock.com

A

Mrs_ya/Shutterstock.com

B

Figure 3-5. Types of fuses. A—In-line fuses. B—Cartridge fuses.

3.2 Safety with Tools and Equipment

A ***hand tool*** is a device or implement used to perform work that is held by the hands and does not use electricity or some other type of power. Some basic hand tools are shown in **Figure 3-6**. Do you recognize and know the proper use of each tool? It is very important for all technicians and engineers to be able to identify tools and know how to use tools required for their jobs safely. Whether or not you use each tool depends on the type of work you do and the environment in which you work.

Syafiq Adnan/Shutterstock.com

Figure 3-6. Basic hand tools.

3.2.1 Organizing Tools

It is important to use some method to organize your tools so that you can find them when needed. Tools should be marked in some way to show where they belong and sometimes to whom they belong. An organized tool storage system provides easy access and a way to protect your tools. Tools can be organized using several different storage devices. Popular storage devices for tools include:

- Tool belts, **Figure 3-7**.
- Mobile storage boxes, **Figure 3-8**.
- Tool chests, **Figure 3-9**.
- Rolling storage cabinets, **Figure 3-10**.

Another helpful method for storing tools is the use of pegboard. This material is available in 4′ × 8′ sheets for mounting on walls or carts. Tool outlines may be painted or otherwise marked on the pegboard to indicate places for proper storage. Organized tool storage can contribute to a safe working environment and prevent loss.

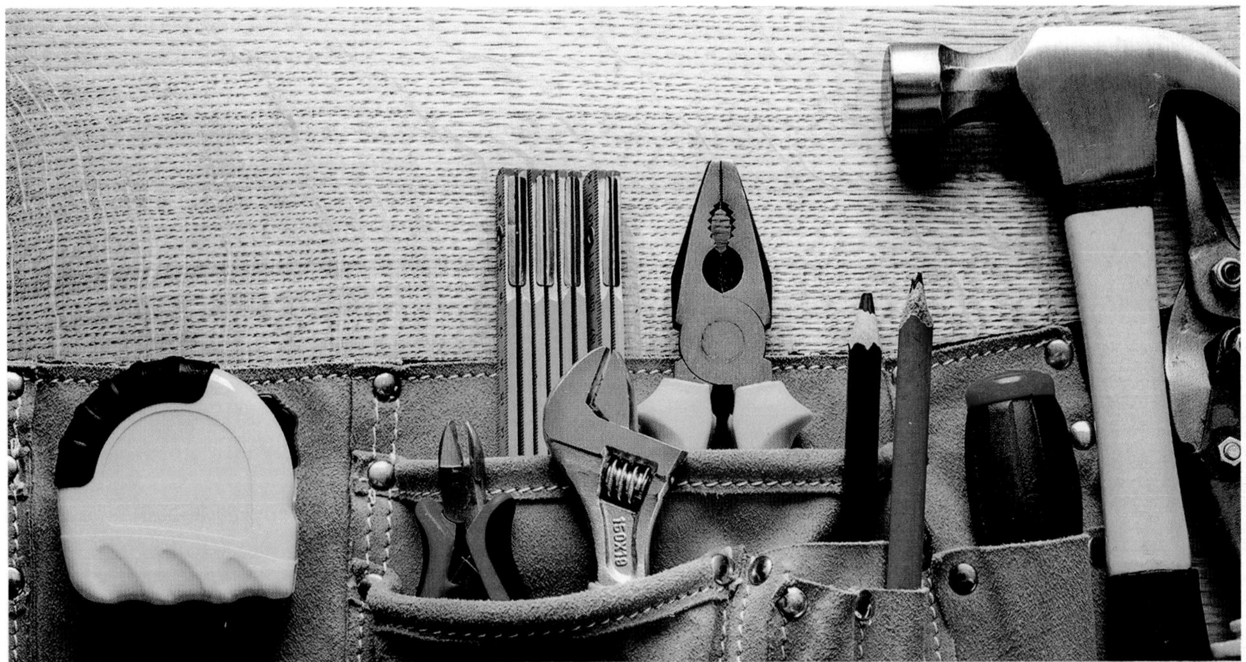

mihalec/Shutterstock.com

Figure 3-7. A tool belt can be used to keep items close at hand.

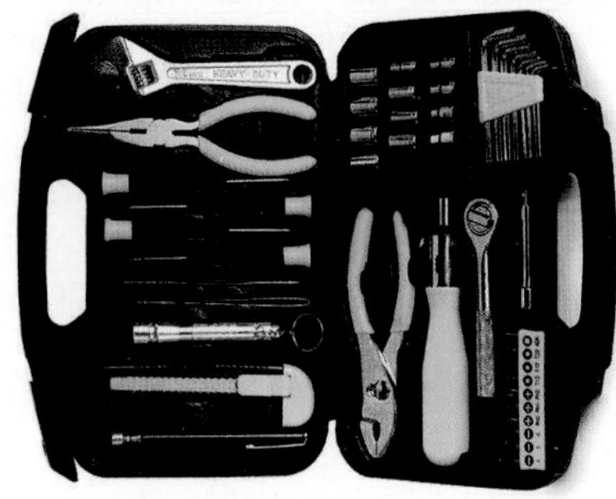

maxim ibragimov/Shutterstock.com

Figure 3-8. A mobile storage box is convenient for moving tools around in the work area.

Yotsatorn Laonalonglit/Shutterstock.com

Figure 3-9. Tool chests come in various sizes for storing a few or several tools.

Baloncici/Shutterstock.com

Figure 3-10. Rolling storage cabinets can be used to organize tools.

Goodheart-Willcox Publisher

Figure 3-11. GFCIs shut off power when a problem occurs in the circuit. The GFCI includes a reset button, which can be used to reset the GFCI when the circuit's problem is corrected.

3.2.2 Grounding Power Equipment

A ***power tool*** is a device or implement that is used to perform work and that uses electricity or some other type of energy. All power tools and equipment should be properly grounded. A three-wire conductor should be used with power tools and electrical equipment. A three-prong plug connects to the equipment or safety ground of the electrical outlet to ensure safe operation. Equipment grounds reduce the possibility of shock due to electrical shorts or leakage. It is dangerous to use a three-prong to two-prong adapter to plug into an outlet unless a separate ground is connected. Never remove the ground prong of a plug.

A ground fault circuit interrupter (GFCI), shown in **Figure 3-11**, is a very important electrical safety device used to turn off an electrical circuit when a safety hazard occurs. Such a hazard might be a person touching an electrical wire or an improperly grounded machine or metal cover used for electrical wiring.

GFCIs may be installed as a rapid-action circuit breaker in an electrical distribution panel in a home, lab, or other building. They may also be installed in an electrical duplex outlet where equipment is plugged in. If a problem occurs in an electrical circuit, the GFCI shuts off the power immediately, avoiding deadly shock. Once the problem with the circuit has been identified and corrected, the GFCI may be reset by pushing a reset button.

3.2.3 Safety Rules

A work area should have posted safety rules that are available to everyone who uses the area. These rules apply to the use of tools and electrical equipment as well as work procedures. Many of them are widely understood by many people. However, workers need to be constantly aware of safety issues and promote the use of safe procedures to others. If you see an unsafe practice, let your coworker know or report it to a supervisor. Some safety rules you should remember include:

- Always be safety aware. Use caution when operating any type of tool or equipment.
- Clean tools after using them to keep them in good working order. Tools that are not properly maintained may not function properly, causing safety hazards.
- Check lubrication on tools and equipment and lubricate as needed.
- Store tools in a safe, protected area.
- Wear personal protective equipment (PPE) as required. See **Figure 3-12**.
- Pay close attention to your work. Do not become distracted and do not distract others.

- Inspect electrical cords on tools and other devices to be sure they are in good condition.
- Remember the OSHA safety color codes shown in **Figure 3-13**. For example, the color red stands for *danger* or *stop*.
- Always use the right tool for the job, and use it correctly.
- Repair or replace damaged tools immediately. Do not use broken or malfunctioning tools and equipment.
- Do not place tools where they may fall and create a potential hazard.
- When repairing electrical tools or equipment, be sure that the power is turned off. Take steps to ensure that no one else can turn on the power while you are working. Use ***lockout/tagout procedures***, which are practices used to disable machinery and prevent injury to workers from hazardous energy releases.
- Wear suitable clothing. Do not wear loose or dangling items, such as a necktie. Clothing should have close-fitting sleeves and trouser legs. Remove jewelry, such as rings and wristwatches.
- Be sure that rags containing oil, gasoline, alcohol, or paint are kept in an approved type of container. The chances of a fire are decreased with good, neat housekeeping.

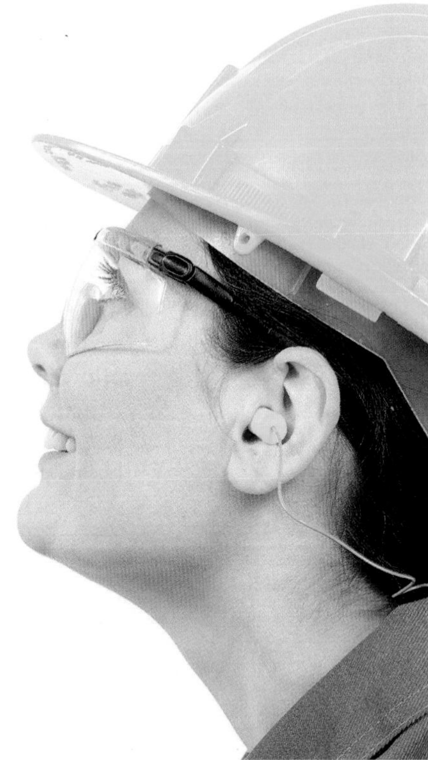

florin operea/Shutterstock

Figure 3-12. Personal protective equipment may include protection for your eyes, ears, and head.

OSHA Safety Color Codes	
Color	**Meaning**
Red	Fire protection equipment and apparatus Portable containers of flammable liquid Emergency stop buttons and switches
Yellow	Caution and for marking physical hazards Waste containers for explosive or combustible materials Caution against starting, using or moving equipment under repair Identification of the starting point or power source of machinery
Orange	Dangerous parts of machines Safety starter buttons Exposed parts (edges) of pulleys, gears, rollers, cutting devices
Purple	Radiation hazards
Green	Safety Location of first-aid equipment (other than fire extinguishers)

Goodheart-Willcox Publisher

Figure 3-13. OSHA safety color codes can be used to help alert workers to hazards.

- Follow recommendations regarding electrical voltages and maintain an awareness of safe practices for electrical tools and equipment.
- If possible, become trained in first aid and safety.

Remember that if you do not follow safe procedures, you can create conditions that are unsafe for others.

3.3 Robotic Safety Considerations

Safety is an important consideration for both installation and operation of robotic systems. A robot may have a large work envelope and degrees of freedom that allow it to move within a large area. Robots can employ a wide range of automated functions, which may include unexpected starting and stopping. Operators must be properly trained to use the robot safely. Operators and other workers who may have occasion to be in the area should be made aware of possible safety hazards.

OSHA offers information and safety guidelines for the safe operation of robots. "Industrial Robots and Robot System Safety" can be found in the *OSHA Technical Manual*. Topics included in the information presented are robot classifications, typical hazards, and safeguarding personnel.

Safety requirements related to industrial robots are controlled by ANSI/RIA R15.06-2012: *American National Standard for Industrial Robots and Robot Systems—Safety Requirements*. This document provides safety requirements for industrial robot manufacturing environments, installation sites, general operation, and personnel safety.

3.3.1 General Safety Guidelines

Safety considerations should begin with research into the characteristics of the specific robot(s) being used. The following questions should be answered:

- What is the size and shape of the robot's work envelope?
- What methods of motion control are used?
- What are the limits for a safe payload (the amount of weight the robot can lift)?
- What is the range of operating speeds?
- What other features are provided by the manufacturer?

The work station should be designed to allow the operator to use the equipment safely and to protect others in the area. Be aware of fixed objects within the work area, such as posts or drains, and leave enough room to pass around these objects to avoid pinch points. Consider the size of the robot, items attached to the arm, and other equipment when designing the space. Use paint or colored tape on the floor to outline the perimeter of the robot's work area. Barriers that cannot be passed through or over may be required for some work stations.

Control panels should be placed outside the work envelope and interlock access to the robot controller may be required. Interlocks are safety devices that are designed to prevent unauthorized access to hazardous areas by requiring a key for entry. Implementing two-step procedures to resume operation after a shutdown is an additional precaution to ensure worker safety. The entire design should be documented, including any changes made during installation and debugging. This is important information for purchasing, operating, and maintenance personnel. Employee training and a conscientious effort to provide safe working conditions are important components of every safety program. Operating personnel,

Robotics Innovations: Drones

Drones, also called unmanned aerial vehicles (UAVs), are flying machines with robotic controls. A quadcopter is one type of drone. These drones are small helicopters that have four motors, and each motor controls one of the four rotor blades. By controlling the four blades independently, quadcopters can move in three dimensions: up/down, left/right, and forward/reverse.

Quadcopters developed by Pennsylvania State University can complete complex maneuvers because of the algorithms they use to control and coordinate flight. They can fly in formation, navigate through openings, such as windows, and work cooperatively to complete tasks. These drones have been programmed to build structures from a set of plans and to lift heavy objects by working together. Individually, these drones are not very strong. Working as a team, however, they can accomplish tasks that larger robots cannot. Their small size also means that they can easily overcome inertia and perform some amazing aerial acrobatics.

Photo courtesy of University of Pennsylvania GRASP Lab, Vijay Kumar, Daniel Mellinger, and Alex Kushleyev

programmers, maintenance workers, and cleaning crews need extensive training. Periodic refresher courses should be scheduled for as long as the robots are in use.

3.3.2 Safety Barriers and Sensors

Safety measures used for working with a robot are more complex than those used for working with other kinds of equipment. Two primary safety devices used are barriers and sensing systems.

One of the safest barriers is a fence, **Figure 3-14**. A safety fence prevents unauthorized entry and provides containment of flying objects. A safety fence is typically 6′ high and constructed of wire mesh, safety glass, or rigid plastic sheets, **Figure 3-15**. Robots that operate at a high speed may require more substantial containment barriers that are sufficient to prevent penetration of a flying object. In toxic applications, such as paint spraying, a solid enclosure will also protect personnel from chemicals.

Fencing can make it more difficult to move or rearrange the robot and other work cell equipment. Fences can also take up valuable floor space and can obstruct the view of critical operations. Fences may also be an obstruction when trying to teach the robot or perform preventive maintenance. Because of these issues, less obtrusive barriers have been developed. One such barrier is the ***light curtain***. This consists of an array of photoelectric sensing devices, **Figure 3-16**. If a worker enters the work envelope, the light beam between two sensors is broken. The sensors send a signal to cut the power to the robot. An

Motoman

Figure 3-14. A steel safety fence surrounds this arc welding robot. The control for the robot is mounted outside the fence. The robot cannot be started until the operator steps outside and closes the gate.

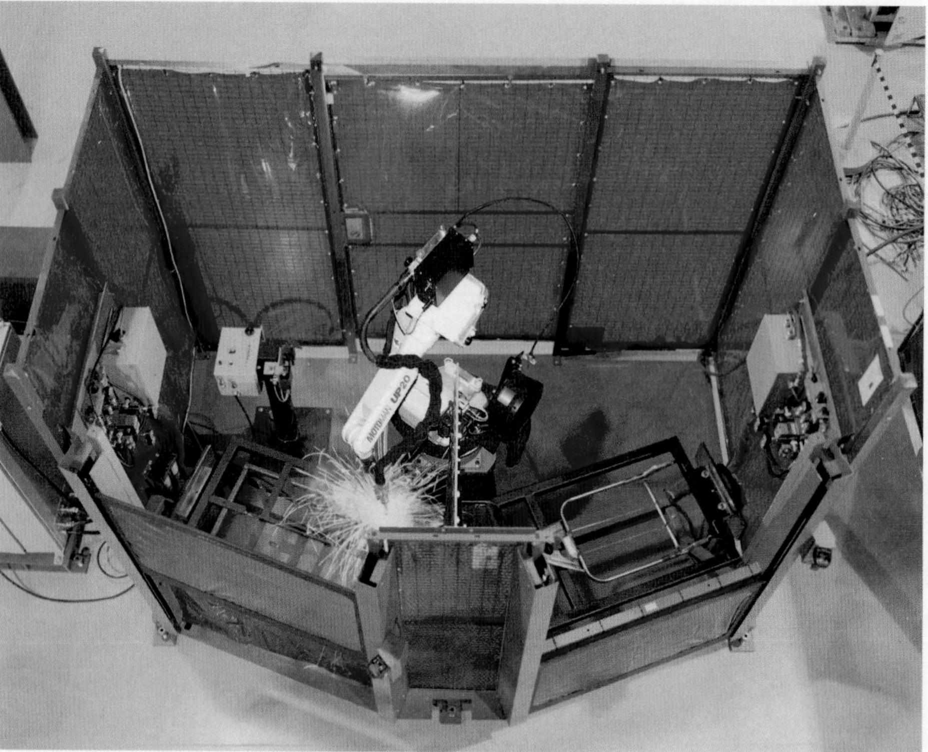

Motoman

Figure 3-15. Glass or rigid plastic panel enclosures are sometimes used to protect personnel.

infrared light curtain is programmable, which allows certain areas to be excluded from the sensing devices. This permits access to equipment that is not within the reach of the end effector of the robot. One drawback of the light curtain is that moving equipment can break the light beam and cause power to the system to be stopped.

One of the simplest sensing systems is a ***pressure sensitive safety mat***. Sensors embedded in the mat send a signal to the controller when weight is placed anywhere on the mat. The controller then sends a signal that shuts down the connected equipment. Pressure sensitive safety mats can be easily placed in dangerous areas within a facility and will automatically turn off dangerous machinery when a worker steps on the mat.

Some sensing systems make use of a combination of capacitive, infrared, ultrasonic, and microwave sensors to detect a worker's presence and track the person within the work envelope. The power is cut off whenever a worker enters the robot's work envelope.

3.3.3 Overload Protection

The end effectors on industrial robots must have some protection against overload. Ordinarily, a feedback signal is sent to the computer system, and the manipulator is withdrawn before damage occurs. Breakaway wrist devices or rapid withdrawal of the manipulator can be accomplished using mechanical fuses, detents, and preloaded springs.

Mechanical fuses are pins or tubes that break or buckle under extreme stress. Mechanical fuses must be replaced after they perform their function, but they are

Motoman

Figure 3-16. This equipment projects a curtain of light. When the light is interrupted, power to the robotic equipment is shut down.

not as expensive as other overload protective devices. ***Detents*** are two or more elements held in position by spring-loaded mechanisms. They move from their original position when placed under excessive stress. When detents move from their original position, power is removed from the machine, providing overload protection. ***Preloaded springs*** may also be used to prevent overload conditions. Excess stress causes the spring to release, and the end effector breaks away from the work area. These devices reset automatically when the overload is removed.

Summary

- General safety awareness is very important. The equipment in a lab or work area should be set up and placed to ensure the safety of operators and bystanders.
- Workers in many jobs need to wear personal protective equipment (PPE), which consists of clothing or devices worn to shield the worker's body and help prevent injuries.
- Accidents and injuries in a laboratory can be related to electric current. Non-conductive floor coverings, fire extinguishers, and hazard warning or safety labels can help prevent accidents and injuries.
- A good first aid kit should be in every laboratory. The phone number of emergency medical services and instruction in the use of 911 should be part of the safety training for workers.
- Circuit breakers and fuses are important safety devices. Both are used in electrical systems to stop the flow of current in the event of electrical shorts or power surges.
- Many types of hand tools are used by workers. Technicians and engineers should know how to use safely the tools required for their jobs.
- Tools can be organized using tool belts, mobile storage boxes, tool chests, rolling storage cabinets, and pegboards mounted on walls or carts.
- Power tools that use electricity may pose a danger of electrical shock. All power tools and equipment should be properly grounded.
- A work area should have posted safety rules that are available to everyone who uses the area.
- Safety is an important consideration for installation and operation of robotic systems. Operators must be properly trained to use the robot safely.
- Safety considerations should begin with research into the characteristics of the robot(s) being used, such as the shape of the work envelope and the methods of motion control used.
- Control panels should be placed outside the work envelope, and interlock access to the robot controller may be required.
- A safety fence prevents unauthorized entry to the work area and provides containment of flying objects.
- Light curtains and pressure sensitive safety mats may be used to shut off power to the robot when a person enters an unsafe area.
- The end effectors on industrial robots must have some protection against overload. Rapid withdrawal of the manipulator can be accomplished using mechanical fuses, detents, and preloaded springs.

Review Questions

Answer the following questions using the information provided in this chapter.

1. How can you avoid accidents or injuries that are the result of taking shortcuts or working with too much haste?
2. What are four examples of personal protective equipment (PPE) that may be used by workers?
3. What is an electrical shock? How can work surfaces and floors in a work area be treated to help avoid electrical shocks?

4. What type of fire extinguisher should be used to put out a fire that involves flammable liquids?
5. Wiring in a laboratory should conform to the specifications of what code?
6. What is the purpose of a first aid kit, and what are some examples of items that a first aid kit could contain?
7. What is the purpose of a circuit breaker or fuse in an electrical system?
8. What are four popular storage devices used for organizing tools?
9. Why should a three-wire conductor be used with power tools and electrical equipment?
10. What are four safety rules related to tools that you should follow?
11. What are three topics related to robots found in the *OSHA Technical Manual*?
12. What are five questions related to the characteristics of a robot that should be answered when planning safety considerations?
13. What is an interlock?
14. What is the purpose of a safety fence, and what are some materials that can be used to construct a safety fence?
15. How does a light curtain function as a safety barrier?
16. How does a pressure sensitive safety mat function as a safety barrier?
17. Describe three devices that are used to provide overload protection for the end effectors on industrial robots.

Learning Extensions

1. Describe some of the unique safety considerations when robots are used in an industrial setting.
2. Locate the document "Industrial Robots and Robot System Safety" at https://www.osha.gov/dts/osta/otm/otm_iv/otm_iv_4.html and discuss the content regarding robotic safety.
3. Visit two websites that list general safety rules for laboratories. List and describe each of these sites.
4. Search the Internet for sample lockout/tagout procedures. Describe the type of information included in one sample document.

Advanced Analysis

1. Select a robot application in a workplace and describe the safety considerations (general and specific) for the area.
2. Develop a list of safety rules for the school laboratory or work area in which you work.
3. Develop a detailed safety plan for the area in which you work that could be used to describe the importance of safety to others in the area.

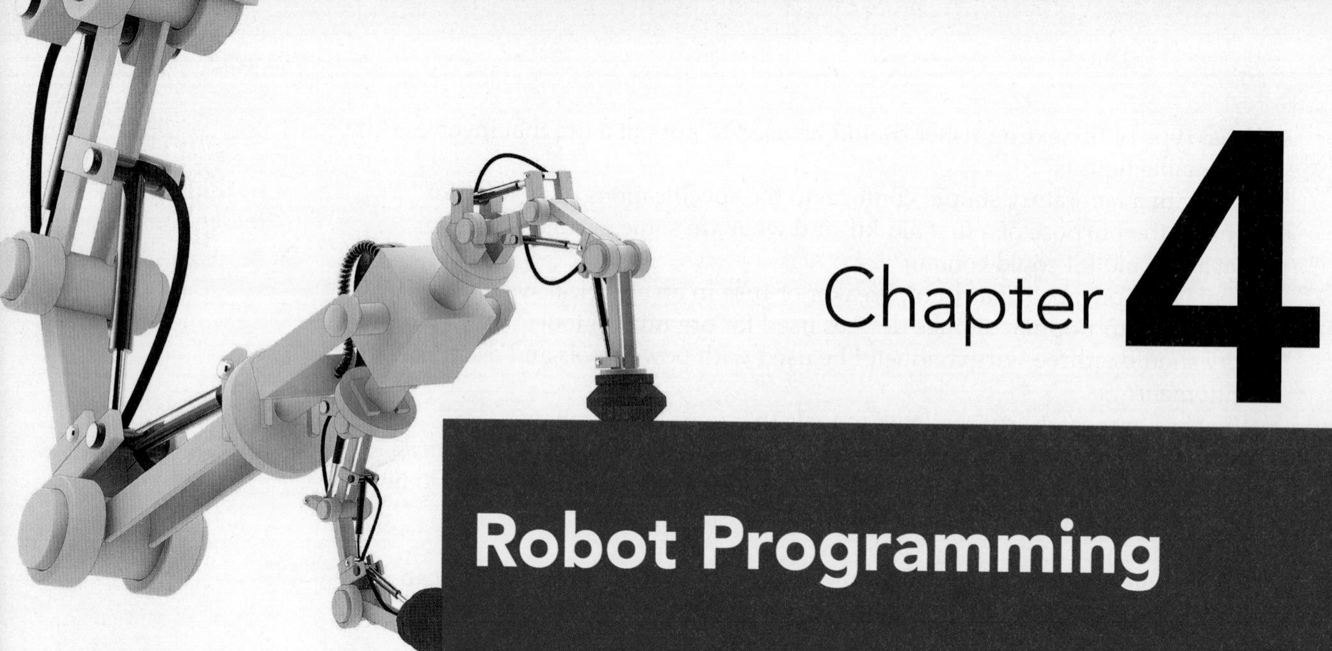

Chapter 4

Robot Programming

Outline

Objectives

Upon completion of this chapter, you will be able to:

- Discuss the three generations in the evolution of programming for robots.
- Identify classifications of robots according to the pattern of motion they use.
- List and describe four programming methods for robots.
- Describe the use of high-level programming languages for robots.
- Discuss the differences between hierarchical control and task-level programming.
- Describe how voice recognition systems can be used with robots.

Technical Terms

compiler
continuous-path (CP) motion
end stop
hierarchical control programming
high-level language
manual programming
manual rate control box
off-line programming
on-line programming
pick-and-place motion
point-to-point (PTP) motion
sensory feedback
subroutine
task-level programming
voice recognition
walk-through programming
WAVE

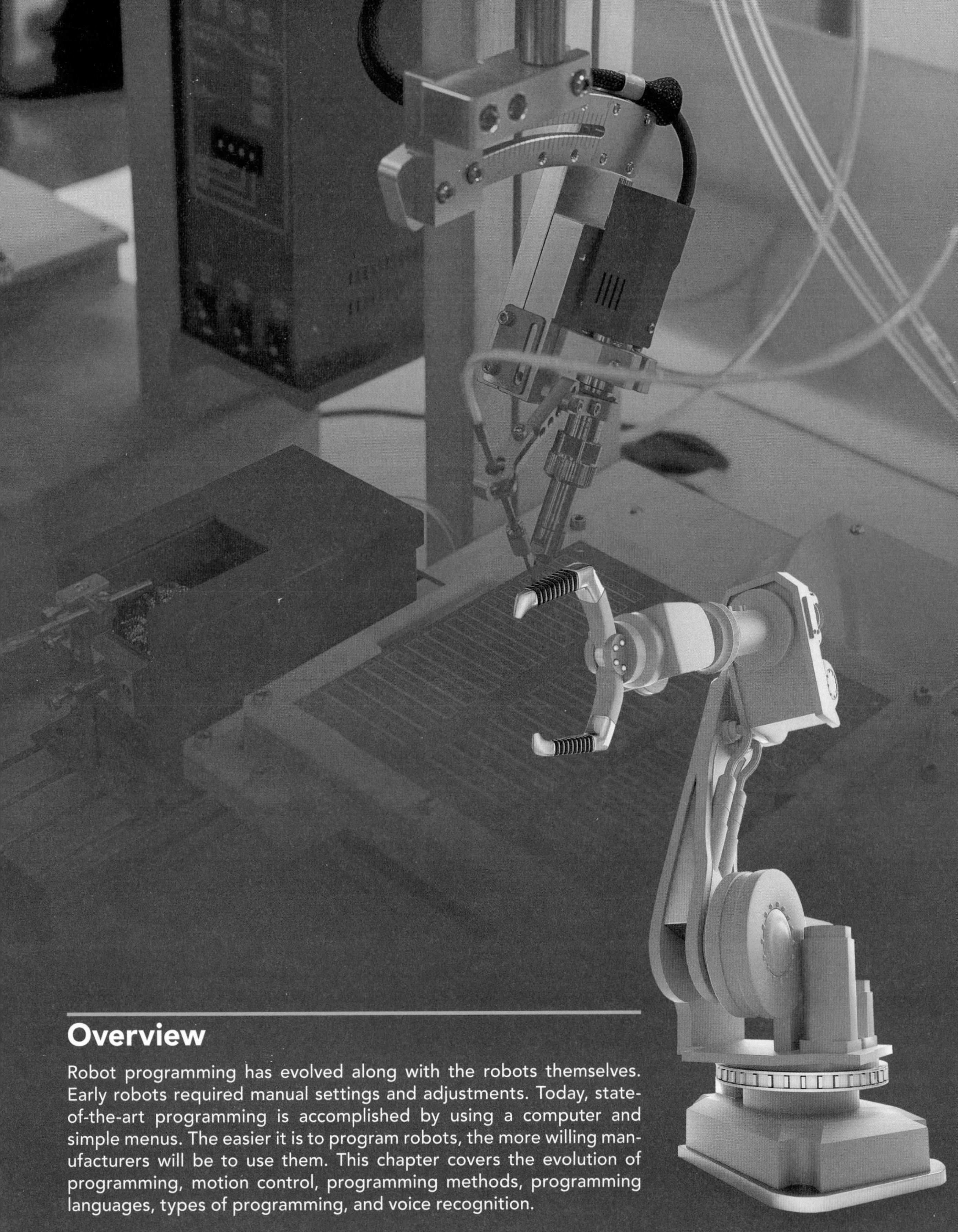

Overview

Robot programming has evolved along with the robots themselves. Early robots required manual settings and adjustments. Today, state-of-the-art programming is accomplished by using a computer and simple menus. The easier it is to program robots, the more willing manufacturers will be to use them. This chapter covers the evolution of programming, motion control, programming methods, programming languages, types of programming, and voice recognition.

4.1 The Evolution of Programming

The evolution of industrial robots consists of three periods called generations. These generations are:

- First generation—Late 1950s through the mid-1970s.
- Second generation—Mid-1970s through the mid-1980s.
- Third generation—Mid-1980s to the present.

4.1.1 First Generation

First-generation robots performed purely repetitive tasks and did not respond to changing conditions. These robots often had only two or three degrees of freedom and used pneumatic or hydraulic actuators for moving objects. Programming these robots involved adjusting mechanical stops and limit switches. This controlled the stroke length of each programmable axis.

More complex programming was done by using a rotating drum that contained actuating switches. A series of sequential moves could be set up and executed. Pneumatic or electric relay logic circuits were also commonly used. This allowed programmed steps to be controlled by a mechanical timer. The total number of programmable steps was generally between 10 and 100, depending on the robot's sophistication.

During the 1960s and 1970s, robotic research was carried out in the laboratories of industry and universities, including Stanford and MIT. A breakthrough in robot design occurred in 1975 with the introduction of robots that used computer-controlled manipulators. Microprocessors provided increased processing power at reasonable cost. This development ushered in the second generation of robot design and programming.

Careers in Robotics: Software Engineer

Software engineers design, develop, and test the computer programming and applications required to operate robotic systems and have them perform the tasks required. The programming developed must meet the specifications provided in both form (platform and language) and function (task performance). The tools software engineers use to do their job change quickly with developing technology and specialized system applications. It is important to remain knowledgeable in the most current technology and trends in the industry.

To enter this field, most employers prefer that workers have a bachelor's degree in computer science or software engineering and knowledge of various computer systems, programming languages, and manufacturing environments. Further education and advanced degrees often open the door to advancement and more complex work.

4.1.2 Second Generation

In the mid-1970s, robot manufacturers began to experiment with more advanced programming. In 1977, Unimation and Olivetti both introduced robots that could be controlled by means of programming languages. A programming language consists of words, numbers, and symbols used in a structured manner to create instructions for a computer. The earliest second-generation robots were not commercially successful. Some companies may have been intimidated by the complex programming required for robots and the high cost of using robots compared to using manual labor. It was not until the early 1980s that second-generation, language-programmable robots began to be used in industry. Robots were used to perform various tasks, such as welding, spray painting, assembling, and machine loading and unloading. The robot had to be programmed to do these activities.

As second-generation robots became more complex, they were equipped with internal sensors and closed-loop control systems. This gave them a limited amount of feedback about their environment. Internal sensors were used to detect the actual position of the robot. That position was compared to the position specified in the program, enabling the robot to move to a correct position if needed. Other internal sensors, such as strain gauges, were used to detect malfunctions. If a malfunction was detected, the problem was automatically corrected, or a warning was sent to the operator.

4.1.3 Third Generation

Third-generation robots were developed through the use of artificial intelligence. Artificial intelligence (AI) is the science and engineering of making machines perform operations commonly associated with intelligent human behavior. Marvin Minsky, one of the fathers of AI, defines it as: *the science of making machines do things that would require intelligence if done by men.* Scientists and engineers have been studying AI since the 1930s. Practical use, however, was largely confined to laboratory research until the late 1980s.

Third-generation robots are capable of sensing the environment and making intelligent decisions to perform tasks more efficiently. Sensing devices, such as vision sensors, provide robots with information about their surroundings, **Figure 4-1**. Robots use this information to determine how to proceed in performing a task. For example, vision systems are used for part recognition, allowing robots to distinguish between good and bad parts on a production line.

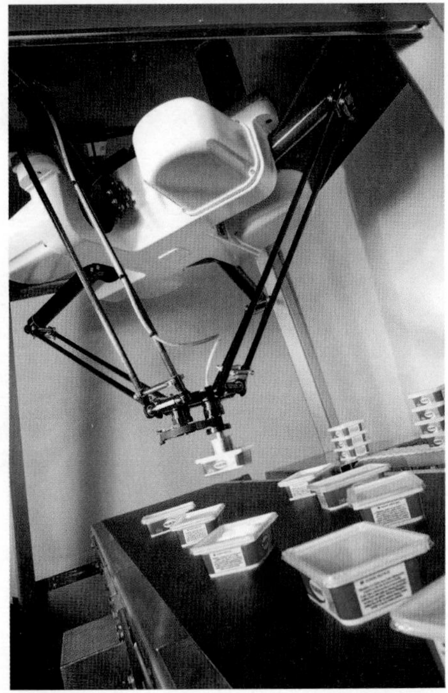

Omron Adept Technologies, Inc.

Figure 4-1. Vision guidance tracks parts on a moving conveyor belt.

4.2 Motion Control

A robot's manipulator moves through a series of points while performing a task. Robots can be classified according to the pattern of motion they use. The three classifications are pick-and-place motion, point-to-point motion, and continuous-path motion.

4.2.1 Pick-and-Place Motion

Pick-and-place motion is movement of the end effector of a robot to the correct position. Pick-and-place motion is often used in manufacturing processes to perform work that is repetitive and does not require many complicated movements to accomplish a task. An example is picking up a part at one location and placing it in another location.

One method of programming the pick-and-place pattern of motion is to manually set mechanical stops or limit switches for each designated point. This method allows for two positions per axis. Programming additional positions per axis requires extra controls and additional time spent in the programming process. Because of the laborious programming, the number of points the robot can move through is comparatively low.

The end effector follows a fixed pattern of movement, as illustrated in **Figure 4-2**. Generally, only one axis of the robot moves at a time. ***End stops*** are devices placed on an axis to physically block movement past a certain point along the axis. They are used to control the length of travel along an axis. Note that the position points used by pick-and-place robots are points along the various axes, not other points in space.

4.2.2 Point-to-Point Motion

Point-to-point (PTP) motion is the movement of a robot end effector through a number of points in space to a desired location. The programmer uses a combination of manipulator axes to position the end effector at a correct spot. The positions are recorded and stored in memory. During playback, the robot moves through the recorded points. The path of motion is a series of straight lines between the points, **Figure 4-3**. As in pick-and-place motion, point location is more important than controlling the path of travel.

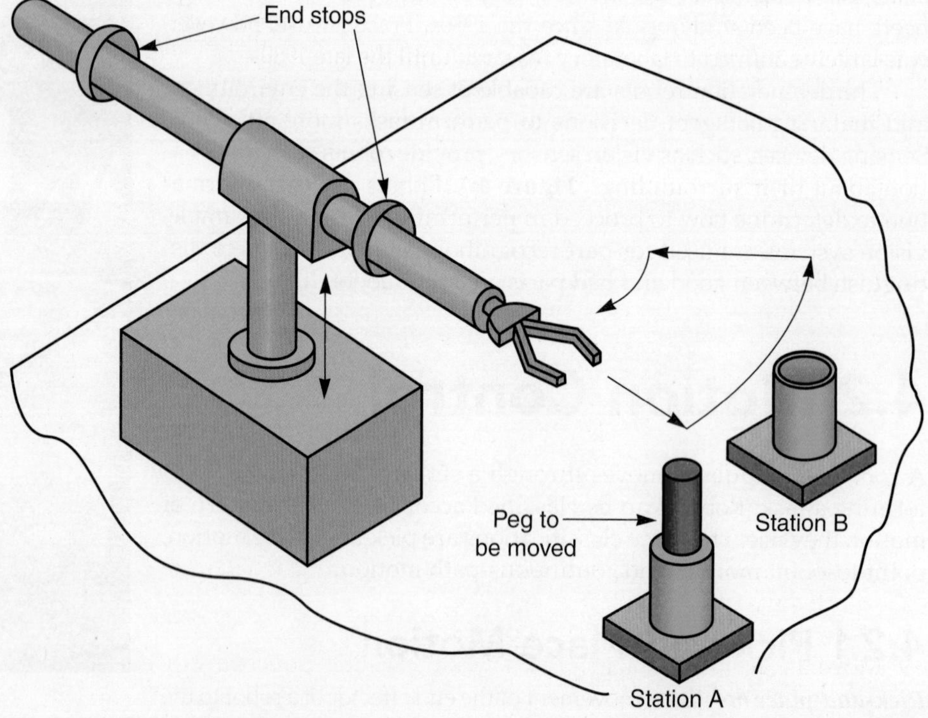

Goodheart-Willcox Publisher

Figure 4-2. Pick-and-place motion involves only two positions per axis.

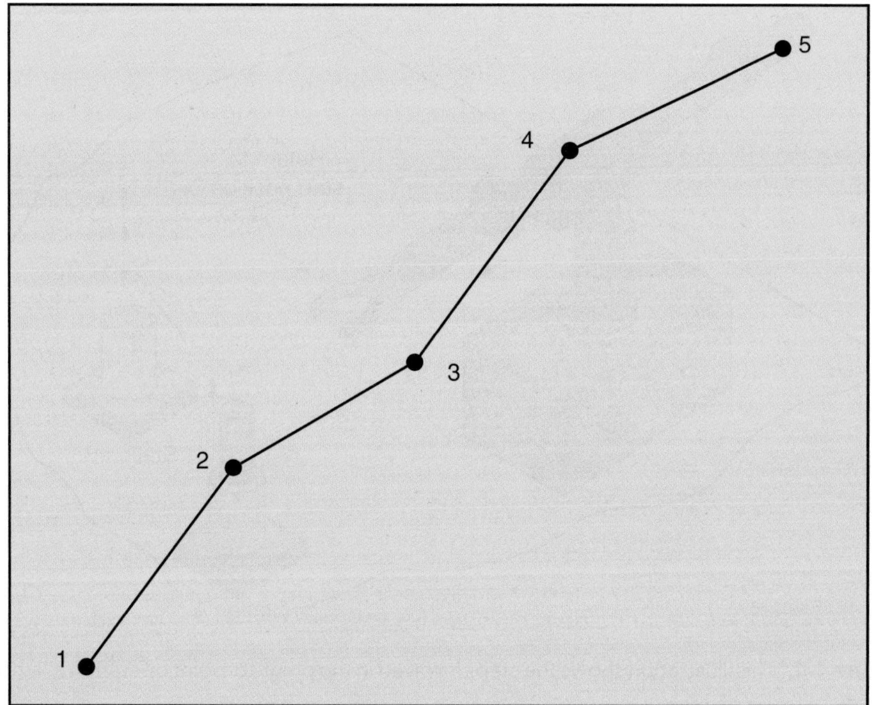

Goodheart-Willcox Publisher

Figure 4-3. Point-to-point motion involves a discrete number of points in space.

To understand point-to-point motion, review the motion illustrated in **Figure 4-4**. Suppose the robot is to take a peg out of its holder and insert it into another holder in a different location. Instead of using end stops to control length of travel, the holder locations are recorded into the robot's memory.

The peg is initially located in a holder at station A. The robot arm is retracted and the gripper is open. The point of insertion is at station B. The steps to be recorded are as follows:

1. Move the arm until the gripper is located above the peg. Record this point into memory.
2. Adjust the wrist joint of the robot until the gripper is properly aligned for grasping the peg. Record this position.
3. Move the gripper down over the peg to the point where the gripper will be able to grasp the peg when closed. Realign the gripper with the peg by adjusting the various joints. Record this point.
4. Close the gripper on the peg and record the point.
5. Carefully lift the peg from the hole vertically. After the peg is out of the hole and is at the desired elevation, record that point.
6. Move the arm until it is approximately over the center of the hole at station B. Record this point.
7. Carefully lower the peg, adjusting the various joints until the peg is properly inserted into the hole. Record this point.
8. Open the gripper to release the peg. Record the point.
9. Move the arm until the gripper is located at some point directly above the peg and record this point.
10. Stop the robot. Move the gripper back to station A so the operation can be repeated with a new peg.

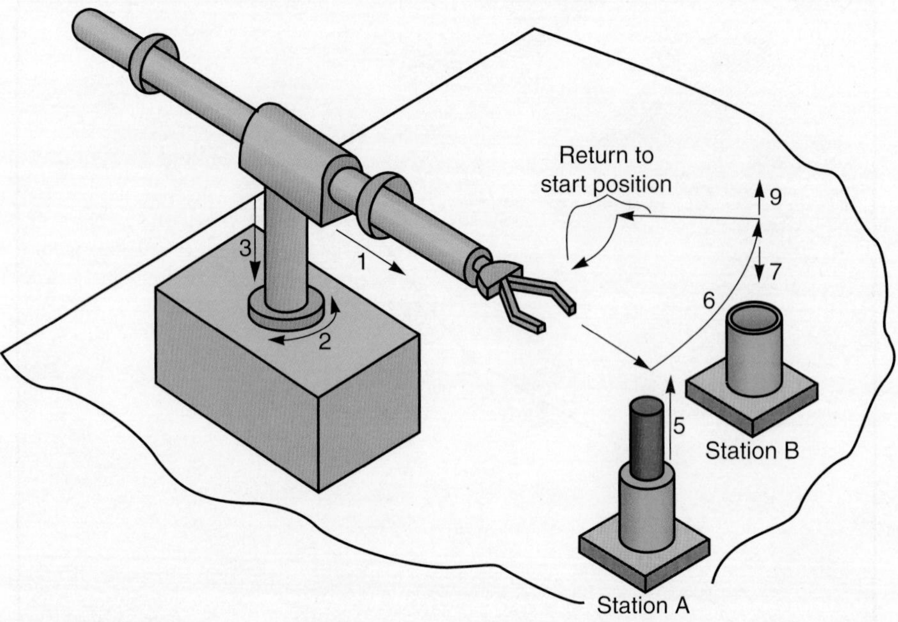

Goodheart-Willcox Publisher

Figure 4-4. This illustration shows the steps involved in the point-to-point operation example.

After the points have been recorded, take the robot out of teach mode. The robot returns to the start position and moves through the sequence using the various points recorded into memory. The robot in this example would have to be stopped after step 9 because no additional pegs were placed at station A. In a real work situation, another peg would appear at station A or additional steps would be added to the program.

During playback, the trajectory of the end effector is generally different from the path used by the operator when establishing the points. This is because the operator must move each axis or joint independently. However, if a joystick is used during programming, the robot will move along a straighter path than it will if the positioning is done manually. Using a joystick also reduces the programming time involved.

Several stops along a given axis can be programmed. Point-to-point servo robots are capable of storing hundreds of discrete points in space, far more than the two stops used in pick-and-place motion. Acceleration and deceleration between points is controlled by a separate device, such as a tachometer.

4.2.3 Continuous-Path Motion

Continuous-path (CP) motion is the smooth and continuous movement of a robot end effector through many points in space to a desired location. This type of motion is an extension of point-to-point motion. The difference is that continuous-path motion can involve several thousand points. Since more points are used, the distance between each point can be extremely small. Using a great number of points results in movement that is smooth and continuous, **Figure 4-5**. With continuous-path motion, control of the path of the end effector is more important than

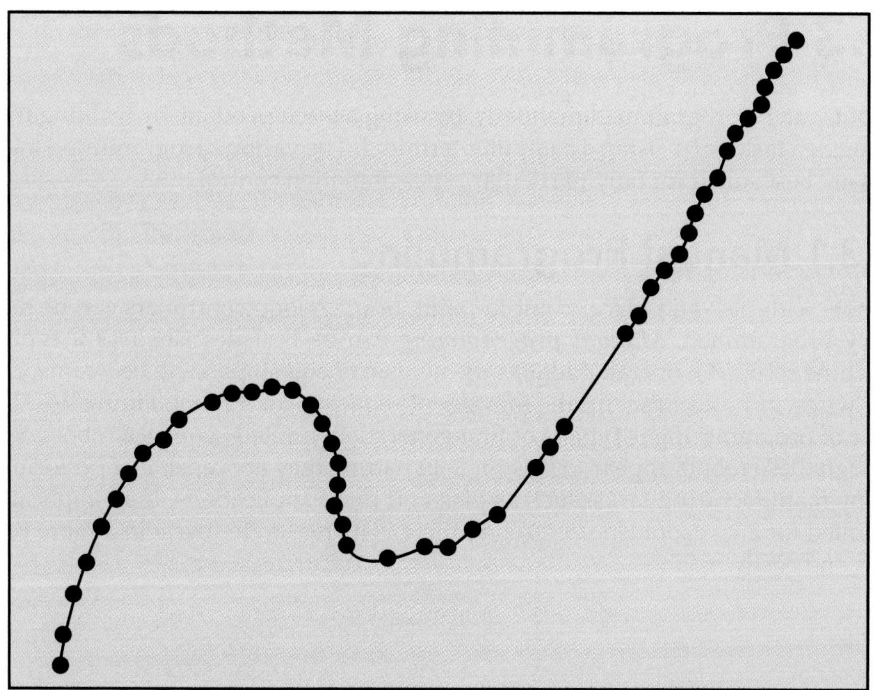

Goodheart-Willcox Publisher

Figure 4-5. Continuous-path motion involves a large number of points and results in movement that is smooth and continuous.

end point positioning. The robot arm generally does not come to rest at various points, as is often required in PTP motion.

Programming is done by an operator who physically moves the end effector through its motions. The positions on the various axes are recorded. Some continuous-path robots record up to 80 points per second.

Playback rate can be changed to provide the best operating speed for the task. For certain applications, such as spraying, it may be better to program the robot at a slow speed and play the program back at a faster speed. Other applications, such as arc welding, require faster programming than playback.

Continuous-path motion offers several advantages, including:

- Movement is smooth and continuous.
- Programming is simple, and no prior knowledge of programming is required.
- Velocity and acceleration of the end effector can often be controlled.
- The operator must simply understand the operation he or she is trying to teach the robot to be successful in programming it.

There are, however, disadvantages to continuous-path motion, including:

- A large amount of memory is required to store all the points recorded for a task. In the past, memory was very expensive. Memory chips are now reasonably priced, so this is not as significant a disadvantage as it once was.
- Since recording during programming is constant, undesirable motions are recorded into memory in addition to intended movements. The robot duplicates all the movements recorded during playback.
- The robot arm must be counterbalanced and able to move freely without power so the operator can produce smooth-flowing motion while programming.

4.3 Programming Methods

Robots can be programmed manually, by using a teach pendant, by walking them through a task, or by using a computer terminal. The various programming methods are best suited for only particular types of motion control.

4.3.1 Manual Programming

Robots with pick-and-place, point-to-point, or open-loop controllers can be manually programmed. ***Manual programming*** can be best described as a type of machine setup. An operator adjusts the necessary end stops, switches, cams, electric wires, or hoses to set up the movement sequence for a robot, **Figure 4-6**. This type of programming is typical of first-generation, limited-sequence robots. Even though these robots appear to be simple in nature, they are capable of performing many manufacturing tasks, such as pick-and-place applications. If an application is suited for a less sophisticated robot, there is no reason to invest in a more complex and costly model.

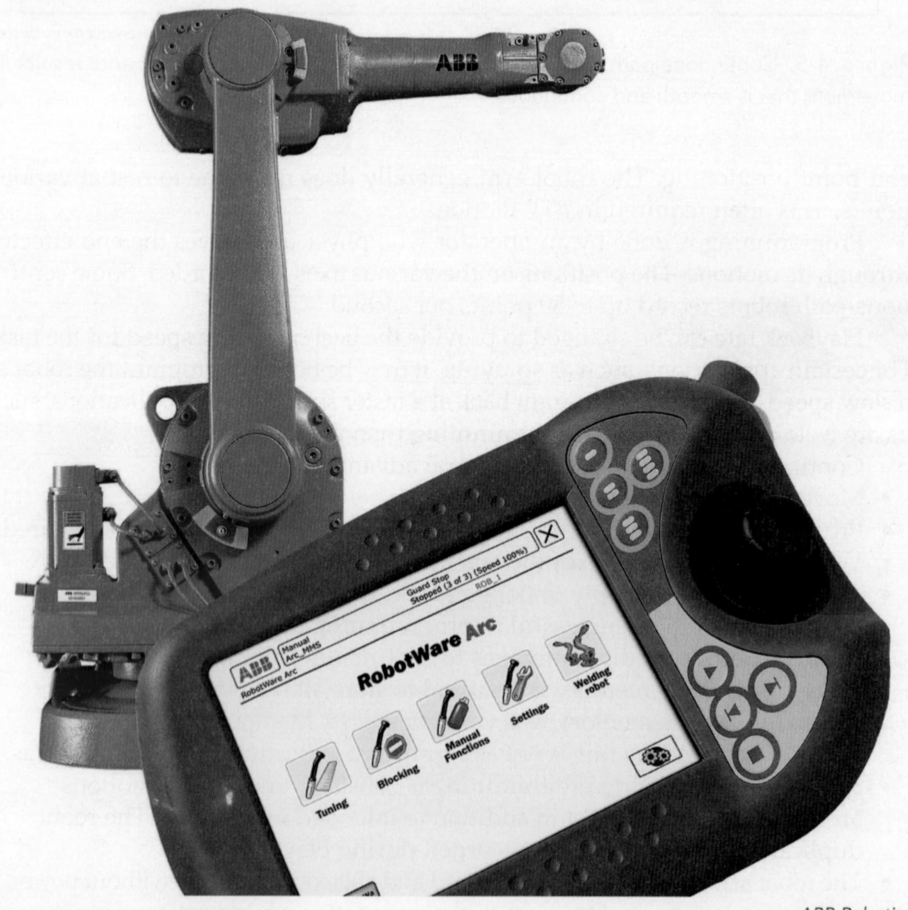

ABB Robotics

Figure 4-6. In manual programming, the mechanical stops and limit switches are manually adjusted to control the movement of the robot.

Manual programming is typically simple and does not require an operator skilled in the use of computers. The capital investment and maintenance costs for manually programmed robots are low. These robots are capable of high operating speeds and have good accuracy and repeatability. However, the flexibility of manually programmed robots is limited. They may have only two or three degrees of freedom. Control of intermediate points along the path is generally not available. Only two positions are programmed for each axis. Depending on the complexity of the robot, the total number of programmable steps is generally 10 to 100.

4.3.2 Using a Teach Pendant

In teach pendant programming, the operator leads the robot through the various positions involved in an operation, **Figure 4-7**. As the end effector reaches a desired point, that point is recorded into memory by pushing buttons on the teach pendant. The recorded points are used to generate a point-to-point path the robot follows during operation.

Using a teach pendant is a popular method of programming because it is convenient, simple to learn, and suitable for programming many tasks found in industry.

ABB Robotics

Figure 4-7. The operator is using a teach pendant to program the robot through various moves.

This method does not require an operator skilled in the use of computers. However, complex motions and applications that require close tolerances may involve a lengthy programming time. The robot must be operating while programming, as the program cannot be entered into the teach pendant while the robot is not operating.

4.3.3 Walk-Through Programming

Walk-through programming is used for continuous-path robots. An experienced operator physically moves the end effector through the desired motions. While the robot moves along the desired path, as many as several thousand points are recorded into memory. The number of points recorded can vary from one manufacturer to another. As the number of points recorded is increased, the movement of the robot becomes smoother and more fluid. However, as the rate of recording increases, so does the need for greater storage capacity. Spraying and arc welding are the most common applications programmed using the walk-through method. Other applications include grinding, deburring, polishing, and palletizing.

Walk-through programming does not require computer experience. However, the person programming the robot must be highly skilled in the precise motion required by the task. The robot must be operating while programming is done using the walk-through method.

4.3.4 Computer Programming

Programming for some robots can be done using a computer. ***On-line programming*** is creating the instructions for a robot on a computer that is connected to the robot console. ***Off-line programming*** is creating the instructions for a robot on a computer that is not connected to the robot, **Figure 4-8**. Final testing of the program is done at the job site.

Computer programming provides greater flexibility than other programming methods. It is not necessary to take a robot out of operation while the program is being written and debugged, so productivity is not affected. Using programming languages makes creating complex operations easier. Using these languages reduces programming time, which increases productivity. The primary disadvantage of programming using a computer is that the operator must be experienced in the use of computers, high-level languages, and programming logic.

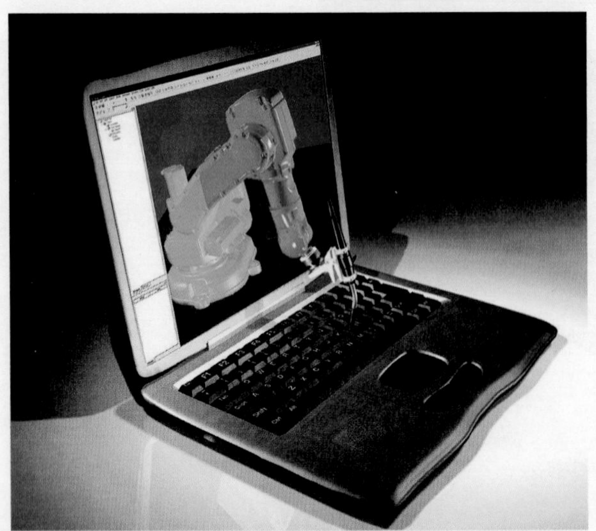

ABB Robotics

Figure 4-8. A personal computer can be used to create instructions for a robot using off-line programming.

4.4 Programming Languages

The first robot programming language, known as *WAVE*, was developed at the Stanford Artificial Intelligence Laboratory in 1973 for research purposes. Most manufacturers that provide off-line programming have developed their own languages because there is not an established set of standards in the industry. Some common languages and their sources are listed in **Figure 4-9**.

Common Programming Languages		
Robotics Programming Language	**Year of Development**	**Originator**
SAIL (Stanford Artificial Intelligence Language)	1968	Stanford University
AL (Assembly Language)	1970s	Stanford University
MCL (Manufacturing Control Language)	1980	McDonnell Douglas Corp.
VAL (Variable Assembly Language)	1980	Unimation Inc.
Karel	1981	Fanuc Robotics America Inc.
AML (A Manufacturing Language)	1982	IBM
RAIL	1982	Automatrix
RPL	1984	Hewlett Packard
RobotBASIC	1984	Intelledex Inc.
Magik	1999	GE Energy

Goodheart-Willcox Publisher

Figure 4-9. This figure shows some common robot programming languages and corresponding originators.

High-level languages are programming languages that more closely resemble standard English than other programming languages, making them easier for people to use. These high-level languages are translated into machine code, which is a language that computers can understand. This is done by means of a program called a ***compiler***.

Even though robot programming languages have been developed by different groups, many are similar to others. The main difference among the languages is in the choice of key words and commands. Each language also has its own syntax, or structure. MELFA BASIC IV, AML, and Karel, for example, are computer programming languages related to BASIC, Pascal, and PL/I languages. Instructions are given in the form of subroutines. A ***subroutine*** is a set of instructions within the program that has a beginning and an end.

Each language defines the number of characters allowed for a program name and use of special characters. With MELFA BASIC IV, for example, the program name must start with a letter and may contain up to twelve characters (letters, numbers, and symbols). Characters are not counted separately. Also, comments may be included within the program code by preceding the comment text with an apostrophe ('). Unlike many programming languages, MELFA BASIC IV does not require a character to signal the end of a line of code. The following sample of MELFA BASIC IV program code provides instructions to move the end effector of a robot from location A to location B.

```
10 'Program Name: Program Exe_2
20 'Author: User
30 'PSAFE is a position clear of all parts and obstructions
40 'P1 is a XYZ location to be taught
50 'P2 is a XYZ location to be taught
60 MOV PSAFE
70 MOV P1 DLY 1.0
80 MVS P2 DLY 1.0
90 MOV PSAFE
100 END
```

With Karel, the program name can contain up to twelve characters, including letters, numbers, and underscores. The program name must start with a letter. Comments can be included within the code by preceding them with two hyphens (--). An identifier follows key words. A semicolon (;) is not required at the end of each line.

The following is a program written in Karel to move the end effector of the robot from location A to location B.

1. PROGRAM Exe_2
2. Variables
3. A: Position --Position location to be taught
4. B: Position --Position location to be taught
5. --Line left blank for separation--
6. Begin --Program execution
7. Move to A
8. Move to B
9. END Exe_2

4.5 Types of Programming

As discussed in Chapter 2, hierarchical control separates instructions for the robot into a number of different levels. The majority of robots in industry today use hierarchical control. However, robots can also be programmed in a manner that focuses on a task as a whole, and this type of control is growing in popularity.

4.5.1 Hierarchical Control Programming

Hierarchical control programming is a method of creating instructions for a robot in which the commands are structured in levels. Each level accepts commands from the level above and responds by generating simplified commands for the level below, **Figure 4-10**. This system uses ***sensory feedback*** (input from the environment) to affect how the robot responds.

First Level

Servo control functions are computed at the lowest level in the hierarchy. This is the level at which most robots in use before 1980 were programmed and controlled. Program commands are compared with feedback from position indicators. If the values are different, a drive signal is generated to move each joint until the correct position is reached.

Programming at the first level does not require the use of a computer. The arm is moved using a ***manual rate control box***. The box consists of a knob and some switches that control the movement of each axis individually. (A rate control box should not be confused with a teach pendant, as it does not have the same capabilities.) The joints are moved one at a time until the robot arm is in the desired location. The values of the position indicators are stored in memory. The arm is moved to another point, and the process is repeated until the desired path is stored in memory. This form of programming is time-consuming and tedious.

Second Level

A computer is required to program a robot at the second and third levels. These levels require real-time interaction of the robot with its environment by means of

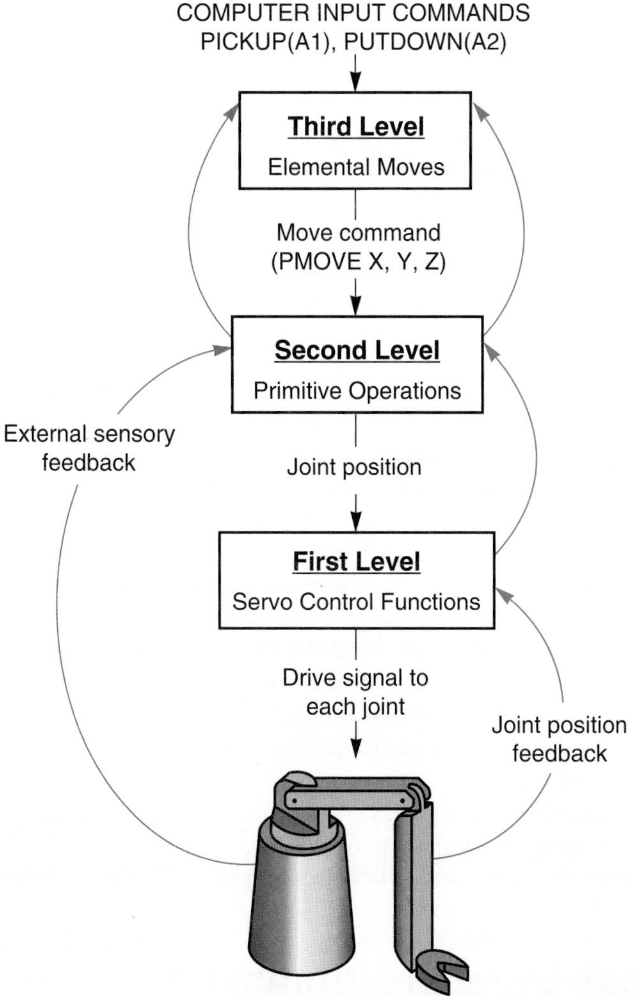

Goodheart-Willcox Publisher

Figure 4-10. This diagram shows a hierarchical control structure.

sensors. Commands are issued for primitive operations, such as a move along a straight line defined by the vectors X, Y, and Z. These commands are translated by the computer into the proper position values.

The second level in the hierarchy receives position feedback and generates the command sequence used by the first control level to accomplish the task. At this level, a joystick can be used for programming instead of the manual rate control box. The operator does not have to move individual joints. Programming tasks are much easier and faster than for the first level.

Third Level

The third level in the hierarchy receives commands for elemental moves. A typical command is MOV P1 or MOVE to A. When these commands are issued by third-level circuitry, the control system monitors feedback signals. It then generates a sequence of commands for primitive operations. These primitive operations become the input to the next level down.

For example, to program a simple transfer task, commands for two elemental moves are typed in the desired sequence into the computer. In the MELFA BASIC IV language this would be:

MOV PSAFE (Safe location away from obstructions)
MOV P1 (X, Y, Z location above position A)
MVS P2 (X, Y, Z location to pick up part)
HCLOSE (Close gripper)
MVS P1 (X, Y, Z location above position A)
MOV P3 (X, Y, Z location above position B to place part)
MVS P4 (X, Y, Z location to place part)
HOPEN (Open gripper)
MOV P3 (X, Y, Z location above placed part)
MOV PSAFE (Safe location away from obstructions)

This set of commands causes the robot to move from a safe location to location X, Y, Z (position two) and pick up an object. It then moves the object through a series of points to location X, Y, Z (position four) and puts the object down. The same sequence is repeated for the next set of positions. These points are previously recorded using a joystick, teach pendant, or other device, moving the robot through the points manually. The X, Y, and Z locations can also be typed directly into memory.

During playback, the robot grasps and releases at the recorded locations. The environment must be tightly controlled. The locations of the objects to be grasped and their orientation must always be the same. A slight misalignment can cause damage to both the robot and workpiece. Controlling the environment often requires additional machinery and equipment that can cost several times more than the basic robot.

Programming even a simple task using hierarchical control and traditional high-level languages can take days. Additional time must be set aside for testing and making changes. Even after the system is up and running, further changes may be required.

4.5.2 Task-Level Programming

In today's manufacturing environment, fast production time and low system costs are critical to a company's success. Yet robot systems are expensive and difficult to install and maintain. A typical robot in an automated work cell might move something, assemble parts, use vision for guidance and parts inspection, control local work cell devices, respond to sensor inputs, keep statistics on processes, and communicate with a host computer or an operator.

Figure 4-11A shows the architecture for a typical custom work cell. It is more complex than it needs to be. Four separate controllers are used. A motion control system runs the robot, while the programmable logic controller (PLC) provides work cell logic. The vision system is for inspection and robot guidance. The personal computer is included for user interface. A simpler system with user-friendly hardware and software would be better. The answer lies in task-level programming.

Task-level programming is a method of creating instructions for a robot in which the user specifies the goals of each task rather than the motions required to achieve those goals. Instructions are entered using simple English-like terms. The details for each and every action the robot is to perform do not need to be specified. This allows the use of instructions at a significantly higher level than those produced by languages such as MELFA BASIC IV, AML, Karel, or RobotBASIC. Many activities are programmed automatically by the computer.

An example of a task-level programming environment is shown in **Figure 4-11B**. In this example, only one controller and one programming system are needed to work with multiple technologies. The need for programming knowledge is reduced because instructions are entered using menu selections or commands stated in English. The operator can focus attention on the task instead of on writing programming code.

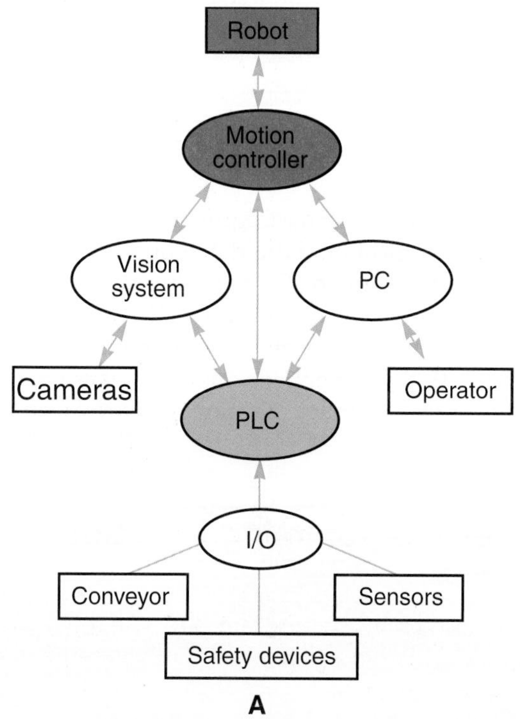

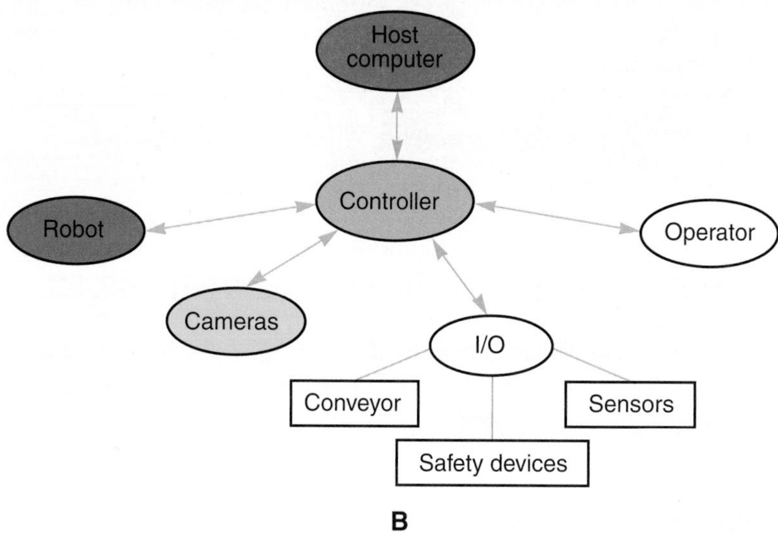

Adept Technology, Inc.

Figure 4-11. Using task-level programming streamlines the programming process. A—This is a typical custom work cell layout. The arrows indicate interactions. B—A single controller and programming system are used to work with multiple technologies.

Interactive screens are used to create, teach, debug, and run application programs. Programming software has become so user-friendly that even inexperienced programmers can program a robot. Some common task-level programming functions are described in **Figure 4-12**.

Task-level programming is a highly efficient method that replaces hundreds, even thousands, of lines of programming code with a small number of statements selected from menus. Representative programming screens are shown in **Figure 4-13**.

4.6 Voice Recognition

Voice recognition, also called speech recognition, is a system that uses recognizable words or phrases as a form of audio data entry for a computer. Using voice recognition can produce the same result as typing words on a keyboard. Only a few years ago, voice recognition was experimental and was used only in the lab. Today, however, voice recognition systems are commonplace. Many organizations use voice recognition with their telephone answering systems. Rather than using a telephone keypad to enter a number selection, you are given the option to speak your selection.

Common Task-level Programming Functions	
Function	**Programming**
Robot Movements	The first step in creating the robot program is to select the sequence of movements from a menu. For example, the MOVE statement tells the robot to pick up parts from one location and place them in another location. This type of sequence may be used to move a part from a conveyor belt to a pallet.
Location Information	A location database stores work cell locations. The height of each location, approach, and departure are taught or modified by making the appropriate selections and entries. The robot's speed, type of motion, and other details are also selected and stored.
Palletizing	The PALLET statement is commonly used in automatic palletizing functions. The operator can define spacing and the number of pallet locations.
Visual Sensors	The type of visual sensors needed can be specified by making the appropriate selections and entries on the program interface. For example, a visual inspection sensor can locate and evaluate parts, while a visual guidance sensor may track and orient the parts along a conveyor belt.
General Control Functions	From the control panel screen, the operator can start the operation, slow the speed of the robot, and step the robot through its motions to be sure the program is performing as intended. The control panel screen may also be used for debugging and cell control.

Goodheart-Willcox Publisher

Figure 4-12. The basic program functions and steps are defined on-screen by the programming software. Programmers use on-screen sequence-editing functions to develop the robot's movement.

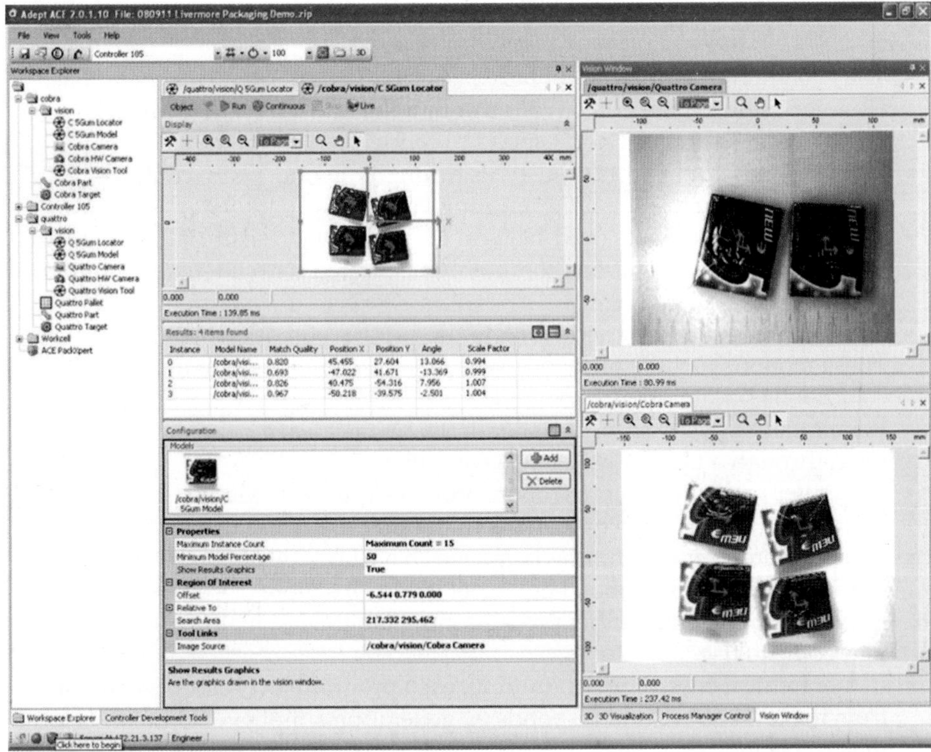

Omron Adept Technologies, Inc.

Figure 4-13. This figure shows representative programming screens.

Voice recognition has improved dramatically over the past few years. Robots have been developed that understand basic phrases in multiple languages and can respond in the appropriate language. Manufacturers are considering placing voice recognition systems in items for use by physically handicapped people. For example, robots equipped with two-way communication can help meet the needs of a person who is completely paralyzed. In this situation, using voice may be the only practical means of communication. Additionally, experts suggest that voice frequency is just as accurate as fingerprints for identification purposes.

Voice communication from a robot may also serve to diagnose and warn of possible problems in a manufacturing operation. If the robot is experiencing trouble with a component, the problem could be verbally communicated to an operator or service person.

One of the biggest hindrances in mainstreaming this technology is likely to be the cost. However, certain key advantages make voice recognition attractive. Verbal communication is inexpensive, and voice recognition can reduce set-up time. This is important since manufacturing trends are shifting away from volume and toward variety. Operators may be intimidated by some of the other set-up methods. With voice recognition, however, anxiety is reduced. Voice recognition can truly be described as user-friendly.

Summary

- The evolution of industrial robots consists of three periods or generations. First-generation robots (late 1950s through the mid-1970s) performed repetitive tasks and did not respond to changing conditions.
- Second-generation robots (mid-1970s through the mid-1980s) could be controlled by means of a programming language.
- Third-generation robots (mid-1980s to the present) use artificial intelligence (AI) and perform operations commonly associated with intelligent human behavior.
- Robots can be classified according to the pattern of motion they use. These motions include pick-and-place motion, point-to-point motion, and continuous-path motion.
- Robots can be programmed manually, by using a teach pendant, by walking them through a task, or by using a computer.
- Computer programming for robots provides greater flexibility for performing higher-level operations than other programming methods.
- Many different programming languages have been developed to provide precise control of the robot. High-level languages more closely resemble standard English than other programming languages, making them easier for people to use.
- In hierarchical control programming, each program level accepts commands from the level above and responds by generating simplified commands for the level below.
- In task-level programming, the user specifies the goals of the task rather than the motions required.
- Voice recognition is a system that uses words or phrases as audio data entry. Robots have been developed that understand basic phrases and can respond in the appropriate language.

Review Questions

Answer the following questions using the information provided in this chapter.

1. The evolution of programming for robots consists of three periods. List these periods and discuss the developments in each period.
2. List three classifications of robots according to the pattern of motion they use and briefly describe the three patterns of motion.

3. What are three advantages of using continuous-path motion control?
4. Identify four methods used to program robots and briefly describe each method.
5. What are some advantages of using a computer to create programs for controlling robots?
6. What distinguishes high-level languages for programming, and why is this distinction important?
7. What method is used to convert high-level programming languages to machine code that a computer can understand?
8. What are some characteristics of high-level programming languages used for robots?
9. What are the main differences between hierarchical control programming and task-level programming?
10. What are some advantages of task-level programming?
11. What are two examples of ways voice recognition can be used with robots?

Learning Extensions

1. Using the Internet, conduct a search to find four robot manufacturers. Identify the programming language each company uses to program the robots they manufacture.
2. Visit the Honda website and find information about the ASIMO robot. View the various videos about the functions of ASIMO. Identify possible uses for this type of robot.

Advanced Analysis

1. For the programming software used in your laboratory, describe the following:
   A. Programming environment.
   B. Axis movement commands.
   C. Circular interpolation commands.
   D. Delay or timer commands.
   E. Speed control commands.
   F. Program control GOTO function.
   G. Program control If…Then…Else command.
   H. Program control subroutines.
   I. Program control repetition.
   J. Robot I/O functions.

Chapter 5

Industrial Applications

Outline

Objectives

Upon completion of this chapter, you will be able to:

- Describe how robots are integrated into the manufacturing process.
- Discuss factors to consider in selecting the proper robot for a given task.
- Identify applications where robots are used in industry.

Technical Terms

accuracy
automated guided vehicle (AGV)
command resolution
dedicated equipment
design for manufacturability
dynamic performance
fixturing
interlocks
operational speed
payload
repeatability
resolution
service robot
spatial resolution

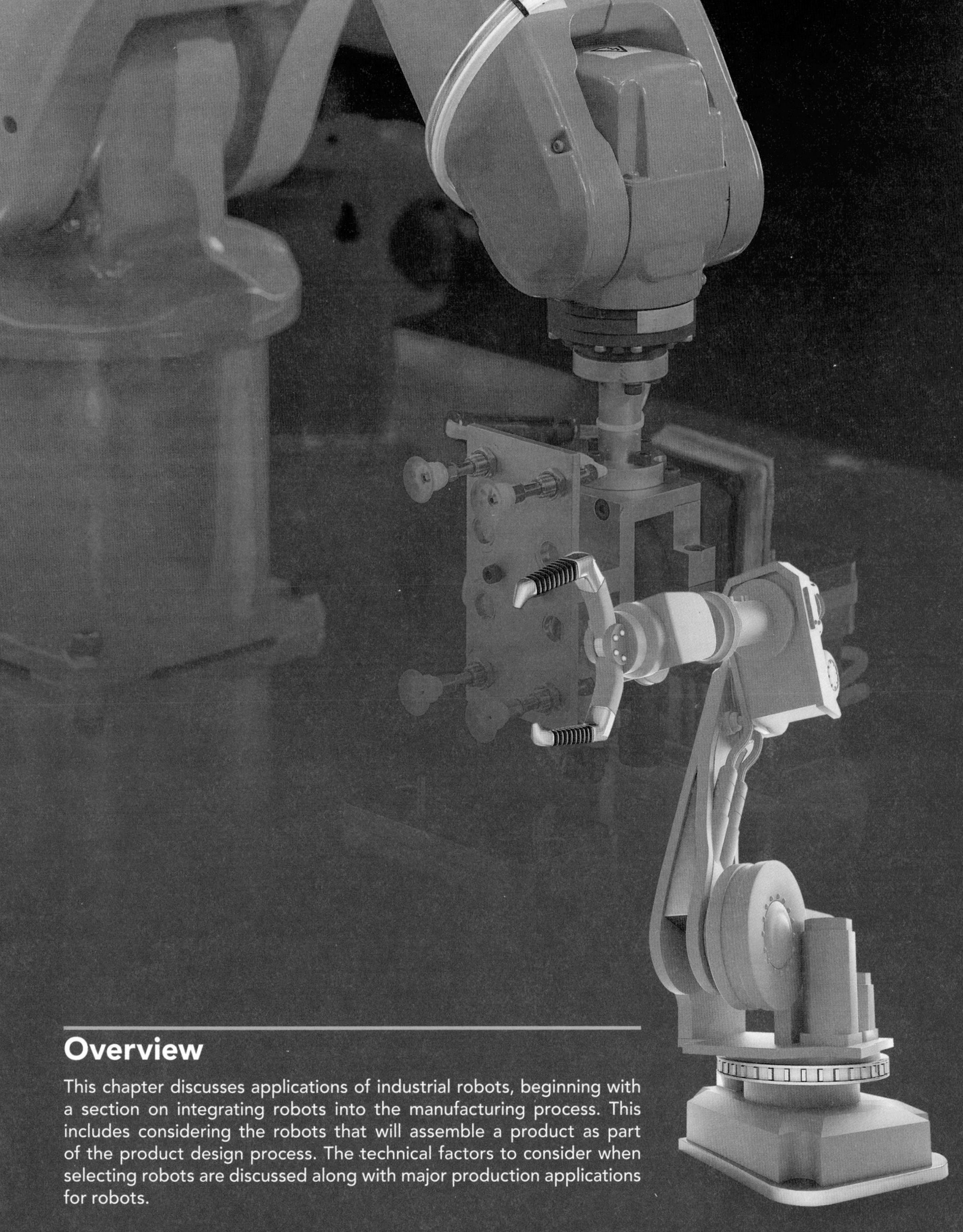

Overview

This chapter discusses applications of industrial robots, beginning with a section on integrating robots into the manufacturing process. This includes considering the robots that will assemble a product as part of the product design process. The technical factors to consider when selecting robots are discussed along with major production applications for robots.

5.1 Integrating Robots into the Manufacturing Process

Automation is a high priority for manufacturing companies worldwide. Competition, improved technology, cost, market conditions, productivity, available manpower, and undesirable work environments are leading manufacturers to install a greater number of robots. How can robots be successfully incorporated into a manufacturing process? The design of the manufactured products and safety considerations are two key factors to consider in answering this question.

5.1.1 Design for Manufacturability

Design for manufacturability involves considering the robots that will assemble a product as part of the product planning process. First, the engineering department must decide which parts of the product can be assembled by robots and which parts can be assembled by people or by using dedicated equipment. ***Dedicated equipment*** is an automated device that is designed to perform only one function. The next step is to design, or redesign, the parts of the product for ease of robotic assembly. The list of guidelines that follows is useful to consider when a robot will be used for assembling a product.

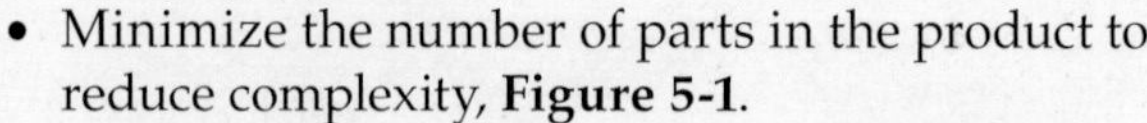

- Minimize the number of parts in the product to reduce complexity, **Figure 5-1**.
- Reduce the directions of approach required to assemble a product. This allows a less sophisticated robot to be used.
- Minimize the number of obstructions in the environment so that the robot can work in a straight line.
- Whenever possible, use subassemblies with components that stack on top of each other or that can be assembled in sequence using downward motion.
- Add chamfers, guide pins, ridges, and other physical characteristics that allow the robot to lock the part into its proper location. The cost of these guides is far less than the cost of robot vision systems.
- Simplify fastening by using tabs, snaps, or other methods that allow parts to be joined in one motion, **Figure 5-2**.
- Eliminate screws, springs, and adjustments.
- Avoid parts that compress or deform easily, such as wires, foils, or foams, which robots do not handle well.
- Give parts common features. For example, all parts might have the same size hole or post in the center. This allows the same end effector to pick up any of the parts.
- Make parts as close to symmetrical as possible to ensure correct assembly no matter how the robot picks them up, **Figure 5-3**. If a part must be asymmetrical, use the asymmetrical features to assist in orientation.

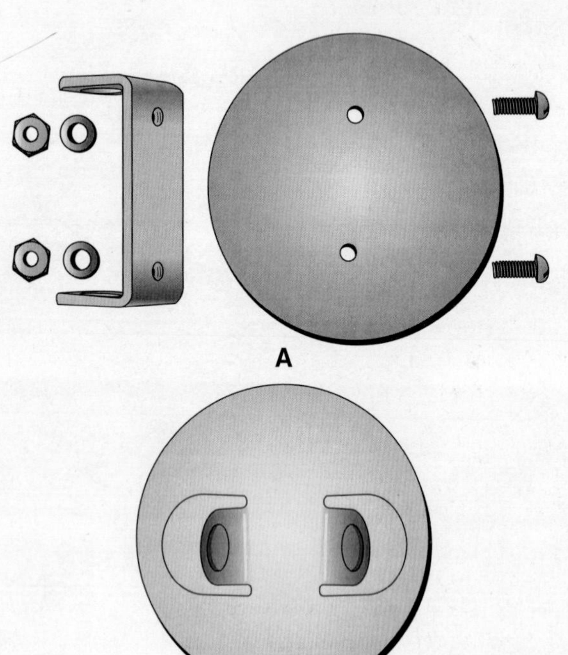

Goodheart-Willcox Publisher

Figure 5-1. A—This subassembly contains a total of eight separate parts, making robotic assembly very difficult. B—The same part is redesigned to accommodate robotic assembly. The brackets are punched out of the original base plate, resulting in the use of one part rather than eight parts.

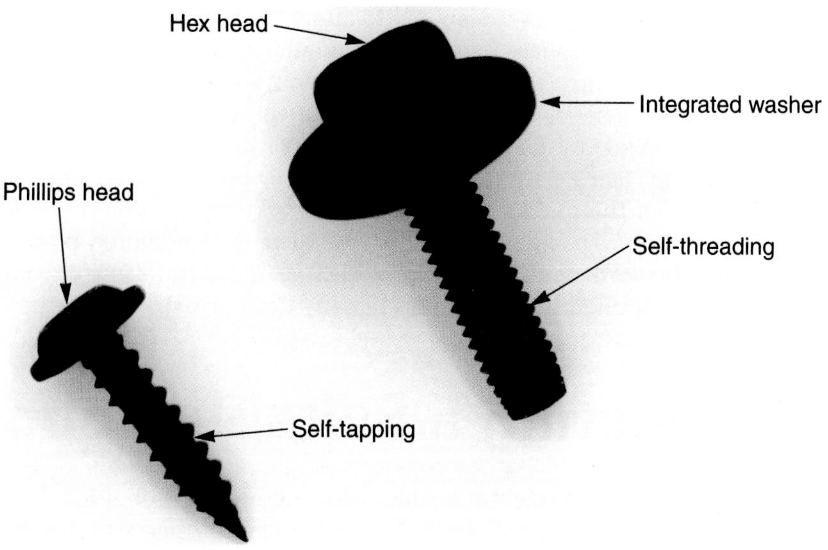

Goodheart-Willcox Publisher

Figure 5-2. These screws have been designed specifically for ease of assembly with features such as self-tapping, self-threading, and a washer attached to the screw head.

Many companies have discovered that when they design for manufacturability, the process becomes so simplified that it is faster and more productive to use people or dedicated machines to produce the products rather than robots. In some cases, all three methods might be used.

5.1.2 Robotic Safety Considerations

Safety is always an important consideration in installing robotic systems. Robots differ from other machinery due to their degrees of freedom and sizable work envelopes. Providing a safe working environment must be considered during the design, installation, maintenance, and operation of robotic systems. A robotic work station can be dangerous if workers are not properly trained or are not made aware of possible hazards.

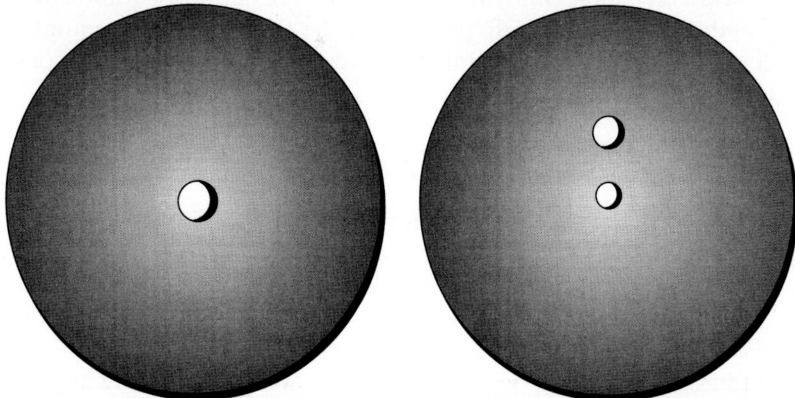

Goodheart-Willcox Publisher

Figure 5-3. The base plate on the right contains asymmetrical holes, which means that it must be installed using one particular orientation. The redesigned base plate on the left is symmetrical and does not require any particular orientation during assembly.

The work station should be designed to allow the robot and operator to work safely. Omit places where space may be too limited by being aware of fixed objects within the work area, such as vertical poles. Keep the end effector in mind when designing the space. The relationship of the end effector to work fixtures and other equipment must be taken into account. Outline the full range of the end effector's motion on the floor using paint or colored tape. This provides a constant, highly-visible perimeter of the robot's work area. The work station should also include barriers that cannot be passed through or over. The design should place control panels outside the work envelope and provide ***interlocks*** (safety devices that are designed to prevent unauthorized access to hazardous areas).

5.2 Selecting a Suitable Robot

With hundreds of robot models available, selecting the best robot for the job can be a major task. Although robots are not as versatile as human workers, they can perform a wider range of tasks than equipment designed to perform one specialized function. Robots can perform repetitive tasks without tiring and can operate in hostile environments. They can maintain greater precision than human workers when performing repetitive work.

There are two types of tasks for which robots are not suited. One such task is extended, high-speed production of many identical items. Dedicated machinery is better suited for that type of work than robots. Robots also are not suited for short, complex tasks that require a high degree of hand-eye coordination. Currently, people are still best suited to performing these tasks. Otherwise, robots are useful for many production jobs. In addition to environmental conditions and coordination with associated equipment, other important considerations when selecting a robot include the work envelope, degrees of freedom, and other measures of performance.

5.2.1 Work Envelope

As discussed in Chapter 2, the work envelope is different for each robot configuration and is defined by the type of joints and degrees of freedom of the robot. All the fixtures and other machinery that will work with the robot must be located within the work envelope. Many times, placement of support equipment dictates the shape of the work envelope. To provide greater flexibility, some manufacturers offer optional wall and ceiling mountings in addition to the standard floor mounting for the robot.

5.2.2 Degrees of Freedom

The number of degrees of freedom determines the ability of the robot to move within the work envelope. Robots with six degrees of freedom can work anywhere within the work envelope. However, not all applications require such a versatile robot. The configuration of each individual work station determines the requirements for degrees of freedom.

Sometimes, an additional degree of freedom can be obtained by mounting the robot on rails. This permits the robot to move horizontally, increasing its overall reach. The six possible degrees of freedom can be placed into two basic categories: body and wrist. The table in **Figure 5-4** provides descriptions of the six degrees of freedom and the associated axes.

Robot Axes and the Six Degrees of Freedom		
Degree of Freedom	**Robot Axis**	**Description**
First	Waist (rotational transverse)	Movement about a vertical axis.
Second	Shoulder (radial transverse)	Extension and retraction of the arm.
Third	Elbow (vertical transverse)	Up and down motion.
Fourth	Pitch	Up and down movement of the wrist.
Fifth	Yaw	Side to side movement of the wrist.
Sixth	Roll	Rotation of the wrist.

Goodheart-Willcox Publisher

Figure 5-4. This table summarizes the six degrees of freedom and the corresponding robot axes.

5.2.3 Resolution

Resolution is the smallest incremental movement the robot can make. Resolution can also be described as the smallest segment into which the work space can be divided. Resolution is determined by the control system for the robot. ***Command resolution*** is the closest distance between movements of a robot. It is calculated by dividing the travel distance of each joint by the number of control increments. For example, a robot travels a distance of 36″ (91.44 cm) along one axis, and the controller is programmed with a total of 8000 points. Divide 36 by 8000 to produce a command resolution of 0.0045″ (0.01143 cm), without taking into account any mechanical inaccuracies.

The manipulator uses mechanical components to position the end effector at the various points in the work envelope. Whenever mechanical parts are used (gears, chains, cables, or ball lead screws), some inaccuracies occur. This happens because chains and cables can stretch and gears are prone to backlash. Also, overweight payloads can create inaccuracies. All these factors can affect resolution.

Tool positioning during programming is an important factor in accomplishing the desired task. However, tool position may vary due to the command resolution and mechanical inaccuracies. ***Spatial resolution*** describes the accuracy of movement of the tool tip of the robot. Spatial resolution takes into account command resolution and mechanical inaccuracy. For example, if the command resolution of 0.0045″ (0.01143 cm) has a mechanical inaccuracy of 0.003″ (0.0076 cm), then the spatial resolution would be 0.0075″ (0.01903 cm). This figure is derived by adding the command resolution to the mechanical inaccuracy.

In the work envelope, spatial resolution varies with tool location. The consistency of the spatial resolution value within a work envelope is affected more by certain robot configurations than by others. For example, the Cartesian configuration offers fairly constant spatial resolution. Rotary joints, however, can impact the spatial resolution value at various points within the work envelope, resulting in necessary adjustments.

5.2.4 Accuracy

Accuracy and spatial resolution are both by-products of command resolution and mechanical inaccuracies. ***Accuracy*** expresses how precisely the end effector of the robot is programmed to reach a predetermined point. The value associated with accuracy is expressed as half of the spatial resolution. A robot with a spatial resolution of 0.010″ (0.0254 cm) has an accuracy of 0.005″ (0.0127 cm).

Large robots with payloads of 100 lb (45.36 kg) or more have an average accuracy of 0.050″ (0.127 cm). Small robots used for tasks where payloads are lighter, such as assembly operations, have an average accuracy of 0.002″ (0.005 cm). Improved accuracy increases with advancements in technology.

The speed of movement and weight of the payload also affect accuracy. As speed increases, accuracy decreases. To stop the manipulator at a given location, speed must be reduced to prevent overshooting the position. If the speed is reduced too much, however, valuable time is wasted. Overcoming the inertia in moving a heavy load can also affect position accuracy. To overcome inertia, it may be necessary to reduce speed. Therefore, there must be a trade-off between speed and accuracy.

5.2.5 Repeatability

It is a common mistake to confuse accuracy with repeatability. Even though both are dependent on spatial resolution and mechanical inaccuracy, there is a difference. Accuracy deals with programming the end effector of the robot to go to a designated position. ***Repeatability*** is a value that expresses the variation in movements that occurs when an action or task is performed several times. For example, you could compare how close the robot comes to a desired position when performing a task each of twenty times.

Good repeatability is more desirable than accuracy because inaccuracies are easier to correct. This is especially true if the inaccuracies are consistent for all movements of the end effector. For example, suppose that a robot is programmed to move its gripper from point A to a target point 30″ (76.2 cm) away. After the robot has made the move, a measurement is taken that shows the robot actually moved 30.10″ (76.454 cm). This represents an inaccuracy of 0.3%. If an inaccuracy of 0.3% is consistent for other command movements, then the programmer can compensate for this error. Adjustments for poor repeatability are more difficult. In fact, repeatability may become so poor that a more sophisticated robot that can perform more consistently is required to complete the task. However, if the error is rather small, additional tooling or alignment devices can be used to compensate.

Repeatability can change with use, especially when robots perform the same task day after day. The mechanical components are subject to wear, which increases mechanical inaccuracies. Mechanical inaccuracies, in turn, reduce the repeatability performance.

5.2.6 Operational Speed

Operational speed, also called ***dynamic performance***, represents how fast the robot can accelerate, decelerate, and come to a stop at a given point. The two most important factors that influence operational speed are the desired accuracy and the payload. The robot configuration and location of the tool in the work envelope also influence operational speed.

Robot manufacturers define robot speed in different ways. Speed may be noted for each joint or for various groups of joints. The range of speed may be given for when there is no load as well as for when there is a full load.

Most people think that robots are more productive than people and assume that robots work at a faster pace. However, a robot's actions often are no faster than those of a person. In fact, a robot's movements can be many times slower than a person's movements. Increased productivity results because the robot works at a constant, steady pace and does not stop for breaks or food. Therefore, basing the total cycle time for a given task on the manufacturer's speed rates may not produce the true cycle time. To arrive at cycle times that are realistic, a prototype layout may have to be built. The prototype layout should consider all the variables in the procedure, such as environment, speed of other machinery, and size of the work envelope. It is important to determine how fast the end effector can move through the total cycle while still maintaining the desired accuracy and repeatability. The robot should produce a quality job in a reasonable amount of time.

5.2.7 Load Capacity

The ***payload*** is the maximum weight or mass of material a robot is capable of handling on a continuous basis. In most cases, the weight of the end effector is included in the payload. Some manufacturers' specifications indicate payload values for both an extended robot arm and for a retracted arm. Other manufacturers may list load-carrying capacity for the wrist joint and end effector in addition to the arm. However, the two factors that greatly affect load-handling capabilities are the type of configuration and the placement of the end effector within the work envelope. The robot's load-handling capability is reduced when the boom is fully extended (compared to when the boom is retracted).

Some robots are capable of lifting as little as 1 lb (0.453 kg), while others may have a lifting capacity of 2000 lb (907.18 kg). In most instances, the weight of the end effector is included in the lifting capacity. If a robot has a lifting capacity of 5 lb (2.268 kg) and the gripper weighs 2 lb (0.907 kg), the weight of a part to be moved or manipulated cannot exceed 3 lb (1.36 kg). Most robot manufacturers list both normal and maximum load-handling capacities.

The average weight carried by robots in US industry is estimated to be 20 lb (9.07 kg). Approximately 50% of robots in industry handle parts weighing less than 10 lb (4.536 kg). The average load-carrying capacity could be even less in the future because smaller robots are being constructed for assembly tasks, which rarely involve heavy payloads. For example, a large portion of the individual parts used in an automobile weigh less than 5 lb (2.268 kg).

5.3 Using Robots in Industry

As robot technology advances, areas where a robot might be used are limited only by our imagination and creativity. A completely automated factory is now possible, and a few are in operation as a result of the advances in computing technology and sensor applications. Even now that the automated factory is a reality, engineers and technicians are needed to ensure that the technology continues to function as designed. As newer robotic equipment is developed and brought into the workplace, the potential uses for robots will increase dramatically.

5.3.1 Pick-and-Place Operations

The process of picking up parts at one location and moving them to another is one of the most common applications for robots. Placing parts or removing them from a uniform series of positions (such as on a pallet) is probably the most common form of pick-and-place operation, **Figure 5-5**. Some pick-and-place operations are used for transferring parts, **Figure 5-6**.

FANUC Robotics

Figure 5-5. This robot palletizes products for storage or shipping.

Motoman *Pacific Robotics*

A B

Figure 5-6. Pick-and-place motion is used to transfer products and materials. A—This robot is loading automotive wheel rims into a storage cart. B—This robot transfers boxes from one conveyor to another.

Using robots that are less sophisticated for pick-and-place operations offers several advantages. They can handle fragile parts made of glass or powder metal. Those used in lightweight pick-and-place applications offer excellent speed while maintaining good accuracy and repeatability. Robots are also useful for handling heavy loads and items that are at extreme temperatures.

5.3.2 Machine Loading and Unloading

Another common use of robots in industry is machine loading and unloading. In production, robots load and unload parts associated with automated machining centers, such as CNC (computer numerically controlled) machines, **Figure 5-7**. A robot's machine loading and unloading capabilities can be applied to several operations, such as forging, injection molding, and stamping.

Industrial operations often expose workers to excessive heat, noise, dirt, and air pollution. Since these operations are considered unpleasant and are frequently dangerous, they are ideal applications for robots. Replacing human operators with robots reduces the need for expensive safety equipment. Workers in unpleasant or dangerous jobs also typically have high absenteeism rates, which contributes to low productivity and poor product quality. Substantial gains in productivity and product quality generally result when robots are properly used.

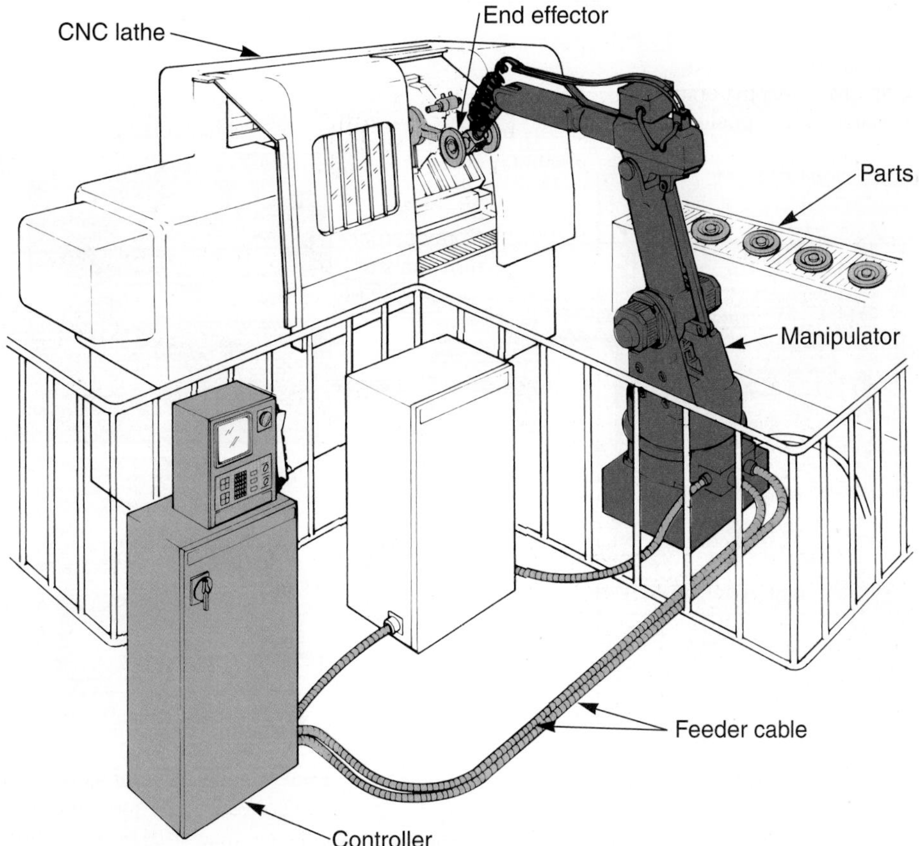

Motoman

Figure 5-7. This robotic system is used for loading and unloading parts to a CNC lathe.

5.3.3 Die-Casting Operations

Die casting is producing metal parts using molds, which are called dies. The first practical application of industrial robotics occurred more than 30 years ago when General Motors installed the Unimate robot in a die-casting operation. The use of robots in die casting has resulted in a 200 to 300% increase in productivity. After robots proved useful in die casting, they were used in investment casting, forging, and welding operations.

Die casting involves pumping molten metals, such as zinc, aluminum, copper, brass, or lead, into closed dies, **Figure 5-8**. After the metal solidifies, the die is opened and the casting is removed. In this operation, the robot removes the hot casting from the die and dips it into a liquid, usually water. This process is called quenching. Next, the robot transfers the cooled part to a trimming press where excess material (flash) is removed and the part is shaped. The robot then transfers the part to a pallet or conveyor belt to be transported out of the work area.

Robots can work with hot castings more consistently than humans. The increase in productivity is the result of decreasing the total time required to produce a finished casting. The decrease in time spent in the die-casting process results in a more uniform die temperature, which reduces the amount of flash material attached to the casting. Therefore, trimming and scraping costs are also reduced.

The robots essentially perform a machine loading and unloading operation. This type of operation does not require critical or complex motions and is ideally suited for low-cost pick-and-place robots.

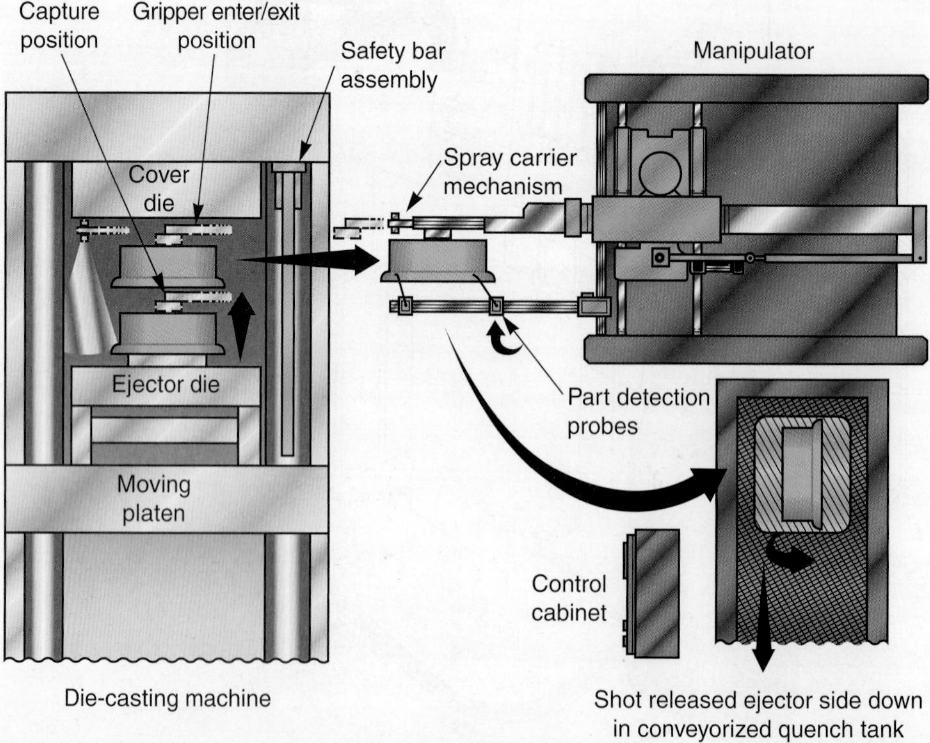

Sterling Detroit Company

Figure 5-8. In a die-casting operation, a robotic arm unloads the hot casting from the press and moves it to the quench tank.

5.3.4 Welding Operations

Robots used in resistance welding and arc welding operations comprise the third largest group of robots in industry. In resistance welding, electric current is passed between two metals, causing them to heat and fuse together at the point of the electric current, **Figure 5-9**. In arc welding, the weld is made along a joint rather than at one spot. The arc-welding process heats the metals until they melt at the joint and fuse into a single piece, **Figure 5-10**.

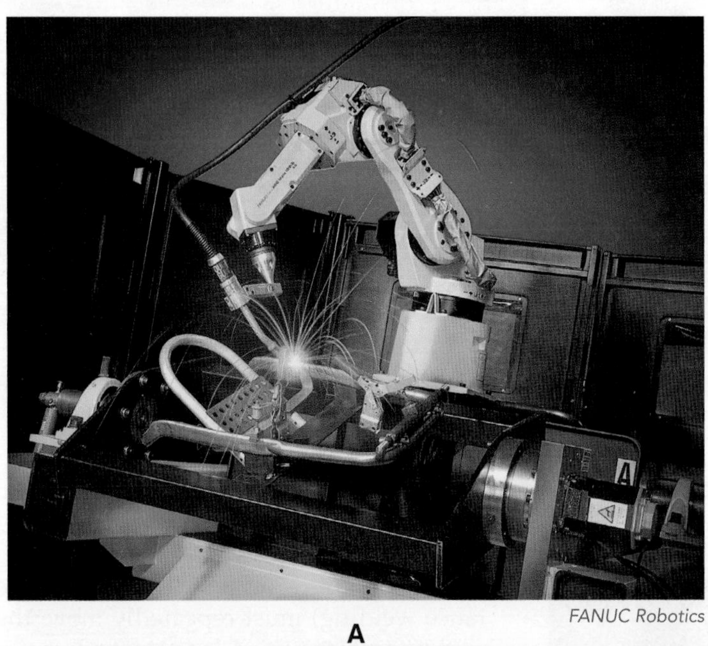

FANUC Robotics

A

Motoman

B

Figure 5-9. Robots are used in industrial resistance welding applications. A—A robotic spot welder is at work in an assembly operation. B—This robot is welding a piece of construction equipment.

Motoman

Figure 5-10. These two robots, mounted on a shuttle track for increased mobility, are being used to perform an arc welding task.

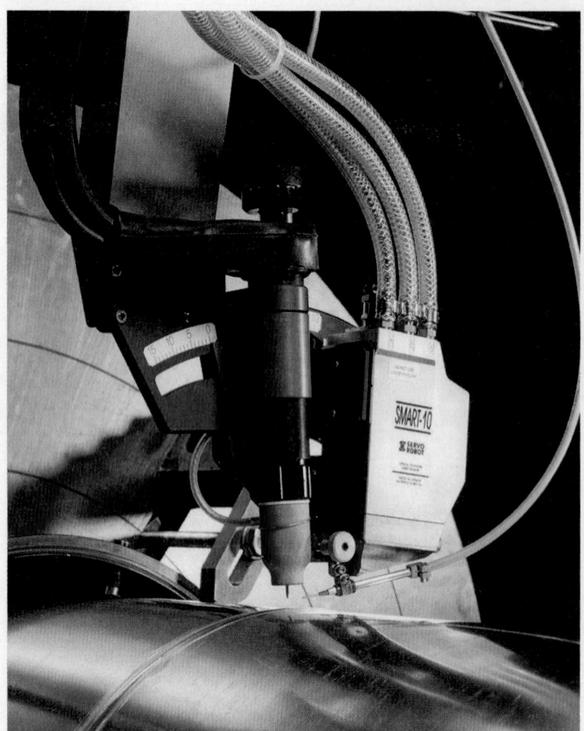

Servo-Robot Inc.

Figure 5-11. Laser seam tracking helps guide the welding laser beam along the seams.

Robots used for spot welding (a type of resistance welding) must repeatedly move the welding gun to each specified location and position it perpendicular to the weld seam. Spot welding is used extensively in the automotive industry to weld body sections and other parts together. The workpieces must be pressed together at the spot where the weld is applied by the welding electrodes. A low-voltage direct current is passed through the parts, causing them to fuse. Robots typically used for spot welding have six or more axes of motion that allow them to approach the weld at any angle. This type of movement would be difficult and awkward for a human operator.

Many problems associated with robotic arc welding are due to poor joint alignment. Robotic arc welding requires more accurate positioning and better-fitting joints than welding done by human operators. A human operator can make adjustments if the location of joints or the gap between parts varies, but most robots cannot. To remedy the problem, robot manufacturers have developed various sensory tracking systems. One example is a robotic system equipped with a visual arc sensor, **Figure 5-11**. The visual sensor tracks the joint position and compensates for deviations.

5.3.5 Spraying Operations

In industrial spraying operations, the spray nozzle is mounted on the robot's wrist. Painting or spray finishing involves the application of a variety of paints, polyurethanes, or other protective coatings to the surface of a part or product, **Figure 5-12**. Spraying operations can also be used to apply sealant or glue to the surface of an object.

Continuous-path programming must be used for spray operations. These operations should be performed by an experienced operator because both intentional and unintentional movements and positions are recorded. For spraying operations, programming the robot to perform in a smooth motion is more important than the actual point locations.

Using robots in spraying applications offers several benefits. Spray vapors are often toxic and explosive, so expensive ventilation systems must be installed for the health and safety of human workers. Since robots do not need fresh air to operate, greater concentrations of solvent can be allowed in the air. With the reduction in ventilation, energy costs are lowered. However, there is the danger of an explosion caused by a spark from the electrical system. In the past, most robots used for painting were hydraulically powered for this reason. Today, spark-free electrical robots are available for use in explosive environments.

The use of robots for sealing and gluing applications has increased as the application methods and the quality of sealing and gluing agents have improved. The components associated with a typical robotic sealing/gluing work station are shown in **Figure 5-13**. Sealing and gluing operations require specialized end-of-arm tooling for three-dimensional surfaces, **Figure 5-14**. To ensure even distribution of the adhesive or sealing product, path control and speed must remain consistent between programmed points.

FANUC Robotics

Figure 5-12. These robots are spray painting a vehicle body in an enclosed work cell.

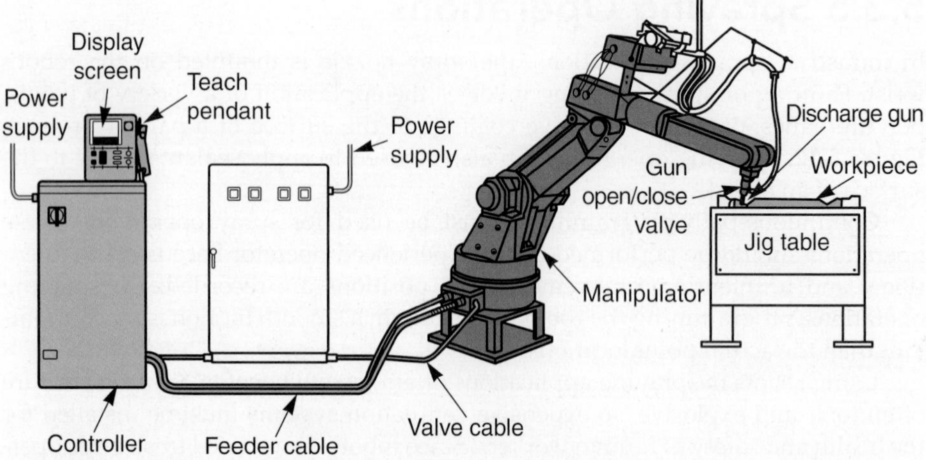

Motoman

Figure 5-13. Components used in a typical sealing/gluing robotic work station.

Motoman

Figure 5-14. This gantry robot applies an adhesive to automotive windshields.

5.3.6 Machining Processes

Some of the machining processes performed by robots include routing, cutting, drilling, milling, grinding, polishing, deburring, riveting, and sanding. Robots used in these applications must have a high degree of repeatability. They must be equipped with quick-change tooling and improved sensory and adaptive control. Robots used in machining processes must also be good at ***fixturing***, which is the ability to provide precise points in the process of accurately locating parts.

Cutting applications require flexible robots with tight servo control loops, directly linked drives, and dedicated cutting axes, **Figure 5-15**. The most common types of robotic cutting applications are gas torch, plasma torch, water jet, and laser. The components used in a typical gas torch cutting robotic work station are illustrated in **Figure 5-16**. Additionally, many industrial cutting operations require special safety precautions to safeguard operators, **Figure 5-17**.

Motoman

Figure 5-15. This robot is using a laser to cut a floorpan for an automobile.

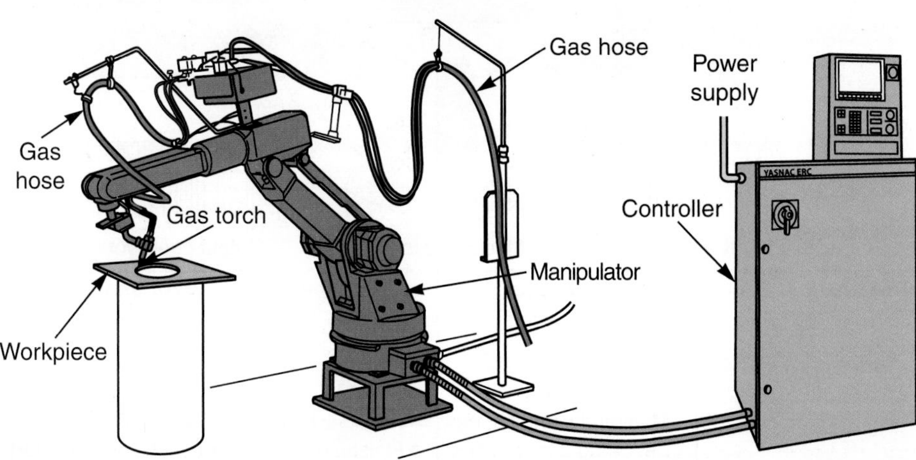

Motoman

Figure 5-16. These components are associated with a typical gas torch cutting robotic work station.

Goodheart-Willcox Publisher

Figure 5-17. This robotic system operates within a plastic safety enclosure.

Deburring and polishing operations require robots with smooth movement and rigid wrist design. The robot control system must be capable of automatically adjusting for various factors, such as tool wear, burr size variation, workpiece variation, and product positioning, **Figure 5-18**. Force-sensing equipment can identify contact with a workpiece and adjust tools for constant pressure, **Figure 5-19**.

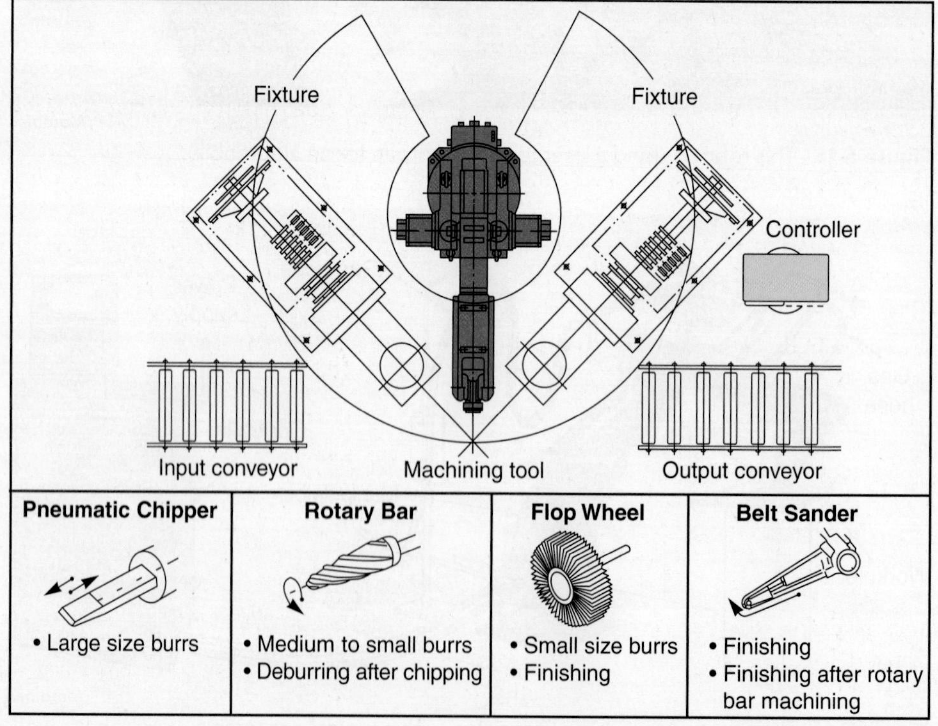

Motoman

Figure 5-18. The components shown here are used for deburring and polishing.

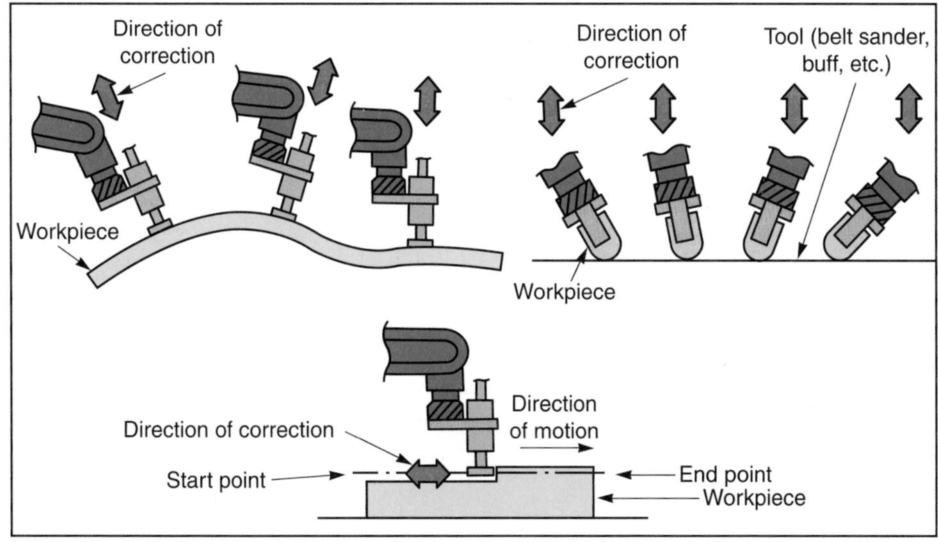

Motoman

Figure 5-19. Force sensors allow for corrections in tool pressure against the workpiece.

5.3.7 Assembly Operations

Making use of robots in some portion of the assembly process has become standard practice in industrial applications. The potential savings and increased efficiency are significant advantages, since assembly operations may account for 50% of the labor cost involved in manufacturing a product. Workers who perform certain kinds of assembly tasks can develop repetitive-task injuries and health problems. Psychological problems due to job stress are also associated with highly repetitive work. The use of automated robotic assembly systems is a desirable alternative to using human workers for highly repetitive work.

Robots used for assembly work are generally small and are designed to move small parts accurately at high speeds, **Figure 5-20**. Most operations involve fitting and holding together parts and assemblies using nuts, bolts, screws, fasteners, or snap-fit joints.

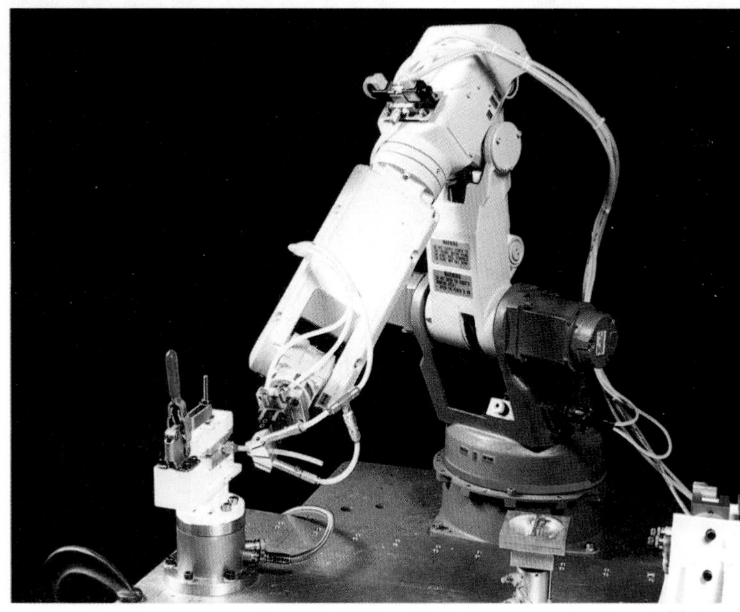

Motoman

Figure 5-20. A robot inserts parts into an assembly.

Newer robots are equipped with sophisticated vision systems that can detect proper orientation of parts. Tactile sensing systems attached to the end effector are used to detect misaligned, missing, or substandard parts. The robot in **Figure 5-21** is equipped with a two-camera vision system. The camera mounted on the manipulator guides the robot around the work cell to gather and assemble parts. The second camera is used for precise placement of a component to within 0.0039″ (0.01 cm) of the programmed point. In this example, the motors being assembled include nearly a dozen different components. The components are fed to the robot by four vibrating bowl feeders and a paternoster (a device that holds trays of components). After the assembly is complete, the robot removes the motors from the assembly area and places them on an outlet tray. This allows the operator to remove the trays without interrupting production.

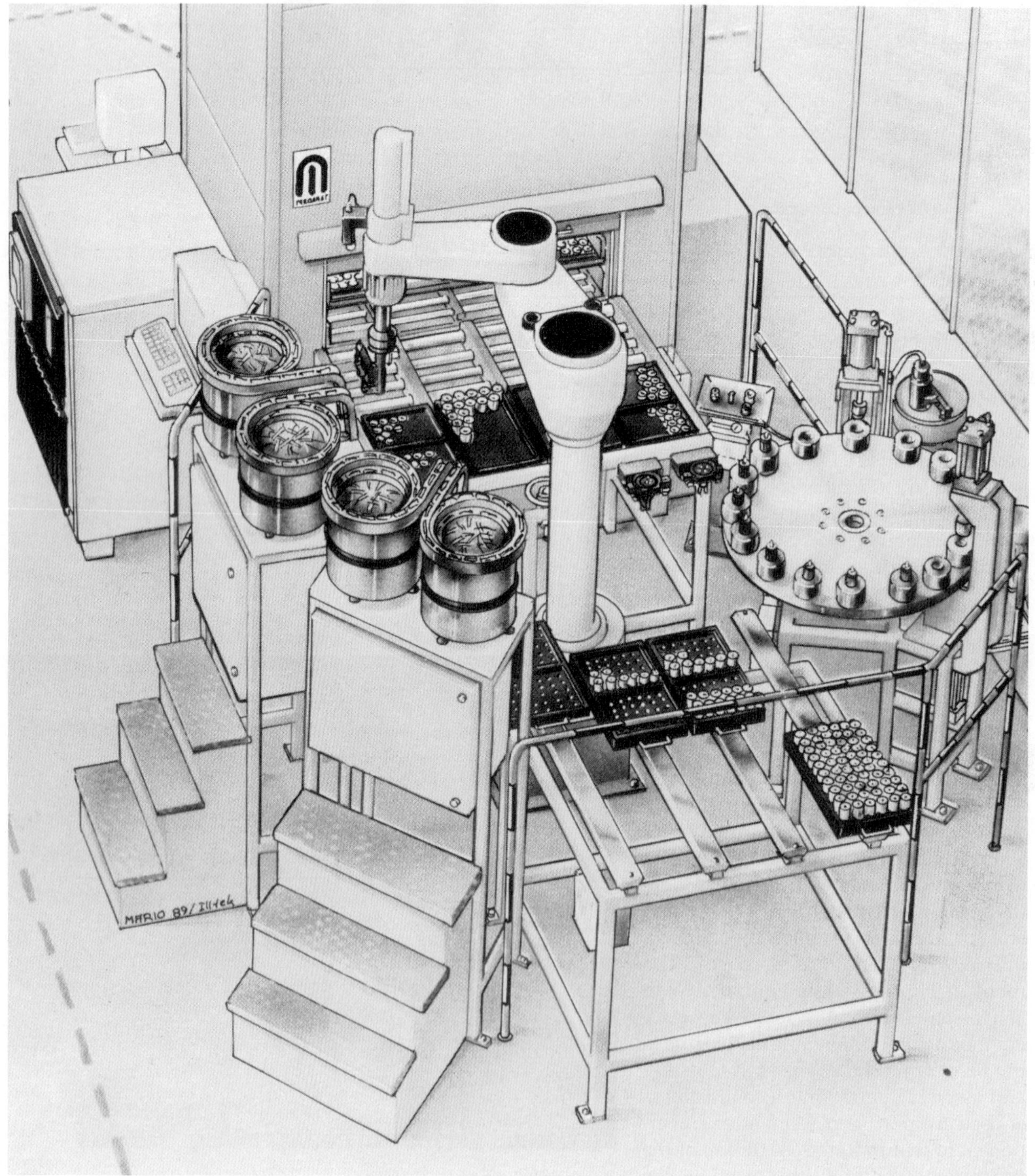

Adept Technology, Inc.

Figure 5-21. A two-camera vision system is used by this robot, which assembles air motors.

To obtain the necessary production mix, the operator enters the model, quantity, and due date for each order. The robot controller uses this information to calculate an assembly schedule. It signals the paternoster to deliver the appropriate trays of components before production of a new batch begins. As a result, no changeover time is required.

5.3.8 Inspection Operations

Quality assurance (ensuring that products are of the highest quality) remains an essential and costly part of robotic manufacturing. The demand for non-contact gauging and inspection systems has grown steadily in recent years. At the heart of these systems is state-of-the-art electronic sensor technology, **Figure 5-22**, which checks tolerances, positioning, fixturing, and defects, among other things.

Robotic inspection systems consist of two subsystems. One subsystem accumulates data, and the other analyzes the data and presents it in a meaningful way. The electronic sensing units used in these systems are set up in one of two ways. They may be mounted on the end effector with the robot programmed to move the sensor along the part. The sensing unit may also be located within the work cell. The unit inspects the parts as they move along the conveyor system.

5.3.9 Material Handling

Automated and robotic systems are often used for material handling. Increased efficiency and flexibility are achieved through automated control and tracking of inventory and handling of parts and products.

Automated guided vehicles (AGVs) are computer-controlled, battery-operated transportation devices (**Figure 5-23**). AGVs operate using one of several navigation options: a buried wire guidepath, a magnetic tape guidepath (inertial guidance), laser target guidance, or a global positioning system (GPS). The type of navigation used depends on the work environment and type of tasks the AGV will perform.

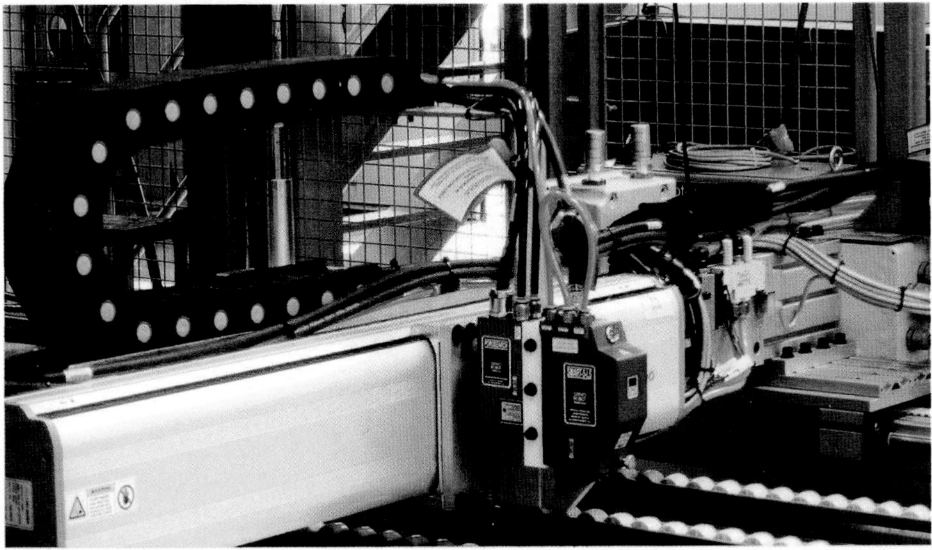

Servo-Robot, Inc.

Figure 5-22. Laser vision cameras inspect welded blanks for defects, such as porosity and undercut.

Daifuku Co., Ltd.

Figure 5-23. This AGV transports palletized materials to a conveyor system within a manufacturing facility.

For safety, AGVs have bumpers, and they will automatically stop if they contact any object in their path. AGVs are bi-directional and emit a beeping sound to warn people of their presence. Some AGVs even signal the command center when their batteries are getting low. The computer then directs the vehicle to a battery charging station.

Careers in Robotics: Robotics Engineer

A robotics engineer designs and maintains robotic systems and is most often employed in the manufacturing industry. In the area of manufacturing, a robotics engineer may design new robotic systems for a production or assembly line and is likely to be responsible for maintaining existing robotic systems. Outside manufacturing, employment opportunities include research and robotic system development for military use, mobile exploration, medical applications, solutions for consumers, and many other areas.

Because robotics engineering relies heavily on computer capabilities and programming, advances in these fields create new opportunities for robotic systems and their applications. From medicine to the military, robots that were once a conceptual design are now a reality due to advances in technology and the creativity of robotics engineers.

Typical education for a robotics engineer includes a graduate degree specializing in electrical or mechanical engineering and knowledge of computer science, applied science, and 3D modeling applications. To work in the manufacturing sector, education in or experience with manufacturing processes is also critical for developing and maintaining the robotics systems in a plant. A doctorate degree opens career opportunities in research and highly specialized system development.

5.3.10 Service Operations

The industrial robots that have been described so far are limited to six or seven degrees of freedom. Their restricted movement means that the materials needed to perform a task must be brought to the work envelope. ***Service robots***, however, are mobile units that can move to a work area to perform the necessary tasks.

One area where service robots are useful and economically feasible is security services, **Figure 5-24**. Mobile robots patrol the hallways of banks, museums, government buildings, and other high-security buildings. The command center can be located in the same building or miles away. When the operator enters the appropriate command, the robot unplugs itself from its charging station and begins its rounds. As it patrols, the robot continuously samples the air and surveys the environment with its sensors. It also automatically responds to any emergency or intrusion. The robot can be programmed to stop and perform automatic surveys when suspicious conditions are detected. The information is transmitted to the command center where the operator can see the situation on a video console.

Service robots are also used in hazardous environments and assist in life-threatening situations. Service robots can be found in such diverse areas as the nuclear industry, security and law enforcement, the military, mining, firefighting, construction and excavation, and chemical or hazardous materials handling.

One day, robots may perform many of the chores in your daily life, such as cleaning your home, cooking your meals, and taking out the garbage. In the retail market today, there are already robots you can program to clean your floors (**Figure 5-25**), mow your lawn, and monitor the security of your home (**Figure 5-26**). In his book *Robots in Service*, Joseph F. Engelberger writes that stand-alone robots that are able to pump gas, fill prescriptions, cook and serve foods, clean commercial buildings, and aid the handicapped and elderly are real prospects. These robots will bring the magic back to robotics. Many robotics experts agree that in the decades ahead service will be the fastest-growing robot application.

MobileRobots, Inc.

Figure 5-24. PatrolBot follows a programmed patrol route or schedule and provides remote video and two-way audio to the command center.

iRobot Corporation

Figure 5-25. The Roomba® can be programmed to vacuum the rooms in a home and automatically adjusts when moving from carpeting to hard floors.

Educational Robot Company

Figure 5-26. The Spy-Cye provides a mobile video monitor in your home. You can log on to your Spy-Cye while away from home and see that everything is secure.

Summary

- Competition, improved technology, cost, and other factors are leading manufacturers to install a greater number of robots.
- Design for manufacturability involves considering the robots that will assemble a product as part of the product planning and design process.
- Safety must always be considered when installing robots. The work station should be designed to allow the robot and operator to work safely.
- Selecting a robot that is appropriate for the job is important. Robots can perform repetitive tasks without tiring and can complete operations in hostile environments.
- The work envelope is different for each robot configuration. All fixtures and machinery that will work with the robot must be located within the work envelope.
- The configuration of each individual work station determines the number of degrees of freedom required for the robot to complete a task.
- Resolution refers to the smallest incremental movement a robot can make. Spatial resolution describes the accuracy of movement of a robot tool tip.
- Accuracy expresses how precisely the end effector of a robot is programmed to reach a predetermined point. The speed of movement and weight of the workpiece affect accuracy.
- Repeatability expresses the variation in movements that occurs when an action is performed several times, such as when the end effector of a robot moves to a designated position each time it performs a task.
- Operational speed (dynamic performance) represents how fast a robot can accelerate, decelerate, and stop movement at a given point.
- The maximum weight a robot is capable of handling on a continuing basis is called payload. Load-handling capabilities are affected by the type of configuration and the placement of the end effector within the work envelope.
- As newer robotic equipment is developed and brought into the workplace, the potential uses for robots will increase dramatically.
- Some current applications for robots include pick-and-place operations, machine loading and unloading, die-casting operations, welding, and spray operations for paint or other liquids.
- Machining processes performed by robots include routing, cutting, drilling, milling, grinding, polishing, deburring, riveting, and sanding.
- Robots used for assembly operations are typically small, and they are designed to move parts accurately at high speeds. Robots equipped with state-of-the art electronic sensing systems are used for inspection.
- Automated guided vehicles (AGVs) are computer-controlled, battery-operated transportation devices used for material handling.
- Service robots are mobile and can move in a work area to perform various tasks, such as security monitoring and handling materials at hazardous locations.

Review Questions

Answer the following questions using the information provided in this chapter.

1. What are five guidelines that product engineers should consider when a robot will be used for assembling a product?
2. Companies will often design parts for robotic assembly. Are there any advantages to this practice, even if robots are not actually used to assemble the parts?
3. What are some factors to consider when designing a work station that allows the robot and operator to work safely?
4. What are two types of work for which robots are not suited?
5. What are some important considerations when selecting a robot to perform an industrial task?
6. What is the difference between command resolution and spatial resolution?
7. What is the spatial resolution of a robot that has a command resolution of 0.006″ (0.0152 cm) and a mechanical inaccuracy of 0.004″ (0.0102 cm)?
8. What is one example of a robot configuration that offers fairly constant spatial resolution throughout the work envelope?
9. Briefly explain the difference between accuracy and repeatability.
10. What factors influence the dynamic performance of a robot?
11. What is the maximum payload of a robot with a 50 lb (22.68 kg) load-carrying capacity and an end effector that weighs 8 lb (3.628 kg)?
12. List seven major production applications for robots.
13. Which robot application has the greatest potential for growth in the future?

Learning Extension

1. Search the US Department of Labor: Occupational Safety & Health Administration website (www.osha.gov) for the robot safety information in the *OSHA Technical Manual*. Review the types of hazards explained in the document and the information presented on safeguarding personnel.

Advanced Analysis

1. For this activity, perform additional research as needed. Describe in detail the selection of a robot to perform the following operations:
   - Pick-and-place.
   - Machine loading and unloading.
   - Die casting.
   - Welding.
   - Spraying.
   - Machining (specify the operation, such as drilling).
   - Assembly (specify type).
   - Inspection (specify type).
2. Describe the use of an AGV for a material handling operation. Perform additional research as needed.
3. Describe the use of a service robot for a specific application. Perform additional research as needed.

Activity 5-1
Control With DC Stepping Motors

Objectives

The dc stepping motor represents a type of rotary actuator designed to achieve automatic position control of industrial processing equipment. The motor shaft will move a specific number of degrees with each pulse of electrical energy. The amount of angular displacement produced by each pulse can then be repeated precisely with each succeeding pulse. The resulting output of this motor is used to locate or position accurately a work surface automatically for some manufacturing operation.

In this activity, you will construct a manual switching circuit for a dc stepping motor. This particular motor is capable of moving in steps of 1.8° in either direction. The direction of rotation is determined by the switching sequence of the motor coils. Only four coils are used in the construction of this motor. As the motor is stepped through a switching sequence, different coil combinations cause the rotor to move to a new location for each step. Only two of the four coils are energized during an operational step.

In Part A of this activity, you will become familiar with the manual switching logic needed to actuate a dc stepping motor to a specific position. You will also reverse the direction of rotation by changing the logic sequence order of the switch combinations. Through this experience, you will be able to observe how switching logic is transformed into rotary motion.

In Part B of this activity, driver transistors are employed as control elements. Toggle switches are used in this case to achieve manual control of the logic sequence through transistor conduction. In an actual stepping motor control circuit, driver transistors are switched by pulses from a logic control center. The next activity is devoted to logic circuitry needed to achieve automatic control of a stepping motor. Remember that dc stepping motors are frequently used in robotic systems.

Equipment and Materials

- DC stepping motor (Superior Electric Co. M-111-FD 310 or equivalent).
- 0–15V dc, 1 A power supply.
- 50 Ω, 20 W resistor.
- 470 Ω, 1/8 W resistors (4).
- 1 kΩ, 1/4 W resistor.
- SPST toggle switches (4).
- IN4004 diodes (4).
- 2N3053 transistors (4).
- SPDT toggle switches (2).
- IC circuit construction board.

Procedure

Part A: Manual Switching Control

1. Construct the manual switching control circuit for a dc stepping motor.

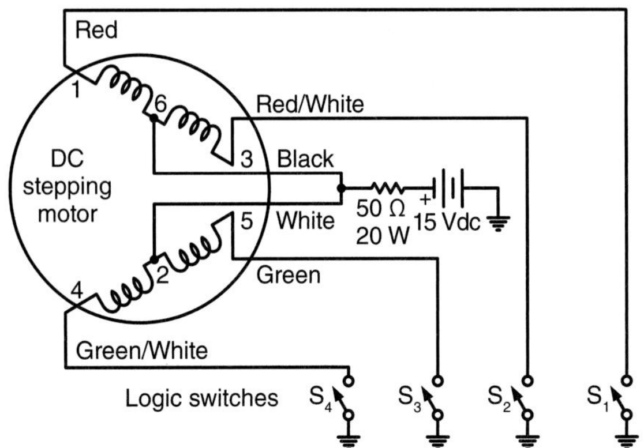

Goodheart-Willcox Publisher

Manual switching control circuit.

2. Turn on the dc power supply and adjust it to 15V dc before closing any of the logic switches. Then turn off the power supply.
3. Set the logic switches to correspond to Step 1A as shown in the following table. Apply power to the motor by turning on the dc power supply.

Step	S_1	S_2	S_3	S_4
1A	on	off	on	off
2	on	off	off	on
3	off	on	off	on
4	off	on	on	off
1B	on	off	on	off

Goodheart-Willcox Publisher

Logic switch settings.

4. What is the initial direction of rotation for the rotor? You may want to place a strip of masking tape near the rotor shaft and mark its initial location. You may need to make a position locating pointer of tape and attach it to the shaft as an indicator of its position change.
5. Set the switches to Steps 2, 3, 4, and finally to Step 1B. Note that Steps 1A and 1B are identical.
6. Starting at Step 1B, set the switches for Steps 4, 3, 2, and 1A in this order.
7. What is the direction of rotation and position of the rotor with reference to its starting location?
8. Steps 1A and 1B in practice are referenced as the same step. They were used in this activity as a basis for establishing the rotational direction control logic sequence. They will now be referred to as Step 1.
9. Starting with the switches at Step 1, sequence the Steps 1, 2, 3, 4; 1, 2, 3, 4; 1, 2, 3, 4; and 1, 2, 3, 4.
10. If each step achieves 1.8° of rotation, how many steps are needed to complete one revolution of 360°?

11. If a protractor is available, you may want to mark the starting position, sequence the switching action, and check the final location.
12. The rotational sequence is cyclic and may be started at any position as long as it is maintained in the same logical order.
13. Turn off the power supply. If you have time to do Part B now, remove only the switches of the circuit. Otherwise disconnect the entire circuit and return all parts to the storage area.

Part B: Manually Controlled Drivers

1. Connect the manually controlled dc stepping motor with the transistor drivers.

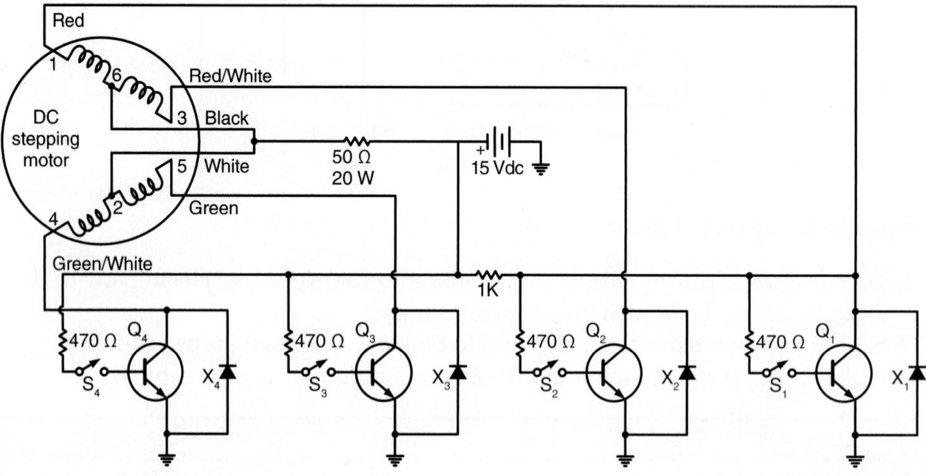

DC stepping motor with transistor drivers.

2. The switching action of this motor can be sequenced to turn in either direction as in Part A of the activity. Repeat the stepping sequence from 1A to 1B. Note the direction of rotation.
3. Reverse the stepping sequence from Steps 1B to 1A and note the direction change.
4. An important aspect of the switching sequence should be noted at this time. Switches 1 and 2 are always in opposite switching states for each step. The same occurs with switches 3 and 4; when one is *on*, the other is in the *off* state. These switches could be easily combined by two SPDT switches for control of all four transistors.
5. Modify the switches so two SPDT switches achieve control of the four transistors, as shown in the following schematic.

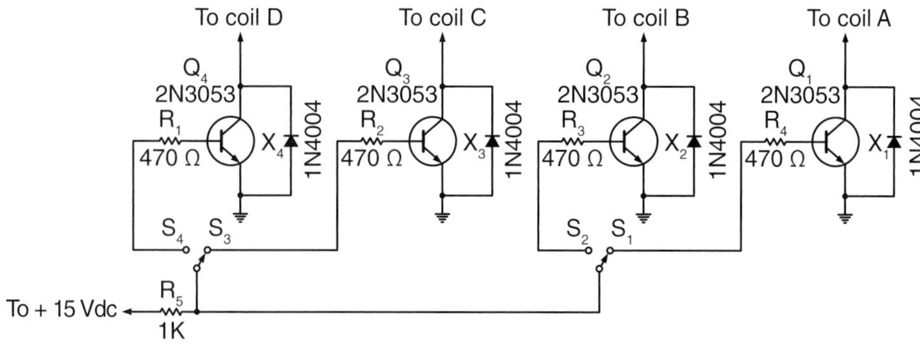

Goodheart-Willcox Publisher

Transistor circuit controlled by two switches.

6. When a particular switch setting applies positive voltage to the base of a transistor, it causes conduction. This, in turn, causes the respective motor coil to be energized. The alternate switch position has no bias voltage. Therefore, reverse bias occurs and the respective coil is off.
7. Cycle the motor through Steps 1A to 1B with the two SPDT switches.
8. Alter the direction of rotation by reversing the switching sequence.
9. It is important to note here that SPDT switches S_1 and S_2 could be achieved by opposite state outputs of a common flip-flop. This would permit the switching action of the motor to be achieved automatically.
10. If at all possible, leave the circuit of this activity constructed and go to the next activity. A logic direction sequencing circuit with automatic binary control will be built for the stepping motor.

Analysis

1. What determines the direction of rotation of a dc stepping motor?
2. If a dc stepping motor in Part A were to form a zero resting position 7.2 cw and 14.4 ccw, what would be the logic switching sequence?
3. Why does forward biasing a transistor cause current to the respective coils of a stepping motor?
4. How is it possible to control the four transistors with flip-flops?

Activity 5-2
Automatic Stepping Motor Control

Objectives

Most applications of a dc stepping motor employ some type of control circuit to energize the motor automatically. Manual sequence switching as a general rule is far too slow for typical applications.

Automatic switching circuits are used to control the rotation of a stepping motor. This can be achieved by a common D flip-flop, as shown in the following. This type of flip-flop has only one input and two opposite state outputs.

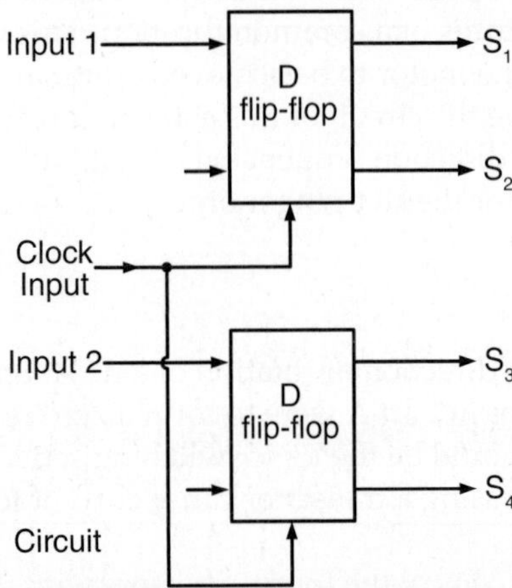

Step Number	Input 1	Input 2	Outputs			
			S_1	S_2	S_3	S_4
3	0	0	0	1	0	1
4	0	1	0	1	1	0
2	1	0	1	0	0	1
1	1	1	1	0	1	0

Goodheart-Willcox Publisher

D flip-flop control circuit and truth table.

Notice that the outputs of switches 1 and 3 follow the inputs of 1 and 2 respectively. You should also note that output switches 1-2 and 3-4 assume opposite states with respect to their corresponding inputs 1 and 2. This means that four valid switch settings are obtained by using only two inputs to the D flip-flops. Column 1 of the truth table refers to the corresponding step sequence number of the manual switching circuit.

The combined inputs 1 and 2 form binary numbers ranging from zero to binary three (00-11). If these numbers are used as the inputs for the flip-flops in proper sequence, the motor will rotate by 1.8° steps. The following table shows a summary of step numbers, binary numbers, and the corresponding decimal equivalents for the stepping sequence.

Step Number	Binary Number of Switches 1 and 3	Decimal Equivalent
1	11	3
2	10	2
3	00	0
4	01	1
1	11	3

Goodheart-Willcox Publisher

Stepping sequence.

The next function of automatic control is to devise a method by which a proper sequence of binary numbers is developed and applied to the inputs of the two flip-flops. The method employed here is one of a decoding process using NAND gates.

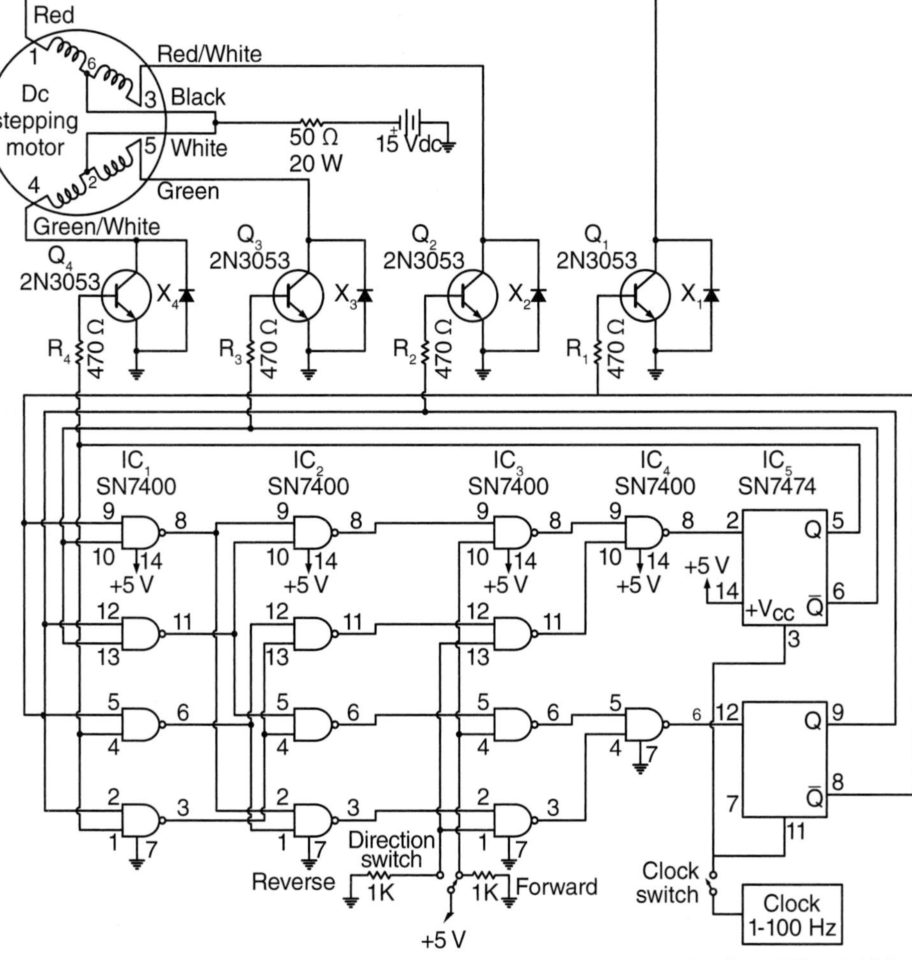

Goodheart-Willcox Publisher

Circuit for applying binary sequence to flip-flops.

When power is applied to an automatic control circuit, the flip-flops will assume some valid state and apply their output to the driver transistors. This state is sensed by the four NAND gates of IC_1, and the next correct step for either direction is set up by the gates of IC_2. IC_3 then operates in conjunction with the forward/reverse enable switch to determine which of the two possible direction sequences will be applied to the flip-flops. The forward/reverse switch is used to enable or disable each half of IC_3, which influences the direction of rotation.

With power applied to the motor, driver transistors, decoding network, and the direction control switch in either position, stepping does not occur until the flip-flops receive a clock pulse. This means that rotation of the stepping motor is directly dependent on the rate by which clock pulses arrive at the flip-flops. Precision control of the motor is based on the availability of a finite number of clock pulses applied to the flip-flop inputs.

In this activity, you will build an automatic directional sequence controller for a dc stepping motor. Through this activity, you will see how a number of ICs are connected together to achieve a specific operational sequence.

You will also be able to test and observe the functional steps of a control circuit in operation. Stepping motor applications of this type are commonly used to control industrial robots.

Equipment and Materials

- Dc stepping motor (Superior Electric Co. M-111-FD310 or equivalent).
- 0–15V, 1 A power supply.
- 0–5V, 1 A power supply.
- 50 Ω, 20 W resistor.
- 470 Ω, 1/8 W resistors (4).
- 1 kΩ, 1/4 W resistors (2).
- Dc voltmeter or multimeter.
- 2N3053 transistors (4).
- SPDT toggle switch.
- SPST toggle switch.
- SN7400 dual inline ICs (4).
- SN7474 IC.
- IC circuit construction board.
- Clock circuit or generator.
- Oscilloscope.
- Electronic multifunction meter.

Procedure

1. Construct the dc stepping motor automatic directional control circuit.
2. Turn on the 5V and 15V power supplies initially with the clock switch off.
3. With a dc voltmeter, measure the driver output signals at the collector of each transistor. When a transistor is conducting, the collector voltage will be low, and when it is nonconducting, it will be approximately 5V. What is the state of the transistor?
   This represents the initial step of the motor.
4. Which step of the truth table is represented by the transistors?

5. Set the clock to produce one pulse per second (1 pps) and momentarily turn on the clock switch for one pulse. Repeating Step 3, what is the state of the transistors? Which step does this represent?
6. Repeat Step 5 for two or three steps to determine the sequential pattern of the steps. Note that a step change cannot be produced until the clock pulse is applied.
7. Place the direction switch in the reverse enable position. Repeat Steps 5 and 6 to determine the operational sequence. How does it compare with the first readings?
8. Turn on the clock switch and adjust the clock to produce one pulse per second. How does the motor respond?
9. Increase the clock pulse frequency to 10 pps and observe the change. To stop the motor at a specific position, simply turn off the clock switch.
10. Sequence the stepping operation so the motor moves one half of a revolution and then returns to its original position.
11. Disconnect the circuit at this time.

Analysis

1. How would the motor respond if the pulses per second from the clock were increased from 100 to 1000?
2. Describe the operation of a D flip-flop.
3. What is the function of IC_3?

Herrndorff/Shutterstock.com

Engineers and technicians are still needed to ensure automated factories run as designed.

Unit II

Power Supplies and Movement Systems

Three types of motion are used in automated applications: rotary, linear, and reciprocating. These types of motion can be produced using either electrical or fluid (hydraulic or pneumatic) powered operating motors, relays, solenoids, actuators, or cylinders. The manipulator is the basic mechanical unit of a robot. It has several moving joints and performs the actual work functions of the machine.

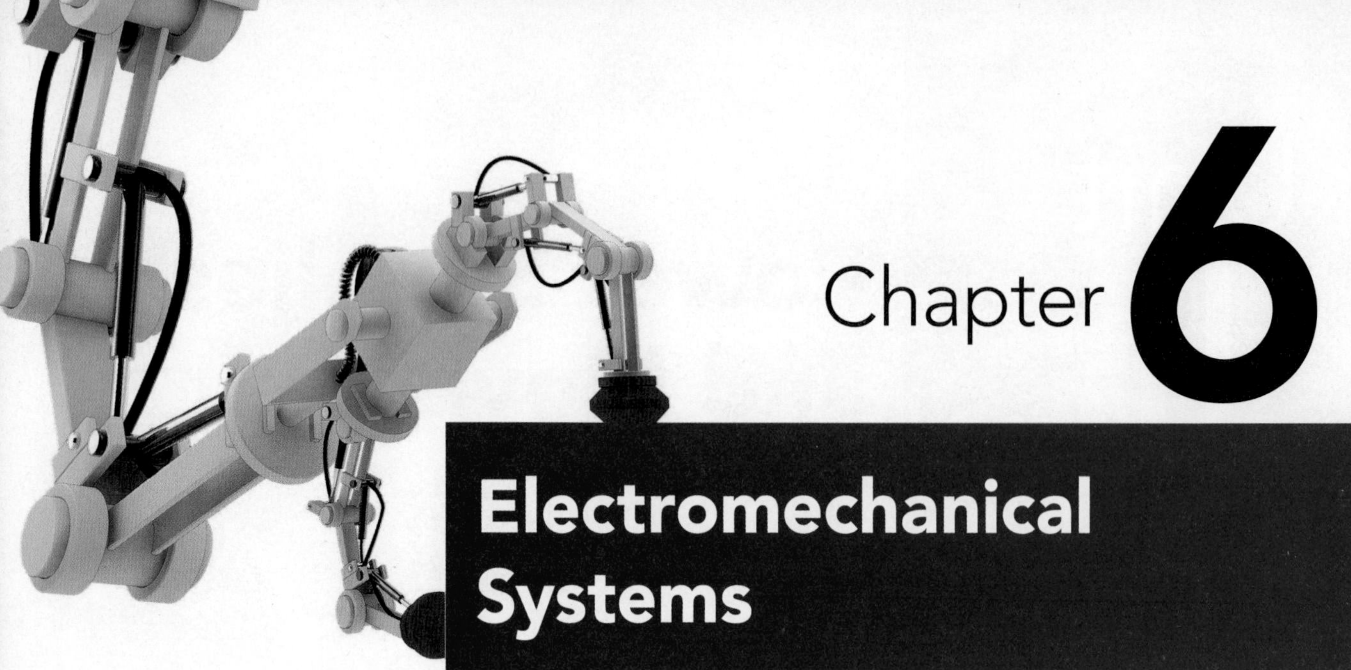

Chapter 6

Electromechanical Systems

Outline

Objectives

Upon completion of this chapter, you will be able to:

- Discuss the five basic subsystems that are common to all automated robotic systems.
- Describe mechanical systems as they relate to robotics.
- Explain how sensing, timing, and control systems are used in the operation of robots.
- Discuss rotary motion systems used for robotics.

Technical Terms

alternating current (ac)
armature
bifilar construction
brushes
commutator
comparator
compound-wound dc motor
control
counter electromotive force (cemf)
cycle
cycle timing system
dc stepping motor
delay timing system
detector
digital system
direct current (dc)
electric motor
electromechanical system
error detector
feedback
field winding
flux
indicator
interval timing system
light pipe
load
permanent-magnet dc motor
rectification
rotor
sensing system
series-wound dc motor
servo system
servomotor
shunt-wound dc motor
single-phase ac motor
single-phase induction motor
slip
squirrel cage rotor
stator
subsystem
synchronous motor
synchronous speed
synthesized system
system
three-phase ac motor
three-phase induction motor
three-phase synchronous motor
timing system
torque
transmission path
universal motor
work

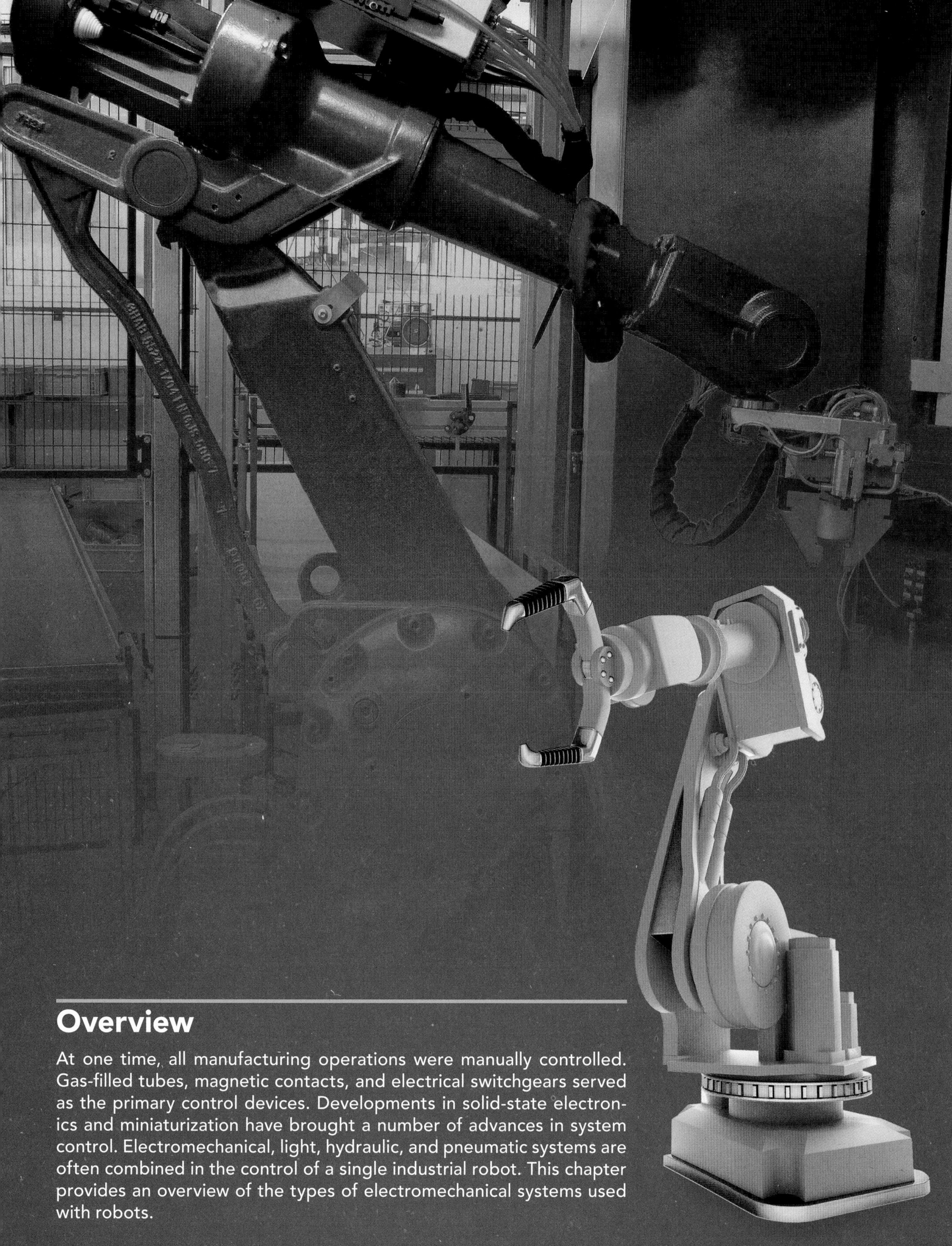

Overview

At one time, all manufacturing operations were manually controlled. Gas-filled tubes, magnetic contacts, and electrical switchgears served as the primary control devices. Developments in solid-state electronics and miniaturization have brought a number of advances in system control. Electromechanical, light, hydraulic, and pneumatic systems are often combined in the control of a single industrial robot. This chapter provides an overview of the types of electromechanical systems used with robots.

6.1 Automated Systems and Subsystems

Nearly all manufactured products are manufactured using some type of electromechanical system. A ***system*** is a combination of components that work together to form a unit. An ***electromechanical system*** transfers power from one point to another through mechanical motion that is used to do work. ***Work*** occurs when energy is transformed into mechanical motion, heat, light, chemical action, or sound. Punch presses that move up and down, rotating machinery, and robots are all examples of electromechanical systems.

A robot is a unique type of system that may require several different types of components for proper operation. These various components are the ***subsystems*** that comprise a complete robot. When subsystems are combined, the result is referred to as a ***synthesized system***. Familiarity with each of the subsystems and the location of each within the overall system is important to understanding the system itself, **Figure 6-1**.

Various subsystems are used in virtually all automated systems. An electrical power system, for example, is needed to produce and distribute electrical energy. Hydraulic and pneumatic systems are used for motion and other functions. Light systems (also called photoelectronic or opto-electronic systems) are used for inspection operations and in many types of sensors. Mechanical systems are needed to hold objects for machining operations and to move parts and assemblies on a production line. Although each system has unique features, these basic subsystems are common to all automated systems: an energy source, transmission path, control, load, and indicators.

6.1.1 Energy Source

The energy source provides power for the system. The most common source of power for synthesized systems is ***alternating current (ac)***. In alternating current

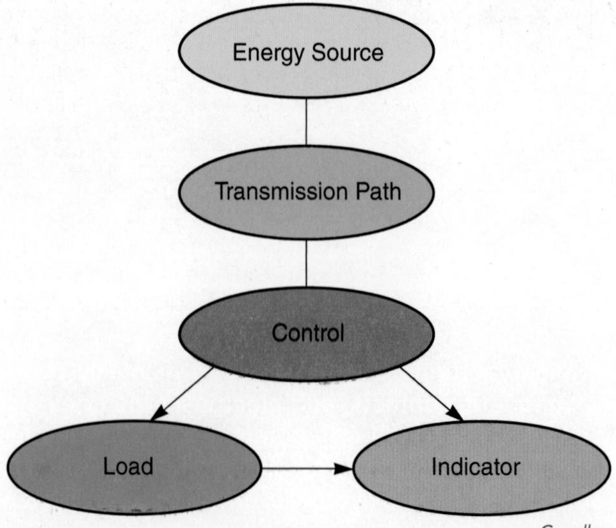

Goodheart-Willcox Publisher

Figure 6-1. The basic parts of an automated system must work together or the system will not function properly.

power, electrons flow first in one direction and then in the opposite direction. Each repeated pattern of direction change is called a ***cycle***, **Figure 6-2**. The alternating current required for an automated manufacturing system may be either single-phase or three-phase. Single-phase ac power consists of one alternating current, **Figure 6-2.** Three-phase ac power consists of three alternating currents that vary by 120°. Three-phase alternating current is ordinarily used for larger systems because it can handle heavier power needs. Three-phase motors are discussed later in this chapter.

Some machines used for automated manufacturing require ***direct current (dc)***. In direct current, electrons flow in only one direction. Many robotic systems have motors and other parts that use direct current. The process of converting alternating current to direct current is called ***rectification***. Rectification is the most convenient and inexpensive method of providing dc energy to machines.

6.1.2 Transmission Path

The ***transmission path*** is a subsystem that provides a channel for the transfer of energy. It begins at the energy source and continues through the system to the load device. The term ***load*** refers to a part (or parts) designed to produce work. This path may be a single feed line, electrical conductor, light beam, or pipe. Some systems may have a supply line and a return line between the source and the load. There may also be a number of alternate transmission paths in the system. These may be connected in series to a number of small load devices or in parallel to many independent devices.

6.1.3 Control

The ***control*** alters the flow of power and causes some type of operational change in the system, such as changes in electric current, hydraulic pressure, light intensity, or airflow. Control devices operate within the transmission path. In its simplest form, control occurs when a system is turned on or off. This type of control can take place anywhere between the source and the load device.

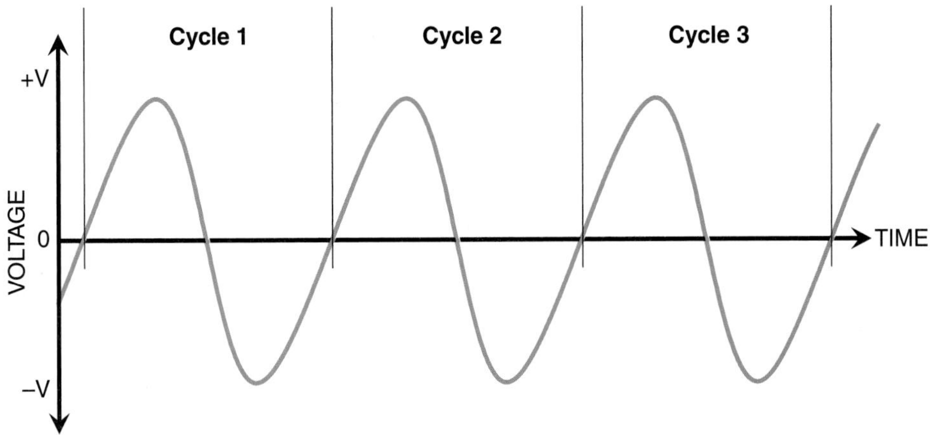

Goodheart-Willcox Publisher

Figure 6-2. This illustration depicts the voltage cycles in ac power. Each cycle contains two changes in direction—a high peak and a low peak.

6.1.4 Load

As stated earlier, the load is a part (or parts) designed to produce work. Work occurs when energy is transformed into mechanical motion, heat, light, chemical action, or sound. Normally, the largest portion of energy supplied by the source is consumed by the load device. Loads include electric motors, heating systems, lighting systems, alarms, horns, and mechanical actuators.

6.1.5 Indicators

The ***indicator*** is a subsystem that displays information about operating conditions at various points throughout the system. Some types of indicators include digital meters, pressure gauges, tachometers to measure speed, and thermometers to measure temperature. In some systems, the indicator is optional. In other systems, indicator readings are essential. When operations or adjustments are critical, indicator readings may affect how the robot functions.

6.2 Mechanical Systems

Mechanical systems transfer power from one point to another through mechanical motion that is used to do work. Automated applications use three kinds of motion: rotary (motors), linear (relays, actuators, and cylinders), and reciprocating (certain types of motors and cylinders). The energy source for these motions can be either electrical power or fluid power (hydraulic or pneumatic). The energy may be used to operate motors, relays, solenoids, actuators, or cylinders. As in other systems, the load is responsible for performing this action. Industrial loads are usually designed for continuous operation over long periods of time. The basic mechanical unit, such as the robot's manipulator, has several moving joints and performs the actual work function of the machine.

The transmission path of a mechanical system transfers power from the energy source to the rest of the system. The transmission path may consist of electrical conductors, belts, rotating shafts, pipes, tubes, or cables. For example, the power from an electric motor may be transferred by a belt or gear to operate a machine tool.

Control is accomplished by changing pressure, direction, force, and speed. Pressure regulators, valves, gears, pulleys, couplings, brakes, and clutches are used to control variables such as force, speed, and direction. Such changes alter the performance of the system, allowing adjustments to be made that optimize the functionality of the system.

In a mechanical system, indicators measure physical quantities. These physical quantities include pressure, flow rate, speed, direction, distance, force, torque, and electrical values. Many of these quantities must be monitored periodically. Some indicators are used to test system conditions during maintenance. Others are designed to measure physical changes that take place. Some examples of indicators are pressure gauges; flow meters; tachometers to measure speed; anemometers to measure wind direction; and multimeters to measure electrical voltage, current, and resistance.

A warehouse conveyor system is an example of a mechanical system. A conveyor moves materials, such as boxes, from one location to another. The energy source for a conveyor system is typically an electric motor. The transmission path includes pulleys, belts, or gears that connect the motor to a conveyor belt. Control is accomplished by changing speed or direction of the motor. Indicators in this type of mechanical system can measure speed or flow rate to determine the movement of boxes along the conveyor belt.

6.3 Electrical Systems

There are many different types of electrical systems used in robotics. Electrical systems include those used for sensing, timing, control, and providing rotary motion (motors). The first three systems are discussed in this section of the chapter. Rotary motion systems are discussed in the following section.

6.3.1 Sensing Systems

A ***sensing system*** consists of components that signal a response to a particular form of energy, such as light. The light source, transmission path, control, and load device are essential parts of the system. The light is produced by electrical energy and may be in the form of incandescent lamps, flames, glow lamps, electric arcs, solid-state light, or laser light. Sensing systems have become one of the fastest growing and most diversified areas of industrial robotics. New systems that combine optics, electromagnetics, and electronics have revolutionized automated manufacturing. Light systems are the most common form of sensing systems. Other sensing systems include sound, pressure, flow, and magnetic sensors.

The transmission path of a sensing system is somewhat unique. In **Figure 6-3**, light energy travels in a straight line in the form of an intense beam of electromagnetic waves. If the light must go around corners or be directed to unusual locations, light pipes could be used as the transmission path. ***Light pipes*** are flexible, fiber-optic rods that may be extended over long distances to transfer light energy from its source to a distant location.

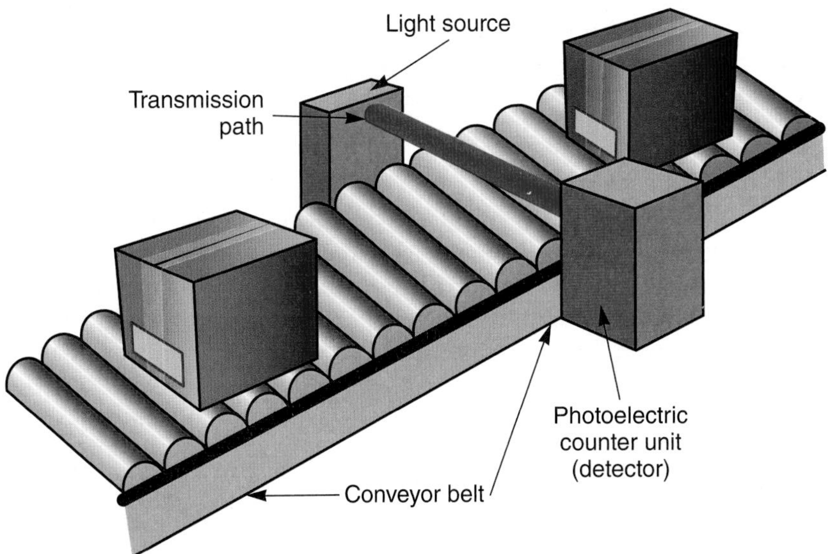

Goodheart-Willcox Publisher

Figure 6-3. This sensing system counts the units moving along the conveyor using system responses to light energy.

The ***detector*** is part of a sensing system that responds to energy from the source and outputs a signal that is used to control the load device. In some sensing systems, the load may be controlled directly by light-source energy. In other systems, the light energy must first be detected and amplified. There are even applications in which the detector itself serves as the load.

Sensing system control may be achieved by interrupting a light beam between the source and detector. Other control methods may alter the intensity, focus, shape, or wavelength of the light source. The detector's sensitivity can be adjusted to adapt it to specific operating conditions.

A common application of light-sensing systems in industrial processes is a counter (detector) on a conveyor line. An example may be a conveyor line that transports boxes from a packing area to a storage area. A light source can be placed along the conveyor line and aimed at the counter. As each box passes the light beam, the detector (counting device) adds another box to the count. Control is accomplished by breaking the light beam between the source and detector to count the number of boxes passing along the conveyor line.

6.3.2 Timing Systems

Timing systems turn a device on or off at a specific time or in step with an operating sequence. Types of timing systems include delay timing, interval timing, and cycle timing. ***Delay timing systems*** provide a lapse in time before the load device becomes energized. ***Interval timing systems*** are used after a load has been energized and operate using specified time periods. For example, an interval timing system may be set to allow the load to remain energized only for a certain period. ***Cycle timing systems*** are typically more complex and may include both interval and delay timing to provide energizing action in an operational sequence.

Timing systems also include thermal devices, motor-driven mechanisms, or other mechanical, electrical, or electrochemical devices. Hydraulic, pneumatic, mechanical, heat, and electrical energy sources may be used in various combinations to power timing systems. A type of timing operation is accomplished by a microprocessor in a computer. A microprocessor continually receives instructions, executes them, and continues to operate in a cyclic pattern. All actions occur at a precisely defined time interval. An orderly sequence of operations, such as this one, requires a type of precision timing system.

6.3.3 Control Systems

Control of an automated manufacturing system can be caused by input from an operator (person). It can also occur automatically due to a physical change. During production, control systems are continually at work making adjustments that alter machine operations. The complexity of an operation determines the number of control functions needed. In many cases, several control components are used in various parts of the system.

The control unit of an industrial robot determines its flexibility and efficiency. Some robots have only mechanical stops on each axis. Others have microprocessor (computer) control with memory to store position and sequence data. Some important factors in the selection of a control unit are speed of operation, repeatability, accuracy of positioning, and the speed and ease of reprogramming.

Non-servo, or open-loop, control systems are the most basic. They use sequencers and mechanical stops to control the end point positions of the robot arm. Pick-and-place (fixed-stop) robots use this type of control. Programmable servo-controlled robots are more complex. These continuous-path robots move from one point to another in a smooth, continuous motion.

Open-loop control systems are used almost exclusively for manual-control operations. There are two variations of the open-loop system: full control and partial control. Full control simply turns a system off or on. For example, in an electrical circuit, current flow stops when the circuit path is opened. Switches, circuit breakers, fuses, and relays are used for full control. Partial control alters system operations rather than causing them to start or stop. Resistors, inductors, transformers, capacitors, semiconductor devices, and integrated circuits are commonly used to achieve partial control.

To achieve automatic control, interaction between the control unit and the controlled element (load) must occur. In a closed-loop system, ***feedback*** is a signal or data that provide information about the interaction between the control unit and the controlled element (load). Feedback can be activated by electrical, thermal, light, chemical, or mechanical energy. Both full and partial control can be achieved through a closed-loop system, **Figure 6-4**. Many of the automated systems used in industry are of the closed-loop type.

In a closed-loop system with automatic correction control, **Figure 6-5**, energy goes to the control unit and proceeds to the controlled element. Feedback from the controlled element is directed to a ***comparator***, which is an element that compares the feedback signal to a reference signal or standard. A correction signal is developed by the comparator and sent to the control unit. This signal alters the system to conform to the data from the reference signal. Systems that function in this manner maintain a specific operating level regardless of external variations or disturbances.

Automatic fabrication methods, packaging, and machining operations have been improved through advances in digital systems. A ***digital system*** is a set of components that work together to process numeric information. Digital (numeric) instructions are supplied by variations in pressure, temperature, or electric current. These instructions are then changed into a series of on/off electrical signals. The signals are processed by the logic circuitry of a computer and are directed to specific subsystems that perform the necessary operations.

Electrical power energizes the load device, which performs the work. The load of a system may be electrical actuators or fluid-power cylinders designed to move parts of the robot. When appropriate signals are received, the robot performs the

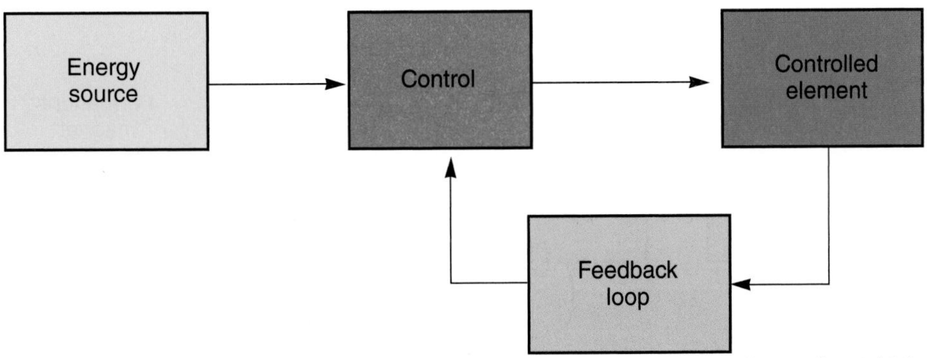

Goodheart-Willcox Publisher

Figure 6-4. A typical closed-loop system, as shown in this diagram, incorporates a feedback loop.

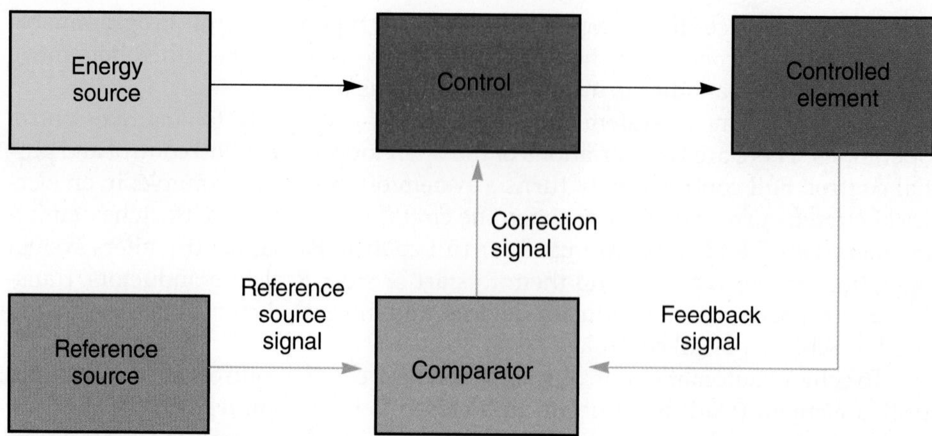

Goodheart-Willcox Publisher

Figure 6-5. The comparator and reference source in this closed-loop control system allow automatic adjustments to be made to a process while it is running.

necessary tasks. Digital systems can control various operations, such as positioning and moving, clamping, and directing material flow.

6.4 Rotary Motion Systems

An ***electric motor*** is a device that converts electrical energy to mechanical movement in the form of rotary motion. Essentially, an electric motor is created by placing an electromagnet, called an ***armature***, between two permanent magnets. The north and south poles of the armature are aligned with the north and south poles of the permanent magnets. When a current is passed through the armature, it becomes magnetized and begins to rotate within the magnetic field of the permanent magnets. Rotation continues until the armature's north pole is opposite the south pole of a permanent magnet and the armature's south pole is opposite the magnet's north pole, **Figure 6-6**. If the current through the armature is reversed, its poles will reverse, and the armature will rotate again. The rotary motion, or turning force, that is produced is called ***torque***. The amount of torque produced by a motor depends on the strength of the magnetic fields and the amount of current

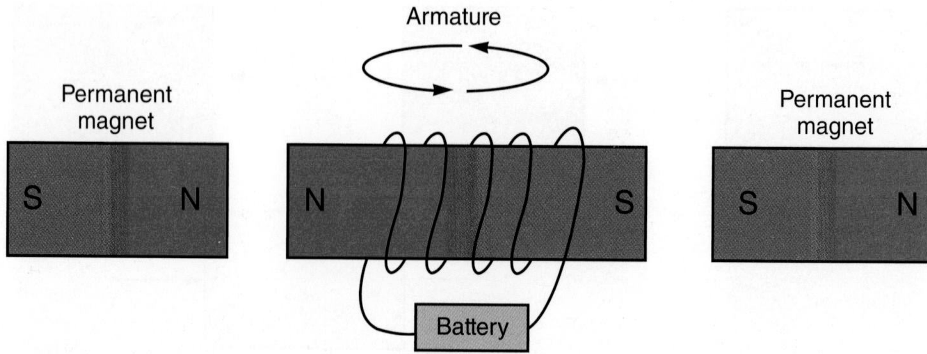

Goodheart-Willcox Publisher

Figure 6-6. Matching magnetic poles that repel one another cause the armature to rotate in a motor.

flowing through the conductors. As either the magnetic field or current increases, the amount of torque also increases.

All motors have several parts in common. The ***stator*** is the stationary portion of a motor and includes the permanent magnets, a frame, and other stationary components. The ***rotor*** is the rotating component of a motor and includes the armature, shaft, and associated parts. Movement of the rotor creates torque. Auxiliary devices for dc motors and ac motors are discussed in the following sections.

6.4.1 DC Motors

DC motors are used when speed control is a critical factor. The basic parts of a dc motor are shown in **Figure 6-7**. A ***commutator*** is a part or assembly used to switch the direction of current flow alternately to produce a one-directional direct current (dc) through the windings (coils of wire) of the armature. The ***brushes*** are carbon devices that rub against the commutator in an electric motor and allow current to flow through them. The flow of current through the brushes to the commutator and armature windings transfers current from the stationary power source to energize the rotating armature conductors.

Most dc motors use electromagnetic windings rather than permanent magnets to create the magnetic field of the stator. The coil wires wrapped around the electromagnets in the stator are called ***field windings***. The interaction of the stationary electromagnetic field of the stator coils and the electromagnetic field of the rotating armature produces rotation of the dc motor.

The operational characteristics of dc motors are shown in **Figure 6-8**. As the armature of a dc motor rotates, it generates its own voltage called the ***counter electromotive force (cemf)***. This voltage flows against the voltage coming into the motor. The amount of cemf depends on the number of rotating conductors and the speed of rotation. As the mechanical load (amount of work effort demand

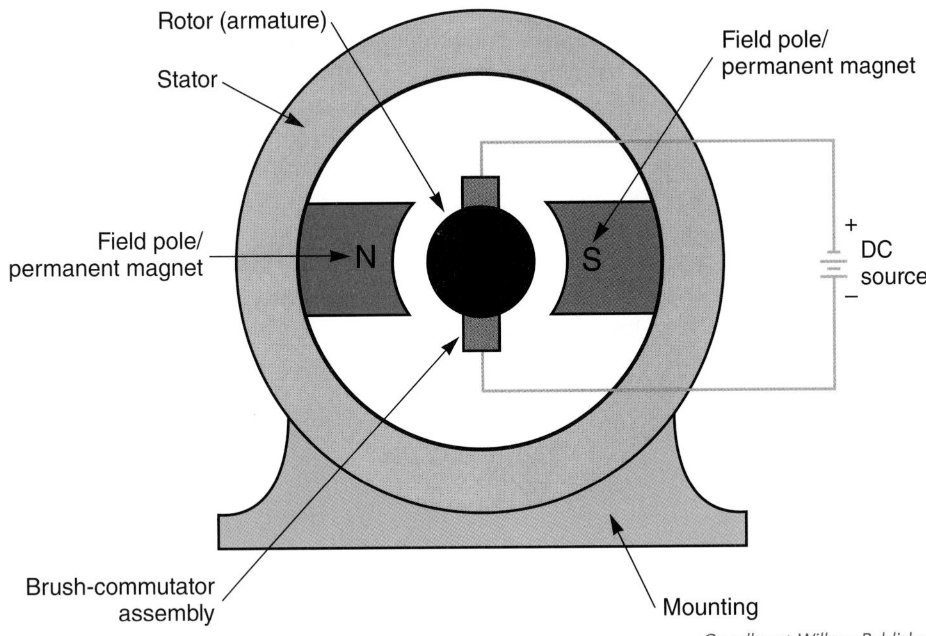

Goodheart-Willcox Publisher

Figure 6-7. The basic parts of a dc motor include a brush-commutator assembly for switching the current flow.

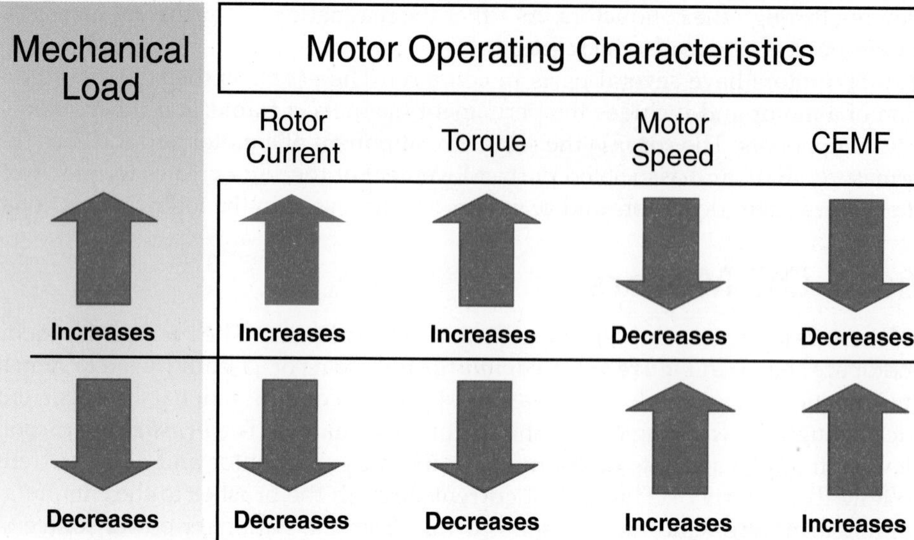

Goodheart-Willcox Publisher

Figure 6-8. This illustration depicts the effect of mechanical load on the operating characteristics of a dc motor.

on the machine) increases, the motor speed decreases. The opposite is also true. As the mechanical load decreases, the motor speed increases. The amount of cemf increases or decreases in direct relation to the speed of rotation.

Since the cemf flows against the supply voltage, the actual working voltage of the motor increases as the cemf decreases. When the working voltage increases, more current flows through the rotor windings. The torque of the motor is directly proportional to the amount of current flowing through the armature. Therefore, torque increases as armature current increases and decreases as armature current decreases. The amount of torque also varies with changes in load. As the load on a motor increases, torque increases to handle the greater load. The increase in torque causes the motor to draw more current from the power source.

When a motor starts, it draws a very large initial current (compared to the current it draws at full speed) because of the absence of cemf. To reduce the starting current of a motor, resistors wired in series with the armature circuit are often used. Once the motor reaches full speed, these resistors are bypassed by an automatic or manual switching system. This allows the motor to produce maximum torque.

Horsepower is a measure of the amount of work performed over a specified amount of time. The horsepower rating of a motor represents the power of a motor. It is based on the amount of torque produced at the rated full-load values and is a very common rating for electric motors used for robotic applications. As torque or the work requirement increases for any application, the horsepower of the motor used to drive the machine must also increase. This can be expressed mathematically as:

$$hp = \frac{2\pi ST}{33{,}000}$$

$$hp = \frac{ST}{5252}$$

(hp = Horsepower rating, π = Constant, S = Speed of the motor expressed in rpm, T = Torque developed by the motor expressed in ft/lbs.)

As noted earlier, an important feature of dc motors is the ability to control speed. By changing the applied dc voltage, speed can vary from zero to the maximum rpm. Some types of dc motors have more desirable speed characteristics than others in terms of the motor's ability to maintain a constant speed under varying load conditions. The difference between no-load motor speed (rpm) and the rated full-load motor speed (rpm) is used to determine the percentage of speed regulation for different dc motors. Lower speed regulation values mean better speed regulation capabilities. Thus, a motor that operates at nearly constant speeds under varying load situations provides good speed regulation.

There are four basic categories of commercially available dc motors: permanent-magnet, series-wound, shunt-wound, and compound-wound. Each type has different characteristics that result from the basic circuit arrangement and physical properties.

The ***permanent-magnet dc motor*** is used when a low amount of torque is needed for applications, such as motor-driven timers and printers connected to computer systems, **Figure 6-9**. The dc power supply is connected directly to the conductors of the rotor through the brush-commutator assembly. The magnetic field is produced by permanent magnets mounted in the stator.

In a ***series-wound dc motor***, the armature and field circuits are connected in a series arrangement, **Figure 6-10**. This is the only type of dc motor that also can be operated using ac power. For this reason, series-wound motors are sometimes called

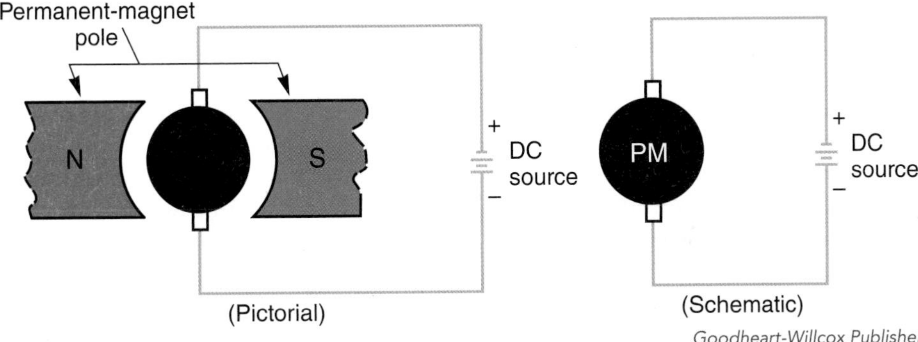

Goodheart-Willcox Publisher

Figure 6-9. A permanent-magnet dc motor is used for applications in which low torque is required.

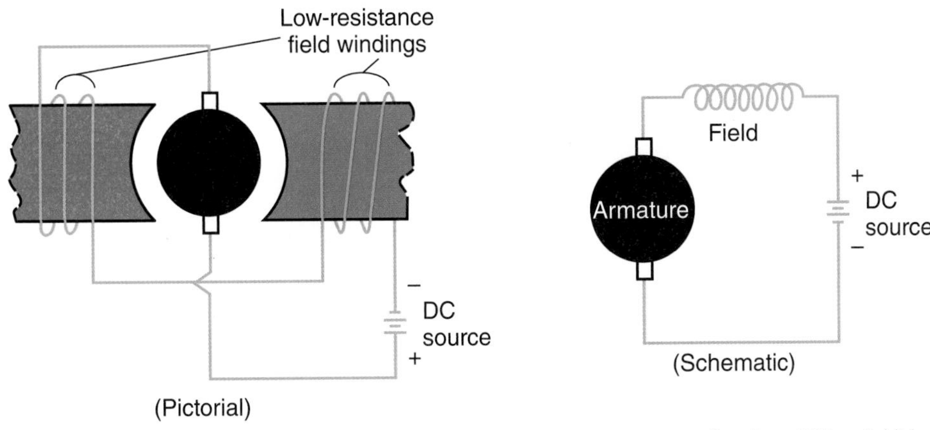

Goodheart-Willcox Publisher

Figure 6-10. A series-wound dc motor produces high torque, but it has poor speed regulation.

universal motors. In this type of motor, there is only one path for current to flow from the dc voltage source. Therefore, the field has a low resistance. Changes in load applied to the motor shaft cause changes in current though the field. If the mechanical load increases, the current also increases. This creates a stronger magnetic field. The speed of a series-wound motor ranges from very fast when there is no load to very slow under a heavy load. Since large currents may flow through the low-resistance field, series-wound motors produce high torque. Series-wound motors are used when heavy loads must be moved and speed regulation is not as important.

The ***shunt-wound dc motor*** is the most commonly used type of dc motor, **Figure 6-11**. The motor's field windings have relatively high resistance and are connected in parallel with the armature. Since the magnetic field is a high-resistance parallel path, a small amount of current flows through it.

Most of the current drawn by a shunt-wound motor flows in the armature circuit. Since the armature current has only a slight effect on the field strength, variations in load have little effect on motor speed. The field current, however, can be controlled by placing a variable resistance in series with the field windings. Since the current in the field circuit is low, a low-wattage resistor (or rheostat) can be used to vary the motor's speed. As the field resistance increases, the field current decreases. The opposite also applies. A decrease in field resistance increases the field current. Changes in the field current result in corresponding changes in the ***flux*** (strength) of the electromagnetic field. As field flux decreases, the armature rotates faster; as it increases, armature rotation slows.

Shunt-wound dc motors are commonly used for industrial applications because of their effective speed control characteristics. Many types of variable-speed machine tools are driven by shunt-wound dc motors.

A ***compound-wound dc motor*** has two sets of field windings. One set of field windings is in series with the armature; the other is in parallel with the armature, **Figure 6-12**. It has the high torque of a series-wound motor and good speed regulation like a shunt-wound motor. However, compound-wound motors are more expensive than the other types.

6.4.2 AC Motors

AC motors are commonly used in industrial settings as well as in commercial and residential applications. ***Single-phase ac motors*** operate using a single-phase ac

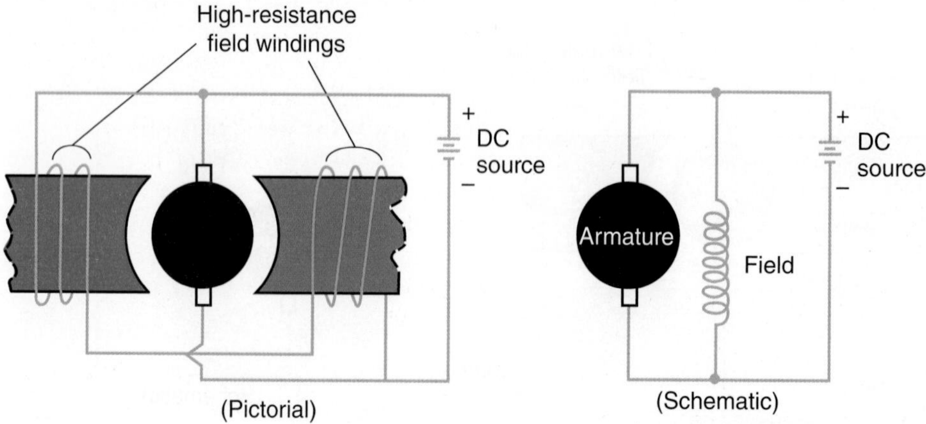

Goodheart-Willcox Publisher

Figure 6-11. Shunt-wound dc motors are commonly used in industry because their speed is easily regulated.

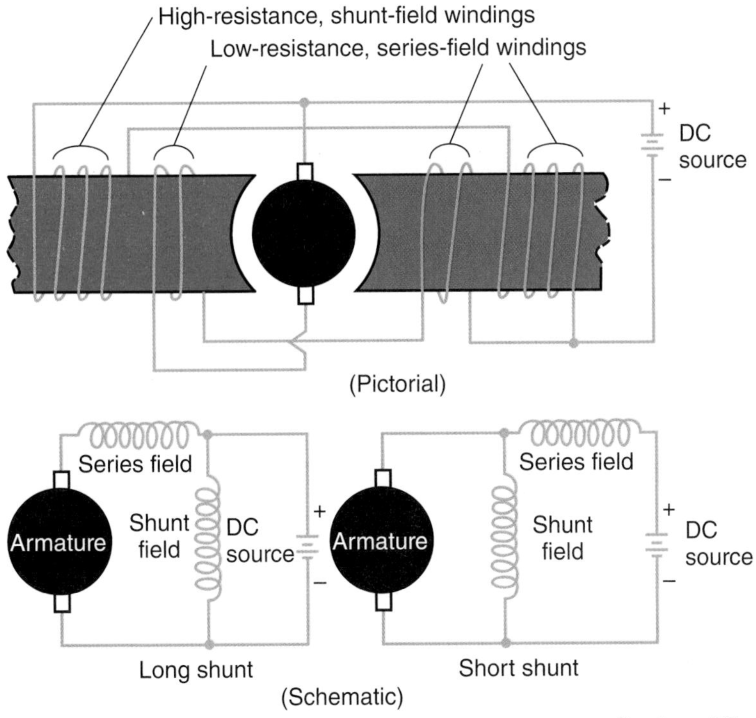

Goodheart-Willcox Publisher

Figure 6-12. The compound-wound dc motor combines the best features of the series-wound and shunt-wound motors.

power source. There are three types of single-phase ac motors: universal motors, induction motors, and synchronous motors.

A ***universal motor*** can be powered by either an ac power source or a dc power source and is built like a series-wound dc motor, **Figure 6-13**. It has concentrated field windings and speed and torque characteristics similar to those of series-wound dc motors. Universal motors are used mainly for portable tools and small equipment.

The ***single-phase induction motor*** has a solid rotor, called a ***squirrel cage rotor*** (**Figure 6-14**). The top and bottom of large-diameter copper conductors are soldered to connecting plates. When current flows in the stator windings, a current is induced in the rotor. The stator polarity changes in step with the applied ac frequency. This develops a rotating magnetic field around the stator. The rotor becomes polarized and rotates in step with the stator's magnetic field. However, due to inertia, the rotor must be set into motion by some auxiliary starting method.

The speed of an ac induction motor is based on the speed of the rotating magnetic field and the number of stator poles. The speed of the rotating magnetic field (rpm) is developed by the frequency of ac voltage (Hz) and the number of poles in stator windings. The stator speed is also referred to as ***synchronous speed***. A two-pole motor operating with a 60 Hz source has a synchronous speed of 3600 rpm. For 60 Hz operation, the following synchronous speeds may be obtained:

- Two-pole, 3600 rpm.
- Four-pole, 1800 rpm.
- Six-pole, 1200 rpm.
- Eight-pole, 900 rpm.
- Ten-pole, 720 rpm.
- Twelve-pole, 600 rpm.

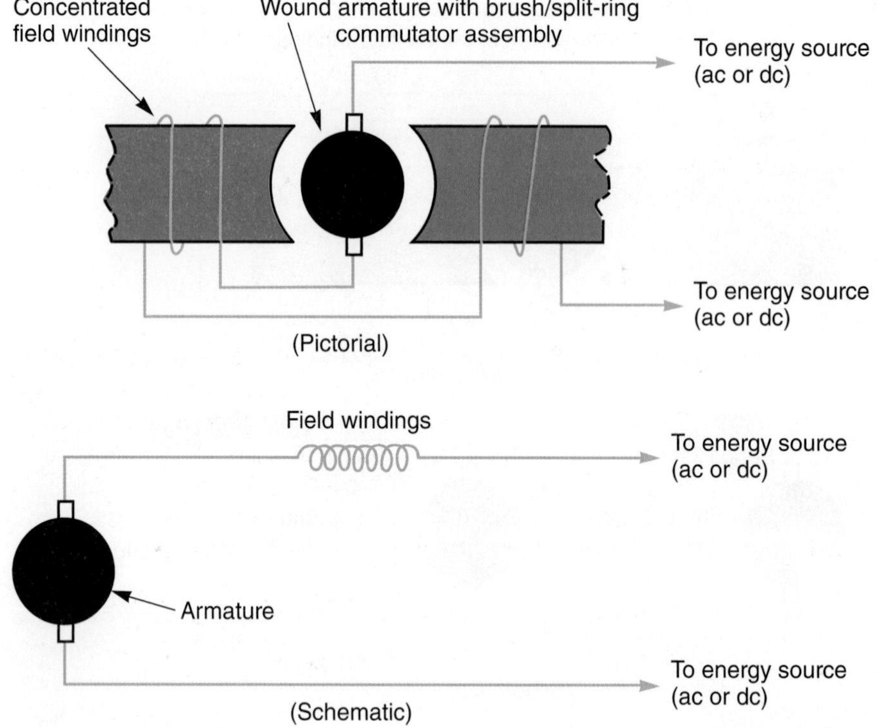

Goodheart-Willcox Publisher

Figure 6-13. Universal motors can operate on either ac or dc energy and are often used to power portable tools.

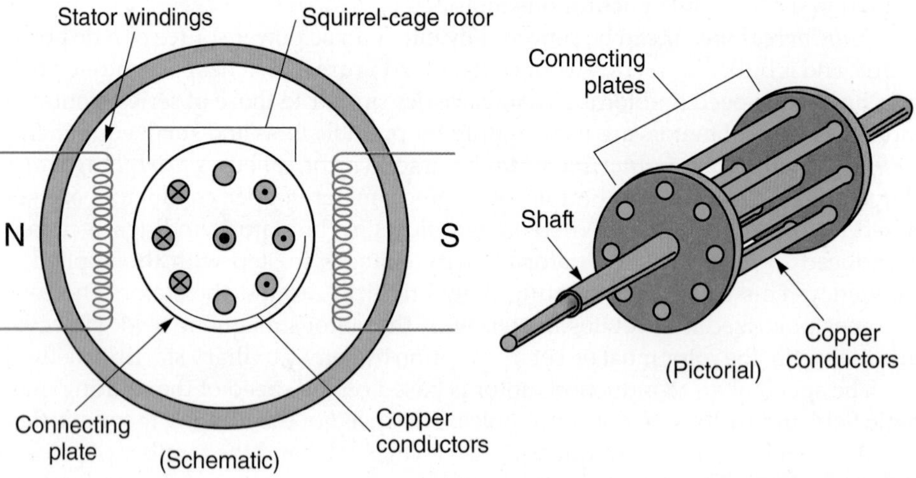

Goodheart-Willcox Publisher

Figure 6-14. The components of a squirrel cage rotor on an induction motor include copper conductors, connecting plates, and a shaft.

The rotor speed must be somewhat less than the synchronous speed in order to develop torque. The difference between the synchronous speed and the rotor speed is called ***slip***. The greater the difference between the rotor speed and the synchronous speed, the more torque is produced. As the difference between the rotor speed and the synchronous speed becomes smaller, the percentage of slip decreases as well.

Most motors used in industry operate using a three-phase ac power source. ***Three-phase ac motors*** are often called the workhorses of industry. The two basic types are induction motors and synchronous motors.

Three-phase induction motors have a squirrel cage rotor. Since three-phase voltage is applied to the stator, no external starting mechanisms are needed. Three-phase induction motors are made in a number of sizes and have good starting and running torque characteristics. Three-phase induction motors are used in industrial applications, such as machine tools, pumps, elevators, hoists, and conveyors.

In **Figure 6-15**, the three-phase power source is connected directly to the windings of the stator. Phase A is connected to windings A1 and A2, which are located 180° apart. Phase B connects to windings B1 and B2, also 180° apart. Phase C follows the same pattern. The windings are distributed 360° around the entire stator. The beginnings of each phase winding (A1, B1, C1) are located 120° apart. This is equivalent to 360° divided by three (or the total number of phases). Likewise, the ends of the windings (A2, B2, C2) are located 120° apart. The stator windings are energized with the three-phase voltages, causing the rotor to rotate in a constant direction. Reversing any two of the winding connections (A and B, B and C, or A and C) of a three-phase ac motor causes the rotor to rotate in the opposite direction.

Three-phase synchronous motors are unique and very specialized. These motors deliver constant speed and can be used to correct power factors of three-phase systems. Direct current is applied to the wound rotor to produce an electromagnetic field. Three-phase ac power is applied to the stator, which also has windings.

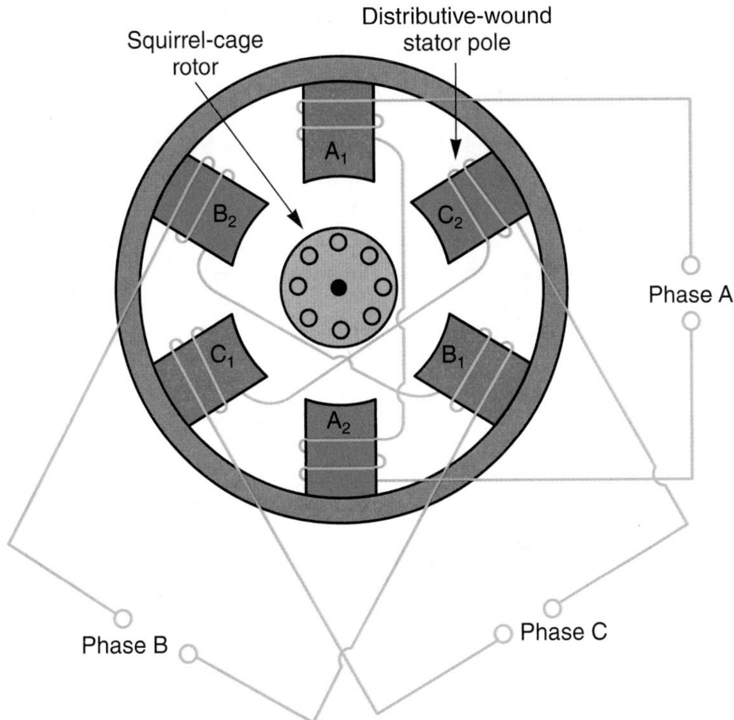

Goodheart-Willcox Publisher

Figure 6-15. Three-phase ac induction motors are widely used for industrial applications and do not require an external starting mechanism.

Robotics Innovations: Edison Robot

The Edison robot is a low-cost robotic solution designed for use in schools. The robot has a surface that is compatible with LEGO® blocks and allows students to build unique robot designs on top of the programmable base. Lesson plans and activities have been developed that are available for download from the Edison website. Programs can be entered with a computer or by simply printing out bar codes that the robot drives over and scans. Onboard sensors can detect light, sound, lines, and infrared signals. Standard TV and DVD remotes can be used to control the robot. The low cost and versatility of the robot make it ideal for schools with small budgets.

Photo courtesy of Edison Robot

A basic three-phase synchronous motor has no starting torque. Some external means must be used to start it. A synchronous motor will rotate at the same speed as the revolving stator field. At synchronous speed, rotor speed equals stator speed, and the motor has zero slip.

6.4.3 Servo Systems

Servo systems are machines that change the position or speed of a mechanical object in response to system feedback or error signals. Positioning applications include numerical control machinery, process control indicating equipment, and robotic systems. Changing the speed of a mechanical object applies to conveyor belt control units, spindle speed control in machine tool operations, and disk or magnetic tape drives for computers. In general, servo systems follow a closed-loop control path, **Figure 6-16**.

The input of a servo system is the reference source to which the load responds. By changing the input in some way, a command is applied to the ***error detector***. This device receives data from both the input source and the output device. If a correction is needed, the signal is amplified and applied to the actuator. The actuator is normally a servomotor that produces controlled shaft displacements. The output device is usually a synchro system that relays information back to the error detector for position comparison.

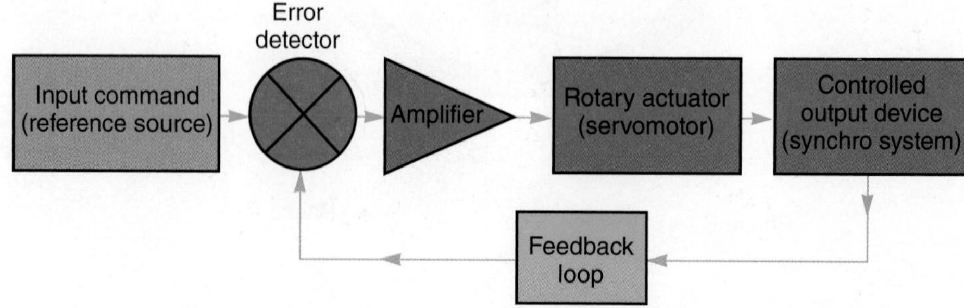

Goodheart-Willcox Publisher

Figure 6-16. A typical servo system uses feedback signals to adjust the position or speed of a mechanical object.

A ***servomotor*** is a component of a servo system that produces controlled shaft displacements used to achieve a precise degree of rotary motion, **Figure 6-17**. Servomotors must respond accurately to error signals and be capable of reversing direction quickly. The amount of torque developed by a servomotor must be high. There are two types of servomotors: the synchronous motor and the stepping motor.

A ***synchronous motor*** contains no brushes, commutators, or slip rings. It is comprised of a rotor and a stator assembly, but there is no direct contact between the rotor and stator poles. An air gap must be carefully maintained in order for synchronous motors to operate.

The speed of a synchronous motor is directly proportional to the ac frequency and the number of pairs of stator poles. Since the number of stator poles cannot be altered after the motor has been manufactured, frequency is the most significant factor in controlling speed. Speeds of 28, 72, and 200 rpm are typical. A speed of 72 rpm is commonly used in numerical control applications.

Single-phase ac synchronous motors are commonly used in low-power applications, **Figure 6-18**. This type of motor is normally limited to low-power applications because it develops excessive amounts of heat during starting conditions. A typical low-power application of single-phase ac synchronous motors is for precision timing circuits that require precise and constant speed.

The stator layout of a two-phase synchronous motor with four poles per phase is illustrated in **Figure 6-19**. This motor can start, stop, and reverse quickly. In this example, there is room for 48 teeth around the inside of the stator. However, one tooth per pole must be eliminated to provide space for the windings. This leaves a total of 40 teeth. The four coils of each phase are connected in series to achieve the correct polarity.

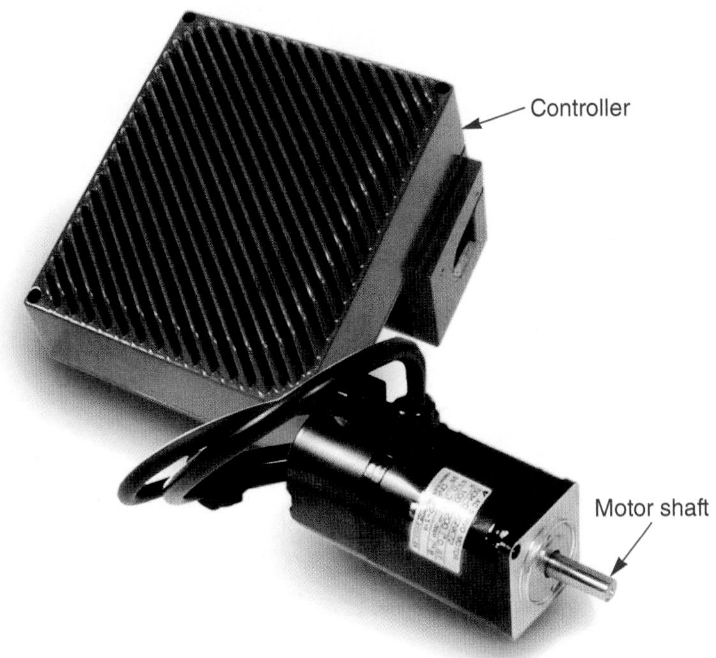

Adept

Figure 6-17. A typical servomotor is used for robot axis positioning. The controller provides signals to position the axis accurately and the motor shaft connects to the robot axis.

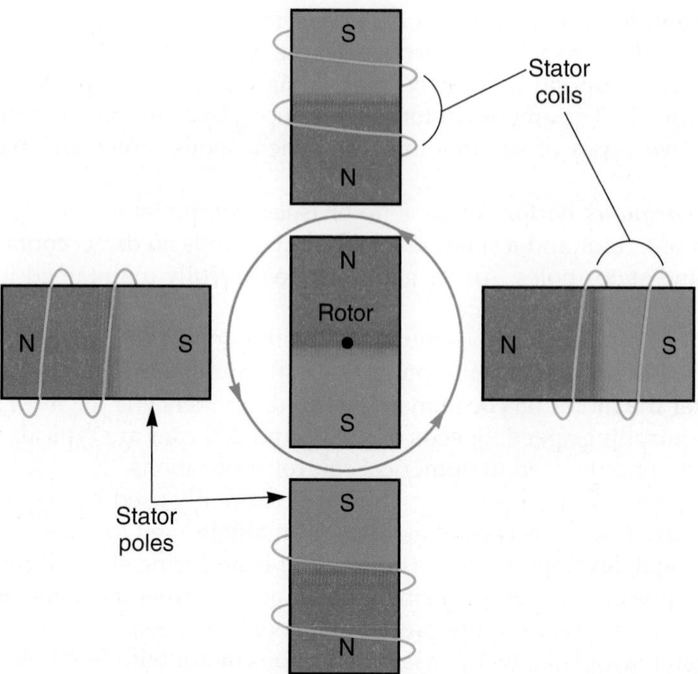

Goodheart-Willcox Publisher

Figure 6-18. A single-phase synchronous motor has a simple construction. The rotor turns without making direct contact with the stator poles.

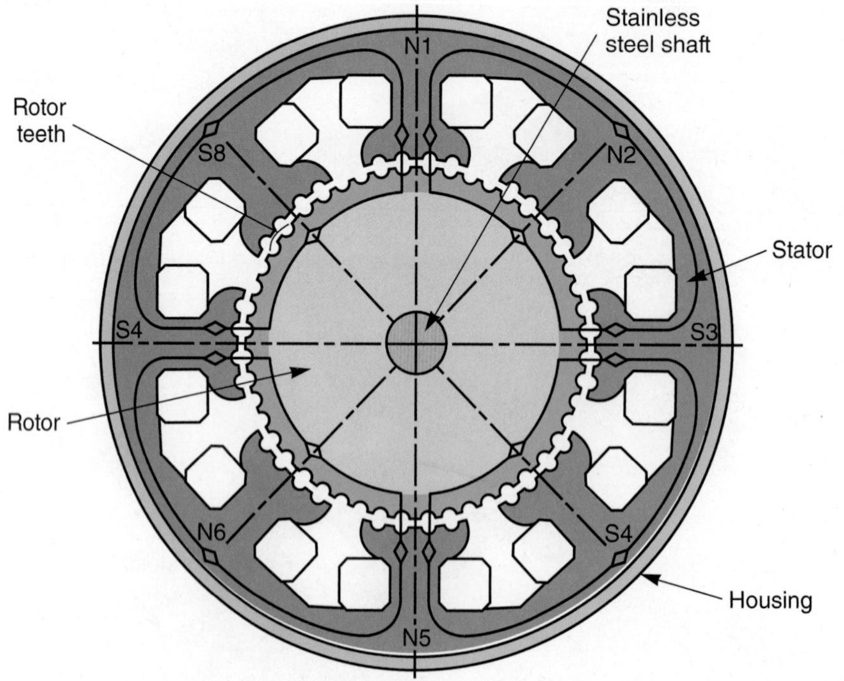

Superior Electric Co.

Figure 6-19. This illustration shows the stator layout of a two-phase synchronous motor with four poles per phase. Poles N_1 to S_3 and N_5 to S_7 represent one phase. Poles N_2 to S_4 and N_6 to S_8 represent the second phase.

The rotor of the synchronous motor has a permanent magnet. There are 50 teeth cast into the form of the rotor. The front section of the rotor has one polarity while the back section has the opposite polarity. The difference in the number of stator teeth (40) and rotor teeth (50) means that only two teeth of each part can be properly aligned at the same time. Since the rotor's sections have opposing polarities, the rotor can stop very quickly and can reverse direction without hesitation.

Because of its gear configuration, the synchronous motor can start in one and one-half cycles of the applied ac frequency and can stop in 5° of mechanical rotation. Synchronous motors of this type draw the same amount of current when stalled as they do when operating. This is very important in automatic machine tool applications that involve heavy mechanical loads.

Nearly all high-power servomechanisms use ***dc stepping motors***, **Figure 6-20**. This is a type of motor in which the rotor has a permanent magnet. These motors are primarily used to change electrical pulses into rotary motion. They are more efficient and develop significantly more torque than synchronous servomotors. The shaft of a dc stepping motor rotates a specific number of degrees with each pulse of electrical energy. The amount of rotary movement, or angular displacement, can be repeated precisely.

The velocity, direction, and travel distance of a piece of equipment can be controlled by a dc stepping motor. Stepping motors are energized by a dc drive amplifier that is controlled by a computer system. The movement error is generally less than 5% per step. The construction of a dc stepping motor is very similar to that of an ac synchronous motor. Some manufacturers make servomotors that can be operated as either ac synchronous or dc stepping.

The stator coils are wound using ***bifilar construction***, in which two separate wires are wound into the coil slots at the same time. The two wires are small, permitting twice as many turns as with a larger wire. This simplifies control circuitry and dc energy source requirements.

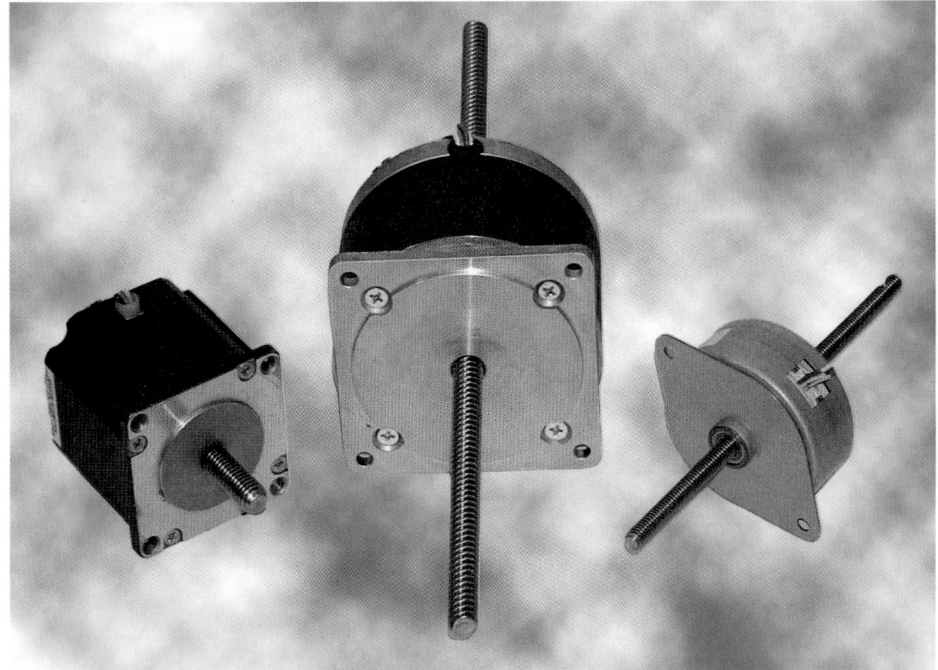

Anaheim Automation, Inc.

Figure 6-20. These stepping motors are used to power linear actuators.

Operation of a stepping motor is achieved using a four-step switching sequence. Any of the four combinations of switches 1 or 2 will produce an appropriate rotor position. The switching cycle then repeats itself. Each switching combination causes the motor to move one-fourth step.

The rotor in the circuit illustrated in **Figure 6-21** permits four steps per tooth, or 200 steps per revolution. The amount of linear displacement, or step angle, is determined by the number of teeth on the rotor and the switching sequence. A stepping motor that takes 200 steps to produce one revolution moves 360°/200 or 1.8° per step. It is not unusual for stepping motors to require eight switching combinations to achieve one step.

6.4.4 Rotary Electric Actuators

Robots require rotary electric actuators to produce rotary motion different from that produced by an electric motor. This type of rotary motion controls the angular position of a shaft. Using rotary electric actuators, rotary motion is transmitted between locations without direct mechanical linkage. Servomotors and synchronous motors are types of rotary actuators. For these applications, computer signals are applied to the actuators and translated into precise amounts of rotary motion. The electrical signals are applied to the motor, causing mechanical (rotary) movement.

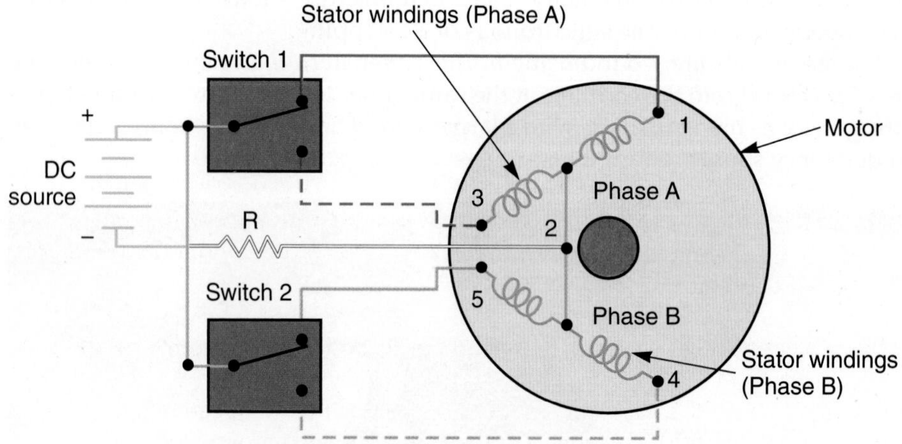

Switching Sequence*

Step	Switch #1	Switch #2
1	1	5
2	1	4
3	3	4
4	3	5
1	1	5

*To reverse direction, read chart from bottom to top.

Superior Electric Co.

Figure 6-21. This dc stepping motor is similar in construction to an ac synchronous motor.

Summary

- Electromechanical systems transfer power from one point to another through mechanical motion used to perform work.
- A robot is a unique type of system that may require several different types of subsystems for proper operation.
- An energy source provides power for a robotic system. The most common source of power for synthesized systems is alternating current (ac).
- A transmission path provides a channel for the transfer of energy in a system.
- A system control alters the flow of power and causes some type of operational change in the system, such as changes in electric current or hydraulic pressure.
- The load of a system is a part or parts designed to produce work. Work occurs when energy is transformed into mechanical motion, heat, light, chemical action, or sound.
- Indicators are used to display information about the operational condition throughout the system.
- Mechanical systems, such as motors, linear actuators, cylinders, and reciprocating devices, produce some form of mechanical motion.
- An electrical sensing system signals a response to a particular form of energy, such as light. Some sensing systems combine optics, electromagnetics, and electronics.
- Electrical timing systems turn a device on or off at a specific time or in step with an operating sequence.
- Control systems are used to make adjustments that alter machine operations. Open-loop and closed-loop systems are used to provide control.
- Electric motors are rotary motion systems that convert electrical energy to rotary motion.
- Direct current (dc) motors are used when speed control is essential. Permanent-magnet, series-wound, shunt-wound, and compound-wound are types of dc motors.
- Single-phase ac motors operate using a single-phase ac power source. Three types of single-phase ac motors include universal motors, induction motors, and synchronous motors.
- Servo systems are machines that change the position or speed of a mechanical object in response to system feedback or error signals.
- Servomotors are used with robotic systems for precise control of rotary motion. DC stepping motors are used with robotic systems for precision motion control.
- Robots require rotary electric actuators to produce rotary motion different from that produced by an electric motor.

Review Questions

Answer the following questions using the information provided in this chapter.

1. What is a subsystem?
2. What is a synthesized system?
3. What are five basic subsystems that are common to all automated systems?
4. What is the function of a mechanical system?

5. What are three kinds of motion used in automated applications of mechanical systems?
6. Give an example of a mechanical system used in industry.
7. How are sensing systems, timing systems, and control systems used in the operation of industrial robots?
8. How do non-servo and servo control systems for a robot differ?
9. Explain the difference between full control and partial control in an open-loop control system for a robot. What methods are used to achieve each type of control?
10. What is the function of feedback in a closed-loop system with automatic correction control?
11. What is an electric motor and what are the basic parts of an electric motor?
12. Describe the relationship of load, speed, cemf, current, and torque in an electric motor.
13. What does the horsepower rating of a motor represent?
14. Identify four types of dc motors and list the characteristics of each.
15. Identify the various types of single-phase ac motors and list unique characteristics of each type.
16. What factors contribute to the speed of an ac induction motor?
17. Describe the purpose of a servo system.
18. Describe the basic construction of a synchronous motor.
19. What is a dc stepping motor?

Learning Extensions

1. Select a piece of equipment or a machine and identify the parts of its system (energy source, transmission path, control, load, indicators).
2. Choose a type of electric motor and search the Internet to find the specs and operational information, such as horsepower ratings available, cost, and common applications.
3. Make a sketch of a type of electric motor that includes illustrations of a simplified rotor and stator. Describe how rotation is achieved.
4. Go to the websites for manufacturers and obtain information on rotary actuators that may be used with robotic systems.

Advanced Analysis

1. What are some typical applications of an industrial environment system? Perform research as needed.

Activity 6-1 Electrical Meters

Objectives

Electrical meters are used to measure voltage, current, resistance, and other electrical quantities. There are two general types of electrical meters: analog and digital. Analog meters have a hand or pointer that moves according to the amount of the quantity being measured. This type of meter is seldom used today. The other type, digital meters, display numbers indicating the amount of the quantity that is measured. Digital meters are by far the most common type of meter used today. Measurements are made in many types of circuits.

You should learn the proper procedure for measuring resistance, voltage, and current for working with robotic systems. These three measurements are the most common for electrical circuits. Meters used for testing electrical circuits are often referred to as multimeters, such as a DMM (digital multimeter), which is used to measure voltage, current, or resistance. A representative type of digital multimeter will be discussed in this activity.

Measuring Resistance and General Meter Use

Many important electrical tests can be made by measuring resistance. Resistance is opposition to the flow of current in an electrical circuit. The current that flows in a circuit depends on the amount of resistance in that circuit. You should learn to measure resistance in an electrical circuit by using a meter.

A multimeter is used to measure resistance, voltage, or current. The type of measurement can be changed by adjusting the function-select switch (or push button) to the desired measurement. Often, the current range is labeled *mA* (milliamperes) for measuring current, *V* for voltage, and the ohm symbol (Ω) for measuring resistance. Also, a function-select switch or push button is pressed so that the meter may be used to measure alternating current (ac) values or direct current (dc) values.

The ohms measurement function of a DMM is typically divided into ranges, such as 200 Ω, 2 kΩ, 20 kΩ and 20 MΩ. This type of meter is called a multirange meter and there are many different types used. The meter may be adjusted to any of the positions for measuring resistance.

The test leads used with the meter are ordinarily black and red. These colors are used to help identify which lead is the positive or negative side. The positive (red) lead is inserted into the meter for voltage, resistance, and current measurements. The black lead (negative) is used as the *common* (COM) lead of the meter. Resistance values are sometimes indicated as *K ohms* (multiply by 1000) on the digital scale of the meter for some of the resistance ranges of the meter.

Another type of digital meter is an autoranging meter. The range values do not need to be changed for larger or smaller voltages, currents, or resistances. To measure resistance with this meter, the negative (black) lead is inserted in the COM jack and the positive (red) lead is placed in the V/Ohm jack. The function select switch of the meter is rotated to the ohm (resistance) symbol. Resistance values are indicated on the digital scale of the meter.

Measuring Voltage

Voltage is applied to electrical equipment to initiate operation. A technician should to be able to measure voltage in order to check the operation of equipment. Many electrical problems develop due to either too much or too little voltage being applied to the equipment. Multimeters are used to measure voltage in an electrical circuit. Typical voltage ranges of a DMM might be 200mV (millivolts), 2V, 20V, 200V and 1000V. When the function-select switch or push button is adjusted to 1000V on the dc volts range, for example, the meter can measure up to 1000 volts dc. The same is true for the other ranges of dc voltage. The voltage value of each range is the *maximum* voltage that may be measured on that range. When making voltage measurements, adjust the function-select switch or push button to the highest range of dc voltage to be measured. The red test lead should be put into the V-Ω-A jack. The black test lead should be put into the COM jack. Voltage values are measured with the meter connected *in parallel* with the circuit or component being measured. Voltage values are indicated in volt units on the digital scale of the meter.

To measure voltage with an autoranging meter, the negative (black) lead is inserted in the COM jack and the positive (red) lead is placed in the V/Ohm jack. The function select switch is placed in one of the voltage settings: mV for millivolts, V for dc voltage, or $\tilde{V}$ for ac voltage. Voltage values are usually indicated directly in volts on the digital scale of the meter for each range.

Measuring Current

Current flows through a complete electrical circuit when voltage is applied. Many important tests are made by measuring current flow in an electrical circuit. The current values in an electrical circuit depend on the amount of resistance in the circuit. Learning to use a multimeter to measure current in an electrical circuit is also important.

To measure the current through a circuit, first make sure no voltage is applied to the circuit while you are connecting the meter. Current is measured with the meter *in series* with the circuit. The first step of the procedure for measuring current is to remove a connecting wire from the circuit at the point where the current will be measured. Set the multimeter to the *highest* current range. Connect the negative test lead to the most negative point of the circuit and the positive test lead nearest the positive power source terminal. After the meter has been connected in this manner, voltage may be applied to the circuit. If necessary, adjust the meter to a lower range to get an accurate current reading. The current value should never exceed the value of range setting of the meter. The current value is read directly on the scale of the meter.

Now look at the controls of a typical multimeter being used for measuring current. The current ranges might be 2 mA (milliamperes), 20 mA, and 200 mA. For example, when the function-select switch or push button is adjusted to 2000 mA on the current range, the meter measures up to 2000 mA (2 amperes). The same is true for the other ranges of current in terms of the maximum current reading. The current value of each range is the *maximum* current that may be measured on that range. When making current measurements, adjust the function-select switch or push button to the highest range of current. The red test lead should be put into the mA jack. The black test lead should be put into the COM jack. Current values are usually indicated in milliampere units on the digital scale of the meter.

To measure current with an autoranging multimeter, the range values do not need to be changed for larger or smaller values. The negative (black) lead is inserted in the COM jack and the positive (red) lead is placed in the mA jack of the meter. The function select switch is adjusted to the mA (milliamperes) or μA (microamperes) setting, either dc or ac. Current values are indicated in milliamperes or microamperes, corresponding with the setting on the digital scale of the meter.

The multimeters discussed are representative of the types of digital meters used. The most common type has separate ranges that need to be set for making resistance, voltage, and current measurements. The autoranging meter needs only to be set for the desired function for measuring voltage, current, or resistance.

Analysis

Obtain a meter to use for completing your lab activities. Study the meter and answer each of the following questions. Indicate N/A (not applicable) if the item does not apply to your meter. Explain at the end of the item.

1. What company manufactured the meter?
2. What is the model number of the meter?
3. The meter will measure up to what A of dc?
4. *True or False?* Ohms-adjust control is used each time the resistance range is changed.
5. To measure dc current greater than1 A, the range switch is placed in what position?
6. For measuring current in a circuit, the meter should be connected how (series or parallel)?
7. For measuring voltage, the meter should be connected how (series or parallel)?
8. To measure 18 mA of current, the range switch should be placed in what position?
9. To measure 10 μA of current, the range switch should be placed in what range?
10. To measure resistance, the red test lead must be placed in the jack with what marking? The black test lead must be placed in the jack with what marking?
11. *True or False?* Polarity must be observed when measuring ac voltage.
12. *True or False?* Polarity is not important when measuring resistance.
13. Up to how many volts dc can be measured with the meter?
14. For measuring a resistor valued at 10 Ω, what range should be used?
15. *True or False?* Polarity should be observed when measuring dc voltage.
16. *True or False?* It is correct to measure the resistance of a circuit with voltage applied.
17. *True or False?* When measuring an unknown value of voltage, one should start at the highest range and work down to the lowest value without exceeding the measured value.
18. *True or False?* Meters should always be handled with care and safety.

Activity 6-2 Basic Electrical Problem Solving

Objectives

Basic electrical problems are often encountered in any area which involves electrical control systems. Robotics is no exception. The most basic problems in electrical systems involve Ohm's law. Ohm's law is a mathematical formula which explains the relationship between voltage, current, and resistance. This relationship must be understood before electrical concepts are meaningful. In this activity, you will complete some practical problems by applying Ohm's law.

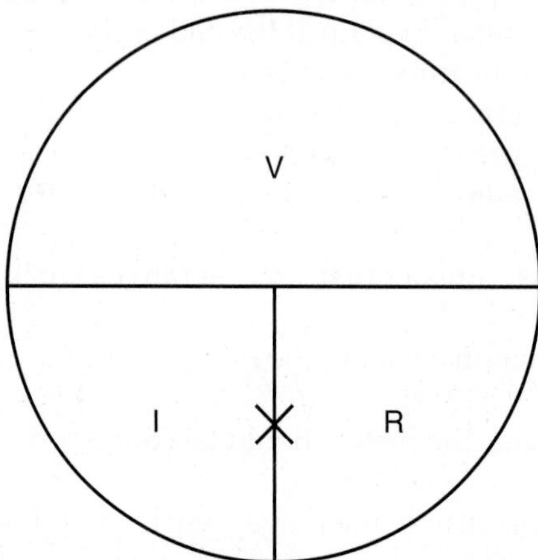

Goodheart-Willcox Publisher

Ohm's law circle: V, voltage; I, current; R, resistance. To use the circle, cover the value you want to find and read the other values as they appear in the formula: $V = I \times R$, $I = V/R$, $R = V/I$.

Analysis

1. A doorbell requires 0.2 ampere of current in order to ring. The voltage supplied to the bell is 120 volts. What is its resistance?
2. A relay used to control a motor is rated at 25 ohms resistance. What voltage is required to operate the relay if it draws a current of 0.25 ampere?
3. An automobile battery supplies a current of 7.5 amperes to a headlamp with a resistance of 0.84 ohm. What is the voltage delivered by the battery?
4. What voltage is needed to light a lamp if the current required is 2 amperes and the resistance of the lamp is 50 ohms?
5. If the resistance of a radio receiver circuit is 240 ohms and it draws a current of 0.6 ampere, what voltage is needed?

6. A television circuit draws 0.15 ampere of current. The operating voltage is 120 volts. What is the resistance of the circuit?
7. The resistance of the motor windings of an electric vacuum cleaner is 20 ohms. If the voltage is 120 volts, what is the amount of current drawn?
8. The coil of a relay carries 0.05 ampere when operated from a 50-volt source. Find its resistance.
9. How much current is drawn from a 12-volt battery when operating an automobile horn of 8 ohms resistance?
10. Find the resistance of an automobile starting motor if it draws 90 amperes from the 12-volt battery.
11. What current is drawn by a 5000-ohm electric clock when operated from a 120-volt line?
12. Find the current drawn by a 50-ohm toaster from a 120-volt line.

Activity 6-3
Control Systems Overview

Objectives

The term *system* is commonly defined as a combination of parts that work together to form a complete operating unit. In this respect, the term can apply to a variety of different systems used by industry to achieve some operation in the manufacturing process. An application of this term is the industrial electronic system. Further application of this term includes a wide variety of subsystems under the heading of industrial electronics. Opto-electronic systems, timing systems, digital systems, and environment control systems are all included in this study of industrial systems.

Each industrial system includes a number of unique features or characteristics that distinguish it from other systems. More importantly, however, there are a number of basic functions common to all electronic systems. Energy source, transmission path, control, load, and indicator are terms commonly used to describe these basic system functions.

In this activity, you will have an opportunity to study a representative system and pick out the basic parts of the system. By doing this, you will become more familiar with the systems concept and be able to apply it to a specific situation.

Procedure

1. Briefly define the function of the following basic system parts. Use general terms that would apply to all basic industrial electronic systems.
   - Energy source.
   - Transmission path.
   - Control.
   - Load.
   - Indicator.
2. Explain how the systems concept could be applied to an electrical power system used to supply an industrial plant site.
3. How could the systems concept be applied to a simple electrical circuit?

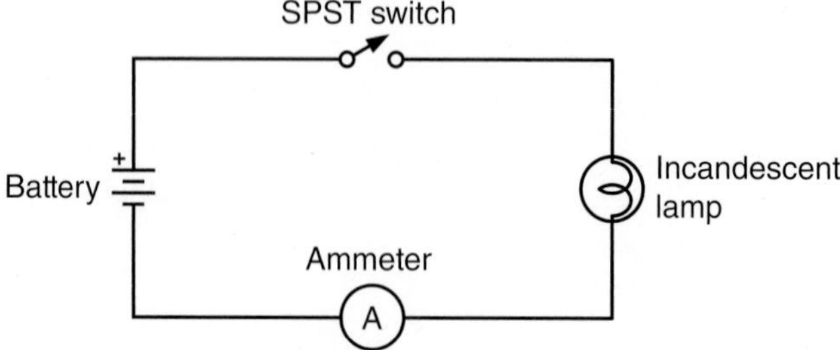

Goodheart-Willcox Publisher

Electrical circuit.

4. The photoelectric box counter is a typical industrial system. What are the basic parts of this system?

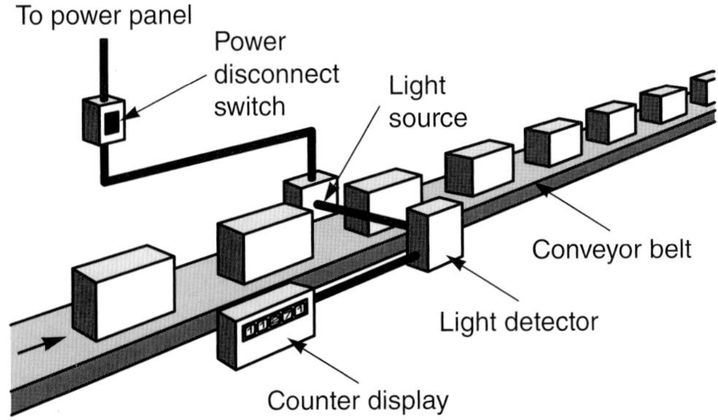

Goodheart-Willcox Publisher

Industrial electronic system.

Analysis

1. Of what significance is the basic systems concept in industry?
2. What is an opto-electronic system?
3. What are some applications of an industrial timing system?
4. What does the term *digital system* mean?

Activity 6-4
AC Synchronous Motors

Objectives

The ac synchronous motor used in robotic systems is commonly classified as a constant-speed device. It has extremely rapid starting, stopping, and reversing abilities. Motors of this type require only a simple clockwise-stop-counterclockwise rotary switch for control. The starting and running currents of a synchronous motor are identical, which is unique for ac motors. This characteristic means that a motor of this type can withstand a high inrush of current when direction changes occur. As a general rule, the synchronous motor can even be stalled without damaging the motor.

Motors of this type are commonly used as drive mechanisms for robotic systems in machinery operations. Starting begins within 1 1/2 cycles of the line frequency, and stopping occurs in five mechanical degrees of rotation. This motor represents a unique part of all servomechanisms used in precision rotary control applications today. In this activity, a simple test circuit will be constructed so you may observe starting and stopping of an ac synchronous motor and the current flow characteristics of the motor.

Equipment and Materials

- AC synchronous motor—Superior Electric Type SS150 or equivalent motor rated at no more than 200 in/oz.
- Capacitor—3.75 μF, 330V ac.
- Resistor—250 Ω, 25 W.
- SPST switches (3).
- Isolated ac power source—120V.
- AC ammeter—0–5 A.
- Piece of wood.

Safety

Use protective eyewear during this activity.

Procedure

1. Construct the ac synchronous motor test circuit shown in the following figure. Ensure that the motor is mounted securely.

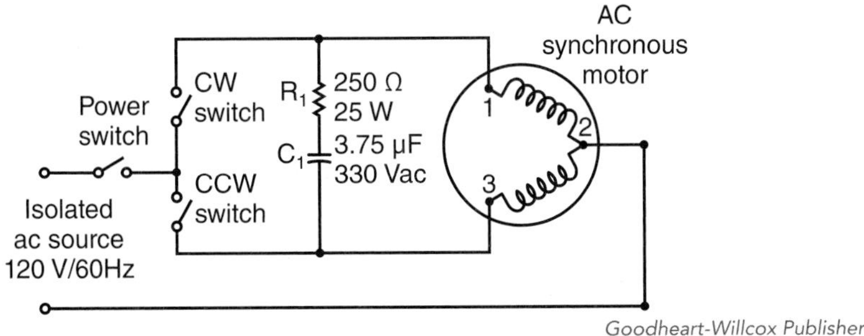

AC synchronous motor test circuit.

2. Turn on the switch for clockwise rotation.
3. Momentarily turn on the power switch and observe the rotation of the motor.
4. With a tachometer, measure and record the speed of rotation.
5. Momentarily turn off the power switch and observe the stopping action of the synchronous motor. You may want to run several trials to see the quickness of the stopping action. How does this compare with other ac motors?
6. Note each time that the motor is turned on, how quickly it comes to speed.
7. With an ac ammeter, measure and record the running current and starting current of the synchronous motor.
8. Carefully wedge a piece of wood between the rotating shaft and the bench while holding the motor. This will provide a simple loading method for test purposes.
9. When the motor is loaded down, how does the running current respond?
10. If the motor is completely stalled, how does the running current respond?
11. Remove the load from the motor, turn off the clockwise switch, and turn on the counterclockwise switch. The motor should rotate equally well in the counterclockwise direction and have the same basic characteristics. Test these again to verify the theory.
12. Switch off the counterclockwise switch, then switch on the clockwise switch. Notice the ease with which direction change occurs.
13. Turn off the ac power source and disconnect the circuit. Return all parts to the storage area.

Analysis

1. Discuss the operation of an ac synchronous motor.

Activity 6-5
DC Stepping Motors

Objectives

DC stepping motors represent a unique electromechanical rotary actuator that is used in robotic systems. These motors are designed to change electrical pulses into rotary motion. The amount of rotary movement or angular displacement produced by each pulse is repeated precisely for each succeeding pulse. The resulting rotation of these motors may be used to locate or position worktables or fixtures accurately for automatic machining operations. Very precise degrees of accuracy can be achieved with these devices.

In this activity, you will construct a simple test circuit for a dc stepping motor. You will be able to produce very accurate shaft movements by using different switching step combinations. Both the four-step and eight-step switching combinations will be tested. Through this activity, you will be able to observe the physical operation and test the accuracy of motor shaft rotation. The switching operation of this activity is achieved automatically by logic devices in actual operating circuits. Precise switching operations obviously improve the accuracy of this circuit.

Equipment and Materials

- DC stepping motor—Superior Electric Type M061-FD02 or equivalent.
- SPST switches (4).
- Resistor—50 Ω, 20 W.
- Resistor—1000 Ω, 1/2 W.
- DC power supply—0–25V, 1 A.
- Diodes—1N4004 (4).

Safety

Use protective eyewear during this activity.

Procedure:

1. Construct the stepping motor test circuit as shown in the next figure.
2. The physical movement of the stepping motor shaft is small and rather difficult to observe. By attaching a simple indicating device, such as a small wire or paper pointer, to the motor shaft the mechanical rotation can be readily observed.
3. The switching sequence in the full-step mode will produce 1.8 steps for each switching combination.

4. Adjust the dc power supply to 15 volts. Set the indicator to a starting position and mark it as the starting reference. Then turn on the switch combinations for Step 1 of the full-step mode.

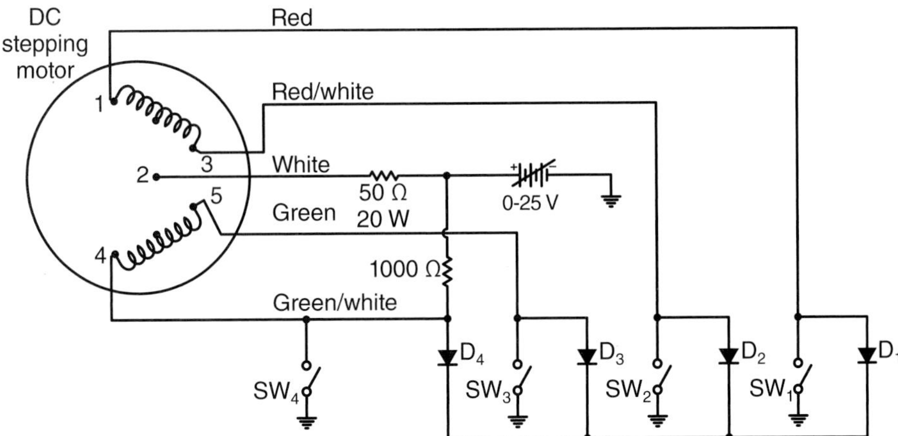

Four-step input (full-step mode)				
Step	SW_1	SW_2	SW_3	SW_4
1	On	Off	On	Off
2	On	Off	Off	On
3	Off	On	Off	On
4	Off	On	On	Off
1	On	Off	On	Off

Eight-step input (half-step mode)				
Step	SW_1	SW_2	SW_3	SW_4
1	On	Off	On	Off
2	On	Off	Off	Off
3	On	Off	Off	On
4	Off	Off	Off	On
5	Off	On	Off	On
6	Off	On	Off	Off
7	Off	On	On	Off
8	Off	Off	On	Off
1	On	Off	On	Off

Goodheart-Willcox Publisher

DC stepping motor test circuit and stepping tables.

5. Go to the next switching combination indicated by Step 2 of the chart. Each combination will produce a rotary step.
6. Follow the stepping procedure through at least two of the four-step cycles. Notice that the switching step combination repeats itself after Step 4.
7. Mark the indicator location as the stopping position for reference.
8. Calculate the rotational degrees by multiplying the number of steps by 1.8.
9. If a protractor is available, measure the number of degrees of rotation between the starting indicator mark and the last switching location mark of the motor shaft.
10. Repeat the procedure outlined in Steps 2 through 9 of this activity for the eight-step sequence. In this stepping mode, each step produces 0.9° of rotation.
11. Disconnect the circuit and return all components to the storage area.

Analysis

1. Explain how a dc stepping motor operates.
2. How many steps are required by a dc stepping motor in the full-step mode to produce one revolution?
3. How many steps are required by a dc stepping motor in the half-step mode to produce one revolution?
4. Describe a typical application of the dc stepping motor.

Activity 6-6 Solenoids

Objectives

A solenoid is an electromagnetic coil with a movable core constructed of a magnetic material. The core or plunger is sometimes attached to an external spring. This spring causes the plunger to remain in a fixed position until moved by the electromagnetic field created by current applied to the coil. This external spring also causes the core or plunger to return to its original position when the coil is de-energized.

Solenoids are used for a variety of applications. For example, most door chime circuits use one or more solenoids to cause the chime to sound. Many gas and fuel oil furnaces use solenoid valves to turn the fuel supply on or off automatically on demand. Most dishwashers use one or more solenoids to control the flow of water. The solenoid is representative of the many types of electromechanical control devices used in conjunction with robotic systems. In this activity, you will learn the basic operation of the solenoid by observing a solenoid as it is used to operate door chimes.

Equipment and Materials

- Digital multimeter.
- Door chimes (low voltage, 6–24V).
- Variable dc power supply.
- Air-core coil (200 turns, No. 24 wire) or equivalent.
- Cold-rolled steel core, 3/4″ diameter, 6″ long or suitable substitute.
- SPST switch.
- Connecting wires.

Procedure

1. Adjust the multimeter to measure resistance. Measure and record the resistance of the air-core coil to be used in this activity.
2. Connect the circuit as shown in the following.

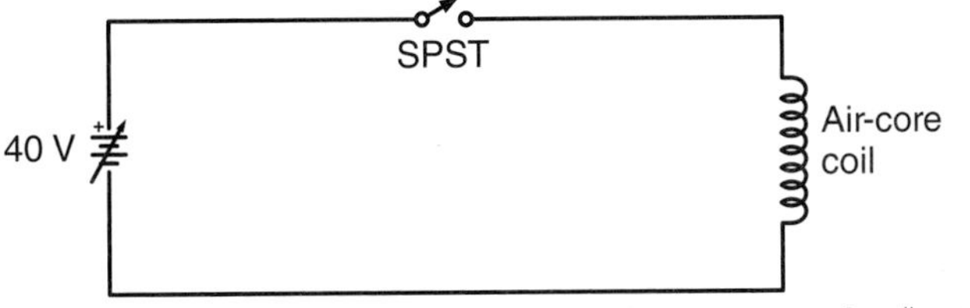

Goodheart-Willcox Publisher

Solenoid test circuit.

3. Place about 2″ of the end of the cold-rolled steel bar into the coil.

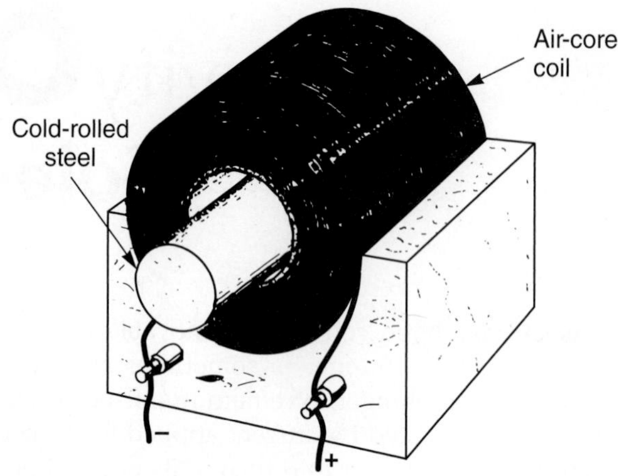

Goodheart-Willcox Publisher

Air-core coil and steel core.

4. Close the SPST switch and describe what happens to the cold-rolled steel core. To prevent the coil from heating, do not allow the SPST switch to remain closed for more than 20 seconds.
5. Place the bar inside the coil and close the SPST switch. Try to remove the bar with the power on. Describe what happens. Do not allow the coil to remain energized for more than 20 seconds.
6. How could the above solenoid be used as a pusher or puller coil?
7. Disconnect the circuit.
8. Acquire a door chime assembly. Examine it to determine if it is to be powered by 6, 12, 18, or 24 volts. Record the voltage used for the chime assembly.
9. Remove the protective chime cover to expose the spring-loaded solenoids and solenoid connections. Measure and record the resistance of the solenoid coil.
10. There should be three solenoid connections. One will be common to the other two. Connect the common connection and either of the remaining connections to the power supply.

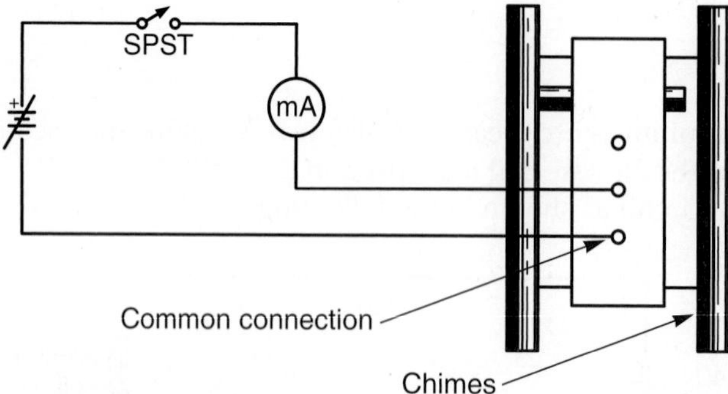

Goodheart-Willcox Publisher

Solenoid circuit with door chime.

11. Close the SPST switch and slowly adjust the power supply from zero until the plunger of the solenoid is drawn into the core. Record the current necessary to pull the plunger.
12. Open the SPST switch; adjust the power supply to the voltage indicated in Step 8.
13. Open and close the SPST switch several times. Describe what happens.
14. Disconnect the circuit and return all components to the storage area.

Analysis

1. What is a solenoid?
2. Where are solenoids used?
3. Why are solenoids sometimes called pusher coils?
4. Why are solenoid plungers spring-loaded?

Activity 6-7
Electromagnetic Relays

Objectives

Relays are electromagnetic switches and are excellent examples of how a magnetic field attracts a magnetic material. These devices contain a coil that creates an electromagnetic field; an armature, which is constructed of a magnetic material attracted by the coil; and a number of contacts or switches that open or close when the magnetic field attracts the armature.

In this activity, you will study the electromagnetic characteristics of a relay. Relays are a popular type of control device that might be used with robotic systems.

Equipment and Materials:

- Multicontact relay.
- 6V lamp with socket.
- Variable dc power supply.
- Resistor—1 kΩ.
- 6V battery.
- Connecting wires.
- Multimeter.

Procedure

1. Prepare the multimeter to measure resistance. Measure and record the resistance of the relay coil.
2. Using the multimeter, determine how many normally open and normally closed contacts are used with your relay.

3. Construct the circuit shown. Be sure that the variable dc power supply is adjusted to zero. The multimeter should be adjusted to measure dc current on the highest range.

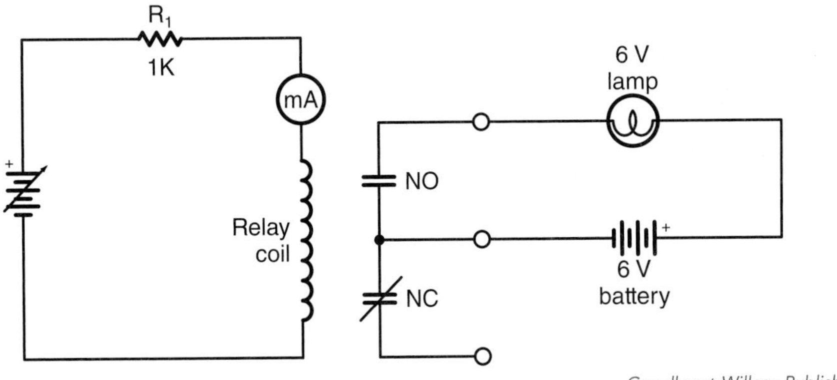

Goodheart-Willcox Publisher

Circuit using electromagnetic relay.

4. Slowly adjust the variable dc power supply from zero until the 6V lamp is turned on. Record the current measured by the multimeter when the relay is energized. This is the pickup current.
5. Slowly adjust the variable dc power supply toward zero until the 6V lamp is turned off. Record the current measured by the multimeter when the relay de-energized. This is the dropout current.
6. Turn the variable power supply off.
7. Alter the circuit as shown.

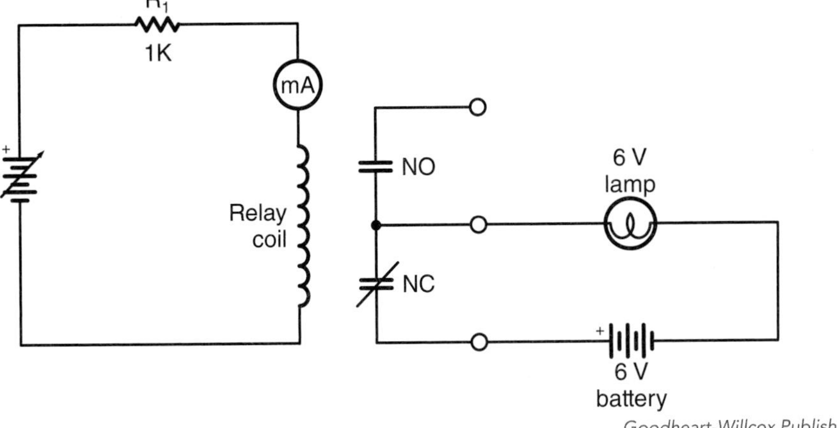

Goodheart-Willcox Publisher

Circuit with relay contacts reversed.

8. You will notice that the only difference in the two circuits is the type of contacts used. In step 3, the normally open contacts were used. In this procedure, the normally closed contacts are used, causing the lamp to remain on until the relay is energized.
9. Adjust the variable dc power supply and record the pickup and dropout currents as you did in steps 4 and 5.

10. How do the currents recorded in step 9 compare with the current recorded in steps 4 and 5?
11. How did the action of the 6V lamp in steps 4 and 5 compare with the action of the lamp in step 9?

Analysis

1. What are normally open contacts?
2. What are normally closed contacts?
3. What is the meaning of the term *pickup current*?
4. What is the meaning of the term *dropout current*?
5. Using Ohm's law, compute the voltage across the relay coil when the relay is energized (see steps 1 and 4).

Activity 6-8 Motor-Driven Timers

Objectives

The motor-driven timer provides a wide variety of timing actions for industrial circuit applications. In its simplest form, this timer has an electric drive motor, a ratchet release coil, and a ratchet dial that is held stationary until released. When the timing cycle reaches its set-time, the ratchet is released and the dial resets itself by spring action. Both on delay and off delay reset timers are available. More sophisticated reset timers permit a wide range of timing operations in a single unit.

In this activity, a simple reset timer will be used to build a load control circuit that produces either interval or delay timing operations. Initially, when the timer is energized by the control switch, load A is turned on and load B is turned off. After the expired time setting or the time out condition has been reached, load A is turned off and load B is turned on. The action of a motor-driven timer is used to represent a type of timing application that might be used in conjunction with robotic system applications.

Equipment and Materials

- Reset timer (Eagle Signal HD-50 Series) or equivalent.
- 7.5 W incandescent lamps with sockets (2).
- SPST toggle switch.

Procedure

1. Refer to the timer manufacturer's product manual before attempting to complete this activity.
2. Wire the reset timer for the on delay operation as shown in the following.

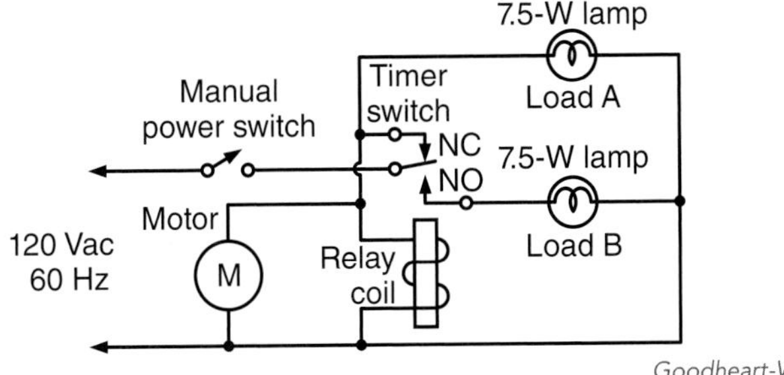

Reset timer with on delay operation.

3. Adjust the time setting dial by pulling it out and turning it to a desired setting. Releasing the dial automatically locks it in the new setting.
4. Turn on the manual power switch and describe the operating condition of loads A and B.
5. When the dial setting trips, describe the operating condition of the load and the motor.
6. Start a new operation cycle by opening the control switch momentarily and then turning it on again. After the cycle has been in operation for a few seconds, momentarily open the control switch and then close it. What action does this initiate?

Analysis

1. What are some industrial applications of a reset timer?
2. Could this timer be modified to achieve a different function of some type? Explain.
3. Explain the difference between interval and delay timers.

Activity 6-9 Digital Timers

Objectives

Digital timers employ a time base generator (oscillator), a counter, and a load-driver circuit. The counter circuit frequently has several output terminals that permit selection of time ranges. Control ranges of 1T, 2T, 4T, 8T, 16T, and 32T are typical. Each output is rated in terms of the time base (T) of the generator circuit.

In this activity, you will employ the SE/NE555 timer IC as the time base generator of a simple digital timing system. The output of the generator will be counted and divided to some extent by the range of the system. Through this circuit investigation, you will see and work with the basic parts of a simple digital timing system. Commercially designed timers of this type are usually housed in an enclosure that does not permit access to component parts. These timers generally do not employ moving parts and are all solid state. The timing period is very accurate, and it usually ranges from microseconds to hours. Again, this type of device is being studied to represent the various types of timing control applications that might be used with robotic systems.

Equipment and Materials

- SE/NE555 IC.
- 200 kΩ, 1/8 W resistors (2).
- 390 Ω, 1/8 W resistor.
- Light-emitting diode.
- 4 μF, 25V dc capacitor.
- 0.01 μF, 100V capacitor.
- 50 μF, 25V capacitor.
- SPST toggle switches or pushbuttons.
- SN7490 IC.
- Circuit mounting board.

Procedure

1. Construct the experimental digital timer circuit of the following by first connecting the time base generator. Do not energize the SN7490 at this time.

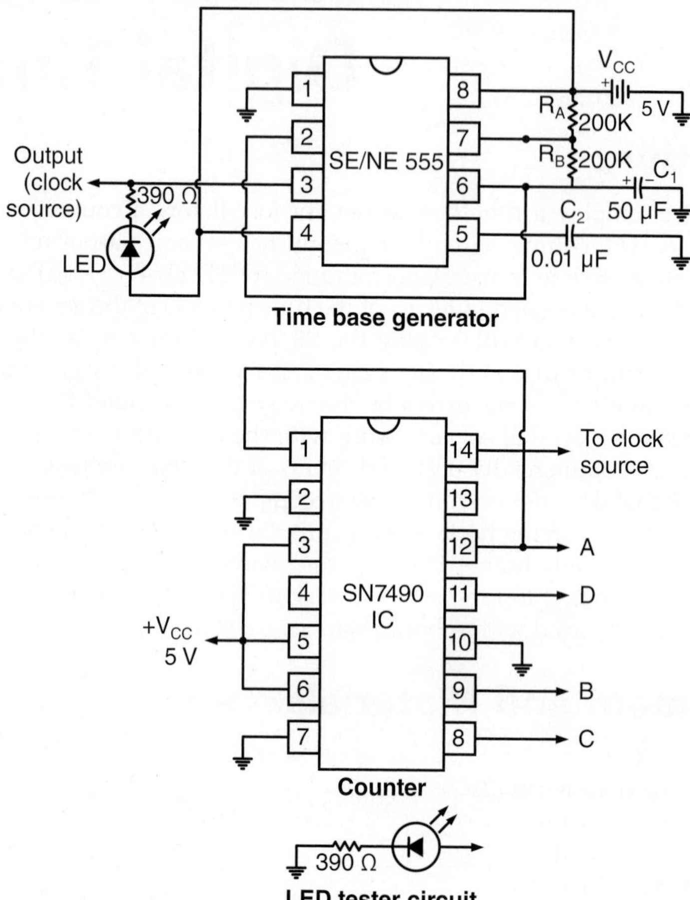

Goodheart-Willcox Publisher

Digital timer circuit.

2. Determine the charging time of the circuit by the $t_c = 0.693\,(R_A + R_B)\,C_1$ formula. Test the accuracy of the circuit. Record your findings.
3. Determine the discharge time of the circuit by the $t_d = 0.693R_B\,C_1$ formula. Test the accuracy of the circuit. Record your findings.
4. Turn off the power supply and complete the SN7490 counter IC. Connect it to the same power source used by the 555. Connect the second LED indicator to the D output of the 7490 IC. Connect the IC clock source to the output (pin 3) of the 555.

5. Turn on the power source and count the number of timing cycles needed to energize the 7490 LED readout. What mathematical function does this represent?
6. Turn off the power source and move the LED readout of the 7490 to output A. Turn on the power again and determine the counting cycles needed to energize the LED. What mathematical function does this represent?
7. Turn off the power supply and alter the SN7490 circuitry. Change C_1 of the time base generator to 4 μF.

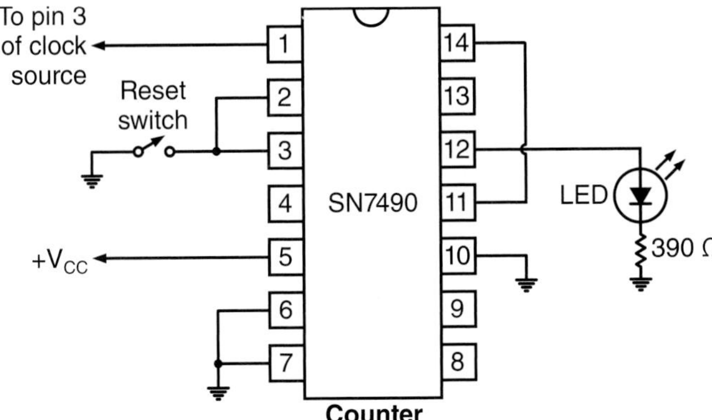

Goodheart-Willcox Publisher

Modified counter circuit.

8. Turn on the power supply and open the reset switch. Close the switch during a time when the time base generator LED is off. Count the number of timing cycles needed to energize the SN7490 LED. Also, count the number of timing cycles needed to turn off the LED. What does this represent as an output of the IC?

Analysis

1. How could the digital timer of this experiment be extended? Name two ways.
2. How can the digital timer of this experiment be made variable?
3. Would it be advantageous to have a reset switch on an actual digital timer? Why?

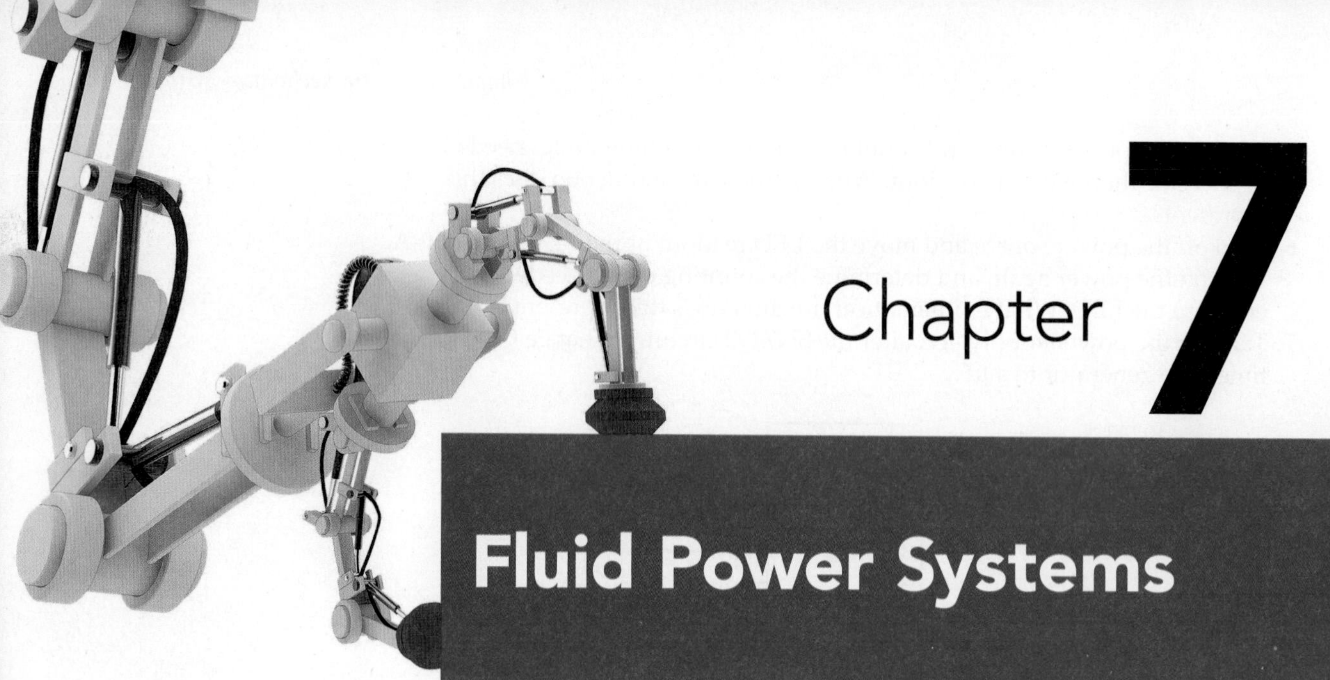

Chapter 7

Fluid Power Systems

Outline

Objectives

Upon completion of this chapter, you will be able to:

- Describe the characteristics of hydraulic and pneumatic systems.
- Discuss the characteristics of fluid flow.
- Explain Pascal's law and how it is applied in fluid power systems.
- Identify the components in a fluid power system and explain the use of each.
- Discuss uses for hybrid power systems.

Technical Terms

centrifugal pump
conditioning
desiccant
direction control device
filter
flow control device
flow indicator
fluid motor
fluid power system
force
FRL unit
heat exchanger
inertia
lubricator
metering
non-positive displacement pump
Pascal's law
positive displacement pump
power
pressure
pressure indicator
pressure regulator valve
pressure relief valve
prime mover
reciprocating pump
resistance
rotary gear pump
rotary vane pump
strainer
turbulence
weight

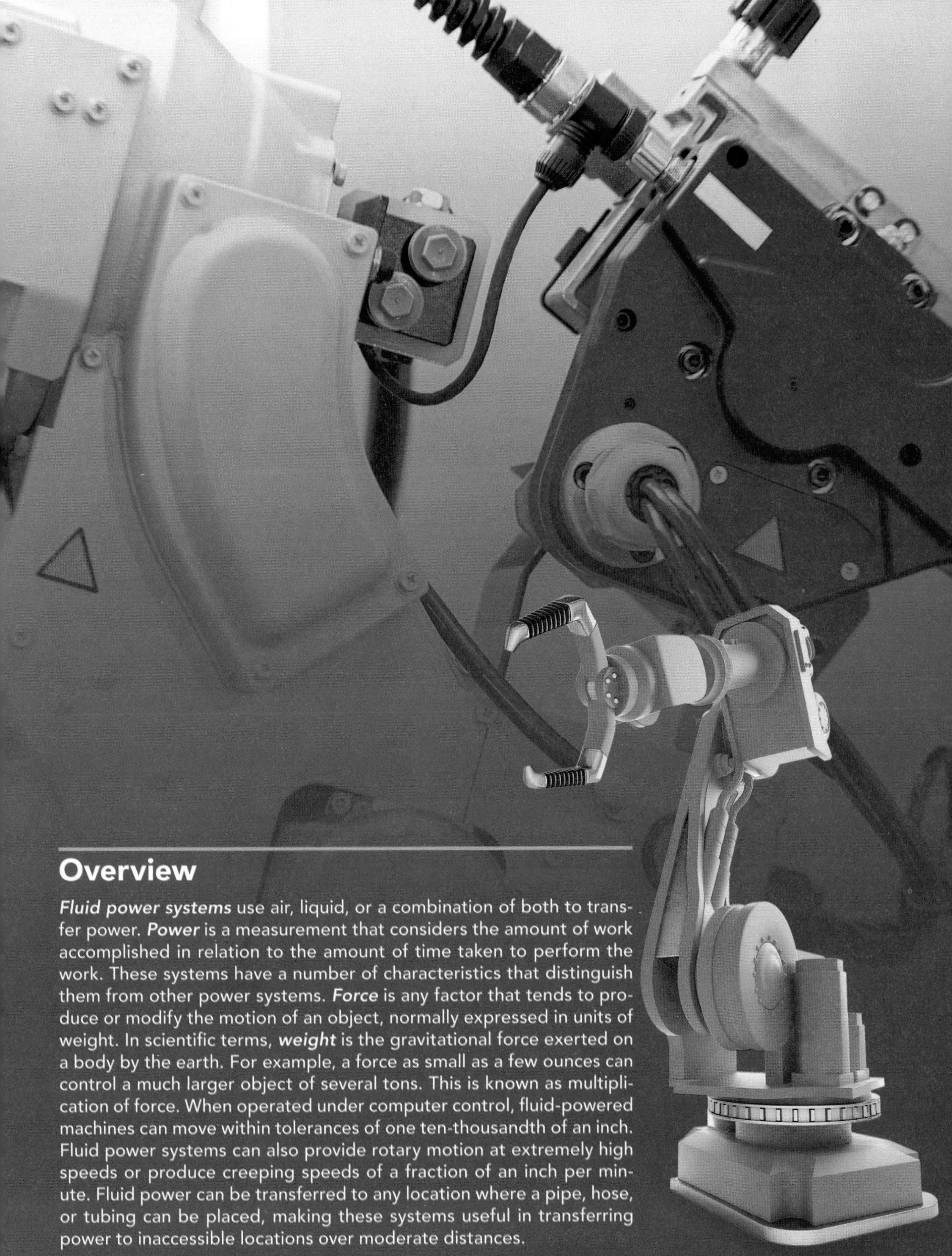

Overview

Fluid power systems use air, liquid, or a combination of both to transfer power. ***Power*** is a measurement that considers the amount of work accomplished in relation to the amount of time taken to perform the work. These systems have a number of characteristics that distinguish them from other power systems. ***Force*** is any factor that tends to produce or modify the motion of an object, normally expressed in units of weight. In scientific terms, ***weight*** is the gravitational force exerted on a body by the earth. For example, a force as small as a few ounces can control a much larger object of several tons. This is known as multiplication of force. When operated under computer control, fluid-powered machines can move within tolerances of one ten-thousandth of an inch. Fluid power systems can also provide rotary motion at extremely high speeds or produce creeping speeds of a fraction of an inch per minute. Fluid power can be transferred to any location where a pipe, hose, or tubing can be placed, making these systems useful in transferring power to inaccessible locations over moderate distances.

Fluid power systems are used to operate a vast array of equipment—from automated manufacturing machinery to power hand tools. They are compatible with electrical, digital, or mechanical systems. Fluid power systems are efficient, dependable, easy to maintain, and economical to operate for long periods of time.

This chapter provides an introduction to fluid power and discusses principles and applications of hydraulic and pneumatic fluid power systems.

7.1 Fluid Power System Models

In fluid power systems, electrical energy is often used to drive a fluid pump. Electrical energy and mechanical motion are converted into the energy of a flowing fluid. Hydraulic systems use oil or another liquid, while pneumatic systems operate with air. The operating principles associated with these systems are similar in many respects. All fluid power systems consist of an energy source, transmission path, load, controls, and indicators.

7.1.1 Hydraulic System Model

Hydraulic systems are used for many applications in robotic system design. Hydraulic systems are relatively simple in operation, but are also heavy-duty and adaptable systems. Hydraulic fluid can be transmitted through a network of hoses to operate motors, actuators, and cylinders (load devices), which then causes a robot to operate. Hydraulic fluid is controlled by control valves and other devices as it is distributed through the transmission network of hoses and tubes. Hydraulic pumps are used as the fluid source to power the system.

The energy source for a hydraulic system typically powers an electric motor, which drives a pump (**Figure 7-1**). In this model, the electric motor acts as a prime mover. A ***prime mover*** is a component of a power system that provides the initial power for movement in the system. The motor receives electrical energy from the source and converts it to rotary energy. The rotary energy is changed into fluid energy by the pump. A vacuum is created in the inlet port while the pump is operating, which draws fluid from the reservoir. The fluid moves through the pump and is forced through the outlet port and into the system.

A typical hydraulic fluid power system includes a number of control devices, **Figure 7-2**. For example, a hand shutoff valve permits control by stopping the fluid flow. A four-way, three-position directional control valve (DCV) can be positioned to allow the cylinder to extend (to restrict the flow of fluid) and lock the cylinder in place, or to shift and reverse cylinder movement. ***Pressure*** is the amount of force applied to a specific area. A ***pressure relief valve*** is a control device that protects the system from stress and damage caused by overpressure. If the hand shutoff valve is closed or the DCV is in the blocked position while the pump is running, for example, the relief valve opens and pressurized fluid returns to the reservoir.

The load device is the system component that is ultimately controlled. In the example illustrated in **Figure 7-2**, the double-acting cylinder serves as the system's load device. This cylinder changes the energy of hydraulic fluid flow into linear movement, which produces the motion of the punch-press ram.

The transmission path of a hydraulic system is often steel pipes and steel tubing, but flexible hose is also used. As the fluid passes through the transmission path, it encounters resistance to flow. ***Resistance*** is friction that forms in a fluid

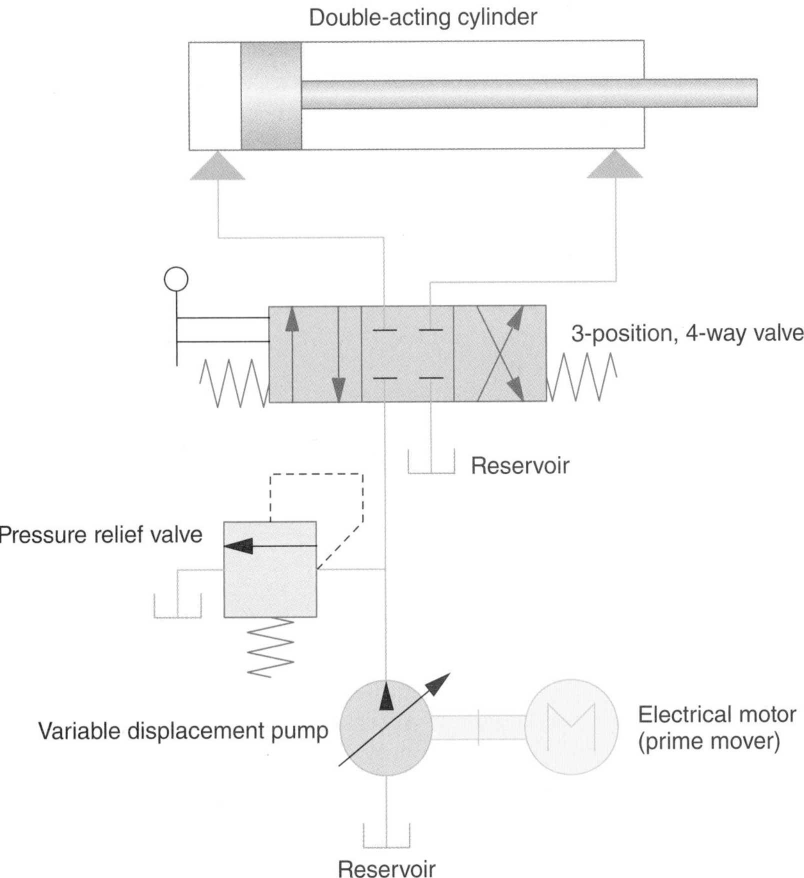

Goodheart-Willcox Publisher

Figure 7-1. In this illustration of a basic hydraulic system, energy moves along the transmission path from the electric motor to the pump. Fluid enters the pump at the inlet port and is forced through the outlet port. The fluid continues through the transmission path to the double-acting cylinder (load device).

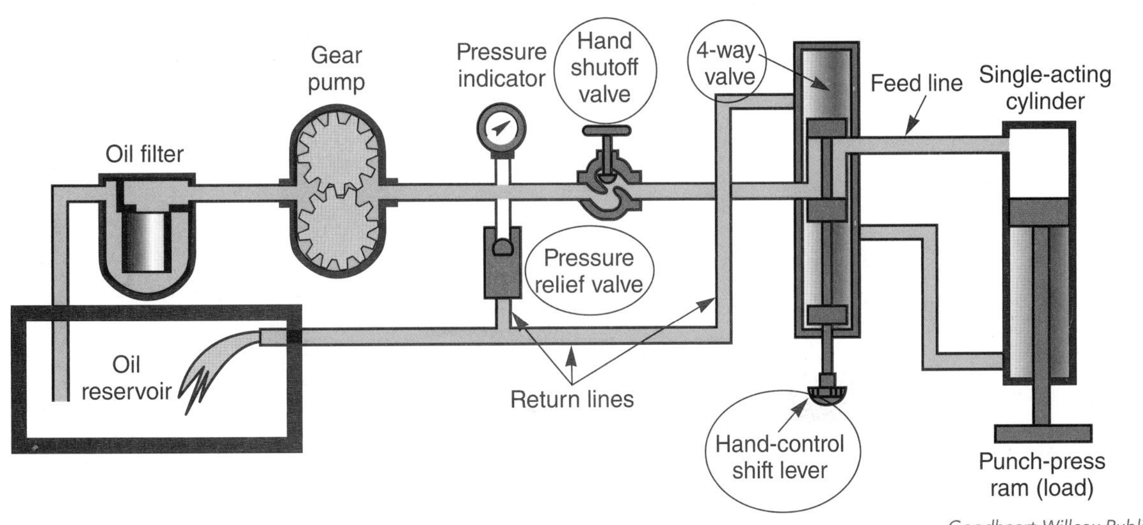

Goodheart-Willcox Publisher

Figure 7-2. This typical hydraulic fluid power system includes several control devices (circled).

system and causes a decrease of power from the input to the output. This causes a buildup of system pressure. Methods in creating resistance to fluid flow in a hydraulic system include friction, control valves, actuators, and piping. Since pressure results from the resistance of fluid traveling through the system, the more resistance produced, the greater the amount of pressure. For example, as the operating speed of a pump is increased, the pressure of the system is also increased due to the greater volume of the fluid entering the system.

Pressure indicators are devices that monitor the fluid pressure within a hydraulic system. Pressure indicators or pressure gauges may be used at several points within a hydraulic system, and may be permanently affixed to certain components. Maintaining appropriate pressure levels within a hydraulic system is an important factor in consistent operation.

A major drawback of hydraulic systems is the use of petroleum-based fluids, which pose environmental dangers. Research and development efforts search for fluids that are more environmentally friendly but still maintain the robust characteristics of petroleum-based fluids.

7.1.2 Pneumatic System Model

In a typical pneumatic system, the energy source powers a compressor, which forces air into a pressurized storage tank, **Figure 7-3**. The compressor is most often driven by an electric motor or by an internal combustion engine. This storage tank holds the pressurized air and serves as the reservoir for the system. Pneumatic systems are used to power hand tools and to provide power for lifting and clamping during machining operations.

Compressed air must be conditioned to remove contaminants that could damage the system components. ***Conditioning*** is the process of removing contaminants, such as dirt and moisture, from hydraulic fluid and air. Conditioning of the compressed air is accomplished with an air filter that uses a condensation trap and drain. A ***filter*** is a conditioning device that removes very small pieces of debris and is typically made of some porous medium, such as paper, felt, or very fine wire mesh. An oiler adds a fine mist of oil to the compressed air, which provides lubrication throughout the system.

A ***pressure regulator valve*** is a device that allows the air pressure in a pneumatic system to be adjusted to a specific level. Constant pressure must be maintained during system operation. Motor-driven air compressors are designed to

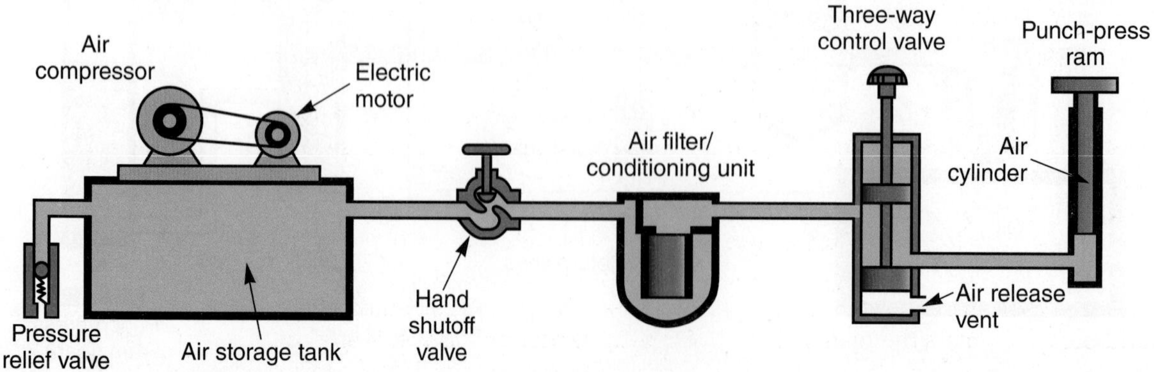

Figure 7-3. This simple pneumatic system powers a punch-press ram actuated by a single-acting cylinder.

start automatically whenever the pressure in the storage tank drops below a certain level.

The transmission path of a pneumatic system consists of piping, tubing, and flexible hoses used as feed lines. Unlike hydraulic systems, pneumatic systems do not use return lines to the storage tank. Return air is simply exhausted into the atmosphere.

A pneumatic system can have several controls throughout the system. Both hand shutoff valves and pressure relief valves control air circulating through the transmission path. Air flow can also be altered by a pressure regulator valve and a three-way valve.

The load device in a pneumatic system changes the mechanical energy of air into linear or rotary motion. An air cylinder, for example, uses linear motion to drive a punch-press ram. The load devices in some pneumatic industrial tools, such as grinders, buffers, drills, and impact wrenches, produce rotary motion.

Pressure indicators are commonly added to monitor tank pressure. The output of pressure regulator valves is also monitored so that exact pressures can be determined. Test indicators are frequently used to troubleshoot faulty components.

7.2 Characteristics of Fluid Flow

In practice, fluid power systems do not achieve 100% power transfer from input to output. The fluid moving along the cylinder walls encounters surface friction or resistance. This is one factor in the decrease of power from input to output. The power loss materializes primarily as heat. Forcing fluid against this resistance develops system pressure. A direct relationship, therefore, exists between the amount of system pressure and surface resistance.

7.2.1 Fluid Pressure

The relationship of resistance and pressure in a static system is illustrated in **Figure 7-4**. The pressure at Point F is zero. A pressure reading of zero could be caused by a break in the system. Other parts of the system show varying amounts of pressure in response to resistance. Point B represents the highest pressure. The full weight of the fluid occurs at this point.

Abrupt changes in the direction of flow create turbulence, which causes pressure drops. ***Turbulence*** is irregular motion or agitation of air or fluids. Conditions such as size and smoothness of internal surfaces, temperature of the fluid, and the location and number of valves and fittings may cause irregular flow characteristics. The corners and components illustrated in **Figure 7-5** are examples of turbulent areas in a system.

Restrictions within the system are also a source of pressure drops. Restrictions can be created by control valves, tubing length, or reduced tubing size. Smaller lines tend to increase the speed of fluid flow, which causes an increase in resistance. A pressure drop is the result of a change in the form of the fluid energy. As fluid pressure enters the system, it has the ability to perform a specific amount of work. However, as fluid passes through the system, the energy changes form. Fluid energy is lost because it is changed into heat due to friction and resistance.

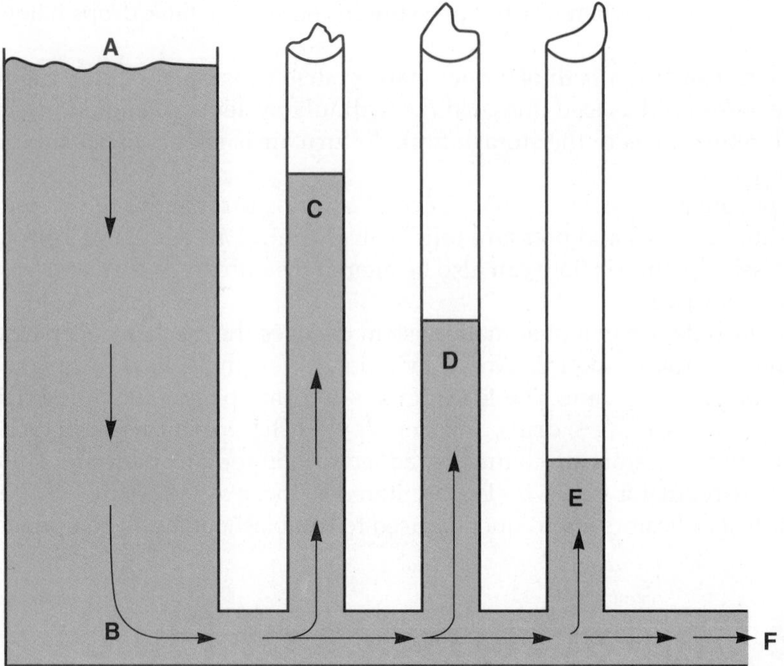

Goodheart-Willcox Publisher

Figure 7-4. This illustration depicts the friction/pressure relationship in a static fluid power system.

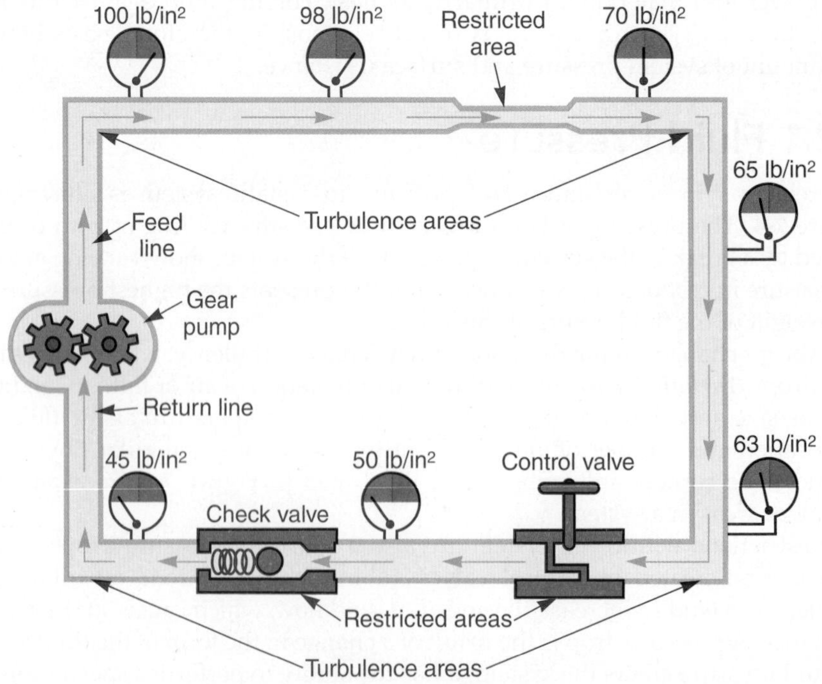

Goodheart-Willcox Publisher

Figure 7-5. In a flowing fluid power system, pressure drops occur either where fluid flow is restricted or where turbulence develops due to a change in the direction of flow.

In a static system, the amount of power loss due to resistance is negligible. In systems that move larger volumes of fluid over longer lines, however, the losses are more considerable. Resistance losses can generally be controlled by reducing the line length, limiting the number of bends, and using appropriately sized lines to prevent high fluid velocity. A properly designed system will help achieve and maintain a high level of efficiency.

When fluid flow ceases, the pressure reaches a stable value throughout the system. A faulty pump or loss of electricity could also cause this to occur. A break in the system line, however, normally causes a complete loss or very pronounced change of pressure.

7.2.2 Fluid Compression

One of the most notable differences between hydraulic and pneumatic systems is the compressibility of the fluids. All gases and liquids are compressible under certain conditions. However, hydraulic fluid (liquid) is considered virtually incompressible, whereas the air used in pneumatic systems is readily compressible. As pressure is applied to a liquid, its volume decreases by 0.5% per 100 psi (6895 kPa) of pressure applied. This change in volume is insignificant under normal operating conditions. Hydraulic fluids exhibit the characteristics of a solid and provide a rigid medium to transfer power through a system.

In pneumatic systems, air must first be compressed before it can be used. When air is compressed, its pressure increases, but its volume decreases. This difference in compressibility results in some differences between hydraulic and pneumatic system components. Typically, however, the basic function of a given component is similar for either system. Air is also affected by temperature. As the temperature increases, so does the volume of the air. Pneumatic systems exhibit spongy and springy movement characteristics due to compressibility of the fluid. Unlike hydraulic systems, pneumatic systems need additional components or considerations to obtain smooth operating movements.

Pneumatic systems offer a number of advantages over hydraulic systems. As the compressor produces pressurized air, it is stored in a tank and released into the system as it is needed. When the task is completed, the air does not need to be reclaimed because it is simply released into the atmosphere.

7.3 Principles of Fluid Power

In 1653, the French scientist Blaise Pascal discovered that pressure applied to a confined fluid is transmitted undiminished throughout the fluid. Also, this pressure acts on all surfaces of the container at right angles to those surfaces. These findings are known as ***Pascal's law*** and are the basis of all fluid power systems. In the illustration shown in **Figure 7-6**, the force applied to piston A creates pressure in the fluid that is instantly transferred to all parts of the cylinder. The pressure exerted on piston B is equal to the pressure originating at piston A. This is true even if both pistons are not the same physical size. The pressure acting on piston B is also applied to the walls of the cylinder. For this reason, the walls of the cylinder must be strong enough to withstand the pressure.

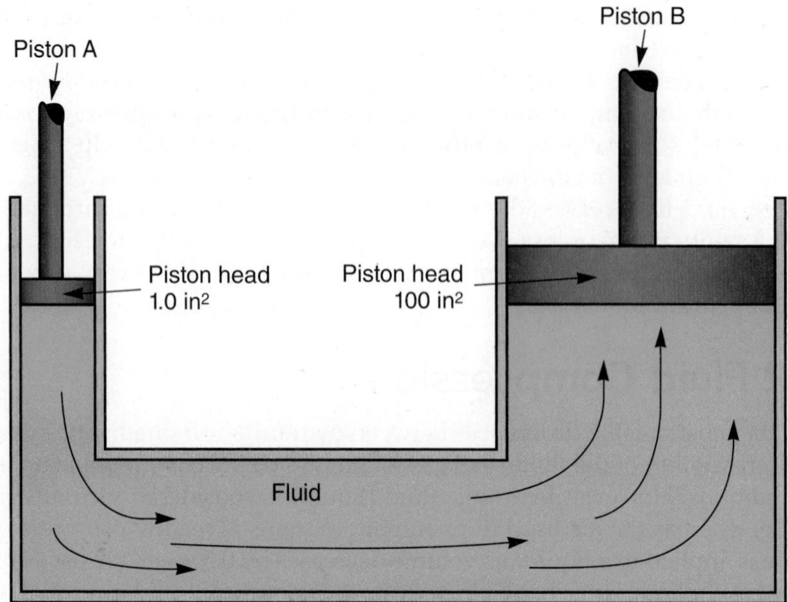

Goodheart-Willcox Publisher

Figure 7-6. This simple example of a hydraulic system illustrates Pascal's law. Pressure exerted by the movement of piston A acts equally on piston B and on all parts of the container.

7.3.1 Force and Pressure

Force is any factor that tends to produce or modify the motion of an object. To move a body or mass, for example, an outside force must be applied to it. The amount of force needed to produce motion is based on the ***inertia*** (resistance to change) of the body to be moved. Force is normally expressed in units of weight.

As stated earlier, weight is the gravitational force exerted on a body by the earth. Since the weight of a body is a force (not a mass), units of force are used to express both weight and force. The basic unit of force in the US customary measurement system is the pound (lb). In the metric system, the basic unit of force is the newton (N).

Pressure is the amount of force applied to a specific area. It is often expressed in pounds per square inch (lb/in^2 or psi) in the US customary system. The pascal (Pa) is the basic unit of pressure in the metric system. Metric pressure measurements are usually expressed in *kilopascals* (kPa); one kPa equals 1000 Pa.

At sea level, the pressure of the atmosphere on the surface of the earth is 14.7 psi (101.3 kPa). In manufacturing, giant hydraulic presses can squeeze metals with a pressure as great as 100,000,000 psi. Force and pressure are simply measurements of effort.

7.3.2 Work and Power

A measure of what a system actually accomplishes is called work. In **Figure 7-6**, work occurs when the force applied to piston A causes it to move a certain distance. Work is expressed in foot-pounds or newton-meters (joules).

A realistic concept of work must take into account the length of time it takes to perform. Power is a measurement that considers the amount of work accomplished in relation to the amount of time taken to perform the work. Horsepower is used to describe mechanical power. One horsepower is equal to moving 33,000 lb a distance of 1′ in 1 minute (or 550 lb a distance of 1′ in 1 second). Motors are rated in horsepower.

In the simple static fluid power system illustrated in **Figure 7-7**, a 100 lb force is applied to piston A, which has an area of 1 in^2. This develops a pressure of 100 psi (690 kPa). The pressure is transferred through the fluid to piston B. The area of piston B is 100 in^2. Since pressure is transferred equally through the fluid to all parts of the cylinder, each square inch of piston B receives 100 lb of force. The result is 10,000 lb (100 psi × 100 in^2) of force applied to piston B.

The distance that piston B moves is directly proportional to the relative area of the two pistons. Moving piston A (1 in^2) 4″ into the cylinder displaces 4 in^3 of fluid. The displaced volume is a product of the area of the piston and the distance it moves. In this example, 1 in^2 × 4″ = 4 in^3 of fluid displacement. Spread over the 100 in^2 surface of piston B, the fluid displacement causes piston B to move only 1/100 of the distance traveled by piston A (1/100 of 4″ = 0.04″ of linear travel). As a result, piston B receives more force because of its size, but travels only a short distance.

The amount of work performed by pistons in a static system demonstrates an unusual relationship. The work done by piston A is 400 in-lb (100 lb × 4″). The amount of work achieved by piston B is also 400 in-lb. This is determined by multiplying the applied force by the distance moved (10,000 lb × 0.04″).

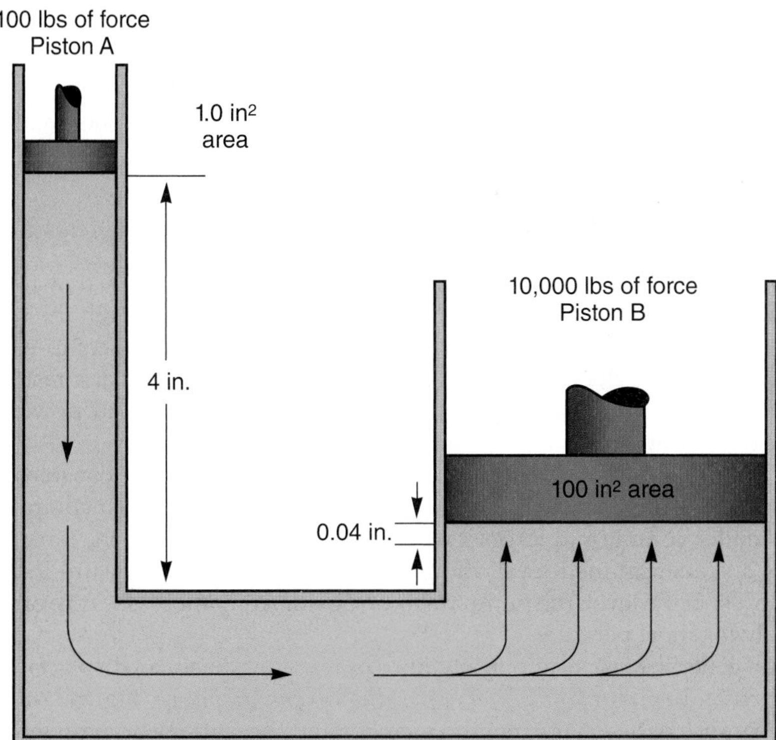

Goodheart-Willcox Publisher

Figure 7-7. In this static fluid power system, movement of the two pistons is proportional to their surface areas.

Robotics Innovations: Security Robot

The company Knightscope Inc. has developed the K5 security robot that can prevent and predict crime. Using an array of data collection sensors, the K5 can send real-time data to a central control center or to a smartphone. The K5 robot is 5′ tall, 3′ wide, and weighs 300 lb, making it an intimidating presence. Monitored in a similar manner as a home security system, this robot can have police, fire, or medical services dispatched where they might be needed. Users of the system can check on their facilities with their phones and monitor video or other sensors at all times. The robot's ability to move to different locations and charge itself make it much more versatile than a typical monitored security system. Marketed for business and industry, the K5 robot is changing the way people think about security.

Photo courtesy of Knightscope Inc.

7.4 Fluid Power System Components

Hydraulic and pneumatic systems use similar, sometimes interchangeable, components. However, hydraulic components are generally larger and more rugged in construction. This is necessary because oil is more dense than air. In addition, hydraulic systems tend to be used at high pressure levels in heavy-duty automated operations. Aside from these differences, hydraulic and pneumatic equipment operate on the same basic principles and function in a similar manner.

7.4.1 Fluid Pumps

The pump is the heart of a fluid system. It provides an appropriate flow to develop pressure. The pump accepts fluid at an inlet port, moves it through the transmission path, and expels it from an outlet port. Gases are typically compressed into a smaller volume, which increases their pressure. Liquids flow at a faster rate to increase the pressure. The specific use of a pump within a fluid power system determines the operation it performs.

Hydraulic pumps operate continuously to keep the fluid in constant motion. A pneumatic pump, on the other hand, operates intermittently. It compresses air into a smaller volume and forces it into a tank for storage. When the tank pressure reaches a predetermined level, the pump turns off. When pressure in the tank drops to a specific level, the pump turns on again. Air compressors are often operated only for short periods.

Two general classifications of fluid pumps are positive displacement and non-positive displacement. A ***positive displacement pump***, **Figure 7-8A**, has a close clearance between the moving member and the stationary components. As a result, a definite amount of fluid passes through the pump during each revolution. With a ***non-positive displacement pump***, **Figure 7-8B**, the fluid is moved by the impeller blades during each rotation. The flow depends on the speed at which the blades are moving. Therefore, the amount of fluid that passes through the pump with each rotation varies.

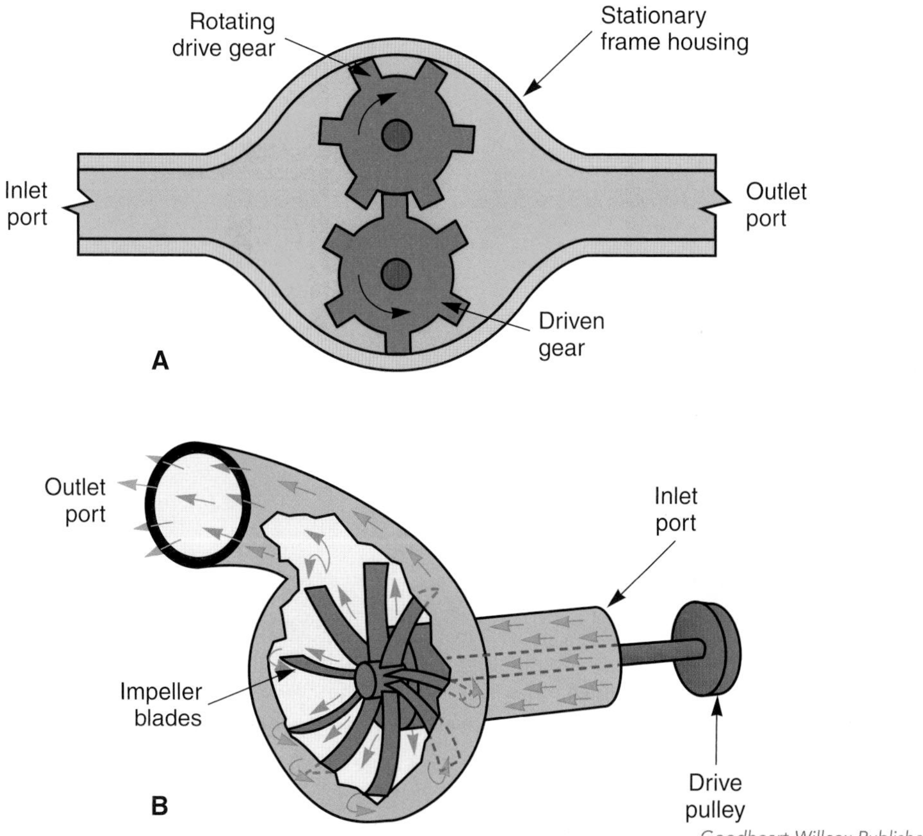

Figure 7-8. Fluid flow in positive and non-positive displacement pumps. A—Positive displacement pumps, like this gear pump, move a specific volume of fluid with each rotation or cycle. B—Non-positive displacement pumps, such as this centrifugal pump, do not move a specific volume of fluid with each rotation. Flow volume depends upon pump speed.

Reciprocating Pumps

Reciprocating pumps are positive displacement pumps that use the reciprocating action (backward and forward motion) of a moving piston to move fluid into and out of a chamber. A partial vacuum is created inside the chamber by the piston as it is pulled to the bottom of its stroke, **Figure 7-9A**. The intake valve opens, admitting fluid (air or liquid) into the chamber. The chamber fills to capacity by the time the piston reaches the end of its stroke. As the piston reaches the bottom, the rotary motion of the drive disk causes the piston to change direction. The discharge valve opens, the intake valve closes, and fluid is forced out of the chamber (**Figure 7-9B**).

For each revolution of the motor shaft, the piston in the reciprocating pump completes both an intake and a discharge stroke. Piston area and chamber volume are the key factors in determining the potential output of this type of pump. In some situations, two or more stages (or cylinders) may be driven by the same motor shaft. Reciprocating pumps are often used to compress air.

Rotary Gear Pumps

Rotary gear pumps are positive displacement pumps that use rotary motion to produce pumping action, **Figure 7-10**. An external-gear rotary pump contains two gears enclosed in a precision-machined housing. Rotary motion from the power

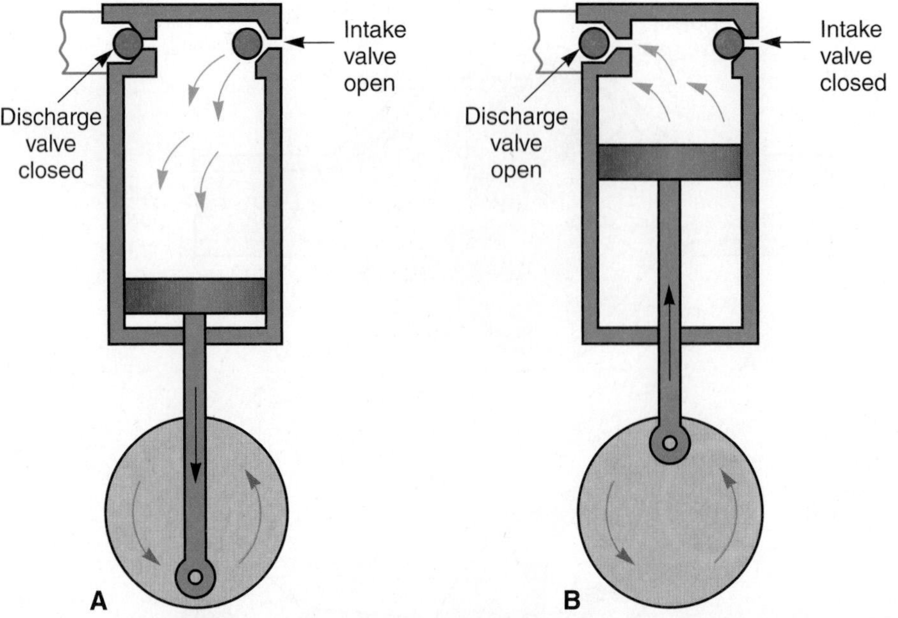

Goodheart-Willcox Publisher

Figure 7-9. Operation of a reciprocating pump. A—During the intake stroke, the intake valve opens and the piston moves down, drawing fluid into the cylinder. B—For the discharge stroke, the intake valve closes, and the discharge valve opens. The piston moves upward to force the fluid out of the cylinder.

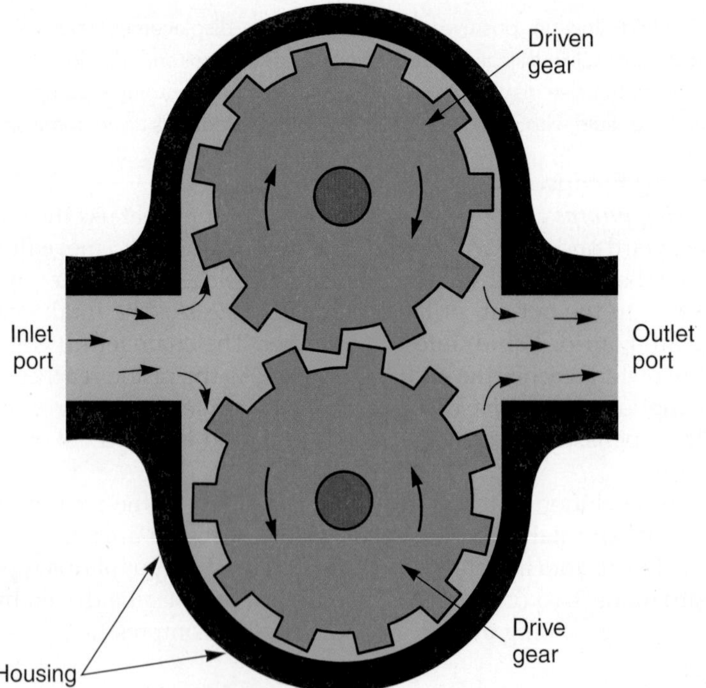

Goodheart-Willcox Publisher

Figure 7-10. Basic construction of an external-gear rotary pump. Spaces between the teeth of the rotating gears carry fluid around the inside of the housing from the inlet port to the outlet port.

source is applied to the drive gear. As it rotates, the drive gear causes the second gear (driven gear) to turn. The teeth of the two gears mesh in the middle of the pump. The rotating gears carry fluid away from the inlet side of the pump and move it to the discharge side. The fluid is ejected through the discharge port. Because the gears mesh tightly, very little fluid returns to the inlet side of the pump.

In an internal-gear rotary pump, one gear rotates within the other (**Figure 7-11**). The inner gear (idler) has fewer teeth than the outer gear (rotor). As the idler gear rotates within the rotor gear (driven gear), the gear teeth separate at the inlet port and mesh at the discharge port. Fluid is drawn through the inlet port, filling the spaces between the teeth. The fluid moves smoothly around the head crescent and is expelled at the discharge port when the teeth mesh. An internal-gear rotary pump operates equally well when rotating in either direction. Output usually ranges from 0.5 to 1100 gpm (gallons per minute).

Rotary Vane Pumps

Rotary vane pumps are positive displacement pumps that use a series of sliding vanes to move fluids. The sliding vanes are placed in slots around the inside of the rotor. As the rotor turns, centrifugal force or spring action forces the vanes outward. These vanes capture the fluid as it passes by the inlet port. As the rotor continues to turn, the fluid is moved to the outlet port.

In an unbalanced (offset) vane rotary pump, **Figure 7-12**, the rotor is offset, and all of the pumping action takes place on one side of the shaft and rotor. As a result, large volumes of liquid can move across the top with little or no return through the bottom. This design causes a side load on the shaft and rotor.

The balanced vane rotary pump is designed with the rotor in the center of its housing. Its elliptical casing forms two separate pumping areas (inlets and outlets) on opposite sides of the rotor and causes the side loads to cancel out. As a result, the flow of air or liquid is smoother than with an unbalanced type of pump.

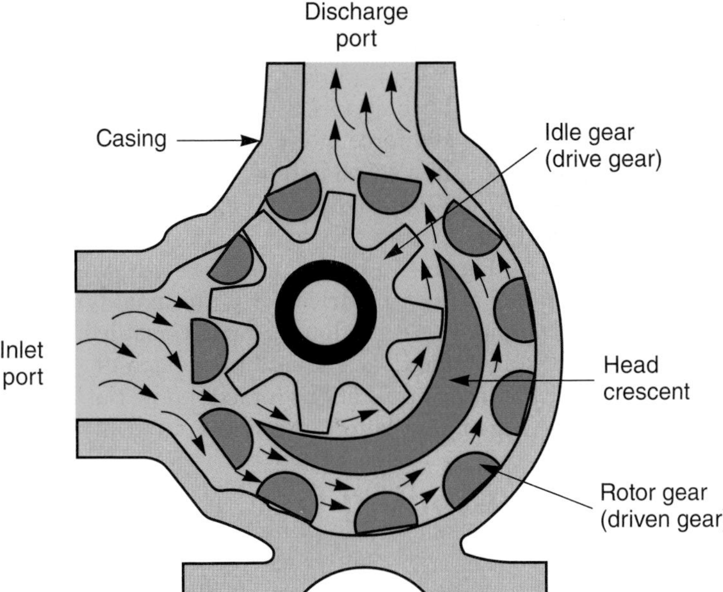

Goodheart-Willcox Publisher

Figure 7-11. Fluid is moved through an internal-gear rotary pump as the gears mesh and separate. Fluid is drawn in at the inlet port, flows around the head crescent, and exits through the discharge port.

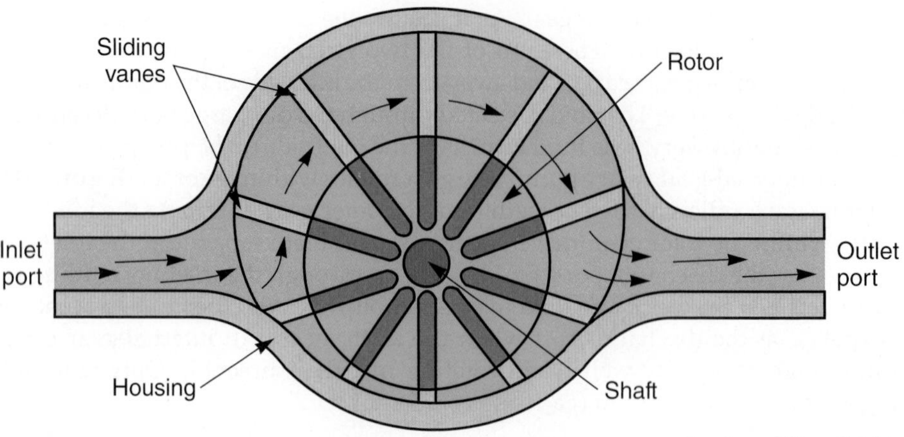

Goodheart-Willcox Publisher

Figure 7-12. An unbalanced straight-vane rotary pump moves a large volume of fluid across the top of the chamber.

Centrifugal Pumps

The ***centrifugal pump*** is a non-positive displacement pump that moves a varying amount of fluid with each rotation using an impeller blade. The clearance between the impeller blade and the housing allows fluid from the inlet port to remain in the pump, even though the impeller blade is rotating. Therefore, the amount of fluid that leaves through the outlet port is not directly related to the amount of fluid that enters through the input port.

Volume depends on the rotational speed of the pump and the amount of resistance in the feed line connected to the outlet port. Increased resistance may cause fluid flow to slow or even come to a complete stop. In this case, any fluid in the pump simply rotates inside without being expelled. When this occurs, the operating efficiency of the pump drops to zero. Increasing the pump speed can solve this problem. As a general rule, centrifugal pumps are only used for transferring large amounts of fluid at low pressure.

A volute pump is a centrifugal pump with a spiral-shaped housing, **Figure 7-13A**. When fluid enters the inlet port, the revolving blades of the impeller cause rotational motion. Centrifugal force moves the fluid out toward the wall of the housing. The fluid circulates in a spiral path to the outlet port. The inlet port continually replaces the fluid expelled from the outlet port to maintain continuous fluid flow.

In an axial-flow pump, **Figure 7-13B**, the blades of the impeller maintain fluid flow along the rotational axis of the drive shaft. The axial-flow pump is also referred to as a propeller pump. The axial-flow pump is versatile because the flow and pressure amounts can be altered by changing the pitch of the propeller to meet various conditions. Common applications for axial-flow pumps include irrigation, drainage, fluid mixing or circulation, and sump pumps.

7.4.2 Fluid Conditioning Devices

Both hydraulic fluid and air must be conditioned before being processed through a fluid power system. Conditioning devices prolong the life of fluid power systems by removing foreign particles and moisture. The number of conditioning devices used depends on the system. A simple hydraulic system may use only a line filter or strainer. A ***strainer*** is an in-line device that captures large particles of

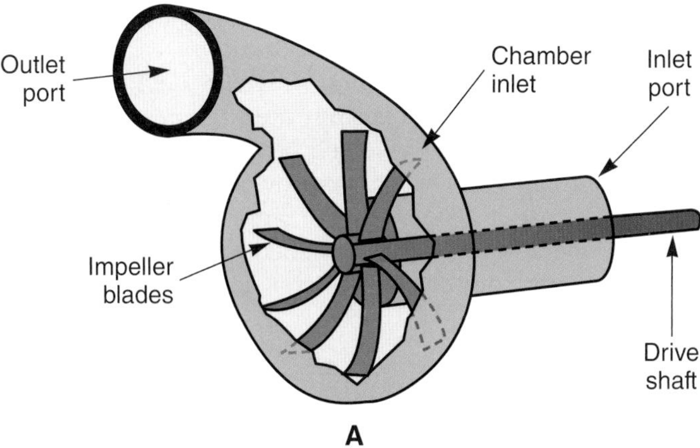

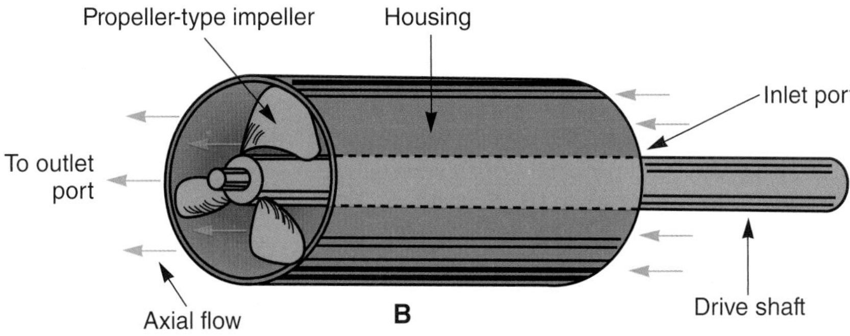

Goodheart-Willcox Publisher

Figure 7-13. Two types of centrifugal pumps. A—A volute pump has a spiral housing that directs the flow of fluid to the outlet port. B—An axial-flow pump has an impeller that moves fluid along the axis of the drive shaft.

foreign matter within hydraulic systems. More sophisticated systems may require filters, strainers, and devices that cool the fluid. Conditioning in pneumatic systems is more complex. Devices in these systems filter the air to remove both dirt and water, regulate the pressure level, and add oil as a lubricant.

Hydraulic Conditioning

The number of components, types of control devices, and operating environment are major considerations in hydraulic fluid conditioning. For systems with manually operated control valves used in a clean environment, a simple intake strainer may be enough. However, systems with precision control valves that operate for long hours in a dirty environment may require micron filters and several strainers.

Strainers can capture large particles of foreign matter. They contain a stainless steel screen with 60 to 200 wires per square inch. Strainers are frequently placed in the filler opening of a reservoir, an air breather, and a pump inlet feed line.

Filters remove very small pieces of debris and provide a finer grade of fluid conditioning than strainers. Filters are typically made of some porous medium, such as paper, felt, or very fine wire mesh. Filter ratings range from 1 to 40 microns. The micron rating refers to the size of particles that can pass through the filter.

In-line and T filters are commonly used in hydraulic systems. T filters have a removable bowl or shell that contains the filtering element. A bypass relief valve opens when the filtering element becomes clogged and restricts flow. In-line filters must be removed when the element is cleaned or replaced. Bypass relief valves can be used with this type of filter, depending on its application.

Some hydraulic systems use ***heat exchangers,*** which are devices that cool the fluid in a hydraulic system to maintain a constant fluid temperature. Hydraulic machinery that operates near a furnace or is used near hot metal often requires heat exchangers to cool the fluid. Heat exchanger devices include forced-air fan units, water-jacket coolers, or gaseous cooling units. Since a hydraulic system produces heat during normal operation, cooling is needed more often than heating. Heating the fluid is required only for portable systems during cold starts.

Pneumatic Conditioning

To condition fluid in pneumatic systems, several different types of devices are used. Filtering must remove moisture as well as foreign particles. In-line and T filters contain chemical elements made of a ***desiccant,*** which is a very dry substance designed to attract moisture. Desiccant elements often require periodic recharging, a heating process that dries the element. Some T filters contain a glass-bowl moisture trap at the bottom of the filter. A drain valve allows accumulated moisture to drain from the trap.

Pressure regulation is also necessary in pneumatic systems. After air passes through a T filter, it moves to a pressure regulator valve. The movement of air through this valve can be controlled with an adjustment screw. The line pressure from the storage tank can be set to the desired operating level.

The pressure regulator valve creates a balance between atmospheric pressure and system line pressure. Atmospheric pressure reaches the top of the diaphragm through the vent, **Figure 7-14A**. System pressure is applied to the bottom of the diaphragm. Turning the adjustment screw adds mechanical pressure to the

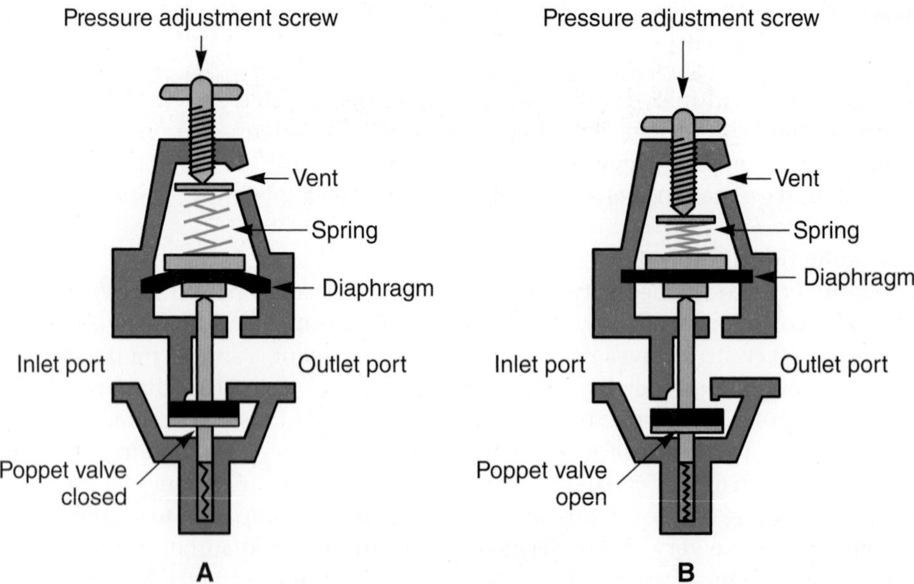

Figure 7-14. Operation of a pneumatic system pressure regulator valve. A—When system pressure exceeds atmospheric pressure, the diaphragm is pushed up, allowing the spring-loaded poppet valve to close. B—If system pressure drops below atmospheric pressure, the diaphragm is pushed down, opening the poppet valve to admit more air from the receiving tank.

atmospheric pressure. When this combined pressure exceeds the system pressure, the diaphragm is pushed down. This action opens the poppet valve and allows air from the storage tank to enter the system lines, When the system pressure becomes greater than the combined mechanical and atmospheric pressure, the diaphragm is pushed up. This action closes the poppet valve to maintain pressure at the set level, **Figure 7-14B**. The pressure adjustment screw can be used to alter the pressure at which the valve opens. Regulators may be used in several places within a system.

Lubricators are conditioning devices that add a small quantity of oil to the air after it leaves the regulator, **Figure 7-15**. This lubrication helps the valves and cylinders in pneumatic systems operate more efficiently and last longer. In many pneumatic systems, an ***FRL unit*** combines the air filter, regulator, and lubricator components, **Figure 7-16**.

When air enters at the inlet port, it flows into the narrowed area, called the venturi, causing the air flow velocity to increase and the pressure to decrease. Pressure in the venturi is lower than in larger areas of the pump. As a result, oil

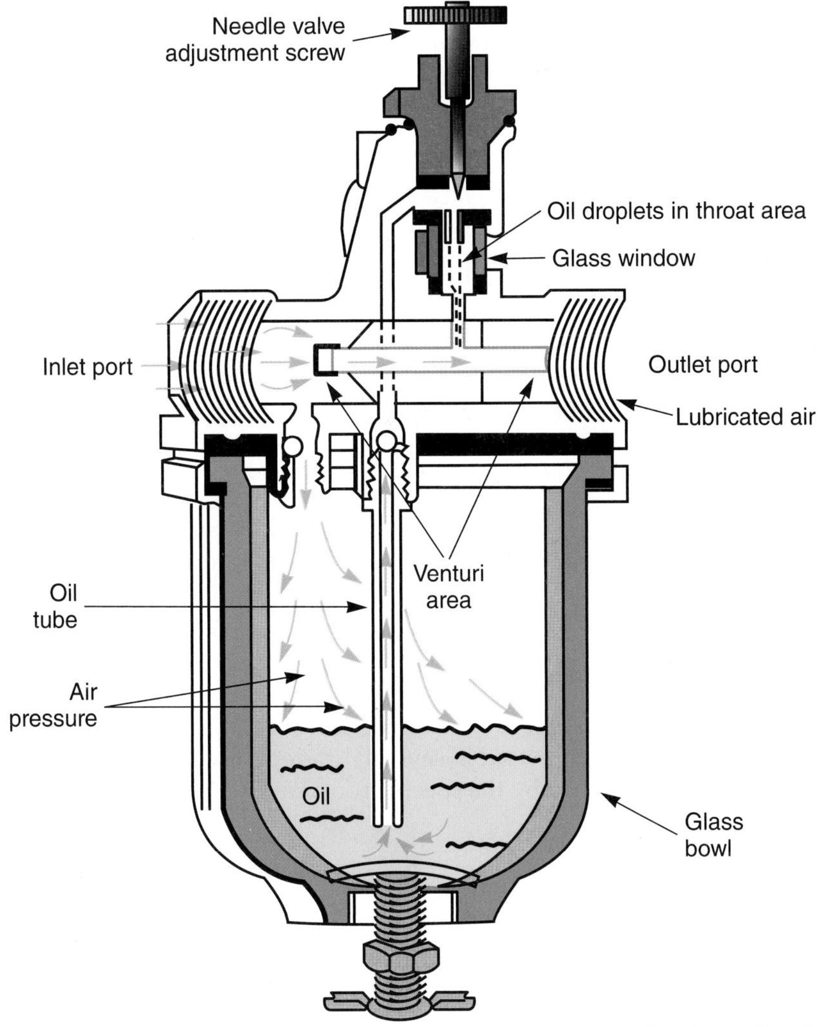

Goodheart-Willcox Publisher

Figure 7-15. A pneumatic lubricator atomizes drops of oil to create a fine mist that mixes with the compressed air supply.

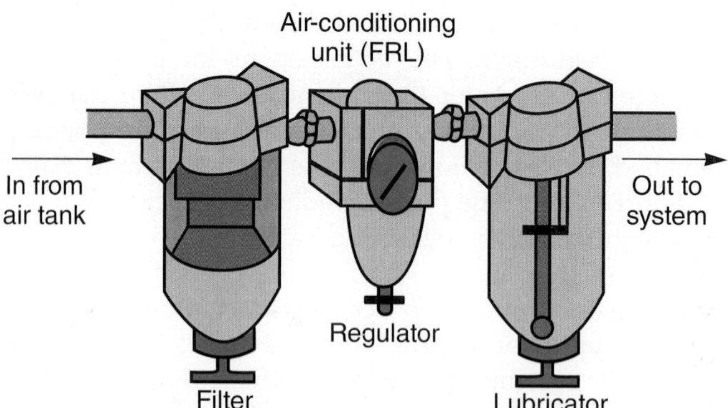

Goodheart-Willcox Publisher

Figure 7-16. This pneumatic fluid power system has an FRL unit that combines the filtering, regulating, and lubricating devices.

is forced from the glass bowl up the oil tube and moves to the top of the unit. The needle valve can be adjusted to regulate the oil flow so that small droplets fall into the throat. Air velocity at the bottom of the throat breaks the droplets into a fine mist that mixes with the air. Finally, the lubricated air passes through the outlet port.

7.4.3 Transmission Lines

Transmission lines can be made of rigid metal piping, flexible metal tubing, or flexible hose. Rigid lines are used for permanent installations where no vibration occurs and are capable of withstanding a good deal of abuse. Rigid lines are economical, but they require additional fittings that add weight to a system.

Steel tubing is used extensively for transmission lines in hydraulic systems. It is lighter than rigid steel piping and can be bent into various shapes, which reduces the number of fittings required. Steel tubing is also more tolerant of vibration, which makes it well suited for automotive and aerospace applications.

Flexible transmission lines are made in a variety of types and sizes. The type of line used depends on the system type and its function. The type of tube (the inner lining reinforcement material and outside cover material) used affects the system pressure limits, the temperature operating range, and the resistance to exposure.

7.4.4 Control Devices

Control is achieved by devices that alter the pressure, direction, and volume of fluid flow. Control devices are used in several different places within a system; the actual location is determined by the specific function of each control device.

Pressure Control

Pressure control functions include relief, reduction, bypass, sequencing, and counterbalancing. Pressure relief valves in hydraulic systems unload the output of a positive displacement pump back into the reservoir when the pressure rises to a dangerous level. In this case, the pressure relief valve serves as a safety device.

In pneumatic systems, pressure relief valves are used to control small amounts of air. Excess air in the system is released into the atmosphere. The output port of a relief valve may be altered in size to maintain the pressure requirements of the system.

Pressure control can also be used to establish operating sequences. Pressure relief valves can direct pressure at specified levels in a predetermined sequence. When the main system pressure overcomes the valve setting, pressure shifts to a different port. A sequencing valve can be used to force two actuators to be operated in sequence. When fluid flow to the first cylinder reaches a preset pressure, the valve allows fluid flow into a branch circuit that extends the second cylinder.

Direction Control

Direction control devices are used to start, stop, or reverse fluid flow without causing a significant change in pressure or flow rate. One-, two-, three-, and four-way valves are commonly used. These valves may be actuated by pressure, mechanical energy, electricity, or manual operation. The control action of a directional valve occurs in different ways.

One-way valves, also called check valves, operate on the seated-ball principle. They permit flow in only one direction. Pressure applied at the inlet port drives the ball away from its seat. This opens the flow path and allows fluid to pass through the valve, **Figure 7-17A**. Pressure applied at the outlet port forces the ball into its seat and prevents flow, **Figure 7-17B**. Valves of this type are often used to permit free flow around controls when the flow direction is reversed.

Two-way valves are installed in transmission lines to permit flow or to shut it off. These valves use gates, plugs, discs, spools, or other precision-machined objects to enable or block fluid flow. For example, the ball in a ball-type control valve can be rotated manually using an outside handle. Flow is enabled when the handle is positioned parallel with the transmission line. Turning the handle 90° stops flow. This type of valve is used primarily for high-pressure applications.

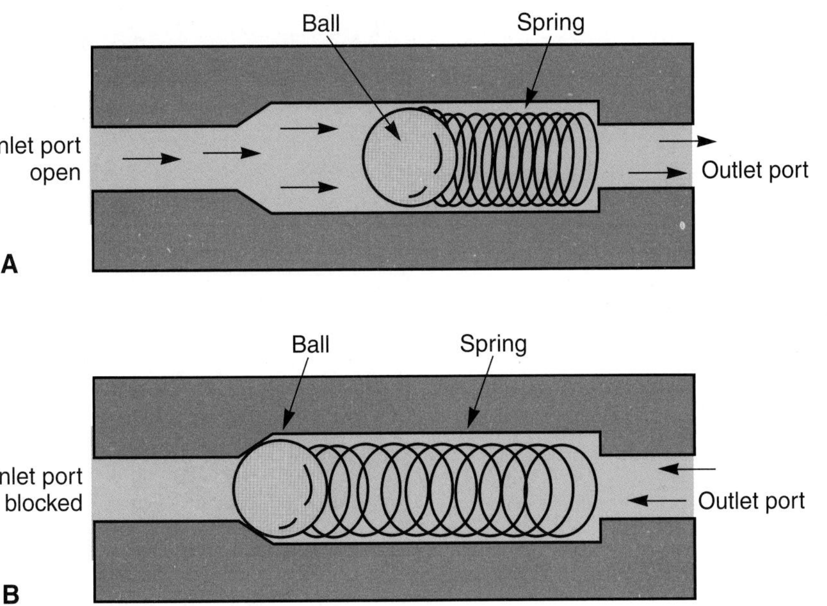

Goodheart-Willcox Publisher

Figure 7-17. A check valve permits flow in only one direction. A—Fluid enters the inlet port and pushes the spring-loaded check ball to allow free flow through the valve. B—Increased pressure from the outlet side pushes the check ball into the closed position, preventing backward flow through the valve.

Three-way valves are used to allow shifting between two different sources of pressure or to direct pressure to alternate devices. These valves are often used to alter cylinder operation or to control hydraulic or pneumatic motors. Valves of this type can be actuated mechanically, manually, or electrically to change hydraulic pressure. Basic designs include shifting spool valves, poppet valves, sliding-plate shear-seal valves, and rotary-plate shear-seal valves (**Figure 7-18**).

Figure 7-19 shows the basic operation of a three-way shifting spool valve. When the manual control shaft is pushed to the right, the spools also shift to the right. Fluid flows through the pressure port and the actuating port, applying pressure to the actuating device controlled by the valve. When the shaft is pulled to the left, the spools shift to the left. This action cuts pressure flow to the actuator and releases it either through the exhaust port or to the tank. Depending on the situation, pressure is directed to the storage tank or vented to the atmosphere through a relief valve. Generally, three-way valves are designed for two-position operation. Some three-way valves have a third position (neutral or off), which increases control capabilities.

Four-way valves are used to start, stop, or reverse the direction of flow, **Figure 7-20**. The valves can be regulated mechanically, manually, or electrically. They are used to control forward and reverse actuation of a double-acting cylinder

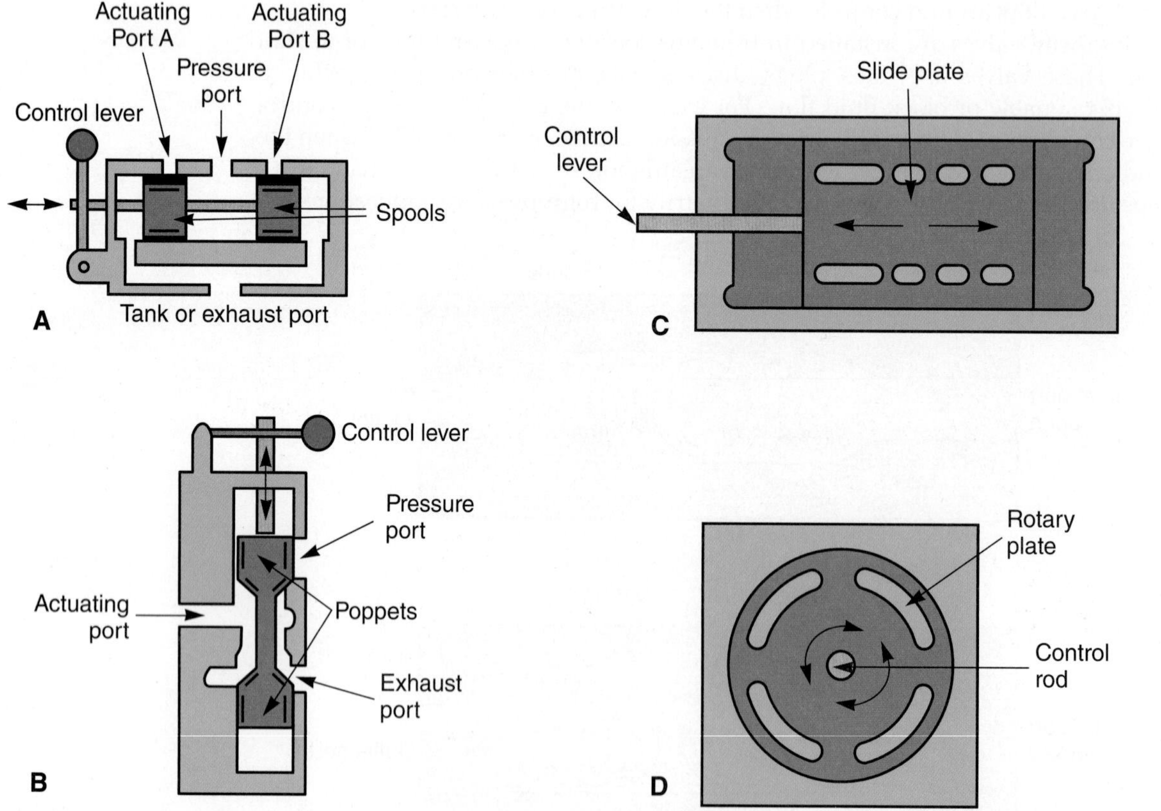

Goodheart-Willcox Publisher

Figure 7-18. Basic types of three-way valves used for fluid control. A—Shifting spool valve. B—Poppet valve. C—Sliding-plate shear-seal valve. D—Rotary-plate shear-seal valve.

or to reverse the rotation of a fluid motor. A ***fluid motor*** is a device that converts the force of a moving fluid into rotary motion through the use of vanes, gears, or pistons.

Flow Control

Flow control devices alter the volume or flow rate of the fluid. The rate at which fluid is delivered to the load of a system determines its operational speed. The speed of an air motor, for example, is dependent on the flow rate of the fluid. By altering the flow rate, the speed of the motor can be controlled.

Cylinder actuating speed is also controlled by the rate of fluid flow. To alter the linear motion of a cylinder, fluid may be controlled at the input feed line or at the return line. Controlling the rate of flow is also known as ***metering***.

The operation of a flow control valve is illustrated in **Figure 7-21**. Fluid flow is controlled from the pressure-flow connection to the free-flow connection. The flow level from pressure flow to free flow is adjusted by a needle valve. Flow from the free-flow connection to the pressure-flow connection forces the check valve ball to move away from its seat, which allows free (uncontrolled) fluid flow.

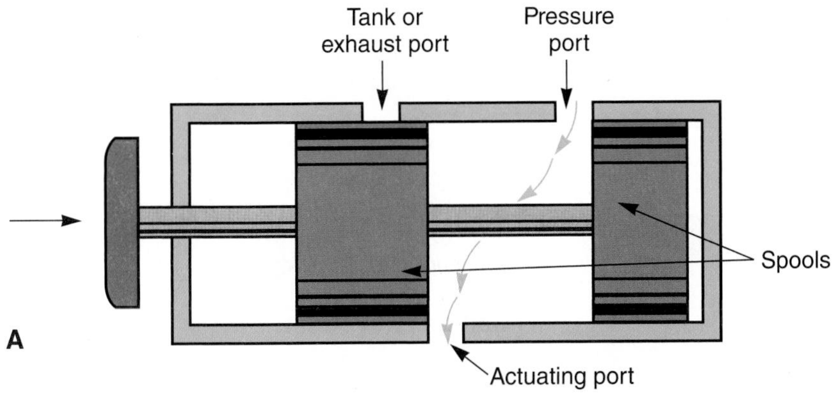

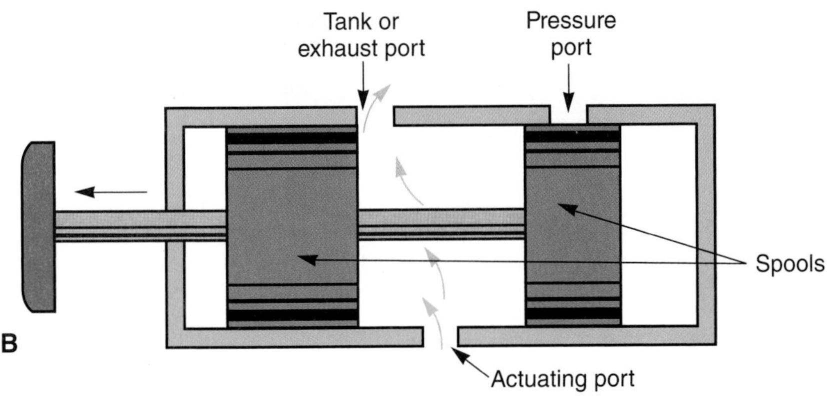

Goodheart-Willcox Publisher

Figure 7-19. A spool-type three-way valve. A—When the spools are moved to the right, pressure flows to the actuating port. B—When the spools move to the left, pressure is directed to an air tank or exhausted to the atmosphere.

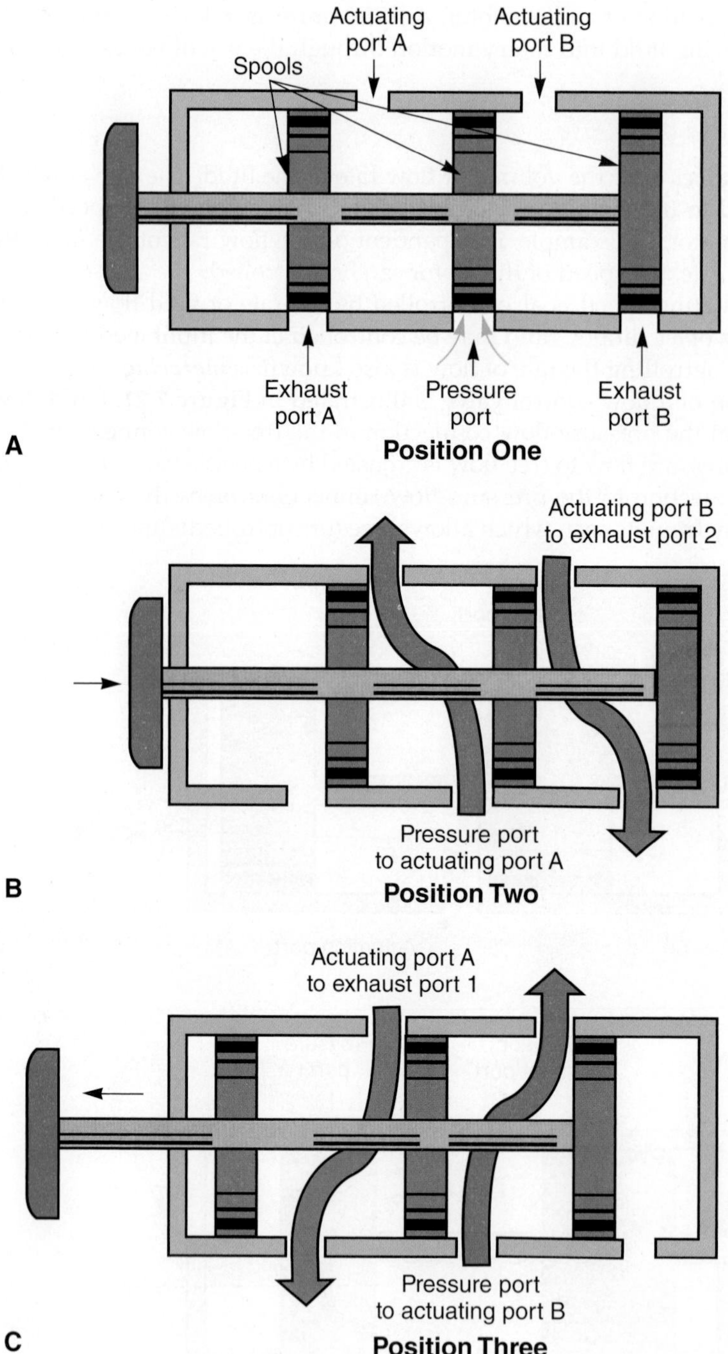

Goodheart-Willcox Publisher

Figure 7-20. In its simplest form, a four-way valve has five working connections: a pressure inlet port, two actuator ports, and two exhaust ports. A—In position 1, the feed line to the pressure port is off. B—In position 2, note the direction of flow from the pressure port to actuating port A. Exhaust from actuating port B is released through exhaust port 2. This moves the piston in a double-acting cylinder in one direction. C—In position 3, note the flow from the pressure port to actuating port B. Exhaust from actuating port A is released through exhaust port 1. This reverses the flow to the cylinder, causing the piston to move in the opposite direction.

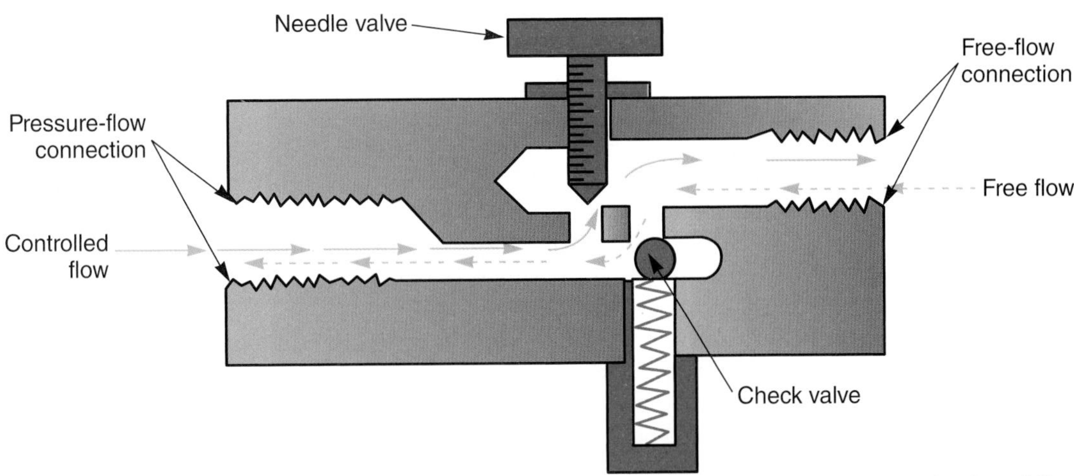

Goodheart-Willcox Publisher

Figure 7-21. In this flow control valve example, fluid flow is adjusted either by the needle (controlled flow) or the seated ball (free flow).

7.4.5 Load Devices

The term *actuator* is often used to identify the load device. Fluid power systems produce work in the form of mechanical motion, provided by either linear or rotary motion. Similar operating principles apply to both hydraulic and pneumatic actuators. Linear actuators, rotary actuators, and fluid motors are three types of load devices.

Linear Actuators

Cylinders used as linear actuators develop the force needed to lift, compress, hold, or position objects (**Figure 7-22**). To produce linear motion, hydraulic fluid or air is forced into a cylinder under pressure. A piston inside the cylinder moves as

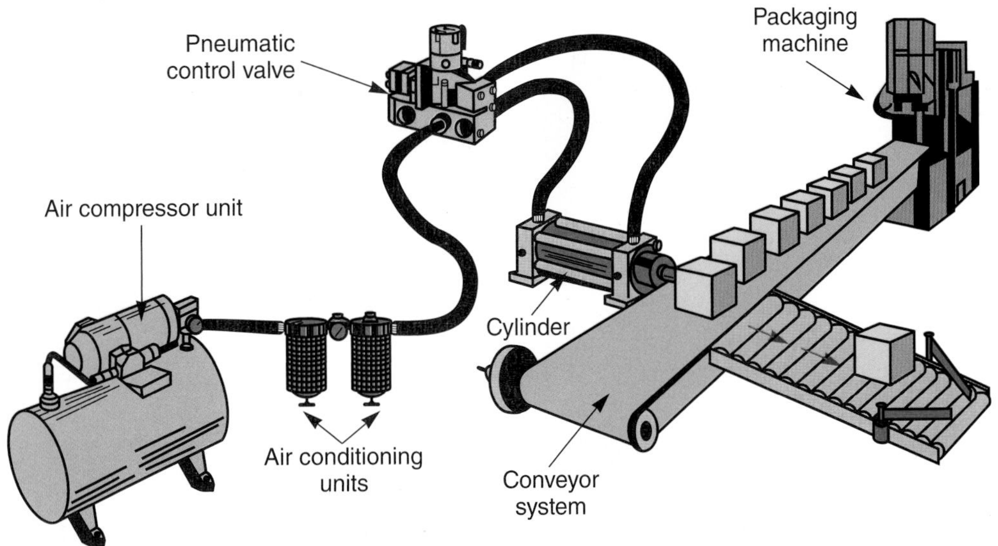

Goodheart-Willcox Publisher

Figure 7-22. In this pneumatic system, the cylinder is the load device and produces linear motion to shift packages from one conveyor to another.

pressure is applied to it. The amount of force created depends on the total area of the piston (in^2 or cm^2) and the amount of fluid pressure (psi or kPa) applied.

Single-acting cylinders have only one input port. The piston moves when fluid is forced into the input port under pressure. A weight (the load) is lifted by the force exerted on the piston by the fluid, **Figure 7-23A**. The combined load includes the weight being lifted, the friction between the piston and cylinder walls, and the heat developed by that friction. To return the piston to its original position, the pressurized flow of fluid is stopped by closing a valve. The fluid under the piston is released by opening the valve's exhaust port. The weight of the load causes the piston to retract, **Figure 7-23B**.

Double-acting cylinders can move in two directions. Two ports are needed because fluid is supplied to one port and is simultaneously expelled from the other. Retracting the piston is accomplished by reversing the fluid flow. Applications for double-acting cylinders include punch presses, rolling mills, machine-tool clamps, paper cutters, and robot actuators.

In **Figure 7-24**, fluid is applied to the right side of the piston and removed from the left side. This forces the piston to move to the left. Switching the fluid flow causes the piston to move to the right. The double-acting cylinder in the illustration is a differential type. The retracting force is somewhat less than the extending force. This is because the area of the piston on the retraction side is reduced by the connection of the piston rod. A non-differential cylinder has rods extending from both ends of the piston. Cylinders of this type can provide an equal force in either direction.

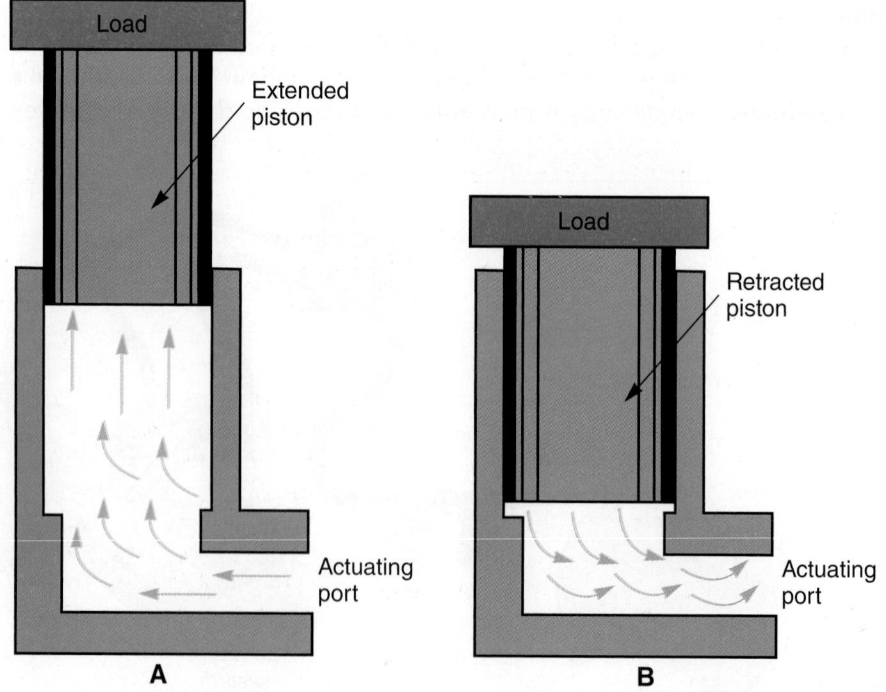

Goodheart-Willcox Publisher

Figure 7-23. Operation of a single-acting cylinder. A—Fluid flowing under pressure through the actuating port forces the piston and its load up. B—The weight of the piston and load causes the piston to retract when pressure is released.

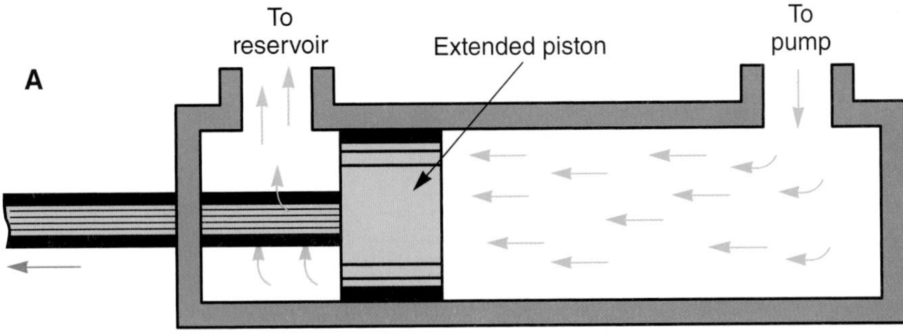

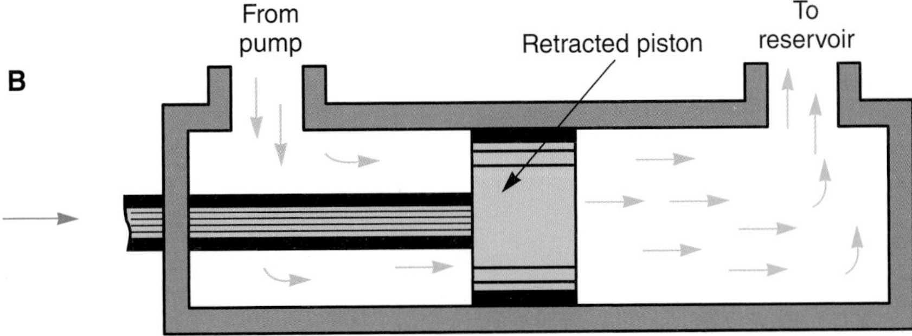

Goodheart-Willcox Publisher

Figure 7-24. Operation of a double-acting cylinder. A—Fluid under pressure from the pump extends the piston. Fluid behind the piston is displaced and flows to the reservoir. B—Reversing fluid flow through the ports retracts the piston.

Rotary Actuators

Rotary actuators produce a limited amount of rotary motion (twisting or turning) in either direction. The rotary actuator shown in **Figure 7-25** applies fluid to port A, which causes the rotor to move in a clockwise direction. Counterclockwise rotation occurs when fluid enters port B and is expelled from port A. A single-vane rotary actuator can turn in either direction through an arc of approximately 280°. A double-vane rotary actuator has twice the turning power, but can only turn approximately 100° in either direction. Actuators of this type are used for lifting or lowering, opening or closing, and indexing operations. One example of this application is the reciprocating operation of a punch press.

Fluid Motors

Fluid motors convert the force of a moving fluid into rotary motion through the use of vanes, gears, or pistons. Fluid motors are classified according to the type of fluid displacement they use. Gear, vane, and piston motors usually have a fixed displacement. They accept and move only a certain amount of fluid with each revolution or cycle. The operating speed depends entirely on the amount of fluid supplied by the source. In variable displacement motors, the amount of fluid circulated can be changed. The piston motor is a variable displacement motor. The length of its stroke is altered to produce variable displacement. Its speed can be changed by an external adjustment. Operating speeds of up to 3000 rpm are typical.

The turning capability of a fluid motor is a measure of torque. The torque is equal to the developed force multiplied by the radius of the rotating arm. The output power of a fluid motor is expressed as a horsepower rating.

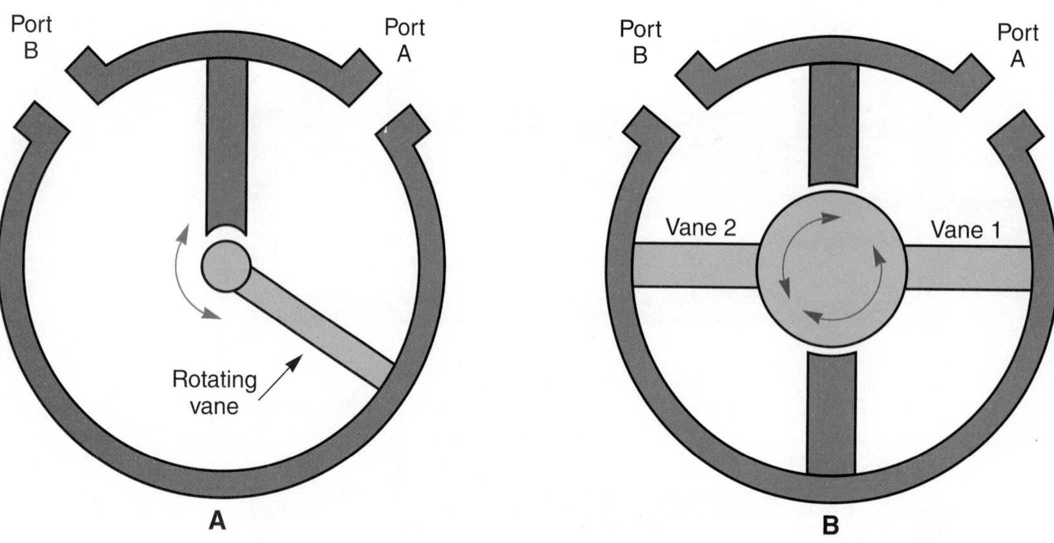

Goodheart-Willcox Publisher

Figure 7-25. Single- and double-vane rotary actuators. A—Single-vane rotary actuators can rotate through a large arc. B—Double-vane rotary actuators have twice the turning power, but permit less movement in either direction.

Gear pumps can be used interchangeably as fluid motors. They are capable of operating at speeds up to 5000 rpm. Both internal and external gear pumps are currently available.

7.4.6 Indicators

The most essential measurement in a fluid system is pressure. Pressure indicators measure and display the pressure at a specific point in the system. Regulators and pneumatic receiver tanks often use permanent pressure indicators or gauges. A wide range of pressures must be measured. Negative pressures (vacuums) as low as 0.00002 psi (0.14 Pa) and positive pressures as high as 1,000,000 psi (6895 MPa) must be measured. Reading such a wide range of pressures requires a number of different devices.

In pressure indicators, an element physically changes shape to show pressure changes. For example, spiral and helix coil elements uncoil when pressure is applied. The Bourdon tube element tends to straighten. The physical change causes movement of an indicator on a scale or a stylus on a paper chart.

Fluid systems also require measurements of temperature and flow. Temperature indicators display the temperature of the fluid at a specific point in the system. ***Flow indicators*** are primarily used to test flow rates from pumps and at the inlet and outlet ports of actuators. By monitoring flow rates, system efficiency is measured and maintenance problems are reduced. Flow and temperature readings are only referenced periodically, so these indicators are not usually permanently installed.

7.5 Hybrid Systems

A number of industrial systems produce mechanical energy by combining fluid power and electrical power systems. Hybrid systems of this type play an important role in automated manufacturing, including robotics. A person working with automated manufacturing systems must be familiar with both fluid and electrical system basics.

An example of a hybrid fluid power system is the hoist used to lift cars in service stations, **Figure 7-26**. Both pneumatic and hydraulic systems produce the power needed to lift an automobile. Adding air under pressure to the top of a long cylinder within an oil-filled tube forces the cylinder to move upward. The tube and cylinder are normally placed in the floor so that the entire unit retracts when air pressure is removed.

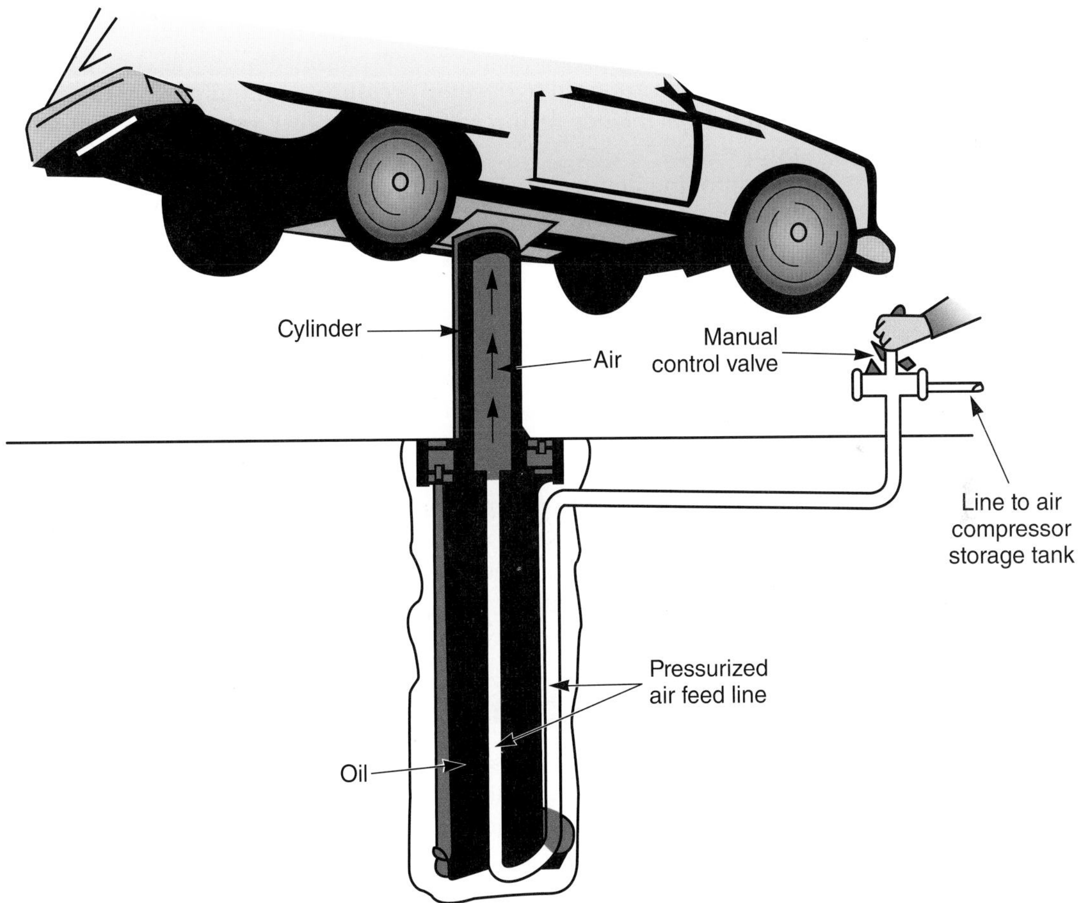

Figure 7-26. A hybrid (air and oil) fluid power system drives this automobile hoist.

Summary

- Fluid power systems use air, liquid, or a combination of both to transfer power. Hydraulic systems use oil or another liquid, while pneumatic systems operate with air.
- A pressure relief valve is used to protect a hydraulic system from damage due to excess pressure. In a pneumatic system, pressure regulator valves are used to adjust pressure.
- In a fluid power system, pressure drops where fluid flow is restricted or turbulence develops.
- Pascal's law, the basis of all fluid power systems, states that pressure applied to a confined fluid is transmitted undiminished throughout the fluid path.
- Force tends to modify the motion of an object. Pressure is the amount of force applied to a specific area.
- A measure of what a system actually accomplishes is called work. Power is the amount of work accomplished in relation to the amount of time taken to perform the work.
- Fluid pumps are used to develop pressure of a fluid power system. Types of fluid pumps include reciprocating, rotary gear, rotary vane, and centrifugal pumps.
- Conditioning devices, such as strainers and filters, are used to prolong the life of fluid power systems.
- Transmission lines of fluid power systems may be either rigid or flexible tubing connected by fittings.
- Control in a fluid power system alters the pressure, direction, or volume of fluid flow. Pressure relief valves, direction control valves, and flow control devices are used.
- Load devices produce work in the form of linear or rotary motion in fluid power systems. Linear actuators, rotary actuators, and fluid motors are types of load devices.
- Indicators are used to measure pressure, flow, and temperature in fluid systems.
- Hybrid systems produce mechanical energy by combining fluid power and electrical power systems.

Review Questions

Answer the following questions using the information provided in this chapter.

1. Briefly describe the characteristics of a hydraulic system.
2. Briefly describe the characteristics of a pneumatic system. What are two applications of a pneumatic system?
3. Define the term *pressure* and explain the function of a pressure regulator valve.
4. What is one reason that fluid power systems do not achieve 100% power transfer from input to output?

5. What are some factors that may cause turbulence in fluid power systems?
6. Describe the differences between hydraulic and pneumatic systems related to the compressibility of the fluids.
7. Explain Pascal's law, which is the basis of all fluid power systems.
8. Define the following terms: *force*, *work*, and *power*.
9. List four types of fluid pumps. Provide a brief description of how each functions.
10. Explain the purpose of fluid conditioning in a hydraulic system. Give examples of conditioning devices used in hydraulic systems.
11. What conditioning devices are used in a pneumatic system?
12. What are some materials used to make transmission lines for a fluid power system?
13. What are some types of control devices used with fluid power systems?
14. How is direction control accomplished in a fluid power system?
15. What are three types of load devices used with robotic systems? Briefly describe the motion each type produces.
16. List some types of indicators used with fluid power systems and briefly describe the use of each.
17. What types of power systems may be combined to form a hybrid system? What is one example of an application in which hybrid systems are used?

Learning Extensions

1. Visit the National Fluid Power Association website. Search the site for information about the fluid power industry and standards development. List three standards that relate to topics presented in this chapter.
2. Visit the Fluid Power Safety Institute website. Review several of the safety products and write a brief summary of the use and features of one of the safety products.
3. Solve the following problem based on the variables given. A cylinder with a 2 in^2 piston head is used to raise a 1000 lb load. What pressure is required to accomplish this task?
4. Solve the following problem based on the variables given. What is the force that can be lifted using a 3 in^2 piston head with a pressure of 500 psi?

Advanced Analysis

1. Prepare a discussion in which you compare and contrast the relationship of pressure and flow in hydraulic and pneumatic systems.
2. Prepare a discussion summarizing pressure, direction, and flow control in a fluid power system.

Activity 7-1
Force through Liquids

Objectives

One of the most significant functions of a hydraulic power system is the transmission of force through a liquid. According to Pascal's law, pressure applied to a confined fluid is transmitted equally through the fluid in all directions, and it is applied at right angles to the containing surfaces. Fluid does not have a specific shape, which allows it to conform to the internal features of the container. Liquid also has poor compressibility, which means it tends to resist compression when force is applied. These ideas are extremely important characteristics of an operating fluid power system.

In this activity, two hydraulic cylinders are connected to form a simple closed-loop system. Through this system you will be able to observe how force is transmitted through a liquid. You will also test the compressibility of a hydraulic fluid.

Equipment and Materials

- Hydraulic fluid or other fluid.
- Double-acting cylinders—1 1/8″ bore (2).
- Assortment of quick-disconnect transparent tubing pieces.
- Safety glasses or goggles.

Safety

Hydraulic power systems operate at high pressure. Always wear protective eyewear and check all fittings before starting any hydraulic system.

Procedure

1. Attach a length of flexible tubing to the port opposite the piston end of one of the 1 1/8″ bore double-acting hydraulic cylinders. Position the unit horizontally with the ram or piston retracted into the cylinder.
2. Place the unattached end of the flexible tubing into a cup of hydraulic fluid. Manually force the piston in and out of the cylinder several times to charge it with oil and clear it of air.
3. Pull the piston all the way out, fully charging it with oil. Set it aside momentarily.
4. Attach a length of flexible tubing to the port nearest the piston end of the second 1 1/8″ bore, double-acting hydraulic cylinder.
5. Using the same procedure outlined in Steps 2 and 3, clear the cylinder of air and charge it with oil. The piston should ultimately be fully retracted when it is charged.

6. Connect the unattached ends of the flexible tubing to the other cylinder.

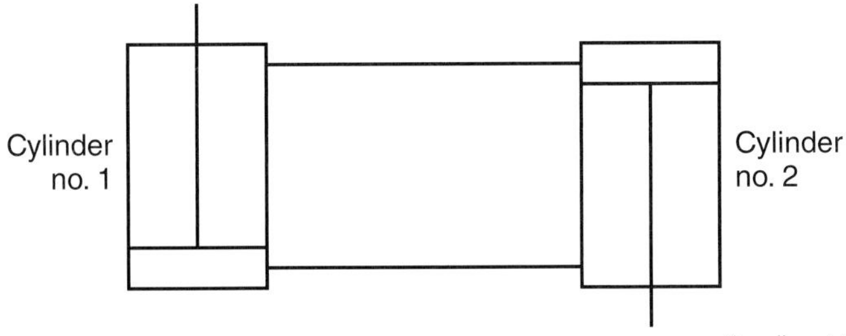

Goodheart-Willcox Publisher

Hydraulic power test circuit.

7. Have a lab partner hold cylinder 1 while you hold cylinder 2. Pushing the ram of cylinder 2 down will transfer the force of cylinder 2 through the fluid to cylinder 1.
8. If transparent flexible tubing is used, you should be able to see the direction of fluid flow by watching any air bubbles trapped in the system.
9. Alter the position of the two cylinders and repeat the procedure. What do the results indicate about a hydraulic system?
10. Adjust the position of the two pistons so they are near the center of the cylinders. Apply force to both pistons. Ideally, there should be no compression of the fluid. Any trapped air in the system will permit some degree of compressibility. How does your system respond?
11. Momentarily remove the flexible tubing from cylinder 1. Force the cylinder down approximately 1/4″, then pull it back to its original position. Return the flexible tubing to the cylinder.
12. Adjust the position of the two cylinders near the center and repeat step 10. How does the system respond?
13. Disconnect the flexible tubing from each cylinder and clear the cylinders of fluid. Wipe excess fluid off all components and return them to the storage area.

Analysis

1. What is the meaning of the term *compressibility*?
2. Explain how force is transferred through a fluid.
3. Describe the mechanical action of a double-acting cylinder.

Activity 7-2
Static Hydraulic System

Objectives

A hydraulic jack is one of the most common static fluid systems in operation today. With this system, a pump cylinder forces fluid into a ram cylinder through check valves. As a general rule, systems of this type are usually self-contained in a single housing.

In this activity, a simple hydraulic jack system will be constructed with discrete components. Through this activity, you will observe fluid flow through check valves that will permit an alternate direction of flow.

A hydraulic jack is often described as a static hydraulic system. This refers to the fact that the system develops its own operating pressure when energized by an outside mechanical force. A distinct mechanical advantage is developed by the two cylinders because of the differences in their size.

Equipment and Materials

- Hydraulic fluid or other fluid.
- Single-acting cylinder—1 1/8″ bore.
- Single-acting cylinder—1/2″ bore.
- Check valves (2).
- Flow control valve (needle type).
- T connectors (2).
- Pressure gauge—160 lb/in^2.
- Assortment of quick-disconnect transparent tubing pieces.
- Safety glasses or goggles.

Procedure

1. Assemble the hydraulic circuit. The 1 1/8″ bore cylinder will serve as the ram of the hydraulic jack, and the 1/2″ bore cylinder will serve as the pump. Mount both cylinders in an upright position.

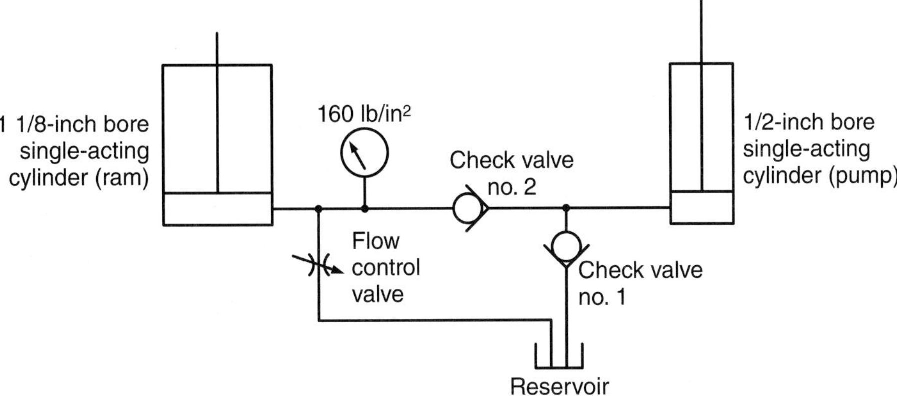

Hydraulic jack circuit.

2. Place the rams of both cylinders in the retracted position before attaching the flexible tubing to the cylinders, and close the flow control valve. Record the stroke length of the ram cylinder and the pump cylinder.
3. Pull the ram of the pump cylinder out of the cylinder, then down. How many millimeters/inches does one downward stroke of the pump cylinder cause the ram cylinder to move?
4. How many strokes of the pump are needed to extend the ram fully?
5. What pressure is indicated on the gauge?
6. Open the flow control valve to drain the ram cylinder. You will need to force the ram slowly to its retracted position. Then close the valve to operate the system again.
7. Attach a weight of approximately 10 lb to the ram cylinder and repeat the operation of the jack.
8. What noticeable changes occur when the system is forced to do work?
9. Disconnect the system and drain the fluid from the cylinders and the tubing. Wipe any excess fluid from the components and return them to the storage area.

Analysis

1. How much work was done by the ram cylinder in step 7?
2. Describe the fluid circulation action that takes place during the operation of this system.
3. Does the pressure gauge change during the pumping action?
4. Why does this system have a mechanical advantage?

Activity 7-3
Hydraulic Ram System

Objectives

Hydraulic presses or rams play a very important role in industry. Mechanical systems of this type are actuated by a pump driven by an electric motor that serves as the source of hydraulic energy. The transmission path of the system distributes energy to the load device, which does some form of mechanical work. Flow and directional control are achieved by special valves in the transmission line. An indicator is used to display system pressure. The hydraulic fluid is sent back to the pump reservoir through a return line. Systems of this type are classified as active units when compared with the static systems of Activities 7-1 and 7-2.

In this activity, you will identify some basic hydraulic components and their associated diagram symbols. You will then assemble these components into a functioning hydraulic ram system. You will be able to observe the change in the direction of fluid flow by valve action. The basic operating principle of this hydraulic system has widespread application in industrial machinery.

Equipment and Materials

- Hydraulic fluid or other fluid.
- Electric hydraulic power unit.
- Flow control valve.
- Two-position, four-way, lever-action control valve.
- Double-acting cylinder—1 1/8″ bore.
- Pressure gauge—160 lb/in².
- T connector.
- Assortment of quick-disconnect transparent tubing pieces.
- Safety glasses or goggles.

Procedure

1. Select the necessary components to construct the hydraulic ram circuit.

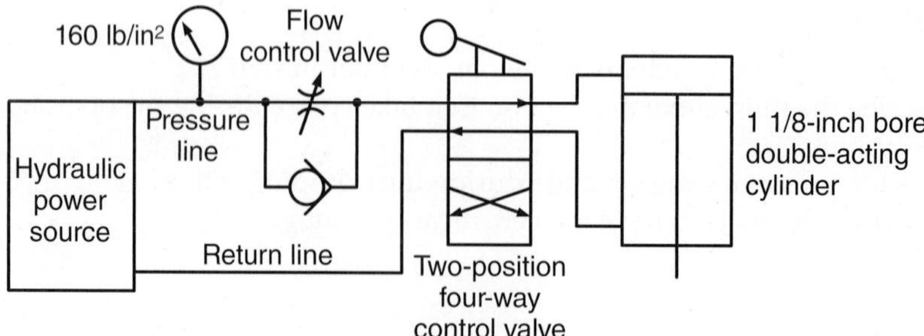

Goodheart-Willcox Publisher

Hydraulic ram circuit.

2. Start the electric pump motor of the hydraulic power source. Slowly open the shutoff control valve of the circuit while observing the pressure gauge. Depending on the initial position of the four-way control valve, the ram of the cylinder may extend itself slowly. Record the system pressure.
3. With the four-way control valve positioned as indicated in the circuit diagram, move the handle to the other position. What influence does this control action have on the ram of the double-acting cylinder?
4. Move the handle of the four-way control valve back to its original position to see the control action on the cylinder. Describe the action of the cylinder ram.
5. The direction of fluid flow can be readily traced by the movement of air bubbles in the transparent tubing. Switch the control valve to extend the ram and observe the direction of fluid flow. If the bubbles are not present, they can be introduced by closing the flow control valve and momentarily disconnecting the return line from the four-way control valve. Depress the check valve in the line and let a small amount of hydraulic fluid drain from the line. Then return the line to the control valve and turn on the flow control valve.
6. What is the direction of fluid flow?
7. Switch the control valve to a position that will cause the ram of the cylinder to retract. What is the direction of fluid flow?
8. Turn off the electric pump motor of the hydraulic power source and disconnect the circuit. Wipe off any excess hydraulic fluid from the components and return them to the storage area.

Analysis

1. What are the primary system functions of the circuit constructed in this activity?
2. What specific component of the circuit is used to achieve each system function?
3. How could the ram in this circuit be used in an industrial application?
4. How does this system differ from the static systems of Activities 7-1 and 7-2?

Activity 7-4
Hydraulic Power Source

Objectives

The power source of a hydraulic system is a pump. This component is responsible for producing a fluid flow that will develop pressure as it encounters resistance as it passes through system components. A hydraulic pump must be in continuous operation whenever the system is operational. Through its action, hydraulic fluid is forced to move at a specific flow rate, which causes a corresponding pressure to occur when resistance is encountered.

In this activity, you will study the basic components of a hydraulic power unit. This will include the hydraulic pump, reservoir, filters, pressure relief valve, pressure manifold, and return manifold. Pressure developed by this power source is variable, with a maximum rating of approximately 300 psi (lb/in^2). For most experimental applications, the power source needs values of 100 lb/in^2 or less. Typical industrial hydraulic power sources have values in the 1000 to 3000 psi (lb/in^2) range.

Equipment and Materials

- Hydraulic fluid or other fluid.
- Hydraulic power source.
- Pressure gauge—160 lb/in^2.
- Assortment of quick-disconnect transparent tubing pieces.
- Measuring container—1/2 gal.
- Open-ended disconnect tubing piece.
- Safety glasses or goggles.

Procedure

Section A. Power Source Analysis

1. A basic hydraulic power source is needed to perform this activity. As a general rule, most of the components of this unit are permanently mounted and ready for operation.
2. Locate the electric motor of the unit and plug it into the appropriate electrical outlet. Trace out the electrical circuit and locate the motor/pump on/off switch. In many units, the pump and motor may be mounted in a single housing.
3. Locate the system reservoir. Trace the fluid flow path from the reservoir to the inlet side of the hydraulic pump. Does your unit employ a fluid strainer or filter in this line?
4. Trace the fluid path from the outlet side of the pump to the pressure manifold. Are there any filters or controls in this part of the system?
5. Where is the pressure relief valve of your system located?

6. Turn on the electric motor. Observe the pressure gauge and record the system pressure. If at all possible, see at what pressure the system returns fluid to the reservoir. This is determined by the pressure relief valve setting.
7. Connect a length of tubing to one of the pressure manifold valves. Run the other end of the tubing to an open tray or to the reservoir return line. Open the manifold valve momentarily until all of the air bubbles pass from the line. Does the pressure gauge change during this operation?
8. Turn off the electrical power to the pump drive motor. Make a sketch of your hydraulic power source, pointing out specific component locations.
9. Draw a schematic diagram of the hydraulic power source using appropriate fluid power symbols.

Section B. Hydraulic Pump Delivery Rate

1. Attach a 160 lb/in^2 gauge to one of the pressure manifolds of the hydraulic power source. Connect a length of plastic tubing to the output side of a second manifold valve. Then connect the other end of the tubing to one of the return manifold connections.
2. Ready the hydraulic power source for operation by applying electricity and turning on the pump motor. Open the pressure manifold valve attached to the pressure gauge. Record the system pressure. This valve also indicates the pressure setting or value of the pressure relief valve. If more than 100 lb/in^2 is indicated, ask the instructor to show you how to adjust the pressure limit.
3. Gradually open the pressure manifold valve with the tubing attached. How does the system pressure gauge respond as this valve is opened?
4. Close the pressure manifold valve and remove the plastic tubing with quick disconnects on each end. Replace it with a piece of tubing with a disconnect on only one end. The open end of the tubing will be used to fill a half-gallon measuring container to determine the pump delivery rate.
5. Open the manifold valve enough to maintain 100 lb/in^2 with the full output flowing through the tube. Fill the container and record the time it takes to pump 1/2 gallon. Run two trial tests at this pressure and average your results. Record your data.

Pressure (lb/in^2)	Half-gallon filling time (seconds)	Pump delivery rate (GPM)	Calculated horsepower*
100			
80			

$$\text{*Horsepower} = \frac{\text{gal/min} \times \text{lb/in}^2}{1714}$$

Goodheart-Willcox Publisher

Hydraulic pump delivery data.

6. Repeat the procedure outlined in step 5 for a pressure setting of 80 lb/in^2.
7. Calculate the horsepower delivered by the pump for each pressure setting, and record the data.
8. Turn off the power unit and disconnect all circuit components. Wipe them free of excess fluid and return them to the storage area.

Analysis

1. Explain the operation of the hydraulic power unit investigated in this activity.
2. Explain how the system pressure is altered.
3. What is the function of the pressure relief valve?
4. What type of displacement does the pump of the hydraulic power unit supply?

Activity 7-5
Hydraulic Pumps

Objectives

A hydraulic pump is designed to provide an appropriate fluid flow that will develop system pressure (based on the encountered resistance to flow). In an operating system, the pump accepts power from a drive motor (prime mover) and converts it into fluid flow.

Fluid is pulled into the inlet port of the pump and is expelled (forced out) from the outlet port. After passing through the pump, fluid flow occurs at a faster rate. This increase in fluid flow causes an increase in system pressure when the flow encounters component resistance. Most hydraulic pumps displace a certain amount of fluid (volumetric displacement) with each revolution.

The gear pump in this activity is used to change rotary motion into fluid energy. This pump contains two gears enclosed in a precision machined housing. Rotary motion from the power source is applied to the drive gear. Rotation of this gear causes the second (driven) gear to turn with the teeth meshing in the middle. The unmeshed gear teeth carry fluid from the inlet port to the center. Here the teeth mesh together and force the fluid to pass from the outlet port.

In this activity, you will construct a simple gear pump circuit and rotate the shaft of the pump manually to observe the direction of fluid flow. This will provide you with an opportunity to see how pump rotation and shaft speed influence the output of a hydraulic pump.

Equipment and Materials

- Hydraulic fluid or other fluid.
- Bidirectional, fixed-displacement, gear motor/pump.
- Flow control valve.
- Double-acting cylinder—1 1/8″ bore.
- Assortment of transparent flexible tubing with disconnects.
- Safety glasses or goggles.

Procedure

1. Connect the simple hydraulic pump circuit. The gear motor in this case will be used as a pump.

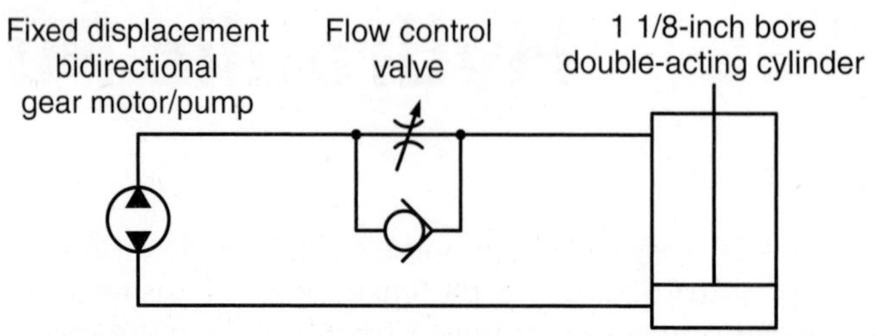

Goodheart-Willcox Publisher

Hydraulic pump circuit.

2. Rotate the shaft of the pump in a clockwise direction with one hand while holding it with the other hand. The tubing pieces should contain enough fluid to make the system operational.
3. Continue to rotate the shaft until the ram of the double-acting cylinder begins to move.
4. Air trapped in the tubing pieces can be used to trace direction of fluid flow. Clockwise pump rotation causes the hydraulic fluid to flow in what direction?
5. Rotate the shaft until the cylinder ram stops moving.
6. Rotate the shaft in a counterclockwise direction. Counterclockwise pump rotation causes hydraulic fluid to flow in what direction?
7. Disconnect the circuit components and wipe off all excess fluid with a cloth. Return all components to the storage area.

Analysis

1. How does the pump speed of rotation influence the flow rate?
2. How does a gear pump produce fluid flow?
3. Explain the basic principle of fluid flow as achieved by a gear pump.
4. Explain how the volumetric displacement of the pump is affected with changes in the pump's rotational speed.

Activity 7-6
Pressure Control Valves

Objectives

When a hydraulic pump is in operation, it supplies a defined amount of fluid to the system with each revolution of the drive motor. With the hydraulic pump operating continuously, there is a need for an alternate flow path for fluid not used by the system. Pressure relief valves placed in the power source are designed to provide an alternate flow path when the need arises.

A pressure relief valve is a type of check valve that permits flow only when a certain pressure is reached. At reduced pressure, the valve checks the return flow and permits the fluid to pass into the system. At higher pressure levels, it passes fluid back into the reservoir. In this application, the relief valve is a safety device.

In this activity, you will investigate the action of a pressure relief valve and check valves more carefully. You will also test the cracking pressure of the pressure relief valve. This refers to the release pressure that causes flow into the reservoir.

Equipment and Materials

- Hydraulic fluid or other fluid.
- Hydraulic power source.
- Pressure relief valve.
- Flow meter.
- Flow control valve.
- Assortment of quick-disconnect hoses.
- Pressure gauge—160 lb/in^2.
- Safety glasses or goggles.

Safety Note

You will be dealing with high system pressures in this activity. Follow all applicable safety procedures. Do not allow system pressure to become excessively high.

Procedure

1. Select the components needed to assemble the circuit. Notice the two paths for the fluid flow in this circuit. This represents a typical power source system divider.

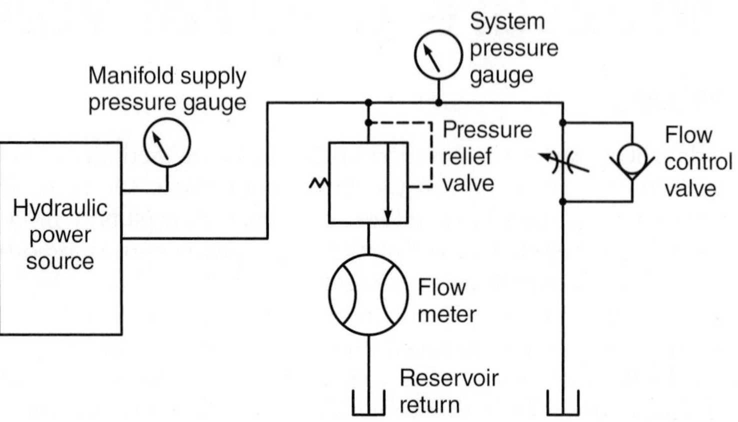

Goodheart-Willcox Publisher

Hydraulic power source system divider circuit.

2. Adjust the pressure relief valve in the counterclockwise direction until it stops. Then back off one turn of the valve control. Close the flow control valve completely.
3. Start the power unit and adjust the pressure relief valve to read 100 lb/in^2 of system pressure.
4. Open the flow control valve completely while observing system pressure. How does the pressure respond? What is the pressure with the flow control valve open?
5. Close the flow control valve again while observing the system pressure gauge. What does the flow meter reading do when the flow control valve is turned off?
6. Adjust the flow control valve to produce 100 lb/in^2 of system pressure. Then note the exact reading where flow first occurs through the flow meter. This is the cracking pressure of the valve.
7. Adjust the pressure relief valve one more turn as instructed in step 2. Then adjust the flow control valve to 150 lb/in^2 of system pressure. Determine the new cracking pressure of the pressure relief valve.
8. Turn off the power source and disconnect the circuit. Wipe all excess oil from the components and return them to the storage area.

Analysis

1. What is the function of the pressure relief valve in a system of this type?
2. Define the term *cracking pressure.*
3. What is the function of a flow control valve in a hydraulic circuit?
4. What accounts for a difference in manifold supply pressure and system pressure?

Activity 7-7
Directional Control Valves

Objectives

A directional control valve (DCV) of a hydraulic system is designed to alter the fluid flow in such a way that it can start, stop, or reverse the flow of the fluid to the system components. Valves that achieve this type of control are described as two-, three-, and four-way directional control valves with different position settings. Controls of this type may be actuated manually, with fluid pressure, with electricity, or by mechanical energy.

In this activity, you will construct a simple hydraulic motor control circuit and test the direction of flow through the circuit with a flow meter. As a general rule, this type of flow indicator will only respond to flow in a single direction. You will also be able to test the direction of flow with respect to different circuit positions in order to see how this device alters the fluid flow. Through this experience, you will become more familiar with the basic components of a hydraulic system and be able to trace the flow path from the manifold source to the reservoir.

Equipment and Materials

- Hydraulic fluid or other fluid.
- Hydraulic power source.
- Pressure relief valve.
- Flow meter.
- Three-position, four-way directional control valve.
- Hydraulic gear motor.
- Assortment of quick-disconnect hoses.
- Safety glasses or goggles.

Procedure

1. Select the components needed to construct the hydraulic circuit. Note the location of the four test points. In this activity, you will be asked to insert the flow meter into each test point and observe the direction of flow. The flow meter will only respond when the flow is from the inlet port to the outlet port. Flow in the reverse direction will not produce an indication on the meter.

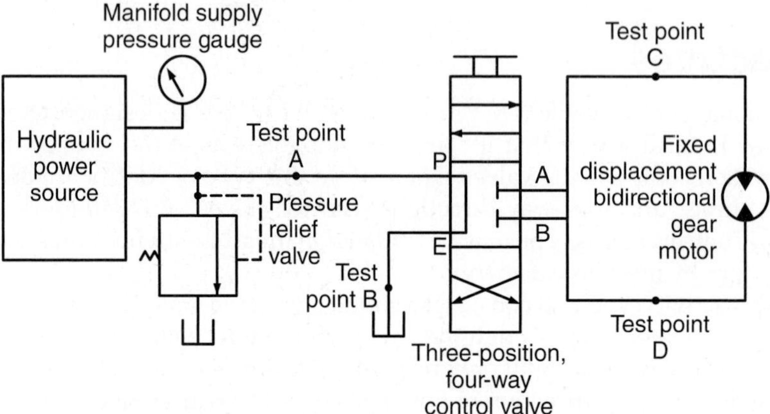

Hydraulic motor directional control circuit.

2. Close the pressure relief valve and back it off one complete turn. Turn on the hydraulic power source and adjust the pressure relief valve for a manifold pressure of 100 lb/in^2 as indicated on the manifold supply gauge.
3. Momentarily turn off the power source and insert the flow meter into test point A. Then turn the power source on again.
4. Pull the lever of the four-way valve to the fully extended position, which should cause rotation of the motor. Record the response of the flow meter and the motor.
5. Shift the directional control valve to the center position and record the flow meter and motor response.
6. Push the lever to the retracted position and note the flow meter and motor response.
7. Move the flow meter to test points B, C, and D, and repeat steps 4, 5, and 6. Record the direction of flow and motor response for each situation.
8. Draw three separate schematic diagrams with arrows showing the direction of fluid flow for each position of the four-way valve.
9. Turn off the hydraulic power source and disconnect the circuit. Wipe all components free of excess fluid and return them to the storage area.

Analysis

1. What do the letter designations P and E on the four-way valve denote?
2. What are the three motor operating conditions of the four-way valve?
3. Would the four-way valve be classified as a full-control or partial-control device?
4. How does the way the hoses are connected from the DCV (ports A and B) affect the motor operation?

Activity 7-8
Hydraulic Linear Actuators

Objectives

The fundamental purpose of a hydraulic power system is to produce some form of mechanical work. The part of the system responsible for achieving this function is called the load. The term *load*, however, is a rather broad term used to describe all components that consume power and do work. The term *actuator* is used more commonly to describe specific components in hydraulic systems that achieve the work function.

Linear actuators are designed to lift, hold, compress, or clamp objects during different manufacturing processes. Hydraulic cylinders are the most common linear actuator. These devices come in a variety of shapes and sizes that range from a fraction of an inch to several feet in diameter and length. The amount of force a cylinder can produce is determined by the surface area of the piston inside it and the pressure applied to it. Therefore, a cylinder with a larger surface area can produce more force than a smaller cylinder with the same pressure applied. Both single-acting and double-acting cylinders are included in the linear actuator classification.

In this activity, you will construct a simple series and parallel linear actuator circuit and test the pressure and circuit operation under load and no-load conditions. Through this experience you will become familiar with different system circuit configurations and observe some of the basic principles of series and parallel connected systems. Each type of circuit has a unique set of characteristics with respect to system pressure.

Equipment and Materials

- Hydraulic fluid or other fluid.
- Hydraulic power source.
- Pressure relief valve.
- Three-position, four-way control valve.
- Pressure gauge—160 lb/in^2.
- Double-acting cylinders—1 1/8″ bore (2).
- T hose connectors (3).
- Assortment of quick-disconnect hoses.
- Safety glasses or goggles.

Procedure

Section A. Series Linear Actuators

1. Select the appropriate components needed to construct the hydraulic circuit. Either a cylinder loading device or a weighted load should be attached to the cylinder in some steps of this activity.

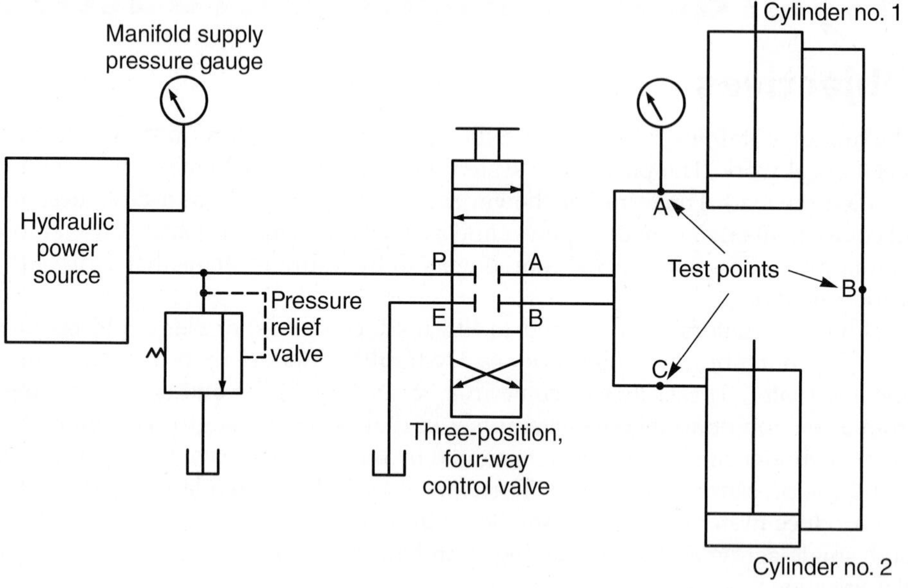

Hydraulic linear actuator circuit—actuator in series.

2. Turn on the power source and adjust the pressure relief valve to 120 lb/in^2.
3. Push the control lever of the four-way valve into the valve while observing the pressure at test point A. Make this reading while the cylinders are moving and record the reading.
4. Move the four-way valve to the center or off position and move the pressure gauge to test point B. Repeat step 3 and record the pressure.
5. Return the four-way valve to the center or off position and move the pressure gauge to test point C. Repeat step 3 and record the pressure.
6. The pressure tests you performed represent the no-load test condition. If a cylinder loading device is available, connect it to cylinder 1 and adjust it to the heavy load condition. If a cylinder loading device is not available, attach a small platform plate to cylinder 1. Place 20 lb of weight on the platform with the cylinder mounted in the vertical position.
7. Repeat steps 3, 4, and 5 with the cylinder loaded. The pressure at test point C should be taken first since the gauge should be at this location from the previous test.
8. Make a chart showing the relationship of pressure between load and no-load conditions at each test point.
9. Increase the system pressure to 160 lb/in^2 and test the operation of the circuit in its loaded condition. How does this added pressure alter the cylinder operation?
10. Reduce the system pressure to 80 lb/in^2 and test the operation of the circuit in its loaded condition. How does it respond to reduced pressure?
11. Turn the power source off and disconnect the circuit.

Section B. Parallel Linear Actuators

1. Select the components needed to construct the hydraulic circuit.

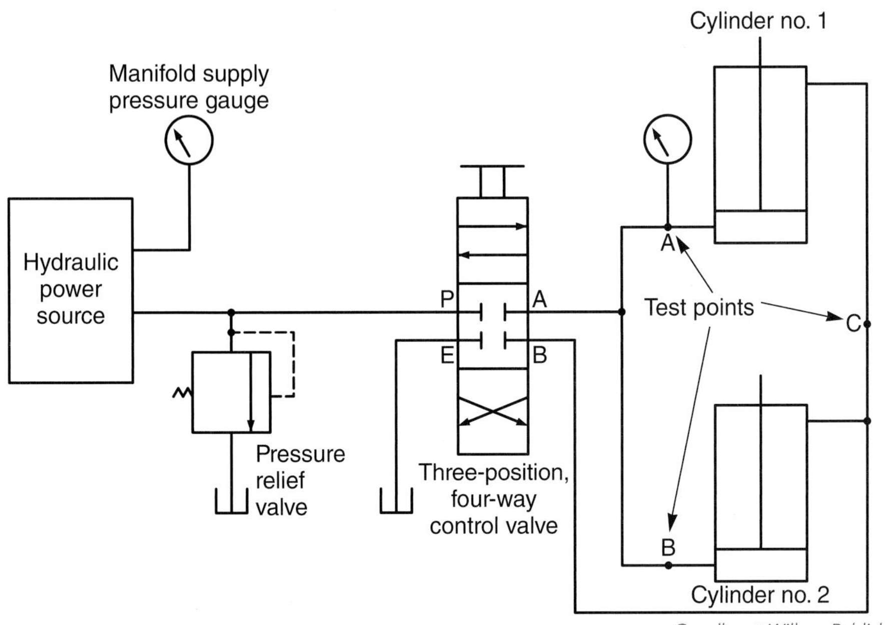

Goodheart-Willcox Publisher

Hydraulic linear actuator circuit—actuator in parallel.

2. Turn on the power source and adjust the pressure relief valve to 120 lb/in^2.
3. Push the control lever of the four-way valve into the valve while observing the pressure at test point A. Record this reading. Now pull the control lever completely out while observing the pressure. Make a pressure reading while the cylinders are in motion. Does the flow direction influence the pressure?
4. Place the four-way control valve in the center or off position. Place the pressure gauge at test points B and C. Repeat the procedure outlined in step 3. Record the readings for test point B and test point C.
5. These tests represent the no-load condition of the linear actuator. If a cylinder loading device is available, connect it to cylinder 1 and adjust it to the heavy load condition. If this device is not available, attach a small platform plate to cylinder 1. Place 20 lb of weight on the platform with the cylinder mounted in the vertical position.
6. Repeat steps 3, 4, and 5 with the cylinder loaded. The pressure at test point C should be taken first since the gauge should be at this location from the previous test.
7. Make a chart showing the relationship between load and no-load pressure conditions at each test point.
8. Increase the system pressure to 160 lb/in^2 and test the operation of the circuit in its loaded condition. How does this added pressure alter the operation of the cylinders?
9. Reduce the system pressure to 70 lb/in^2 and test the operation of the circuit in its loaded condition. How does it respond to reduced pressure?
10. Turn off the power source and disconnect the circuit. Wipe all components free of excess fluid and return them to the storage area.

Analysis

1. Why is this circuit described as a series linear actuator?
2. What influence does loading have on the pressure of a series linear actuator circuit?
3. What is a characteristic of the series circuit with respect to pressure throughout the system?
4. What general influence does system pressure have on the operational speed of a series linear actuator circuit?
5. Why is this circuit classified as a parallel linear actuator?
6. What influence does loading have on the system pressure of a parallel linear actuator system?
7. What is a characteristic of the parallel circuit with respect to pressure throughout the system?
8. What influence does the system pressure have on the operational speed of a linear actuator?

Activity 7-9
Hydraulic Rotary Actuators

Objectives

Hydraulic rotary actuators are designed to produce an appropriate form of rotary motion in either direction. Fluid under pressure flows through the rotating member and causes it to move before it is expelled from the outlet side. Counterclockwise rotation is achieved by reversing the direction of fluid flow. These devices are used in industry for opening and closing gates, for indexing operations, and for lifting operations. The speed of hydraulic rotary actuators is rather slow as compared to electric actuators, but they develop a great deal of torque.

In this activity, you will use a hydraulic gear motor/pump device as a rotary actuator. The turning ability of this device is a measure of the work the device is designed to do. Torque is used to describe the developed amount of work. It is determined by the developed force multiplied by the radius of the rotating arm. You will also observe the influence of pressure and loading on the rotational speed of the actuator. Rotary actuators represent a very important hydraulic system load device.

Equipment and Materials

- Hydraulic fluid or other fluid.
- Hydraulic power source.
- Pressure relief valve.
- Three-position, four-way control valve.
- Flow meter.
- Pressure gauge—160 lb/in^2.
- Bidirectional, fixed-displacement, gear motor/pump.
- T connector.
- Assortment of quick-disconnect hoses.
- Tachometer.
- Safety glasses or goggles.

Procedure

1. Select the appropriate components needed to construct the hydraulic circuit.

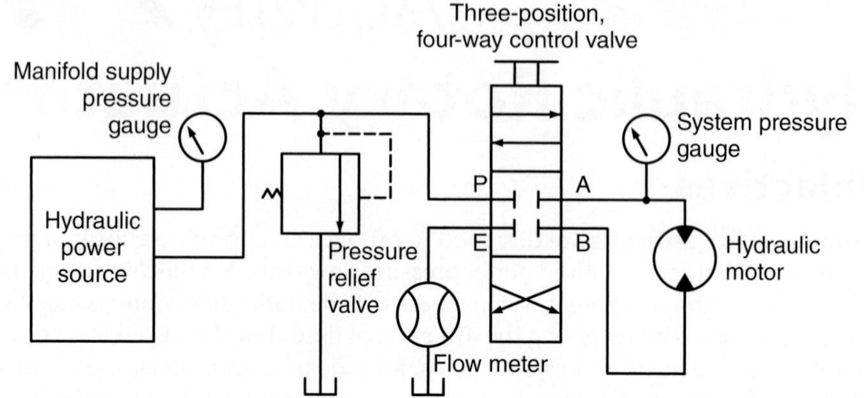

Goodheart-Willcox Publisher

Hydraulic rotary actuator circuit.

2. Turn on the power source and adjust the pressure relief valve for 120 lb/in^2.
3. Push the control lever of the four-way valve into the valve while observing the direction of motor rotation. Is the direction of rotation clockwise or counterclockwise?
4. Pull the valve control lever to its most extended position. What is the direction of rotation?
5. Perform this test again and note the system pressure needed to produce rotation.
6. The flow meter indicates how many gallons per minute (gpm)?
7. If a tachometer is available, measure and record the rotational speed of the motor. You can also measure speed when it is slow enough by marking the shaft and counting the number of revolutions per minute. What is the reading?
8. This represents the unloaded speed and pressure of the motor. Increase the pressure to 160 lb/in^2. Record the influence this has on the speed and flow rate.
9. Reduce the pressure to 80 lb/in^2 and record the influence this has on the speed and flow rate.
10. Attach a rotary load of some type to the mounted hydraulic motor.
11. Repeat steps 2–10 with a substantial load on the motor. Record your findings.

Pressure (lb/in^2)	Speed (RPM)	Flow rate (GPM)
80		
120		
160		

Goodheart-Willcox Publisher

Hydraulic rotary flow rate.

12. Turn off the power source and disconnect the circuit. Wipe all components free of excess hydraulic fluid and return them to the storage area.

Analysis

1. What influence does the system pressure have on the motor speed and flow rate?
2. What influence does loading have on the pressure and speed?
3. Explain how a hydraulic motor produces rotation.
4. What does the term *displacement* refer to in hydraulic motors?

Activity 7-10 Compressibility of Air

Objectives

The most significant difference in a hydraulic and a pneumatic power system is the compressibility of air with respect to fluid. Pneumatic systems are designed to compress air into smaller volumes and increase its pressure. Compressed air can then be stored in a receiver tank and released when the system demands it. This removes the necessity of continuous pump operation that is characteristic of hydraulic systems.

The pressure associated with a pneumatic system can be expressed in a number of ways. Barometric, or absolute pressure, is 14.7 lb/in^2 at sea level. Gauge pressure is the difference in absolute pressure and an unknown system pressure. A gauge reading of 100 lb/in^2 represents an absolute pressure of 114.7 lb/in^2. Pressure below atmospheric pressure is commonly considered as a negative pressure or vacuum.

When a certain volume of air or gas is compressed, it causes a corresponding increase in pressure. Essentially, a decrease in container volume causes a corresponding increase in the density of the molecules. This increases the rate at which the molecules strike the walls of the container, which causes a corresponding increase in pressure. Likewise, when the volume of a container is increased, it causes the gas molecules to spread apart. This causes a corresponding reduction in total container pressure. This relationship is normally expressed by the ratio of: $P_1 \times V_1 = P_2 \times V_2$. P_1 and V_1 refer to the pressure and volume of the original container, while P_2 and V_2 refer to its reduced or expanded condition.

In this activity, you will construct a simple pneumatic system and test the compressibility of air while observing the changes in pressure. The internal, or blind end, volume of a cylinder is based on the area of the piston times the length of the stroke. Cylinder area is measured in square inches or square centimeters and is determined by this formula: Area = $D^2/4$, where *D* is the diameter.

The blind end volume of a cylinder is determined by the cylinder area multiplied by the stroke length, and it is expressed in cubic inches or cubic centimeters.

Equipment and Materials

- Air supply source—0–100 lb/in^2.
- Check valve.
- System pressure gauge—100 lb/in^2.
- Double-acting pneumatic cylinder—1 1/8″ bore.
- Assortment of quick-disconnect hoses.
- Safety glasses or goggles.

Safety Note

Pneumatic systems, like hydraulic systems, operate at high pressure. Care should be exercised while working with pneumatic systems. Follow all safety procedures and use protective eyewear.

Procedure

1. Construct the pneumatic circuit. Note that this circuit includes several items that are part of the primary air supply. These items are optional and may not be included in the air supply feeding each work area. The part of the circuit that you are primarily concerned with begins at the manifold supply valve.

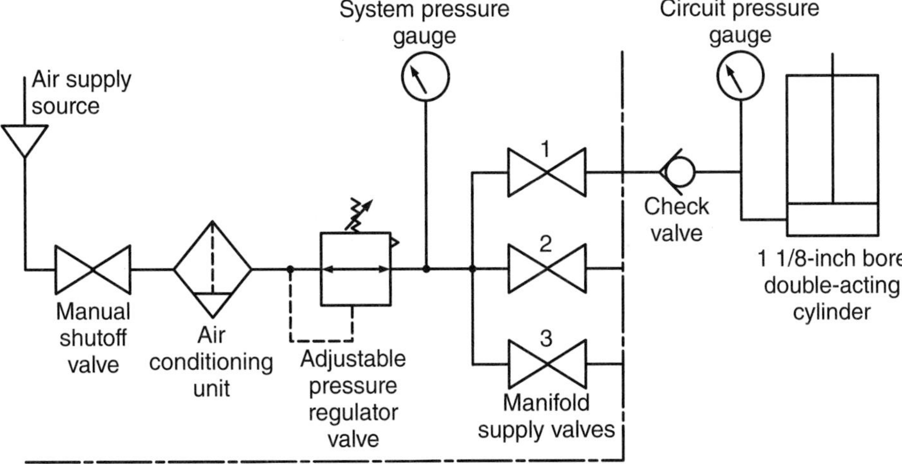

Goodheart-Willcox Publisher

Pneumatic power test circuit.

2. Close manifold supply valves 1, 2, and 3 and open the manual shutoff valve of the air source. Adjust the system pressure regulator to read 15 lb/in^2 on the system pressure gauge.
3. Measure the length of the cylinder shaft when it is fully retracted. Record the reading.
4. Slowly open manifold supply valve 1 while observing the action of the gauge and the cylinder. If the circuit does not respond, turn off the manifold valve and check the flow direction of the check valve.
5. When the cylinder ram is fully extended, measure and record the circuit pressure. Measure and record the extended length of the cylinder ram. Then calculate the blind end volume of the cylinder.
6. Manually force the piston back into the cylinder one-half of its extended distance while observing the circuit pressure gauge. Record the reading. Measure the extended rod length and calculate its blind end volume.
7. Release the cylinder piston and let it extend again.
8. Adjust the supply regulator to 25 lb/in^2 and repeat steps 3, 4, 5, and 6. How does this action differ from the preceding steps?
9. Close the manifold supply valve and the main air supply shutoff valve. Disconnect the circuit and return all components to the storage area.

Analysis

1. Describe the relationship between blind end cylinder volume and pressure.
2. What is the absolute pressure of step 5?
3. As pressure increases, how does its compressibility change?
4. Why does pneumatic pressure increase when the volume of a container is reduced?

Activity 7-11 Pneumatic Pressure and Flow

Objectives

When air passes through a pneumatic system, it normally encounters resistance that tends to reduce or lower the pressure between different locations. This lowering of pressure is primarily due to a form of friction that air encounters as it flows through the system. When airflow is reduced or stopped, it causes a change in friction. In a closed system without an airflow, the pressure is equalized throughout. When the flow of air increases, there is a corresponding increase in pressure drop. The length of the transmission flow path, the diameter of the tubing, the number and type of direction changes, and the path restrictions all have some influence on the amount of pressure drop developed within a system.

In this activity, you will construct a simple pneumatic system and alter the flow of air to see how it influences the system pressure. You will observe both an equalized system condition and the pressure drop of a heavy flow circuit. These conditions are very important considerations in the operation of a pneumatic system.

Equipment and Materials

- Air supply source.
- Flow control valve.
- Pressure gauge—0–100 lb/in^2.
- Manual shutoff valve.
- Muffler.
- T connector.
- Assortment of quick-disconnect hoses.
- Safety glasses or goggles.

Procedure

1. Construct the following pneumatic circuit.

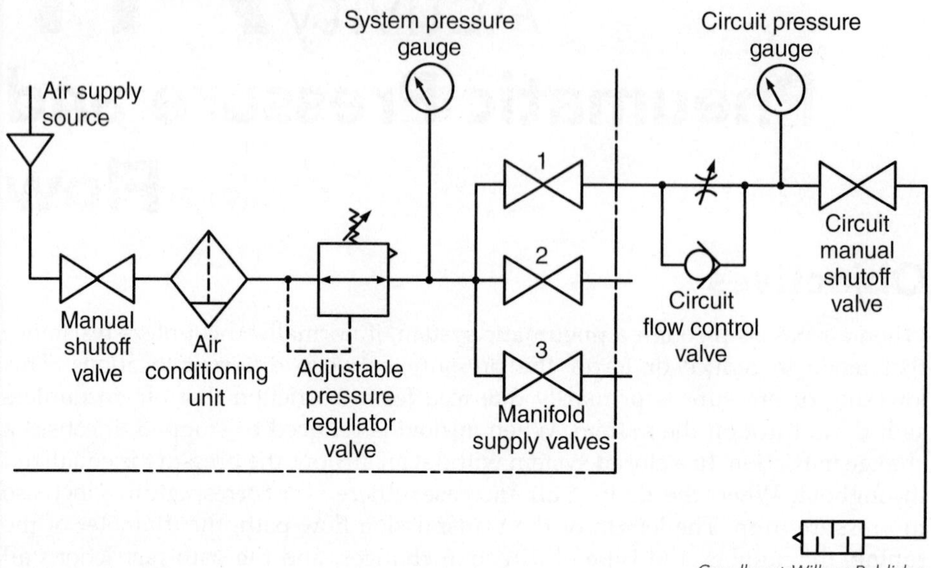

Pneumatic pressure and flow test circuit.

2. Close the manifold supply valves and turn on the manual shutoff valve. Adjust the pressure regulator valve to 15 lb/in^2 of system pressure. Close the circuit flow control valve and the circuit shutoff valve, then slowly open the manifold supply valve. The circuit pressure gauge should read zero at this time if everything is adjusted properly. If not, check the two circuit valves and make certain they are off.
3. Gradually open the flow control valve to produce a reading on the circuit pressure gauge. How does this reading compare with the system pressure?
4. Open the circuit manual shutoff valve. How do the two gauge pressures compare?
5. Turn off the circuit shutoff valve and add approximately six feet of hose between the muffler and the shutoff valve.
6. Turn on the manual shutoff valve again and observe the circuit pressure. What influence does the additional hose have on the pressure reading?
7. Turn off the circuit shutoff valve and remove the muffler with the six-foot length of hose remaining.
8. Turn on the shutoff valve and observe the pressure reading. How does this compare with the value of step 6?
9. Turn off the manifold supply valve and the manual shutoff valve of the air supply. Disconnect the circuit and return all components to the storage area.

Analysis

1. What causes the pressure of a pneumatic circuit to drop at different locations?
2. What is the meaning of the term *system resistance*?
3. What are some of the things that alter the resistance to the flow of fluid in a pneumatic system?

Activity 7-12 Pneumatic Linear Actuators

Objectives

Pneumatic linear actuators are primarily used in industrial applications to lift, compress, hold, or position objects during manufacturing processes. In order for this action to be achieved, air is fed into a chamber under pressure. A piston or ram in the chamber is then forced to move, depending on the applied pressure. Pneumatic cylinders range in size from a fraction of an inch to several inches in diameter. Both single-acting and double-acting cylinders are used to achieve linear mechanical action in pneumatic circuits.

In this activity, you will construct a simple linear actuator circuit to test the lifting capabilities of air. You will observe this operation with no load and under a loaded condition. Through this experience, you will become more familiar with basic pneumatic circuits and their operation. You will also employ a flow control valve, which serves as a throttle that alters cylinder operating speed. This circuit modification is very important in many industrial applications.

Equipment and Materials

- Air supply source.
- Three-position, four-way control valve.
- Single-acting cylinder—1 1/8″ bore.
- Double-acting cylinder—1 1/8″ bore.
- T connectors (2).
- Pressure gauges—0–100 lb/in^2 (2).
- Flow control valve.
- Assortment of quick-disconnect hoses.
- Platform plate and 10 lb weight.
- Safety glasses or goggles.

Procedure

Section A. Single-Acting Linear Actuators

1. Construct the linear actuator circuit. Note that the four-way valve has a plug attached to part of the valve that is not used.

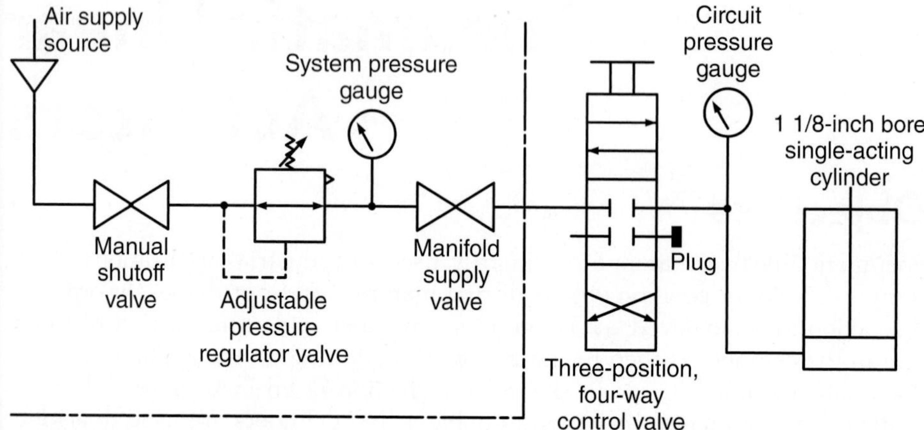

Single-acting pneumatic linear actuator circuit.

2. Turn off the manifold supply valve. Energize the air supply source and adjust the pressure regulator valve for 15 lb/in^2.
3. Turn on the manifold supply valve and switch the four-way valve through all three positions. Describe the action of the cylinder in each position.
4. Place a weight on the cylinder or load it by holding it down with your hands. Run through the switching sequence again with an increased load. How does this influence the action of the cylinder?
5. Increase the system pressure to 40 lb/in^2 and repeat steps 3 and 4. How does increased pressure alter the operation of the cylinder?
6. Turn off the manifold supply valve and insert the flow control valve.

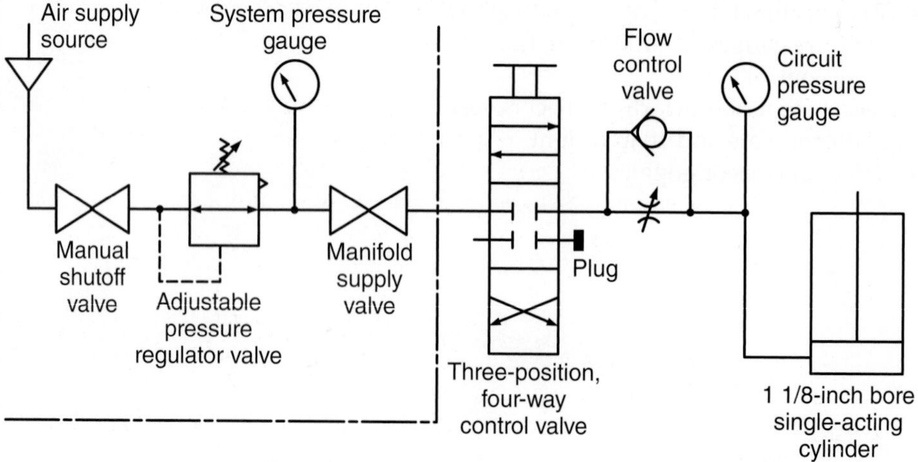

Circuit with flow control valve added.

7. Turn on the manifold supply valve and switch the four-way valve through all three positions. In the retracting position, the flow control valve should be altered to regulate retracting speed. This type of valve serves as a throttle to regulate cylinder retraction.
8. Turn off the manifold supply valve and reverse the flow direction of the flow control valve.
9. Turn on the manifold supply valve and switch the four-way valve through its three positions. How does this alteration change the operation of the circuit?
10. Turn off the manifold supply valve and the air supply shutoff valve. Disconnect the circuit and return all parts to the storage area.

Section B. Double-Acting Linear Actuators

1. Construct the double-acting linear actuator circuit.

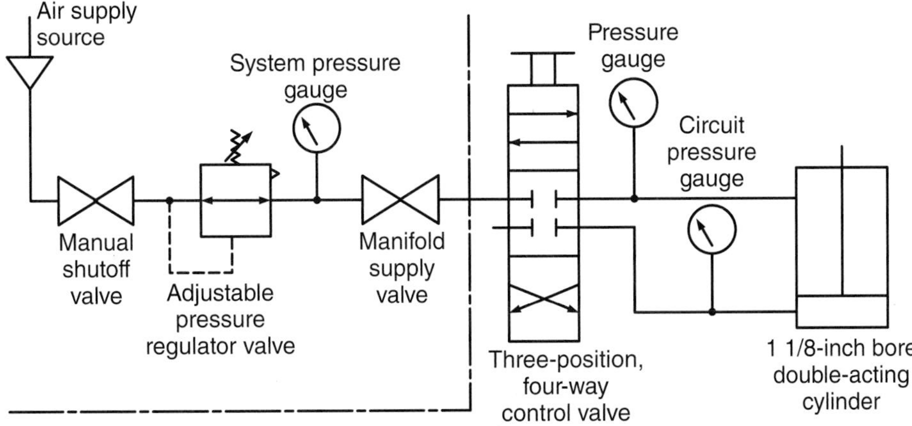

Double-acting pneumatic linear actuator circuit.

2. Turn off the manifold supply valve of the air supply and energize the source. Adjust the regulator pressure to 40 lb/in^2.
3. Turn on the manifold supply valve and switch the four-way valve through all three positions. Make a note showing the extended and retracted positions of the four-way valve. How do the pressure readings respond during operation of the cylinder?
4. Attach a platform to the top of the cylinder with the cylinder extending upward. Place a 10 lb weight on the platform and repeat step 3. What influence does an increased load have on the operation of the circuit?
5. Turn off the manifold supply valve and place a flow control valve in the top cylinder line of the circuit. Connect it so the free flow is from the four-way valve to the cylinder with the throttled flow and from the cylinder to the four-way valve.
6. Turn on the manifold supply valve and switch the four-way valve through all three positions. In the retracting position, the flow control valve should be adjusted to produce a slow retracting action.
7. Turn off the manifold supply valve and switch the location of the flow control valve to the bottom line.
8. Turn on the manifold supply valve and switch the four-way valve through its three positions. How does the control valve alter the operation of the cylinder in this circuit location?

9. If two flow control valves are available, you can place one in each cylinder line and alter the operational speed of the valve in both the retracting and extending positions.
10. Turn off the manifold supply valve and the air supply shutoff valve. Disconnect the circuit and return all parts to the storage area.

Analysis

1. What is an industrial application of a pneumatic circuit of the type studied in this activity?
2. What is a unique difference in this circuit operation and that of a similar hydraulic circuit?
3. Explain the operation of the throttling valve technique of cylinder speed control.
4. Make a circuit diagram for each of the three operating positions of the four-way valve with arrows showing the path of the airflow.
5. Why is the double-acting linear actuator considered to have better control than a single-acting cylinder?
6. Why does the four-way valve *not* require modification in this circuit?

Activity 7-13 Pneumatic Rotary Actuators

Objectives

Pneumatic rotary actuators are designed to produce rotary motion from air pressure. Air under pressure is applied to the rotating member, causing it to move. The air is then expelled from the outlet side. Reversing the direction of airflow causes the motor to rotate in the opposite direction. Pneumatic motors are lightweight and compact. They have a wide range of variable speed characteristics when compared with electric or hydraulic motors. These characteristics are well-suited for hand power tools, hoists, mixers, and explosion-proof applications.

The horsepower and speed of a pneumatic motor can readily be changed by throttling the applied airflow. As the load applied to a pneumatic motor is increased, it normally causes a reduction in operational speed. This behavior is similar to that of an electric motor. With an increased load to an electric motor, however, the speed of rotation must quickly return to the normal running speed, or it will be damaged by excessive heat. Excessive loading of a pneumatic motor will not adversely damage the motor because the airflow has a significant cooling effect. It can be run at alternate speeds for prolonged periods of time without adverse damage to the motor.

Pneumatic motors are energized by air, which means they can be used in potentially explosive environments. This characteristic is extremely important in many industrial applications.

The motor used in this activity produces up to 1/3 horsepower and operates from 500 to 10,000 rpm. It is primarily designed to respond to pressures below 100 lb/in^2. The amount of airflow needed to operate this motor is quite high, however. As a rule, the overall operating efficiency of a pneumatic motor is less than 20% for continuous operation. Air motors are rarely used when operational efficiency is important. They are used where the variable speed characteristic or explosion-proof advantage outweighs efficiency.

In this activity, you will construct a simple pneumatic motor circuit and test the loading effect on applied pressure and speed. You will then test the direction of rotation and see if loading and speed are altered. Through this activity, you will become more familiar with the operation of a pneumatic motor and see how rotary actuators can be harnessed to do work.

Equipment and Materials

- Air supply source.
- Flow control valve.
- Three-position, four-way control valve.
- Reversible pneumatic motor—Gast No. 1AM vane type or equivalent.

- Muffler.
- Assortment of quick-disconnect hoses.
- Pressure gauges—0–100 lb/in^2 (2).
- Safety glasses or goggles.

Procedure

Section A. Rotary Actuator Characteristics

1. Construct the rotary actuator circuit.

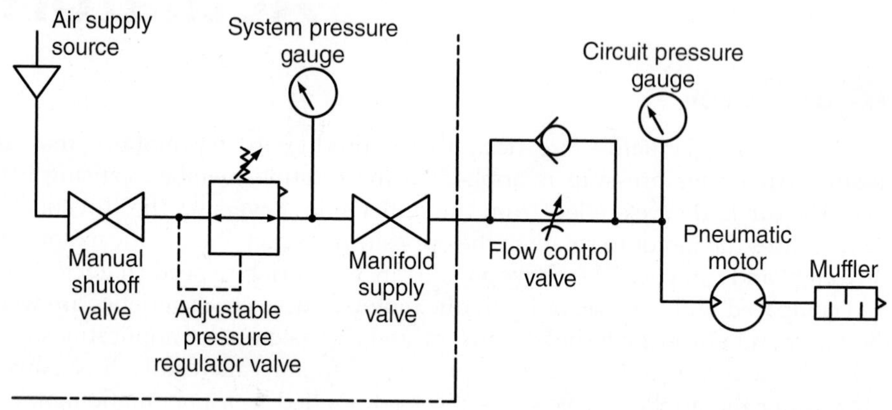

Pneumatic rotary actuator circuit.

2. Turn off the manifold supply valve, then energize the air supply. Adjust the pressure regulator valve for 30 lb/in^2 of system pressure.
3. Close the flow control valve, then open the manifold supply valve. Slowly open the flow control valve while observing the circuit pressure gauge. At what pressure does rotation occur?
4. What is the direction of motor rotation?
5. Gradually increase the flow while observing the circuit pressure. What is the maximum pressure observed? Note that when the pressure increases, the speed increases accordingly.
6. Connect a tachometer to the motor shaft and test the no-load speed. What is the speed reading?
7. Load the motor carefully by wedging a piece of wood between the rotating shaft and the base of the motor. How does loading influence the speed and circuit pressure?
8. Momentarily turn off the manifold supply valve and wait for the motor to stop rotation.
9. Adjust the pressure regulator valve to 70 lb/in^2 of system pressure. Close the flow control valve.
10. Turn on the manifold supply valve and gradually turn on the flow control valve. The motor should be loaded with the wooden wedge again. Increase the flow until rotation occurs. What circuit pressure is needed to produce initial rotation of the motor under a loaded condition? How does this compare with the unloaded condition of step 3?
11. Turn off the manifold supply valve and turn off the air supply source. Disconnect the circuit and return all components to the storage area.

Section B. Rotary Actuator Directional Control

1. Construct the rotary actuator circuit.

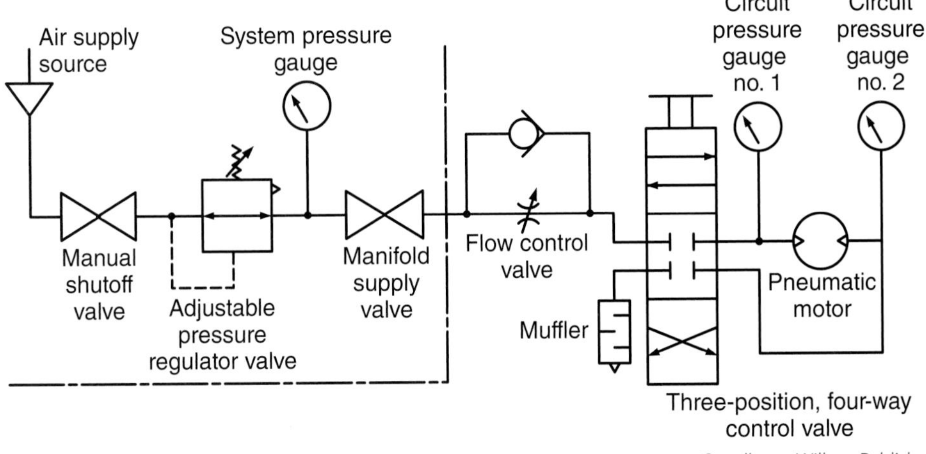

Pneumatic rotary actuator directional control circuit.

2. Turn off the manifold supply valve and energize the air supply source. Adjust the pressure regulator valve to 70 lb/in^2 of system pressure.
3. Turn off the flow control valve, then turn on the manifold supply valve. Switch the four-way directional control valve through each of the three positions. Then switch the four-way valve to one of its extreme positions.
4. Carefully adjust the flow control valve to produce a slow rotation of the motor. Make a note indicating the direction of rotation produced by this switch position.
5. Record the pressure indicated by gauges 1 and 2.
6. Switch the four-way control valve to the center or off position. Wait momentarily until rotation stops. Then switch the four-way control valve to the other extreme position. Make a note indicating the direction of rotation produced by this switch position.
7. Record the pressure indicated by gauges 1 and 2.
8. Increase the speed of rotation by opening the flow control valve. With a tachometer, measure the rotational speed. (Note: Some tachometers will only indicate speed in one rotational direction. Test the tachometer before making this measurement.)
9. Switch the four-way control valve to the off position until the motor stops. Then switch it to produce rotation in the opposite direction. Observe the note in step 8. Does the motor run equally well in either direction?
10. If time permits, repeat steps 1–9 after wedging a piece of wood between the motor shaft and its base to serve as a loading device.
11. Turn off the manifold supply valve and the supply shutoff valve. Disconnect the circuit and return all parts to the storage area.

Analysis

1. What influence does loading have on the motor starting torque and the applied starting pressure?

2. How does loading alter the rotational speed of a pneumatic motor?
3. What significant advantage would a pneumatic motor have in industry use?
4. How would the pressure gauges of this circuit respond if the muffler were removed?
5. What kind of pneumatic motor is used in this activity?
6. How does a pneumatic motor compare with an electric motor?

Activity 7-14
Reed Switches

Objectives

Reed switches are devices that respond to a controlled magnetic field. They are designed to close or open when exposed to either a permanent magnetic field or to an electromagnetic field. The reed switch is representative of various types of switching devices that may be used with robotic systems.

The contacts of a reed switch are housed inside a hermetically sealed glass tube. When actuated, contact sparks are isolated from the outside environment. Industrial applications of this device are quite numerous in explosive areas and dirt-prone environments. They are used to verify Z axis movement on some robots and for other applications involving robotic systems.

A reed switch contains two flat metal strips, or reeds, housed in a hollow glass tube filled with an inert gas. When the reeds are exposed to a magnetic field, they are forced together. This causes them to make or break contact, depending on their design. The normally closed switch breaks contact when exposed to a magnetic field. Normally open contacts, by comparison, are forced closed when exposed to a magnetic field.

In this activity, you will construct a simple electrical circuit and control its action by changes in the magnetic field strength. Through this experience, you will gain some insight into the operation, sensitivity, and control action of a reed switch.

Equipment and Materials

- Dc power supply—0V–5V, 1.0 A.
- Ac power source—6.3V, 60 Hz.
- Reed switch—GE-X7 or equivalent.
- Bobbin-wound, reed switch coil.
- Reed switch magnet.
- No. 47 lamp.
- SPST switch.
- Circuit construction board.

Procedure

1. Construct the following reed-switch circuit.

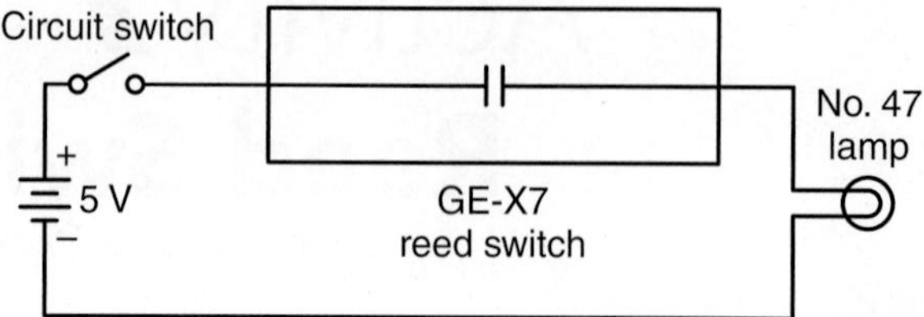

Reed switch test circuit for observing permanent magnet actuation.

2. Turn on the circuit switch and move the small permanent magnet near the reed switch. Try the magnet at different orientations with the reeds. Which type of magnet orientation produces the best control capability?
3. Place the reed switch in the center of the bobbin-wound coil. Connect the following circuit to test the electromagnetic action of the reed switch.

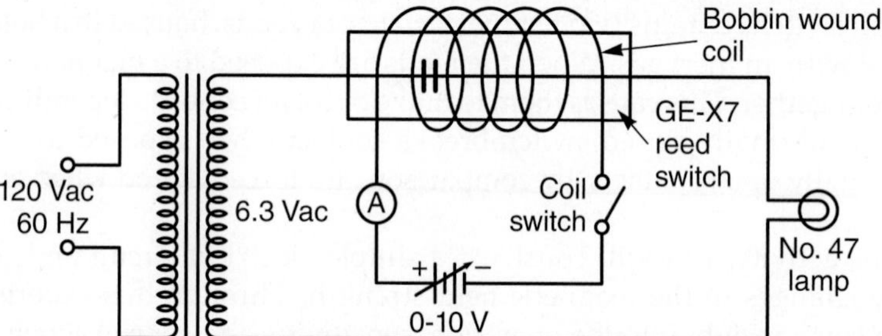

Reed switch test circuit for observing electromagnetic actuation.

4. Turn on the ac power source and the dc coil-control circuit.
5. Starting at 0V dc, gradually increase the dc source voltage until the reed switch is actuated. The meter should be set at a high range until an approximate actuating value is determined. Run at least two trial tests to determine the actuating current needed to energize the switch. What is this value?
6. In some applications, the actuating coil is used to increase the sensitivity of the reed switch by producing a partial field. Increase the dc coil current close to the actuating value. Then place the permanent magnet near the coil to actuate the reed. In this case, the switch can have some degree of variable sensitivity.
7. Turn off the ac and dc power sources for the circuit. Use 5V dc to supply the coil. The primary side of the coil should be connected to a variable ac transformer.
8. Turn on both the ac and dc sources, then increase the ac voltage applied to the coil. Measure the applied ac voltage. Do not increase the voltage to a value that will cause the reed coil to overheat. How does the reed switch respond to ac compared with dc?

9. Turn off the ac and dc power sources and disconnect the circuit. Return all components to the storage area.

Analysis

1. How could a reed switch be used to control a pump motor in a liquid sump tank? Perform additional research as needed.
2. Make a sketch of this circuit.
3. Why could a reed switch be used to control a circuit in an explosive area without danger of an explosion?

DuxX/Shutterstock.com

Hybrid power systems are used in automated manufacturing.

Unit III

Sensing and End-of-Arm Tooling

The control of an industrial robot often depends on the sensing system. Sensing systems use transducers, which are devices that convert light, heat, or mechanical energy into electrical energy. The signal output of the transducer affects the operation of the robot's end effector (end-of-arm tooling). End effectors are attached to the wrist of the manipulator and can grasp, lift, transport, maneuver, or perform operations on a workpiece.

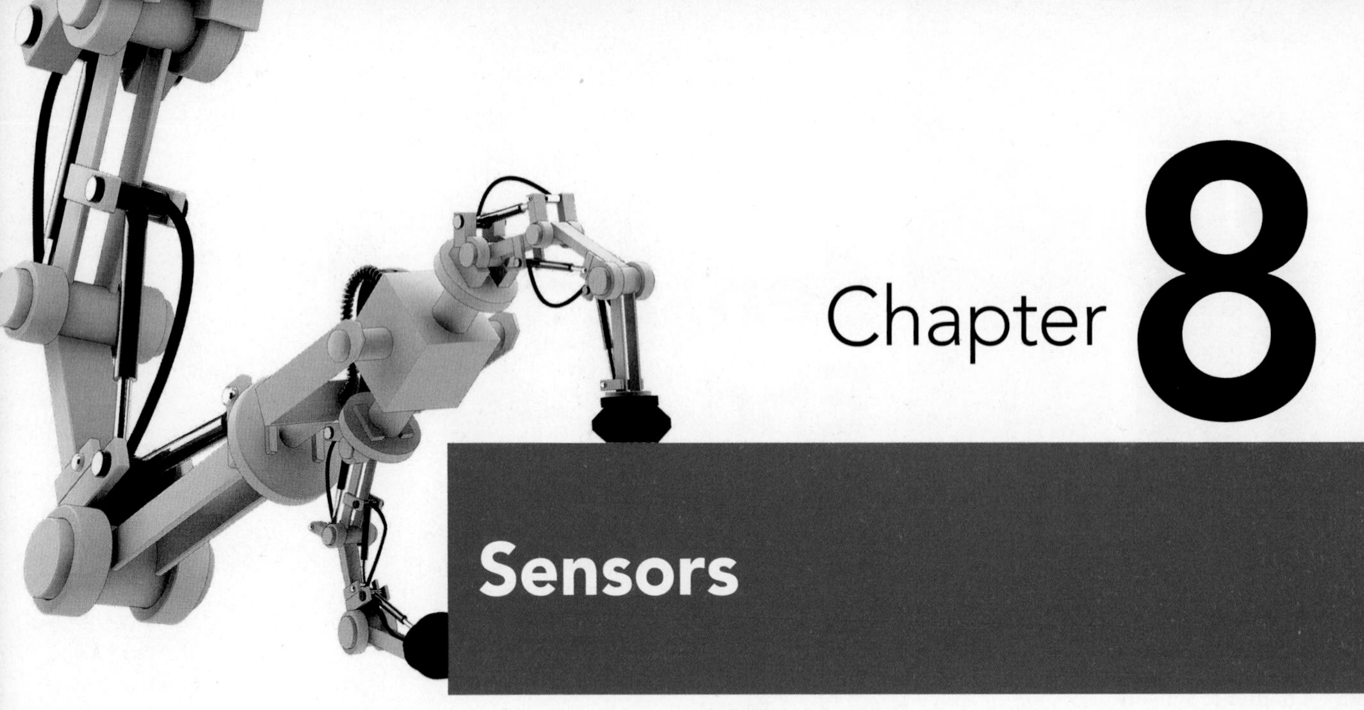

Chapter 8

Sensors

Outline

Objectives

Upon completion of this chapter, you will be able to:

- Explain the function of transducers in the operation of sensors.
- Identify the types of sensors used in an automated system.
- Describe how sensors are integrated into an automated system.

Technical Terms

acoustical proximity sensor
angstrom (Å)
capacitance
capacitive transducer
computer vision sensor
eddy current proximity sensor
electromagnetic spectrum
inductance
inductive transducer
infrared sensor
laser
laser interferometric gauge
light-emitting diode (LED)
limit switch
magnetic field sensor
microswitch
nanometer (nm)
optical fibers
optical proximity sensor
opto-electronic
photoconductive device
photoemissive device
photovoltaic device
piezoelectric effect
proximity sensor
range sensor
reed switch
resistive transducer
sound sensor
speed sensing
stadimetry
strain gauge
tactile sensor
thermistor
thermocouple
thermoelectric sensor
touch-sensitive proximity sensor
transducer
triangulation
ultraviolet sensor
X-rays

Overview

The control of many industrial robots depends on various types of sensors. A sensor operates by converting light, heat, or mechanical energy into electrical energy. The resulting signal affects the operation of the machine. Sensors give robots a higher level of decision-making capability. This chapter discusses some of the sensors used with robotic systems in industry.

8.1 How Sensors Work

Sensors allow robots to interact with their environment, respond to changes, and determine a course of action. Sensors help robots respond to different positions and orientation of parts, variations in the shape and dimensions of parts, and unknown obstacles in the work environment. ***Transducers*** are devices that convert light, heat, or mechanical energy into electrical energy. The conversion of physical quantities to electrical quantities is a basic sensing function. The signal output by the transducer is used to affect the operation of the machine. For example, a ***thermocouple*** is a transducer that converts heat energy into electrical energy. A microphone is a transducer that converts sound energy into electrical energy. Transducers may be resistive, capacitive, or inductive, **Figure 8-1**.

8.1.1 Resistive Transducers

Resistive transducers are devices that convert a variation in resistance into electrical variations. One type works on the potentiometer principle, **Figure 8-2**. This transducer changes resistance when the position of its movable contact is changed. Increasing the length of wire between terminals increases the resistance between those two points. Resistive transducers are often used to sense physical displacement. Displacement moves the sliding contact, which changes the resistance in the control circuit.

8.1.2 Capacitive Transducers

Capacitive transducers measure a change in ***capacitance***, or the ratio of charge on a conductor to the potential difference between conductors. Capacitance exists when two conductive materials (plates) are separated by an insulating material. Capacitance can be increased by increasing the area of the plates or by decreasing the thickness of the insulation. Changing electrical current is a result of physical alteration or movement (mechanical energy) by either affecting plate area or insulation thickness.

One use of capacitive transducers is sensing fluid pressure, **Figure 8-3**. In this application, the transducer is placed into the fluid line. One capacitor plate is a conductive diaphragm that senses any variation in fluid pressure. The other capacitor plate is stationary. When the fluid pressure in the line increases, the conductive diaphragm plate moves closer to the stationary plate. When the distance between capacitor plates decreases, the capacitance between the terminals increases. When the fluid pressure decreases, the distance between the plates increases and capacitance decreases.

8.1.3 Inductive Transducers

Inductance is a property of electrical circuits caused by the magnetic field that surrounds a coil when current is flowing. In ac circuits, inductance opposes changes in current and increases as frequency increases.

An ***inductive transducer*** is a device that measures movement and creates signals that affect current flow, **Figure 8-4**. Inductive transducers usually have a stationary coil and a movable core. The movable core is connected to an object whose movement is to be measured. As the core changes position within the coil, the inductance of the coil varies. Current flow through the coil drops as inductance increases.

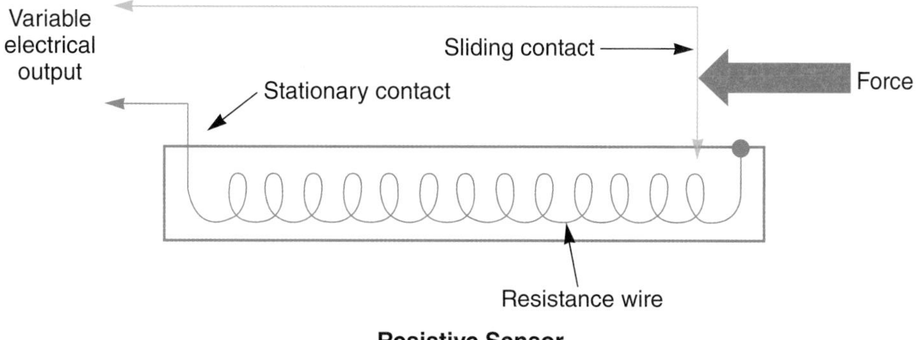

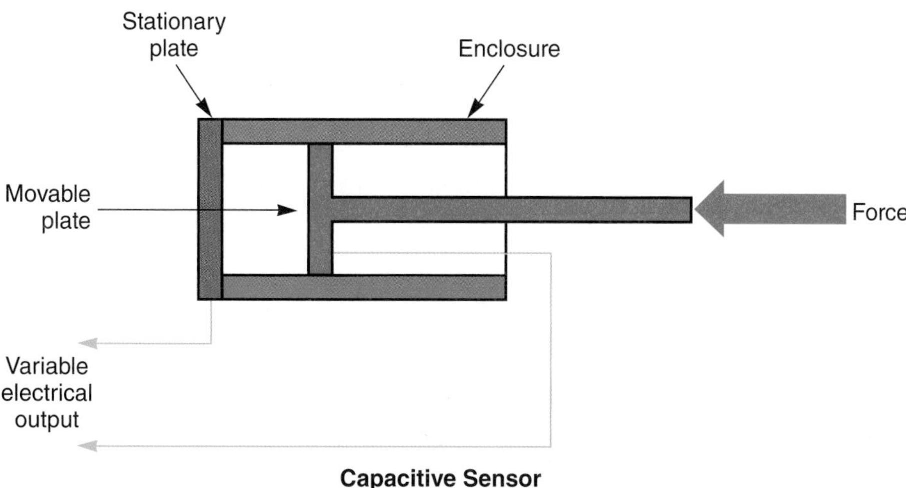

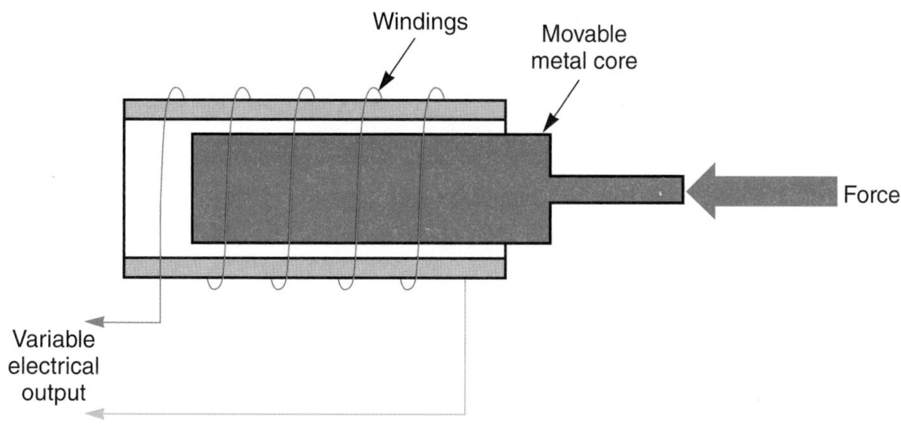

Goodheart-Willcox Publisher

Figure 8-1. Resistive, capacitive, and inductive sensors can be used to create electrical output that shows displacement.

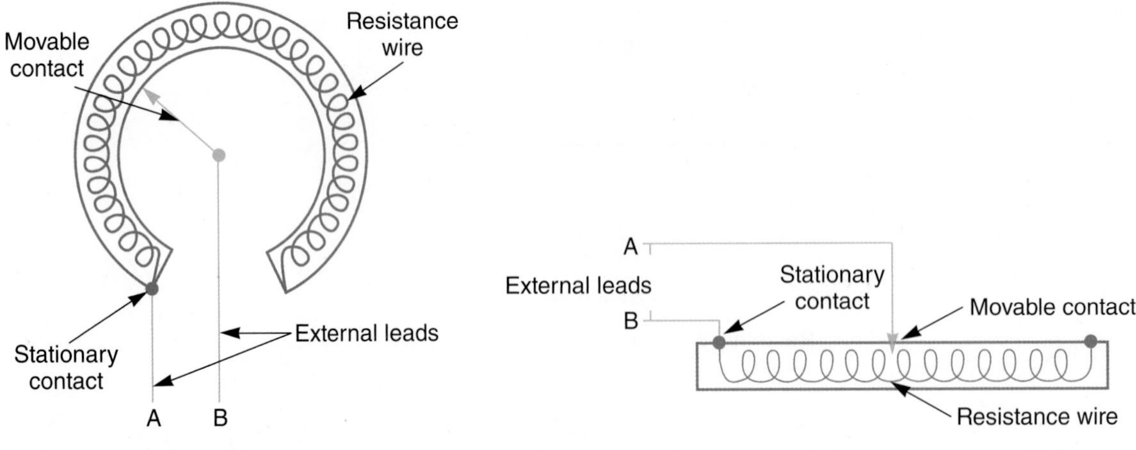

Goodheart-Willcox Publisher

Figure 8-2. Resistive transducers.

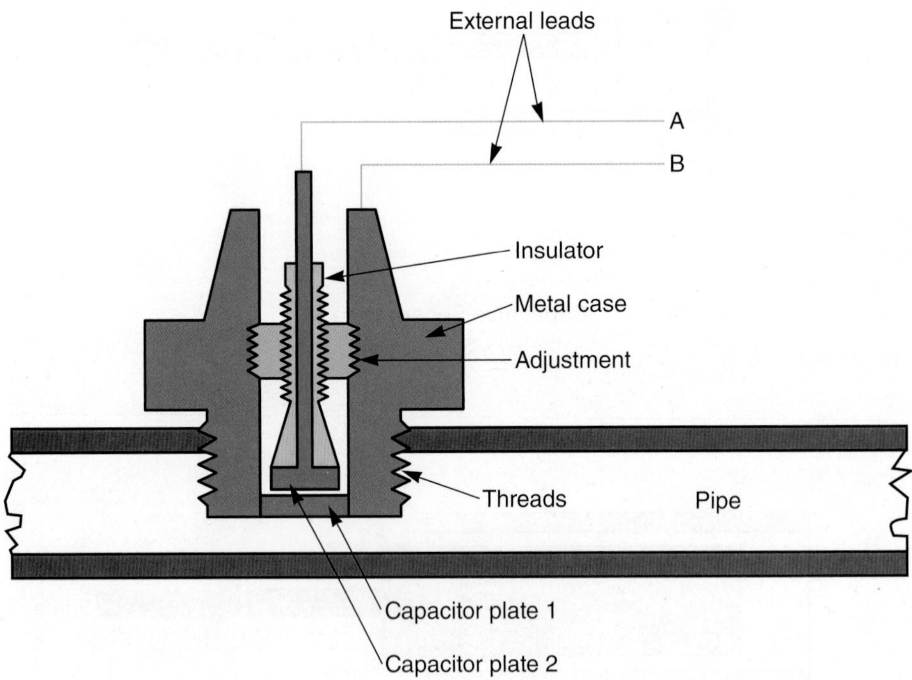

Goodheart-Willcox Publisher

Figure 8-3. This capacitive transducer senses fluid pressure. Pressure changes cause movement of capacitor plate 1, changing the capacitance of the circuit and affecting the control signal emitted by the transducer.

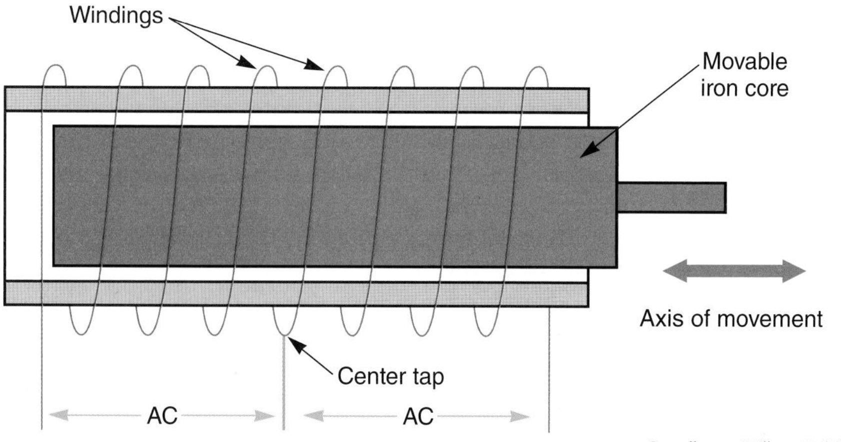

Goodheart-Willcox Publisher

Figure 8-4. A movable-core inductive transducer permits measurement of movement.

The linear variable differential transformer (LVDT) is a common type of inductive transducer, **Figure 8-5**. A movable metal core is placed in a tubular housing that has three windings. The center winding (primary) is connected to an ac source. Voltage is induced in the two outer windings by the primary winding. Initially, the voltage induced in each of the two outer windings is equal. Any movement of the core causes the voltage in one winding to increase and the voltage in the second winding to decrease. The difference between the induced voltages depends on the amount of core movement.

8.2 Types of Sensors

Many different sensing techniques and kinds of transducers are used with robotic systems. The type of sensor used depends on what is to be sensed, the accuracy required, and the environment. Sensors that are part of robotic systems use light, computer vision, sound, temperature, and magnetic fields.

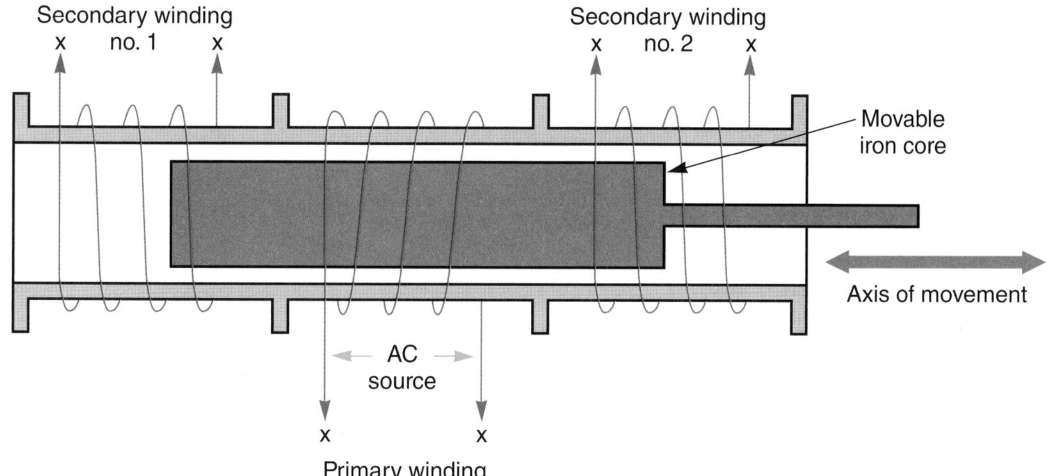

Goodheart-Willcox Publisher

Figure 8-5. A linear variable differential transformer (LVDT) measures movement based on changes of the voltage.

8.2.1 Light Sensors

Visible light is a form of radiation that occupies a narrow band of frequencies within the electromagnetic spectrum. The ***electromagnetic spectrum*** is a range of visible and invisible light, which includes radiation frequencies for radio, television, radar, infrared radiation, visible light, ultraviolet light, X-rays, and gamma rays, **Figure 8-6**. ***X-rays*** are an invisible band of radiation within the electromagnetic spectrum between ultraviolet and gamma rays. The different types of radiation differ only in frequency or wavelength.

The human eye responds only to light in the visible frequencies, **Figure 8-7**. Each color of light has a different frequency and wavelength. Wavelengths of light are measured in ***nanometers (nm)***, which is one billionth of a meter, and the ***angstrom (Å)*** unit, which is one-tenth of a nanometer. Visible light wavelengths range from violet (400 nm/4000 Å) to red (700 nm/7000 Å).

Light sensors respond to changes in light energy using various opto-electronic devices. ***Opto-electronic*** devices use a combination of optical and electronic components. These devices, along with lasers and X-ray devices, are important components in control circuits. A ***laser*** is a device that produces a highly focused or intense beam of light. Opto-electronic devices fall into three categories:

- ***Photoemissive devices*** emit electrons in the presence of light. Phototubes are a type of photoemissive device.
- ***Photoconductive devices*** vary in conductivity according to fluctuations in light. Their electrical resistance decreases when light is more intense and increases when light intensity decreases, **Figure 8-8**. A cadmium sulfide cell is an example of a photoconductive device. Such devices are also called light-dependent resistors (LDRs) since their resistance varies in inverse proportion to their conductivity.
- ***Photovoltaic devices***, or solar cells, convert light energy into electrical energy, **Figure 8-9**. When light energy falls on a photovoltaic device, it creates an electrical voltage. Although their electrical output is low, they are used with amplifying devices to drive a load.

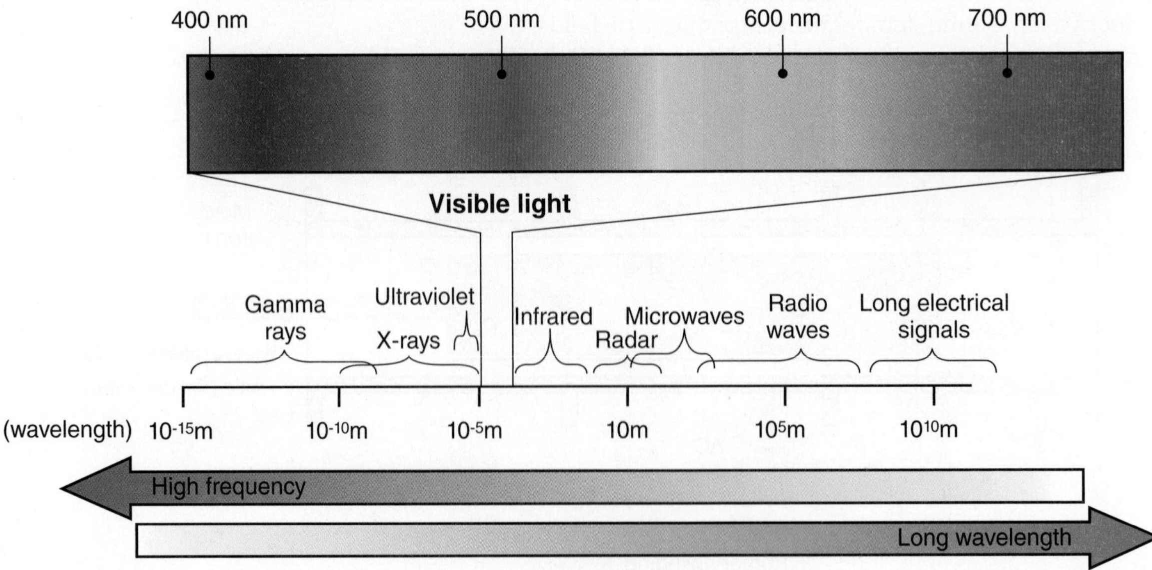

Figure 8-6. The electromagnetic spectrum includes only a narrow band of visible light.

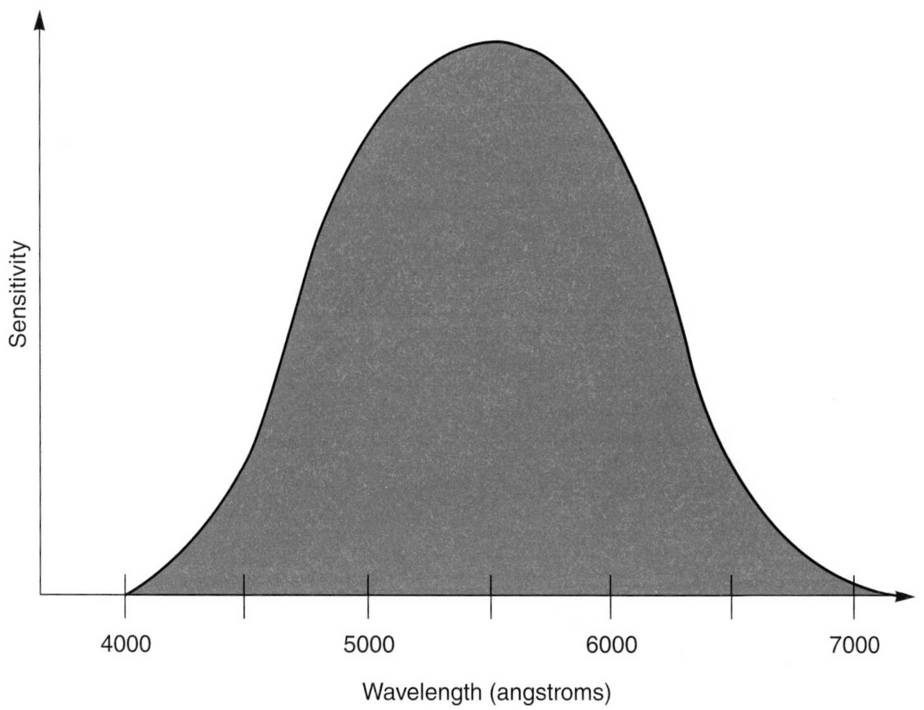

Goodheart-Willcox Publisher

Figure 8-7. The human eye responds to visible light at wavelengths between 4000 and 7000 angstroms, with peak sensitivity in the middle wavelengths (shades of green and yellow).

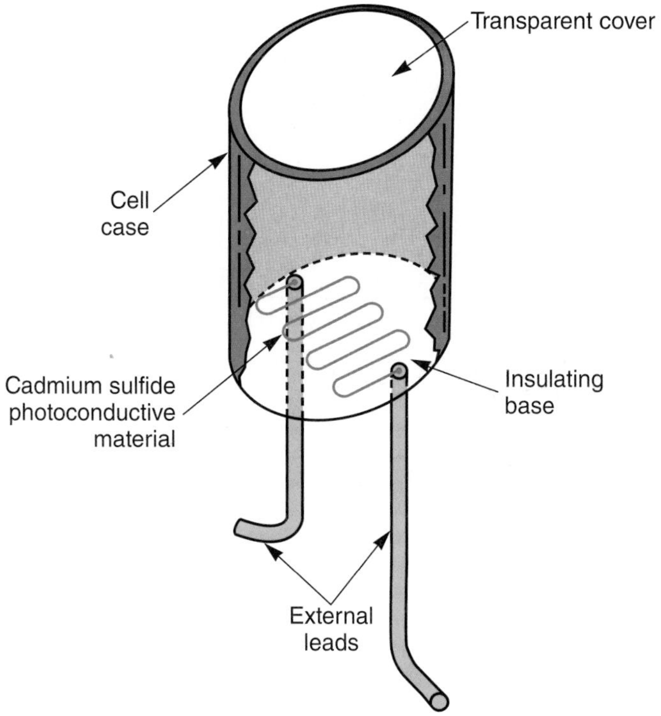

Goodheart-Willcox Publisher

Figure 8-8. The conductivity of a cadmium sulfide cell (cutaway view) varies with the amount of light energy striking it.

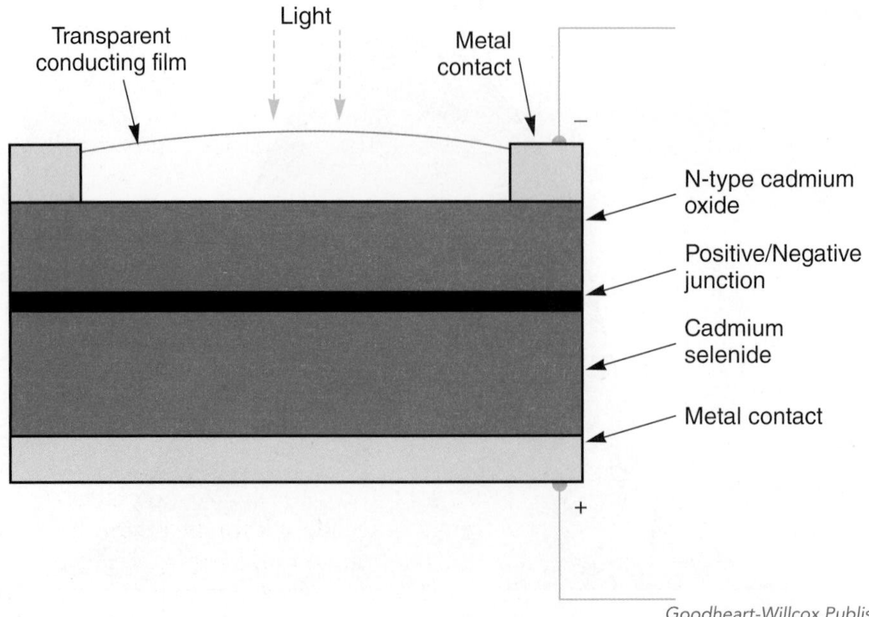

Goodheart-Willcox Publisher

Figure 8-9. Electrons in this selenium photovoltaic cell move from layer to layer, generating electrical energy.

Various types of opto-electronic devices are designed to sense the position of a light beam. They are used with digital control systems to produce an electrical output, **Figure 8-10**. The output is based on patterns created by the photoconductive material onto which the light beam is focused.

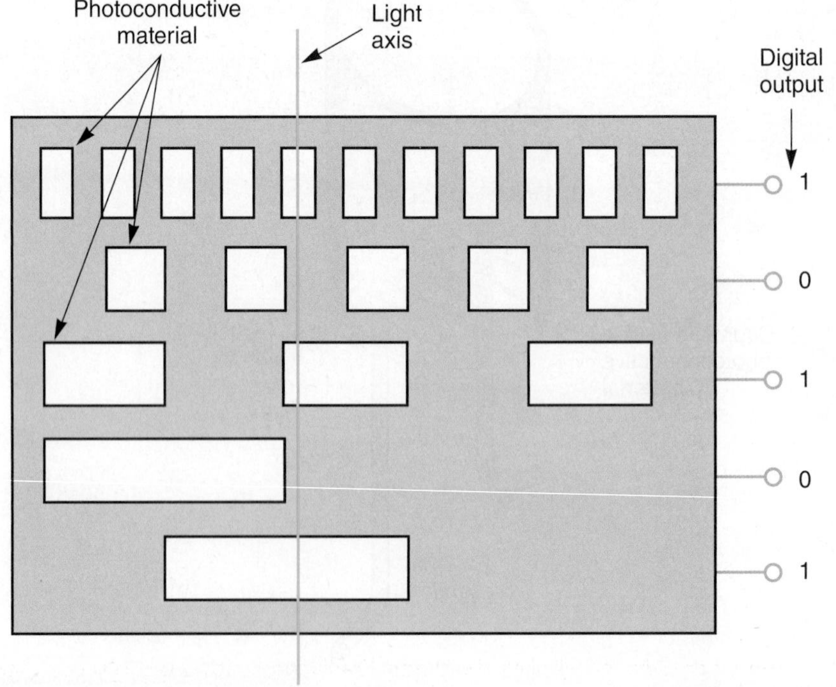

Goodheart-Willcox Publisher

Figure 8-10. This opto-electronic digital readout indicates the position of a light beam.

Infrared

Infrared sensors are devices that respond to radiation in the infrared region of the electromagnetic spectrum. All objects emit infrared thermal radiation. Infrared camera systems, **Figure 8-11**, can detect this radiation even in darkness. Industrial uses include heat-sensitive control systems, optical pyrometers, and infrared spectroscopy for gas analysis.

Ultraviolet

Ultraviolet sensors are devices that respond to electromagnetic radiation in the ultraviolet range. Ultraviolet sensors are used for many applications, such as color measurement, leak detection, impurity detection, and other precision measurement applications. These sensors can also be used in machine vision applications.

Fiber Optics

Optical fibers are strands of material made of glass or plastic that can transmit light from one point to another, **Figure 8-12**. The cladding (cover) material of the fibers is reflective, so light bounces from side to side within the optical fiber light pipe as it moves. Light travels through the optical fibers regardless of how the material is bent or shaped. The light can travel around corners, within a limited space, and over long distances. Fiber-optic sensors have replaced other sensor types in some applications, such as rotation, acceleration, magnetic field measurements, temperature, pressure, and sound.

Infrared Cameras, Inc.

Figure 8-11. An infrared image displays the thermal radiation emitted by objects. In this image, you can see the hot liquid (red and yellow) moving through the cooler piping (blue and purple) of this processing operation.

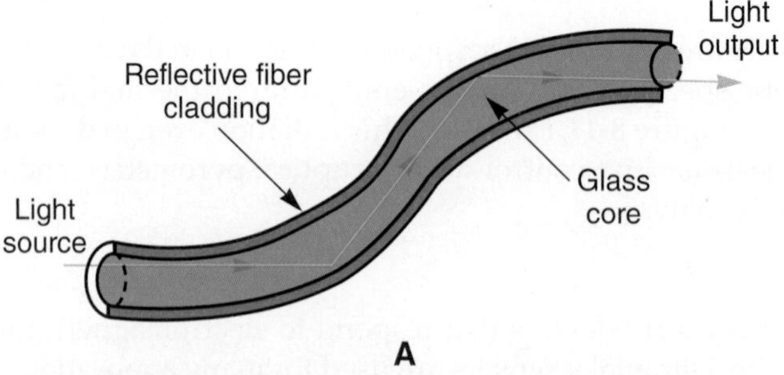

Goodheart-Willcox Publisher

Figure 8-12. Light can be bent around corners using optical fiber, sometimes called a light pipe. A—The components of an optical fiber. B—Fiber-optic strands.

Laser

The term *laser* stands for *light amplification by stimulated emission of radiation*. The development of the laser had a significant and continuing impact on industrial control systems, **Figure 8-13**. A major advantage of the laser is its tightly focused beam, which can travel long distances with relatively little spreading.

To produce a ruby laser beam, a xenon tube in the laser assembly flashes, and chromium atoms in the ruby rod absorb photons of light. The chromium atoms then emit their own photons. The photons travel along the axis of the ruby rod and are reflected by the mirrors on each end, which amplifies the light. The photons align with one another and produce only one wavelength (color) of light. The result is an intense beam of light energy.

Gas Lasers. Gas lasers are often used with sensing and control systems. A common type is the helium-neon laser, **Figure 8-14**. Many other types of gas lasers are available, but they all operate on basically the same principle.

A high dc potential is applied to a plasma tube using a voltage multiplier circuit and a pulse transformer. The filament within the plasma tube is heated by a 6.3V ac source. As electrons from the filament accelerate, they strike helium-neon

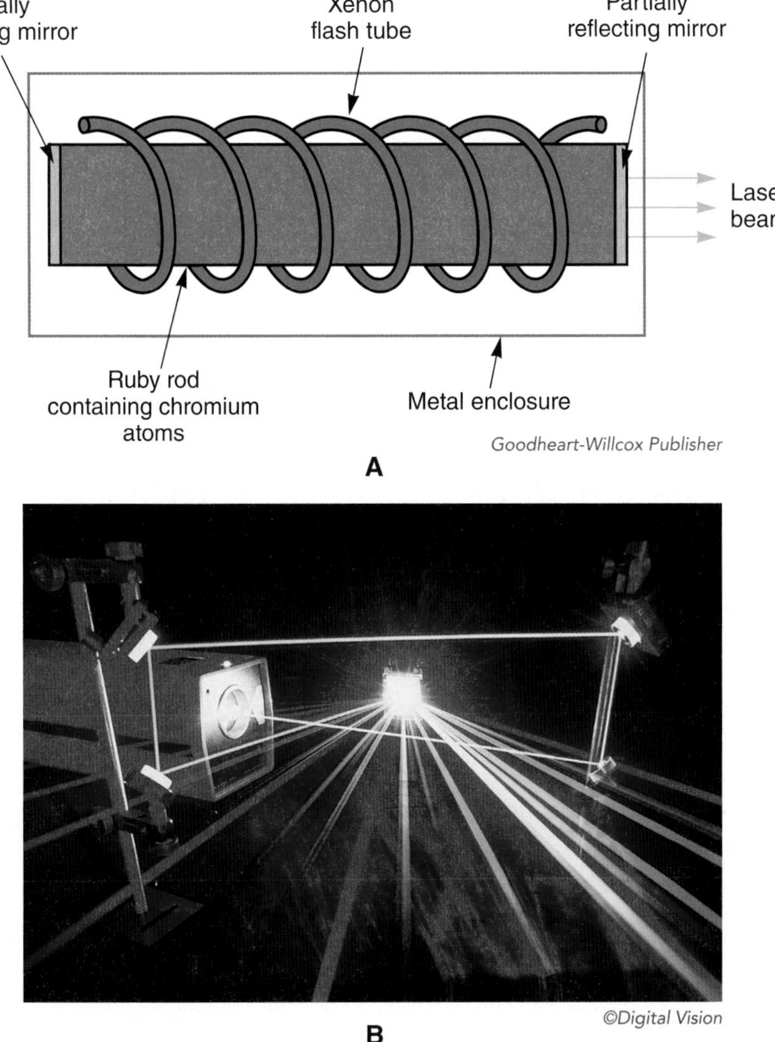

Figure 8-13. A ruby laser produces a beam of photons that do not spread out and scatter as ordinary light does. A—The components of a ruby laser. B—The focused beam of a ruby laser stays intact, even when reflected off of many surfaces.

gas atoms in the tube, which causes them to ionize. The ionized gas emits light. The light reflects from a flat, fully reflective mirror at the top. The plasma tube is cut at a precise angle to control the reflection. The light is reflected back and forth several times to the partially-reflective, spherical mirror at the bottom and is concentrated into a laser beam that is emitted through the spherical mirror.

Semiconductor Lasers. Semiconductors can be used to generate laser beams. Semiconductor injection lasers are very efficient and extremely tiny in size compared to other lasers, **Figure 8-15**. These lasers have a resonant cavity similar to other lasers. However, the cavity is formed on the chip of semiconductor material, which is typically made of gallium arsenide chips. The end faces of a gallium arsenide chip are parallel and flat. Since gallium arsenide is a reflective material, mirrors are not needed for the laser. This is a distinct advantage in terms of complexity and cost.

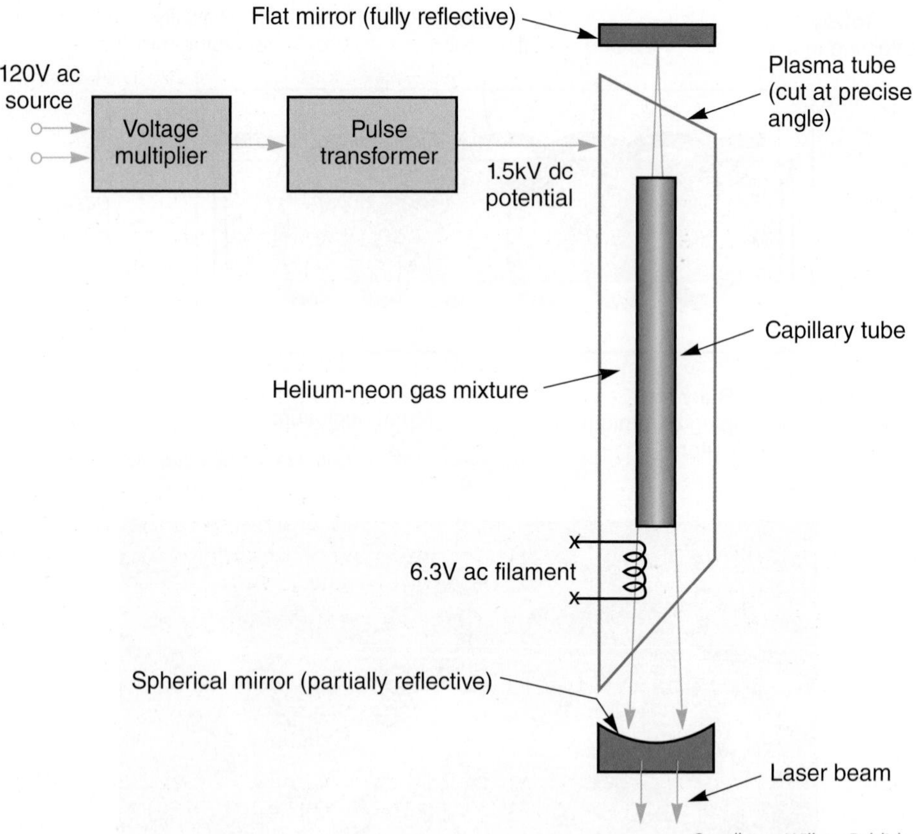

Goodheart-Willcox Publisher

Figure 8-14. A laser using helium-neon gas is one of the types used in industry for sensing and control applications.

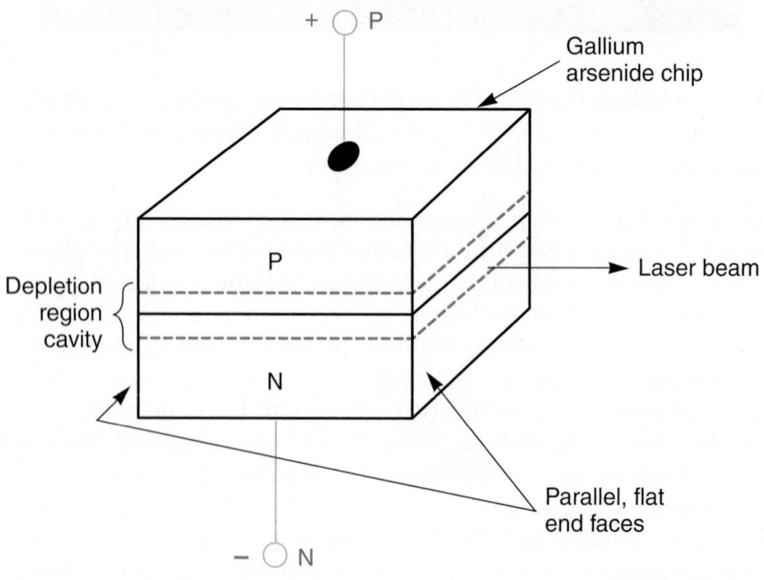

Goodheart-Willcox Publisher

Figure 8-15. This semiconductor injection laser requires no mirrors. It is tiny and very efficient.

As current flows through the chip, light is emitted from the gallium arsenide. Atoms collide in the area where positive and negative semiconductor materials meet, releasing additional photons. Due to the reflective properties of the gallium arsenide, a wave of photons is developed between its flat surfaces. The back-and-forth movement of this wave creates the resonant action required to produce a laser beam.

X-rays

X-rays are invisible rays within the electromagnetic spectrum between ultraviolet and gamma rays. Their short wavelength allows X-rays to pass through the human body. This property makes them valuable in medical treatment and analysis.

A vacuum tube can be used to produce rays similar to those emitted by radium, **Figure 8-16**. The cathode in an X-ray tube is heated using a filament voltage. The anode is constructed of a heavy metal. A high positive potential is applied to accelerate the electrons emitted from the cathode at a very rapid rate. The electrons strike the anode with such velocity that X-rays are created. If the potential is increased, the frequency of the X-rays increases, and the wavelength decreases.

X-ray tubes can operate with applied potential in excess of 1 million volts dc. The resulting X-rays are similar to the high-frequency gamma rays emitted by radium. In industry, X-rays are used to control processes that involve metals. The short wavelength of X-rays allows them to pass through metals and reveal structural characteristics.

8.2.2 Vision Sensors

Vision sensors may be used to recognize objects or to measure the characteristics of objects. Camera-equipped computer systems are able to identify a specific part

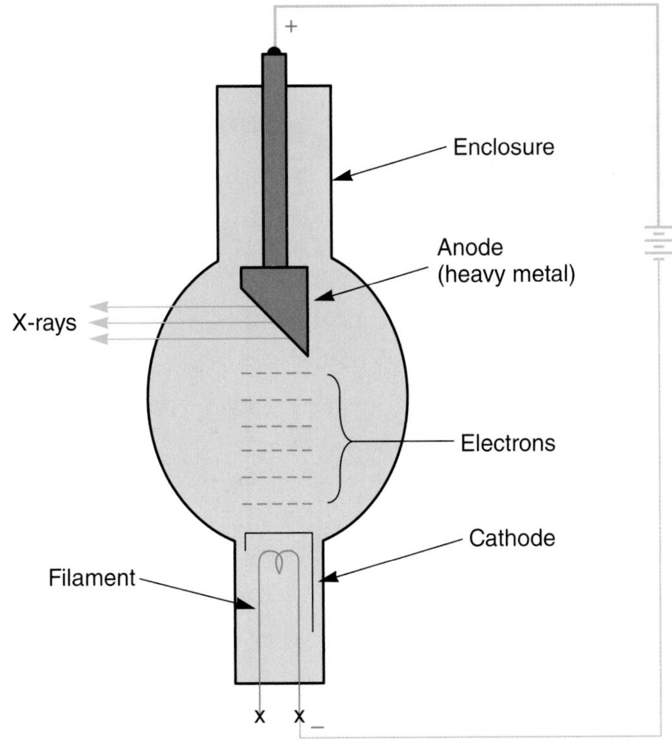

Goodheart-Willcox Publisher

Figure 8-16. The components of an X-ray tube.

using a video camera and can distinguish that particular part from any other. An object may be identified by its shape, outline, or area, regardless of its orientation to the camera.

Computer vision sensors are devices that detect spatial relationships and provide depth information using stadimetry and triangulation. ***Stadimetry*** determines the distance to an object based on the apparent size of the camera image. ***Triangulation*** involves measuring angles and the base line of a triangle to determine the position of an object.

Position detection can be accomplished by placing a camera on the robot's end effector. The feedback is used to guide the end effector to a specific location. This process is referred to as *visual servoing*, and it is used to move material from one spot to another. Servo movement may be used with either stationary or moving objects.

8.2.3 Sound Sensors

Sound sensors are devices that rely on the piezoelectric effect to convert sound to electrical energy. The ***piezoelectric effect*** occurs when certain crystals are subjected to mechanical stress and an electrical potential is developed in them. Rochelle salt and quartz are examples of these crystals.

A common application of the piezoelectric effect is the operation of crystal microphones. When sound waves (vibrations in air) strike a piezoelectric crystal, **Figure 8-17**, a voltage develops across the crystal. The voltage is amplified by the control system. In this way, mechanical energy (vibration) is converted to electrical energy.

8.2.4 Temperature Sensors

Thermoelectric sensors are devices that produce a change in electrical output due to a fluctuation in temperature. They are used to control processes in which temperature must be held at a given level or must not exceed a specific point.

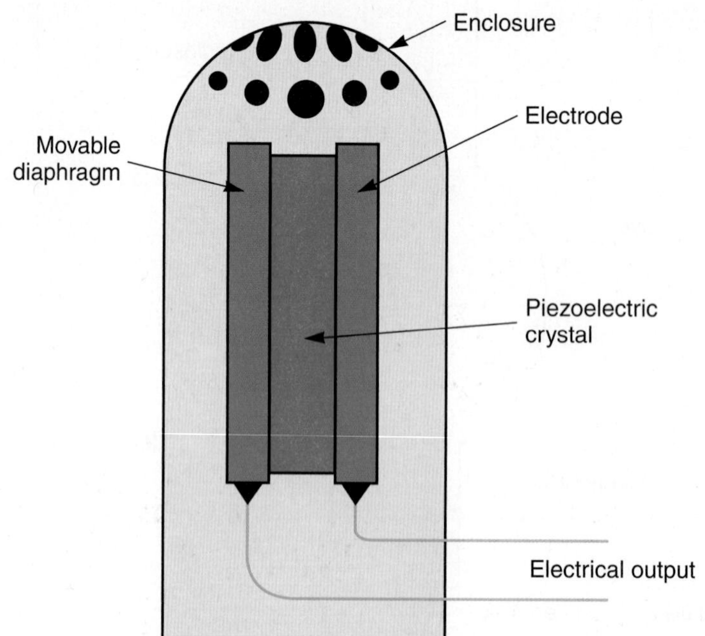

Goodheart-Willcox Publisher

Figure 8-17. One common application of the piezoelectric effect is the crystal microphone.

One device that is commonly used for heat sensing is the ***thermistor***. The resistance of this sensor decreases as temperature increases (and vice versa). This phenomenon is referred to as a negative temperature coefficient of resistance. Various metal-oxide semiconductor materials are used to construct thermistors, **Figure 8-18**. Thermistors are manufactured in a wide range of resistance characteristics and temperature coefficients.

Thermocouples are devices that convert heat energy into electrical energy. A thermocouple, **Figure 8-19**, consists of two dissimilar metal strips fused together at one end. The metals are usually a combination of iron-constantan, copper-constantan, and platinum-rhodium. Different metal combinations respond to different ranges of temperature. When the fused end is heated, a voltage develops at the ends that are not connected. This voltage exists because of the differing coefficients of expansion in the two metals. The voltage produced by a thermocouple is usually in the millivolt range.

8.2.5 Magnetic Field Sensors

Magnetic field sensors are devices that identify a change in an existing magnetic field without making physical contact with objects in the environment. This sensor detects changes in the magnetic field and sends a signal indicating the disturbance. Magnetic field sensors can be used in positioning operations, control

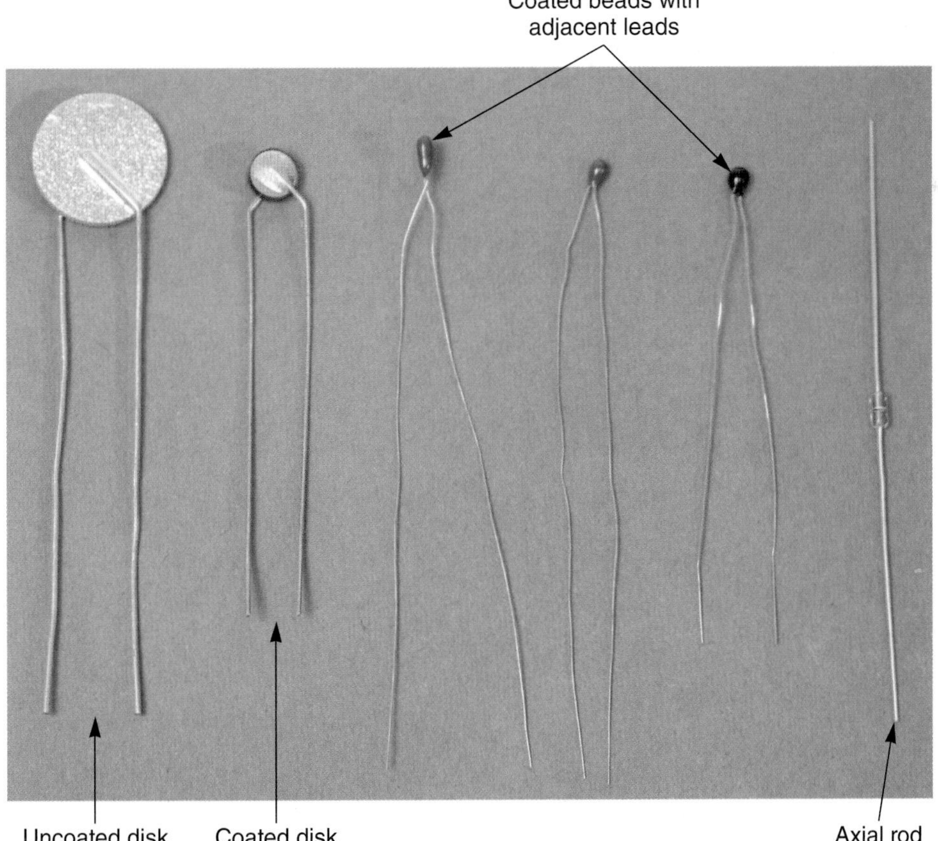

Goodheart-Willcox Publisher

Figure 8-18. In a thermistor, a decrease in temperature causes increased resistance. These are some of the common forms of the device.

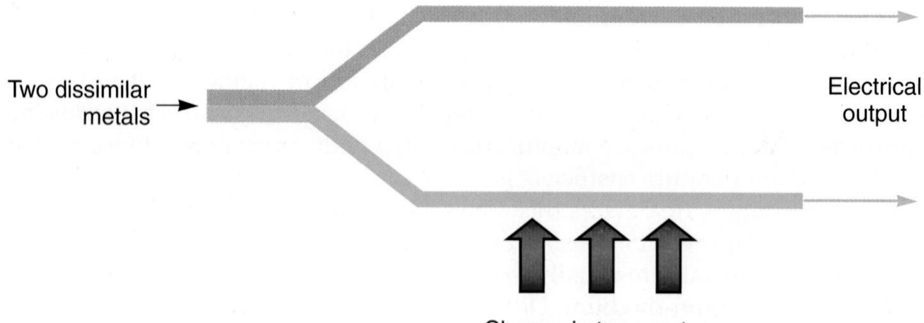

Goodheart-Willcox Publisher

Figure 8-19. A thermocouple converts heat energy into electrical energy.

tasks, range detection, and many other applications. Magnetic field sensors used in industrial applications typically use a permanent magnet as the constant magnetic field. Permanent magnets are not as susceptible to interference from other devices in the work area or from the earth's magnetic field.

Reed switches are simple magnetic field sensors that either make or break contact in response to changes in a magnetic field, **Figure 8-20**. Two flat metal strips called reeds are housed in a hollow glass tube filled with an inert gas. When the reeds are exposed to a magnetic field, they are forced together and complete an electrical circuit. When the magnetic field is turned off, the reeds spring open, and the circuit is broken. The reeds are housed inside a hermetically sealed glass tube, so contact sparks are isolated from the outside. The reeds are also isolated from outside dust and corrosion, which improves the life expectancy of the switch.

8.3 Sensor Applications

There is a wide variety of sensor types available and an equally wide variety of industrial and robotic applications for sensors. Sensor applications for robotic systems include sensing speed, mechanical movement, proximity, and range (distance).

8.3.1 Speed Sensors

Speed sensing involves measuring the rotary motion of shafts, gears, pulleys, and other rotating components of industrial equipment. A dc tachometer system is one method used for speed sensing, **Figure 8-21**. The tachometer is connected directly

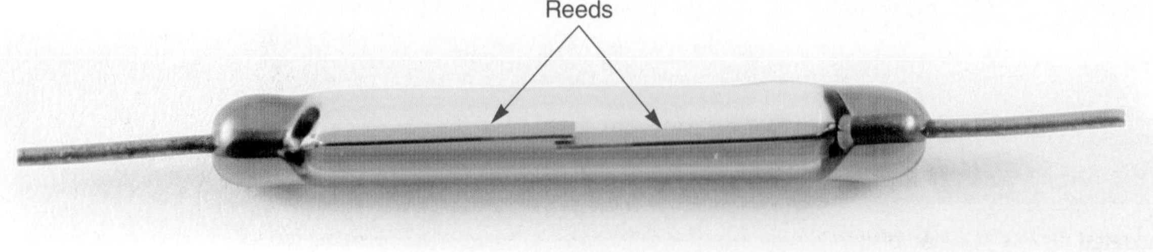

Goodheart-Willcox Publisher

Figure 8-20. A reed switch responds to changes in a magnetic field.

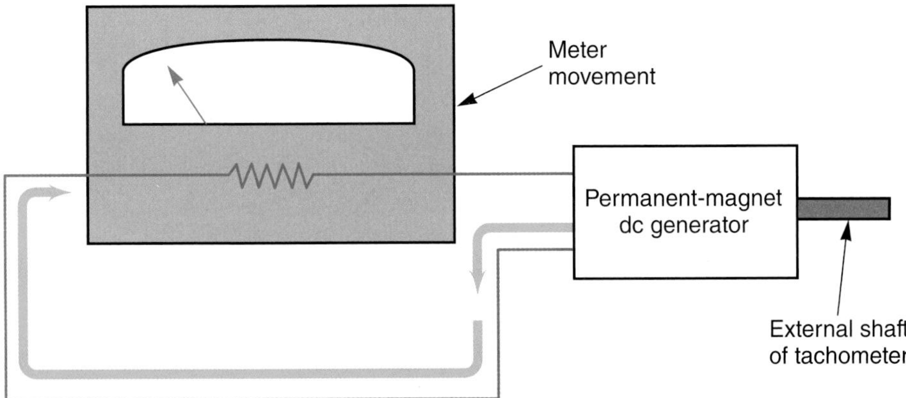

Goodheart-Willcox Publisher

Figure 8-21. A dc tachometer senses speed changes and displays them as meter fluctuations.

to a rotating piece of equipment. The rotation turns the shaft of the small dc generator, creating voltage. The generator's voltage output is directly related to the rotation rate of the shaft. Voltage output is translated into speed readings, which are used to control equipment operation.

Another method of speed sensing is done by using an electronic tachometer, which offers precision and ease of use. A reflective material is placed on the surface of the rotating portion of the equipment. Light emitted from the tachometer is reflected back when it encounters the reflective material. The reflected light reaches a photocell on the tachometer, which converts the pulses of reflected light energy into electrical signals. The electrical signals are used to measure the speed of rotation.

8.3.2 Mechanical Movement Sensors

Mechanical movement is a change in dimension resulting from an applied force. This change can be sensed using a ***strain gauge***, **Figure 8-22**. A strain gauge is made of fine-gage resistance wire, about 0.001″ (0.025 mm) in diameter, and is mounted on a strip of insulation. The wire is flexible, so it stretches when subjected to stress. As the wire stretches, its cross-sectional area is reduced, and its resistance changes. When used in a robotic system, strain gauges emit electrical signals (feedback) based on the amount of pressure the robot exerts to lift an object.

Semiconductors are commonly used in place of traditional metal strain gauges. The rate of change in resistance is approximately 50 times higher than for metal strain gauges, and semiconductors are more sensitive to small changes. A semiconductor strain gauge is as stable as the metal type in terms of consistency of signal changes caused by resistance variation. However, semiconductor strain gauges typically produce a higher electrical signal (feedback voltage) output.

8.3.3 Proximity Sensors

Most ***proximity sensors*** are devices that detect either the absence of an object or the presence of an object within a certain distance. Some proximity sensors provide feedback about the distance between the end effector and an object.

Optical proximity sensors are devices that measure the amount of light reflected from an object. They can respond to visible or infrared light. Incandescent lights can be used as the light source. However, light-emitting diodes are generally

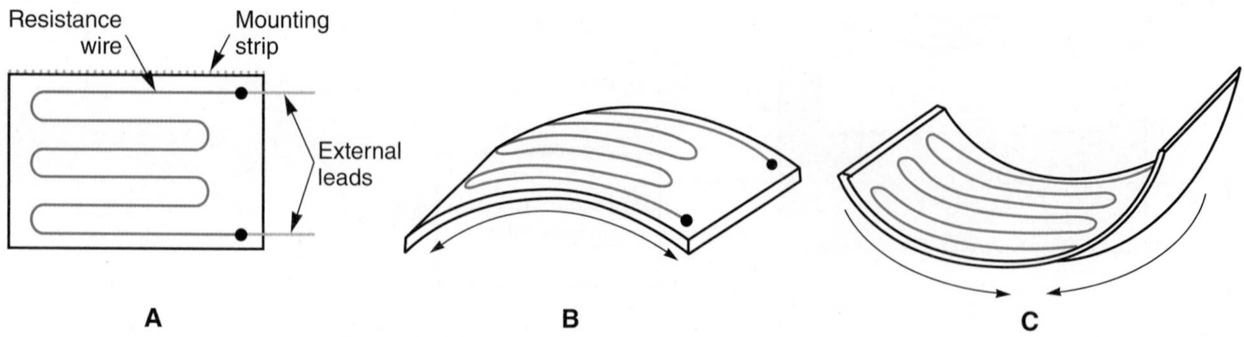

Goodheart-Willcox Publisher

Figure 8-22. A strain gauge responds to mechanical energy by changing its resistance. A—The components of a strain gauge. B—Resistance increases when the strain gauge is under tension. C—Resistance decreases when the strain gauge is under compression.

preferred. ***Light-emitting diodes (LEDs)*** are small, lightweight opto-electronic devices (solid-state lamps) that are made to produce different colors of light. LEDs are more reliable than incandescent lights and are not sensitive to shock and vibration. The semiconductors used in LEDs produce light when an electric current is applied. The response varies according to the type of semiconductor used. The wavelength of radiated energy from an infrared LED is beyond the visible range and, therefore, cannot be seen by the human eye.

Nick Holonyak Jr. invented LEDs.

Inventor Spotlight: Nick Holonyak Jr.

Light-emitting diodes (LEDs) can be found in everyday objects, such as alarm clocks, flashlights, and children's toys. They are also used in complex systems, such as infrared sensors and fiber-optic communication technology. These tiny devices were invented by Nick Holonyak Jr. in 1962 while working in the Advanced Semiconductor Laboratory at the General Electric Company. Though the first visible LED emitted red light, LED lights are now available in an array of colors.

Nick Holonyak Jr. has been involved in many other inventions and developments that have made an immediate impact on technology and modern society.

- In 1958, he helped develop the first light dimmer switch while working at General Electric.
- In 1960, Holonyak created the red-light semiconductor laser (laser diode), which is used in CD players, DVD players, and CD-ROM drives.
- In 1990, he developed the technology that spawned the creation of the lasers used in copy machines, laser printers, computer mice, and communication and data transmission.

Nick Holonyak Jr. is also the co-inventor of the infrared light-emitting transistor (LET). This recent development has the potential to revolutionize telecommunication systems and computer systems in terms of both size and capability.

Photo courtesy of Goodheart-Willcox Publisher

Eddy current proximity sensors are devices that produce a magnetic field in the small space of a detector unit, which can be mounted in a probe. The magnetic field induces eddy currents into any conductive material that is near the probe. A pick-up coil senses a change in magnetic field intensity when an object enters the field.

Acoustical proximity sensors are devices that react to sound. Standing sound waves are generated within a cylindrical, open-ended cavity inside the sensor. The presence of a nearby object interferes with these waves, which alters the sensor output. A microphone may be used to detect a change in sound pressure and measure the distance of the object from the sensor.

Touch-sensitive proximity sensors operate on capacitance developed by a large conductive object (such as the human body). This capacitance changes the frequency of an electronic circuit. A conductive plate or rod may be used to sense contact.

Tactile sensors are devices that use touch to determine whether an object is present. Stress sensors produce a signal that indicates the magnitude of the contact made. A simple type of touch sensor is a microswitch. ***Microswitches*** are electrical switches that are turned on or off with a very small amount of force. They are small, durable, and easy to activate. Common applications for microswitches include computer mouse buttons and joystick controls. Limit switches also respond to contact with an object. A ***limit switch*** is an electrical device that has an actuator that is mechanically linked to a set of contacts, which are either normally open or normally closed (**Figure 8-23**). These switches may be used with robotic systems to sense the presence (or absence) of an object. When activated, contacts that are normally open are closed and contacts that are normally closed are opened. This causes a change of state to an electrical circuit. Limit switches are rugged, simple, easy to install, and reliable. They are used for more heavy-duty applications than microswitches, including conveyor systems, milling, and drilling machinery operations.

Range sensors are devices that determine the precise distance from the sensor to an object. Such devices are useful for locating objects near a work station or for controlling a robot manipulator. A ***laser interferometric gauge*** is a range-sensing system that is sensitive to humidity, temperature, and vibration. Some range-sensing systems use a video camera that operates with sonar (sound).

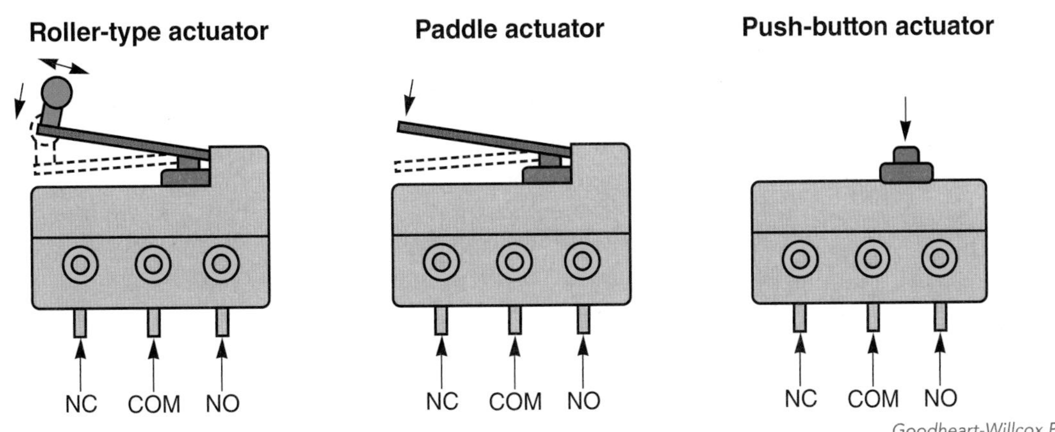

Figure 8-23. Types of limit switches. NC = normally closed contact, NO = normally open contact, COM = common.

Summary

- Sensors (transducers) operate by converting light, heat, or mechanical energy into electrical signals to affect the operation of a robotic system.
- Resistive transducers are devices that convert a variation in resistance into electrical variations. Resistive transducers are often used to sense physical displacement.
- Capacitive transducers measure a change in capacitance, or the ratio of charge on a conductor to the potential difference between conductors.
- An inductive transducer is a device that measures movement and creates signals that affect current flow. The linear variable differential transformer (LVDT) is a common type of inductive transducer.
- Opto-electronic devices used as light sensors include photoemissive, photoconductive, and photovoltaic types.
- Infrared sensors, ultraviolet sensors, optical fibers, and laser devices may be used in robotic sensing applications.
- X-rays are used to control processes that involve metals. The short wavelength of X-rays allows them to pass through metals and reveal structural characteristics.
- Vision sensors may be used to recognize objects or to measure the characteristics of objects.
- Sound sensors use the piezoelectric effect to convert sound to electrical energy.
- Temperature sensors, such as thermistors and thermocouples, produce a change in electrical output due to fluctuating temperatures.
- Magnetic field sensors, such as reed switches, identify a change in an existing magnetic field without making physical contact with an object in the environment.
- Sensor applications for robotic systems include sensing speed, mechanical movement, proximity, and range (distance).

Review Questions

Answer the following questions using the information provided in this chapter.

1. What is the purpose of a sensing system for a robot?
2. What is a transducer? Identify three types of transducers and explain how each one works.
3. What are three categories of opto-electronic devices? Describe each category and give an example of the device.
4. What is the difference between an infrared sensor and an ultraviolet sensor? What are some uses for each type of sensor?
5. Explain the basic principle of how a gas laser operates.
6. How are X-rays used in industrial control?
7. How are sensors used to measure depth?
8. What is visual servoing?
9. Explain the piezoelectric effect.
10. What are thermoelectric sensors? Describe two of these devices.
11. What is a magnetic field sensor, and what are some applications for this type of sensor?
12. Describe how reed switches operate.

13. What are four broad areas of sensor applications for robotic systems?
14. How is a strain gauge used to sense mechanical movement?
15. What are range sensors? Give an example of an application for a range sensor.

Learning Extensions

1. Make a list of five different types of sensors. Search magazines or the Internet for examples of how the sensors are used in the robotics industry. Identify three different applications for each of the sensor types you listed.
2. For each of the sensors identified in the previous activity, note the type of transducer used by the sensor (resistive, capacitive, or inductive).

Advanced Analysis

1. Prepare a discussion in which you compare and contrast the operation of a thermistor and a thermocouple to perform temperature measurements.
2. Refer to **Figure 8-6**, and perform an Internet search. Prepare a simplified discussion of the sections of the electromagnetic spectrum, including radio waves, microwave, radar, infrared, visible light, ultraviolet, X-rays, and gamma rays as they relate to frequency and wavelength.

Activity 8-1 Proximity Detectors

Objectives

Proximity detectors are used to trigger an alarm or turn on a circuit by detecting a physical change within a designated area. These circuits are designed to respond in some way to a change in capacitance. This change usually alters the frequency of an oscillator so the output can be amplified and used to trigger a load control device. Typically, the circuit contains an oscillator, a tuned circuit, an amplifier, and a trigger device. This type of sensor is representative of the type of sensors that might be used in robotic system applications.

One type of proximity detector uses the loaded oscillator principle. In this circuit, the LC components (inductor and capacitor) of the oscillator are shunted to ground by an external capacitance. This capacitance is usually called the sensor plate. Moving a hand or producing a physical change near the sensor plate upsets the electrostatic field of the capacitance. This, in turn, alters the frequency of the oscillator or may load down the circuit in such a way as to quench the oscillation process. Liquid level detection, automatic thickness gauge testers, intruder alarm systems, counters, and switching circuits employ this type of detector.

In this activity, you will construct a loaded oscillator proximity detector. The oscillator is a Colpitts type with split capacitors in the transistor collector circuit. The output of the oscillator is rectified and applied to a voltage comparator circuit where it is amplified by an op amp. The amplified output is used to trigger a load control device. The 5 kΩ resistor in the comparator circuit is used to establish a balanced condition for detection. When the oscillator senses a change, it loads down and stops oscillating. This creates an imbalance, which is then detected by the comparator circuit.

Through this activity, you will test the operation of a loaded oscillator circuit and observe typical waveforms. The test procedure used to analyze circuit operation is also used to troubleshoot the circuit. Through this testing experience, you will become familiar with the oscillator loading principle and be able to follow a logical troubleshooting procedure.

Equipment and Materials

- 0–10V, 1 A dc power supply.
- LM3900 op amp.
- 2N3397 transistor.
- 10 mH coil.
- 1N4004 diode.
- Light-emitting diode.
- 0.1 μF, 200V dc capacitor.
- 0.01 μF, 100V dc capacitor.
- 100 pF, 100V dc capacitor.
- 47 pF, 100V dc capacitor.

- 1 MΩ, 1/4 W resistors (2).
- 68 kΩ, 1/4 W resistor.
- 10 kΩ, 1/4 W resistors (2).
- 6.8 kΩ, 1/4 W resistor.
- 5 kΩ, 2 W potentiometer.
- 1 kΩ, 1/4 W resistor.
- 470 Ω, 1/8 W resistor.
- 500 kΩ, 2 W potentiometer.
- SPST toggle switches (2).
- 6″ × 6″ sensor plate (metal).

Optional according to the selected load control circuit.

- 2N3397 transistor.
- 12V, 125 Ω relay (Guardian 1335-2C-120D or equivalent).
- No. 47 lamp with socket.
- 47 Ω, 1/2 W resistor.
- SCR (MCR 12-D or equivalent).
- 220 Ω, 1/2 W resistor.
- 6.3V, 60 Hz ac source.

Procedure

1. Construct the proximity detector as shown.

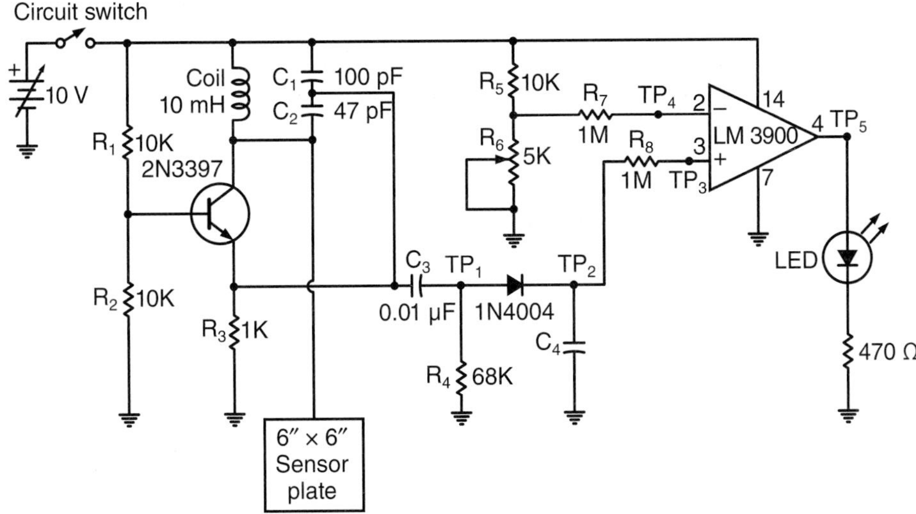

Proximity detector.

2. Close the circuit switch and adjust the 5 kΩ potentiometer for illumination of the LED. There should be a specific point where the LED turns on. Increasing resistance will cause it to remain on, while decreasing the resistance causes it to turn off. The point just before turn-on represents the most sensitive detection setting of the potentiometer. Adjust the potentiometer several times to locate the most sensitive point.
3. If the circuit is not operating properly, perform steps 4, 8, 9, 10, and 11 first to determine which part of the circuit is malfunctioning. If the circuit is

operating satisfactorily, follow the procedure steps in order and record the indicated data.

4. Prepare an oscilloscope for operation and connect it to test point 1. Make a sketch of the observed waveform.

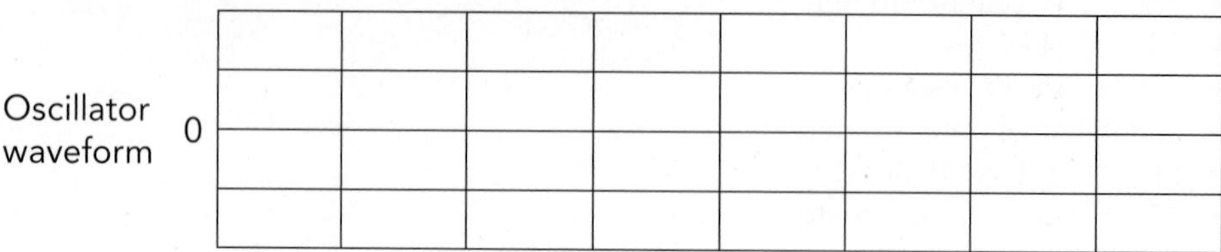

Goodheart-Willcox Publisher

Observed waveform at test point.

5. While observing the oscilloscope, touch the sensor plate. What influence does this have on the waveform?
6. Move your hand over the sensor plate while observing the oscilloscope. What influence does this have on the waveform?
7. Determine the frequency of the oscillator with the oscilloscope or a frequency meter.
8. Measure and record the voltages at the emitter-base and collector of the oscillator transistor Q_1. Compare your measurements with those indicated on the schematic diagram.
9. Measure and record the dc output voltage at test point 2.
10. Touch the sensor plate and record the voltage at test point 2. This represents the loaded output voltage of the oscillator. The output of the oscillator will change approximately 2V between the loaded and unloaded conditions of operation.
11. Measure and record input voltages at TP_3 and TP_4 when the LED is on. This indicates a balanced condition.
12. Measure and record the voltages at TP_3 and TP_4 when the circuit is unbalanced.

13. Measure and record the output voltage of the circuit at test point TP_5 for the balanced and unbalanced conditions. What is the total change in voltage?

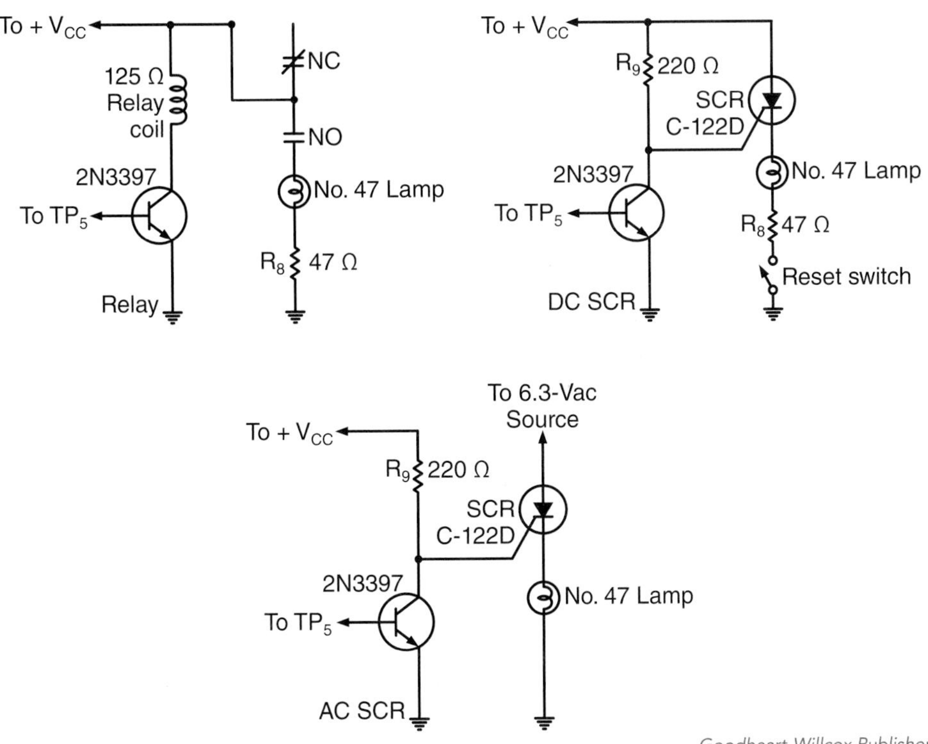

Alternate load control circuit.

14. Select one of the alternate load control circuits and construct it. Connect it to the output of the proximity circuit.
15. Close the circuit switch, adjust the balance potentiometer, and test the control capability of the completed load control circuit.
16. Open the circuit switch. Disconnect the circuit and return all parts to the storage area.

Analysis

1. Describe what is meant by the terms *loaded* and *unloaded* as they relate to oscillator conditions.
2. What is the function of the IC op amp in the proximity circuit?
3. Draw a block diagram representing the major parts of the proximity circuit constructed in step 15.

Activity 8-2 LVDT Detector

Objectives

Frequently, it is not enough to measure only the amount of the displacement of an object. Often the direction of displacement, as well as the amount of displacement, must be indicated. When this becomes the case, a detector circuit similar to that shown below can be used in conjunction with the output of a linear variable displacement transformer (LVDT).

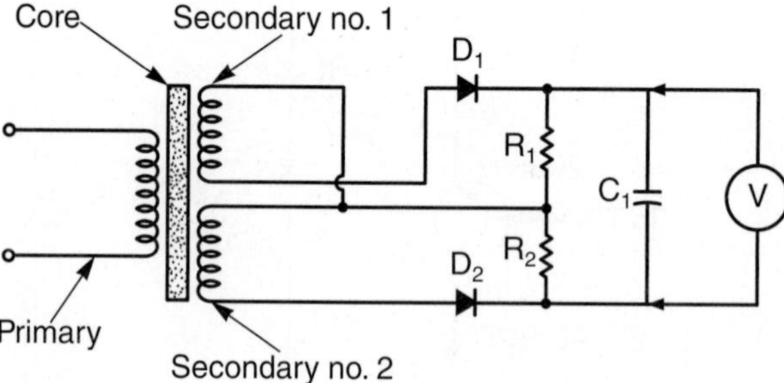

Goodheart-Willcox Publisher

LVDT detector circuit.

In this sensor circuit, the diodes conduct equally when the movable core is in the null position. This results in an equal and opposite voltage drop across R_1 and R_2, causing the readout device (voltmeter) to indicate zero displacement. When displacement occurs, one diode will conduct more than the other, causing an upset in the balanced voltage across R_1 and R_2. The result is an output voltage, as measured by the voltmeter, with polarity that indicates the direction of displacement and with magnitude that indicates the amount of displacement. In this activity, you will examine a detector circuit that could be used with robotic systems or other applications requiring sensors.

Equipment and Materials

- Digital multimeter.
- Air-core coils, 100 turns, A.S. No. 16 wire (2).
- Air-core coil, 200 turns, A.S. No. 24 wire.
- 6″ (15 cm) steel core, 3/4″ (1.91 cm) diameter.
- 6V ac power supply.
- 1N4004 diodes (2).
- Resistors—1 kΩ (2).
- Capacitor—1 μF, 25V dc.
- Connecting wires.
- SPST switch.

Procedure

1. Construct the circuit as shown.

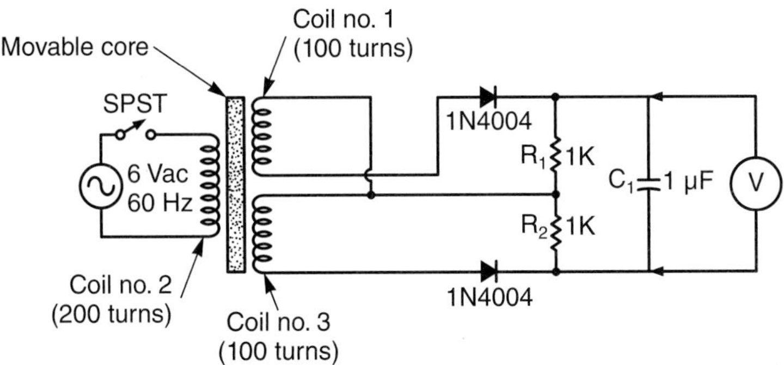

Goodheart-Willcox Publisher

LVDT detector test circuit.

2. Close the SPST switch and slide the movable core until the voltmeter indicates zero.
3. Move the core in the direction and by the amount indicated. Record the voltage output of the detector circuit along with the proper voltage polarity.

Direction of movement from null position	Amount of movement		Voltage output	Polarity (+ or –)
	Inches	Centimeters		
Left	1/16	0.159		
Left	1/8	0.318		
Left	3/16	0.477		
Left	1/4	0.635		
Left	5/16	0.793		
Left	3/8	0.953		
Return to null position				
Right	1/16	0.159		
Right	1/8	0.318		
Right	3/16	0.477		
Right	1/4	0.635		
Right	5/16	0.793		
Right	3/8	0.953		

Goodheart-Willcox Publisher

Output polarity and magnitude of LVDT test circuit.

4. How do the voltages generated by moving the core to the left compare with the voltages generated by moving the core to the right?
5. Ideally, what is the minimum displacement of the movable core that would result in a voltage output?

Analysis

1. Describe how coils of an LVDT can be connected in series to oppose.
2. Explain how the detector circuit used in conjunction with the LVDT enables the direction of displacement to be measured.
3. When might it be necessary to measure the direction of displacement as well as amount of displacement?
4. How would the null position of the core of the LVDT be affected if R_1 of the detector circuit were a value different from R_2?

Activity 8-3
Temperature Sensors

Objectives

Temperature-sensing circuits are widely used in industry as system control elements. Heat sensors, which are by far the most popular of all sensors, are commonly found in alarm circuits that detect changes in temperature at remote locations. Machinery, electronics equipment, and measuring instruments also employ temperature sensors to detect unusual operating conditions. The thermistor is a very popular sensor element for temperature detection that might be used with robotic applications.

Thermistors are primarily classified as temperature-sensitive resistors with an operating range of from –382°F (–230°C) to 1202°F (650°C). When used in bridge circuits with high-gain amplification, thermistors can detect temperature changes as small as 0.001°.

The physical makeup of a thermistor includes a mixture of nickel, manganese, and cobalt oxides formed into a piece of semiconductor material. These oxides are mixed together and fired to form a coherent nonporous material. The mixture is then formed into a variety of shapes. In this activity, a tiny piece of the ceramic material is formed into a bead and enclosed in glass. Changes in temperature cause the resistance of the thermistor to produce a wide range of values. Typically, thermistors have a negative temperature coefficient. This means that an increase in temperature causes a decrease in resistance. Metal, by comparison, has a positive temperature coefficient. This means that a rise in temperature causes a corresponding increase in resistance.

In this activity, you will build a temperature-detecting bridge circuit. The output of the bridge is fed into an op amp for high-gain amplification and increased sensitivity. Any imbalance in the bridge is detected by the amplifier and is used to trigger an output circuit. Circuits of this type are typically found in alarm circuits that function as heat detectors and might be used with robotic and assembly applications in industry.

Equipment and Materials

- Split dc power supply: ±9V at 1 A, or two 9V batteries.
- μA741C op amp.
- Thermistor (Fenwal gB 32J2 or equivalent).
- 200 kΩ, 1/4 W resistor.
- 10 kΩ, 1/4 W resistors (2).
- 5 kΩ, 2 W potentiometer.
- 2.7 kΩ, 1/4 W resistor.
- 1 kΩ, 1/4 W resistor.
- 47 Ω, 1/4 W resistor.
- No. 47 lamp with socket.
- IC circuit construction board.

- SCR (GE C-122D or equivalent).
- Electronic multifunction meter.
- Decade resistance box (optional).
- SPST toggle switch.

Procedure

1. Construct the thermistor bridge circuit. Note that a voltmeter is used to detect the amplified output of the circuit.

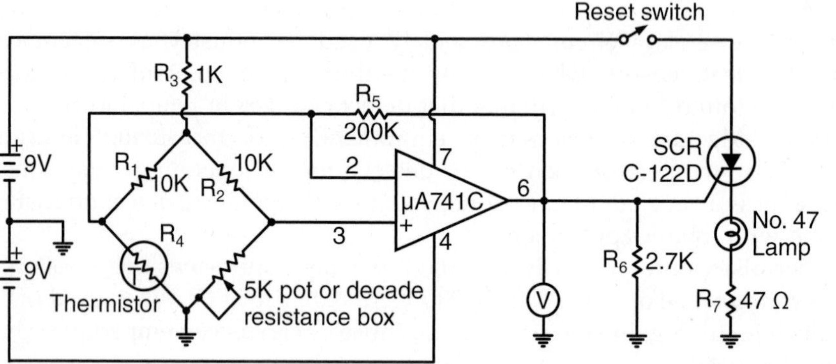

Thermistor bridge circuit.

2. Before connecting the thermistor into the circuit, measure its resistance. Try to avoid touching the glass part of the thermistor when attaching the test leads. With the ohmmeter connected, the resistance should stabilize after a few seconds. What is the stabilized resistance?
3. While observing the ohmmeter, grasp the glass bead between your index finger and thumb. How does body temperature influence the resistance of the thermistor?
4. Connect the thermistor to the bridge circuit as indicated. The reset switch should be in the off position, and the electronic multifunction meter should be adjusted to the zero center position. It should be in the 50V or equivalent range.
5. Turn on the split power supplies and energize the circuit. Balance the bridge by adjusting the potentiometer to produce a zero indication on the meter. A decade resistance box can be used in place of the potentiometer.
6. After the bridge has been nearly balanced, you may switch the voltmeter to a lower range to improve the balancing accuracy.
7. To test the sensitivity of the circuit, place your finger near the glass bead of the thermistor while observing the voltmeter. If the circuit is working properly, the meter should deflect upscale or in the positive direction. Within a few seconds, it should return to the balanced indication when the thermistor has reached its stabilized resistance. If you touch the glass bead with your finger, it usually takes longer for it to return to the balanced state.
8. You may want to try blowing on the thermistor or placing it near some heat-producing source. Avoid temperatures over 302°F (150°C).

9. Close the reset switch of the SCR. Place your finger near the thermistor. While observing the voltmeter, see how much voltage is needed to trigger the SCR into conduction. You must wait a few seconds for the thermistor to stabilize before resetting the SCR.
10. Try several sources of heat to turn on the sensor circuit.
11. Turn off the power supply and disconnect the circuit. Return all components to the storage area.

Analysis

1. What type of temperature coefficient does the thermistor used in this activity have?
2. Explain how a change in thermistor resistance causes the bridge circuit to be imbalanced.
3. Why is it advantageous to use an op amp with the bridge circuit of this activity?

Activity 8-4
Thermocouple Applications

Objectives

Thermocouples are frequently used to measure temperatures in an industrial setting. Due to the relatively low voltage output associated with most thermocouples, amplification circuits are used to increase this output as well as to increase sensitivity. The resulting output of the amplification circuit is used to drive or activate a readout device.

In this activity, you will see the thermocouple used to control the conductivity of an FET. The conductivity of the FET, in turn, controls the action of a single-stage transistor amplifier. Therefore, it controls the current flow through a multimeter that is used as the readout device.

Equipment and Materials

- Type J thermocouple.
- Digital multimeter.
- Resistors—560 Ω, 5.6 kΩ.
- 0–5V dc power supply.
- 10V dc power supply.
- GE-FET-1 field-effect transistor.
- 2N2405 npn transistor.
- Connecting wires.
- 660 W heat cone.
- 120V ac power supply.

Procedure

1. Construct the circuit.

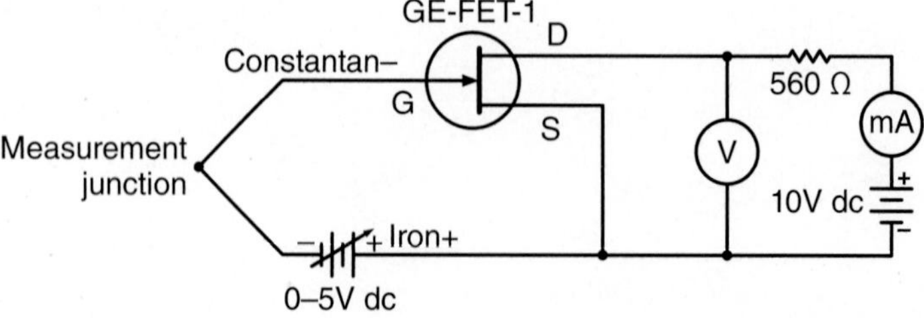

Goodheart-Willcox Publisher

Type J thermocouple FET circuit.

2. Allow the measurement junction of the thermocouple to remain at room temperature. Alter the gate voltage of the FET to equal the listed setting. Record the source drain current and voltage for each FET gate voltage value.

Gate voltage (V)	Source-drain current (mA)	Source-drain voltage (V)
0.2		
0.4		
0.6		
0.8		
1.0		
1.5		
2.0		

Goodheart-Willcox Publisher

Thermocouple-controlled FET circuit data.

3. Adjust the gate voltage to 0.2V.
4. Grasp the measurement junction of the thermocouple between your thumb and forefinger. Describe how this action affects the source drain current and voltage of the FET as compared to the data gathered in step 2 when the gate voltage was 0.2V.
5. Connect the 660 W heat cone to 120V ac, and allow it to warm up for about three minutes.
6. Position the measurement junction of the thermocouple inside the heat cone for a period of 3 minutes and record the source drain current and voltage of the FET.
7. How do the values of source drain current and voltage recorded in step 6 compare with the data gathered in step 2 when the FET gate voltage was 0.2V dc?
8. Disconnect the heat cone from the 120V ac power supply. What effect does the removal of the heat cone have on the source drain current and voltage?
9. Construct the circuit illustrated below.

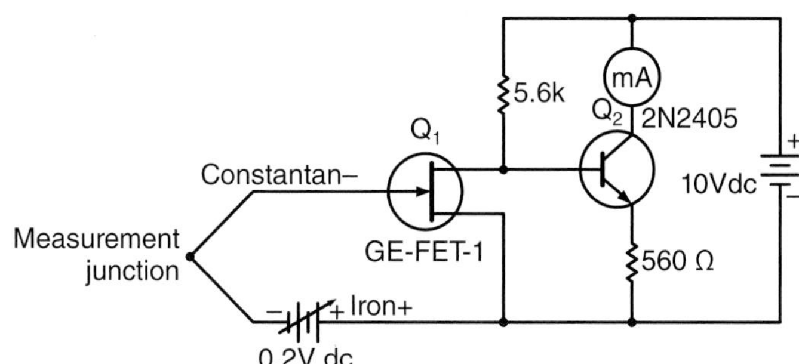

Goodheart-Willcox Publisher

Type J thermocouple transistor circuit.

10. Record the collector current of Q_2 as displayed by the digital multimeter with the measurement junction of the thermocouple at room temperature.

11. Grasp the measurement junction of the thermocouple between your thumb and forefinger. Describe how this action affects the collector current of Q_2 as displayed on the digital multimeter.
12. Place the measurement junction of the thermocouple inside the cone of the *cool* 660 W heater. Connect the 120V ac to the heat cone and allow it to warm up for five minutes.
13. Record the collector current of Q_2 as displayed on the digital multimeter after the heat cone has warmed up.
14. How does the current recorded in step 10 compare with the current recorded in step 13?
15. How can the difference be explained?
16. If the digital multimeter used in the circuit shown in step 9 were calibrated in degrees Fahrenheit, how could the circuit be used to measure temperature?
17. Disassemble the components and return all components to the storage area.

Analysis

1. Why is it sometimes necessary to use an amplifier when a thermocouple is used to measure temperature?
2. What was the readout device used in the circuit in step 9?
3. What would determine the maximum temperature that could be measured by the circuit in step 9?
4. How did the action of the circuit used in step 9 differ from that used in step 2?
5. Explain how the circuit in step 2 could be used to measure temperature.

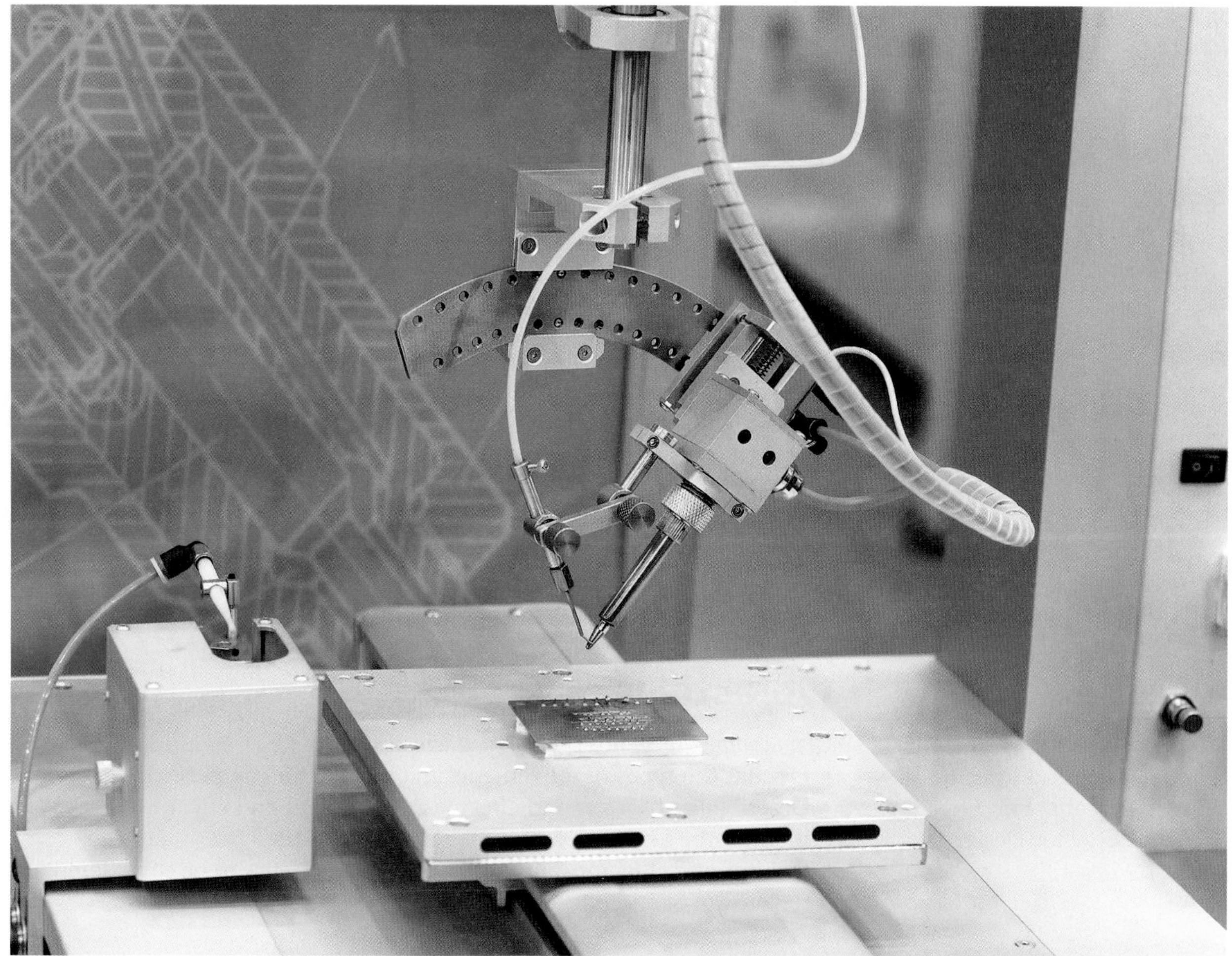

asharkyu/Shutterstock.com

Automated manufacturing may require robots to learn how to interact with their environment. Different types of sensors help robots adjust as needed to complete tasks.

Chapter 9

End Effectors

Outline

Objectives

Upon completion of this chapter, you will be able to:

- Discuss the types of movements an end effector can perform.
- Describe the types of end effector grippers and end effector tools.
- Identify the benefits of changeable end effectors.
- List important factors and desirable characteristics to be considered in the design of end effectors.

Technical Terms

automatic tool changer
collet gripper
compliance
cylindrical grip
electromechanical gripper
expandable gripper
gripper
hook movement
lateral grip
magnetic gripper
mechanical finger gripper
nonprehensile movement
oppositional grip
overload sensor
palmar grip
prehensile movement
remote center compliance (RCC) device
spherical grip
spread movement
tool
vacuum gripper

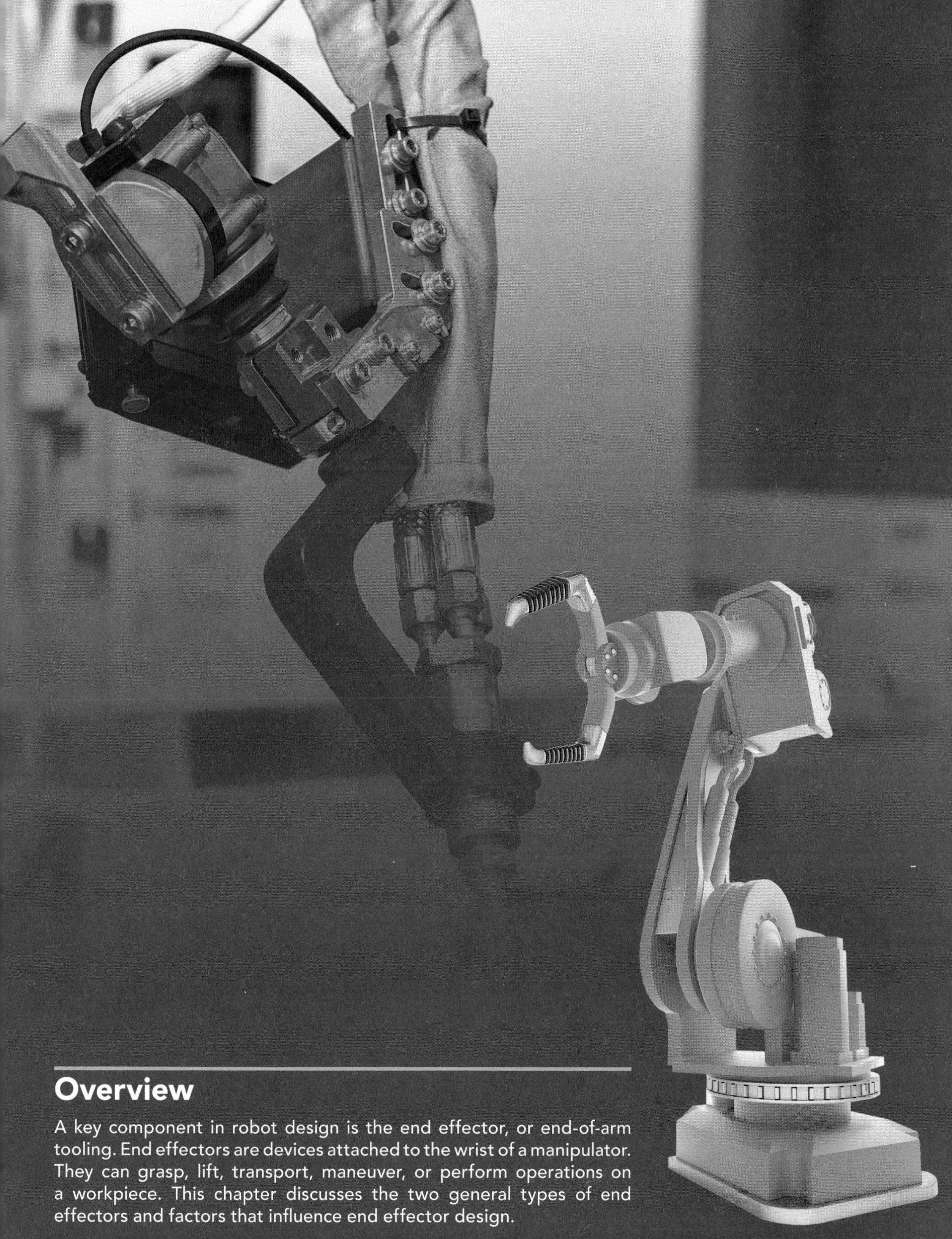

Overview

A key component in robot design is the end effector, or end-of-arm tooling. End effectors are devices attached to the wrist of a manipulator. They can grasp, lift, transport, maneuver, or perform operations on a workpiece. This chapter discusses the two general types of end effectors and factors that influence end effector design.

9.1 End Effector Movement

The robot's end effector can perform many of the same movements as the human hand. The human hand and an end effector have the ability to adjust, grasp, pick up, and rotate objects. Both prehensile and nonprehensile movements are used. ***Prehensile movements*** are actions that require the use of the thumb and fingers to grasp objects. ***Nonprehensile movements*** are actions that do not require particular finger dexterity or use of the thumb.

9.1.1 Prehensile Movements

Fingers that can curl and an opposable thumb provide the dexterity needed for prehensile movements. The opposable thumb is very important because it enables humans and robots to pick up and manipulate very small objects.

People have the ability to use their hands for various grips, **Figure 9-1**. The hand makes the following five basic prehensile (gripping) movements:

- ***Palmar grip.*** Wrapping the fingers and thumb around an object, such as a highlight marker, to grasp the object.
- ***Cylindrical grip.*** Forming a C shape with the fingers and thumb to grasp a cylindrical object, such as a drinking glass or water bottle.
- ***Spherical grip.*** Using the fingers and thumb to hold round objects, such as holding a ball.

Palmar Grip

Cylindrical Grip

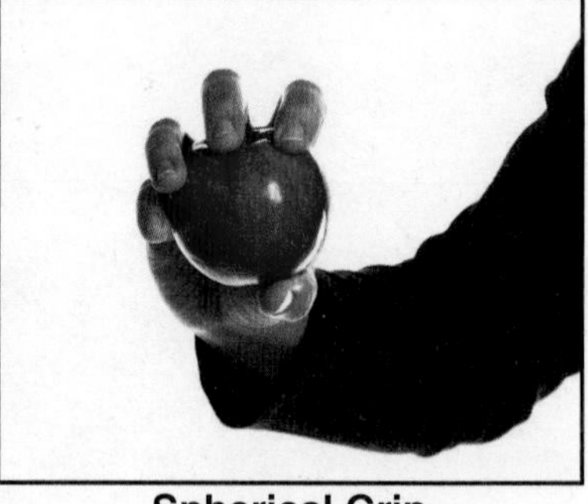

Spherical Grip

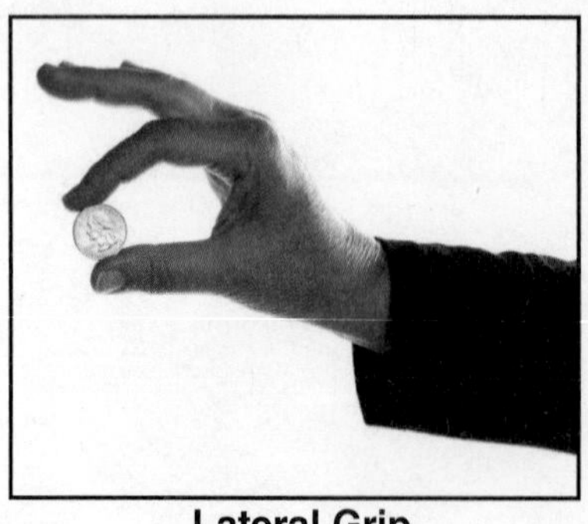

Lateral Grip

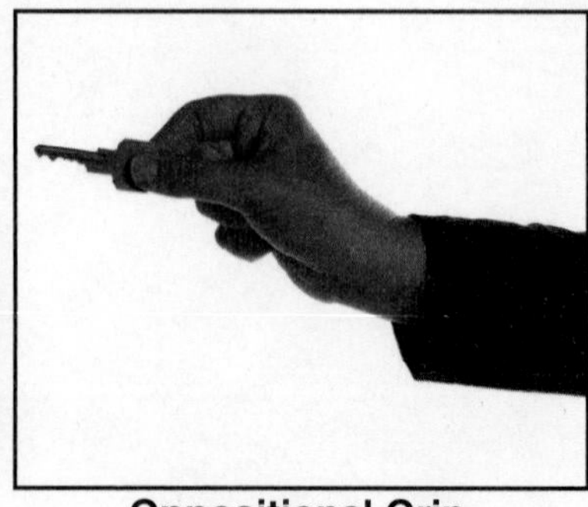

Oppositional Grip

Goodheart-Willcox Publisher

Figure 9-1. The human hand is capable of five basic prehensile grips: palmar, cylindrical, spherical, lateral, and oppositional.

- *Lateral grip.* Grasping flat objects from the sides with the fingers and thumb rather than around the object. Holding a quarter by the edges is an example of using a lateral grip.
- *Oppositional grip.* Using the tip of the index finger and the thumb to hold an object, such as a key.

9.1.2 Nonprehensile Movements

Nonprehensile movements do not require finger dexterity or use of the opposable thumb. These movements include pushing, poking, punching, and hooking. Two specific types of nonprehensile movements are hook and spread, **Figure 9-2**. The ***hook movement*** involves curling the tips of the fingers to pull or lift objects. To perform the ***spread movement***, the fingers and thumb are extended outward until they make contact with the interior walls of a hollow object. The force of the fingers against the walls of the object allows it to be picked up and carried.

9.2 Types of End Effectors

End effectors can be classified as grippers or tools. ***Grippers*** are end effectors that perform prehensile movements by grasping objects and moving them. ***Tools*** are end effectors that execute nonprehensile movements to perform specific tasks, such as welding or painting.

9.2.1 Grippers

Robots use a variety of grippers to grasp, handle, and transport parts, **Figure 9-3**. Some common types of grippers are presented in the sections that follow.

Hook

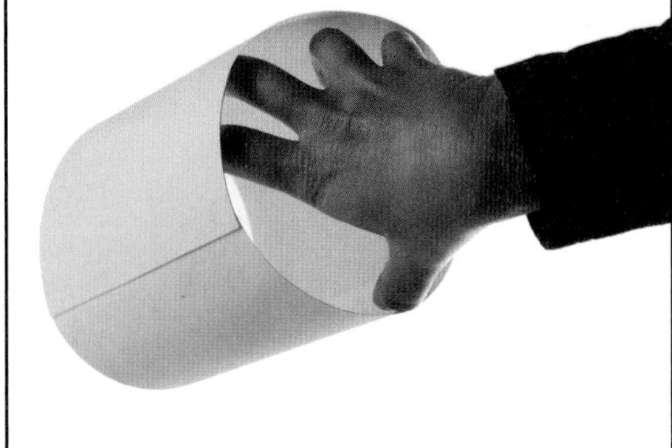

Spread

Goodheart-Willcox Publisher

Figure 9-2. Nonprehensile movements, such as hook and spread, do not depend on movement of the thumb.

Gripper End Effectors

Gripper Type	Gripper Configuration	Gripper Movement	Internal/External Gripping
Mechanical finger	Two-finger Three-finger Four-finger	Parallel or angular	Internal and external
Collet	Round Square Hexagonal	360° clamping contact	Internal and external
Vacuum	One or more suction cups	Vacuum/suction	External
Electromechanical	Permanent magnet Electromagnet	Magnetic attraction	External

Goodheart-Willcox Publisher

Figure 9-3. This classification scheme can be used for mechanical grippers.

Mechanical Finger Grippers

Mechanical finger grippers are end effectors that are most commonly used for grasping objects. Mechanical finger grippers are generally used for grasping parts within a confined space, reaching into channels, or picking and placing any object that has a simple shape. The fingers typically move in a parallel or angular motion to grasp an object, **Figure 9-4**. The gripper fingers are opened and closed using mechanical linkages, gears, cables, chains, or pneumatic actuators. Grippers can have two, three, or four fingers.

Two-Finger Grippers. Two-finger grippers have two stiff fingers that simulate the motions of the human thumb and index finger, **Figure 9-5.** One or both of the fingers may move. For an additional degree of freedom and more flexibility during operations, a two-finger gripper may be equipped with a rotating joint, **Figure 9-6**.

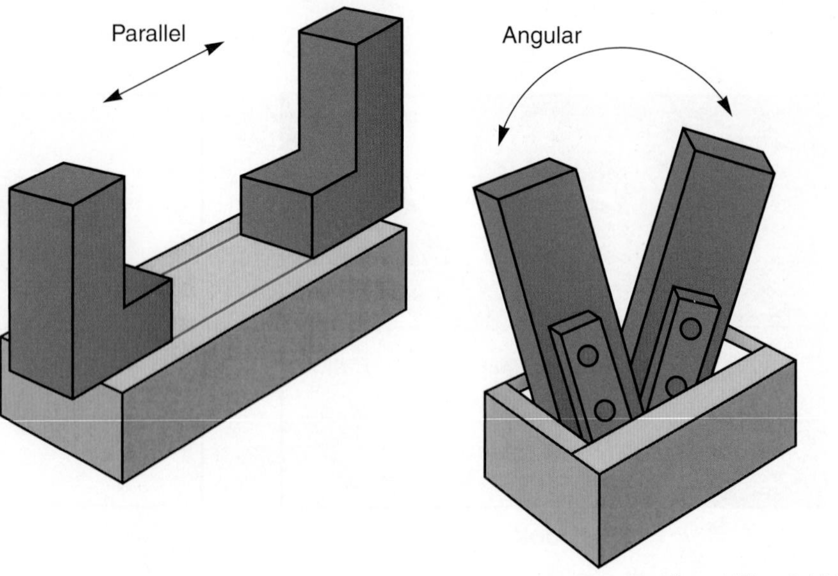

Goodheart-Willcox Publisher

Figure 9-4. Two types of motion are made by mechanical grippers: parallel and angular.

SCHUNK Intec Inc.

Figure 9-5. This two-finger internal gripper is hydraulically operated with parallel motion and is designed to grasp heavy loads.

To handle objects with different shapes, fingers can be interchanged. For grasping cylindrical objects, V-shaped fingers are recommended, **Figure 9-7**. The V shape has two points of contact on each finger, ensuring that the object is centered. Self-aligning, padded fingers are used to grip flat objects. Fingers may also have cavities of more than one size or shape. Multi-cavity fingers may be necessary if an object changes shape or size during processing.

Three-Finger Grippers. Three-finger grippers simulate the action of the human thumb, index finger, and third finger, **Figure 9-8**. This configuration is better than that of two-finger grippers for grasping curved, spherical, or cylindrical workpieces.

Four-Finger Grippers. Four-finger grippers can grasp square and rectangular parts. The four fingers close simultaneously to permit easier part orientation. Four-finger grippers are often constructed from a pair of two-finger grippers.

Collet Grippers

Collet grippers are end effectors used to pick and place cylindrical parts that are uniform in size. Unlike finger grippers, collet grippers deliver 360° of clamping contact. They have a strong clamping force that allows rapid part

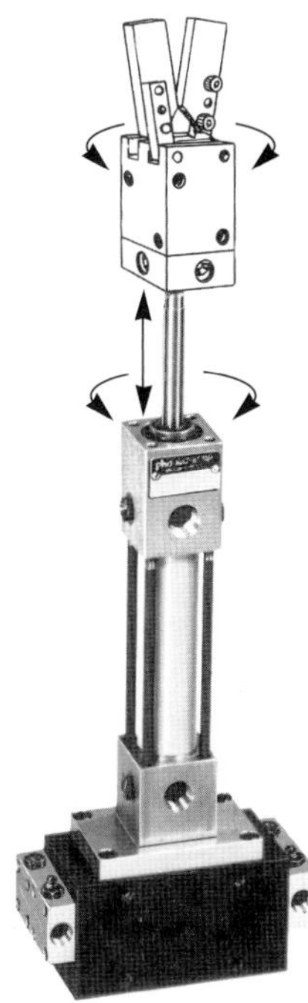

PHD, Inc.

Figure 9-6. A rotary joint gives this gripper one additional degree of freedom.

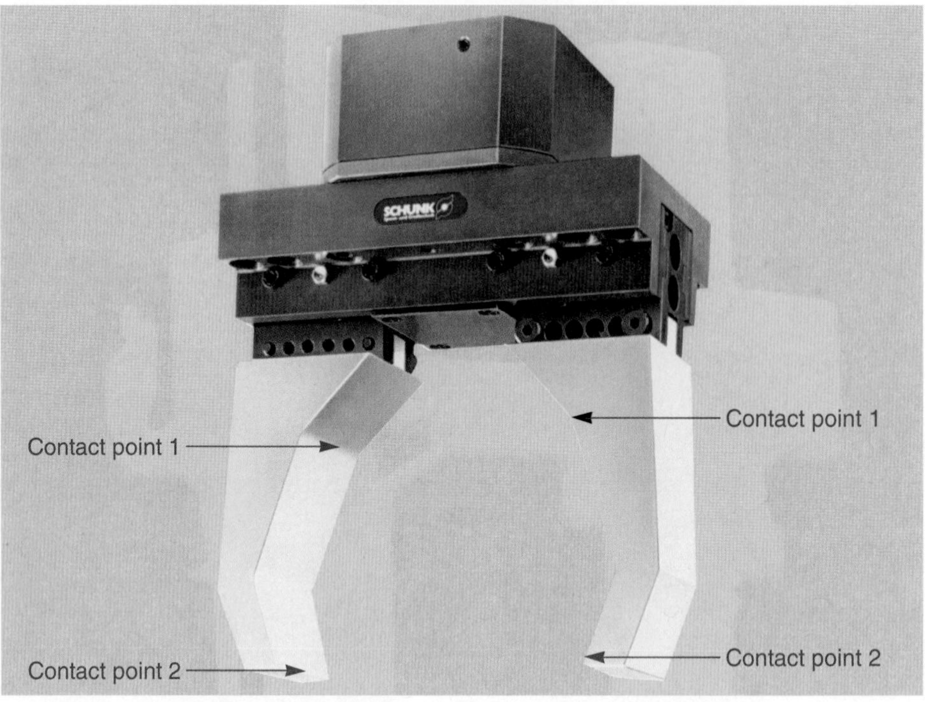

Schunk-USA

Figure 9-7. This two-finger gripper has two points of contact per finger, which allows it to grasp large cylindrical objects securely.

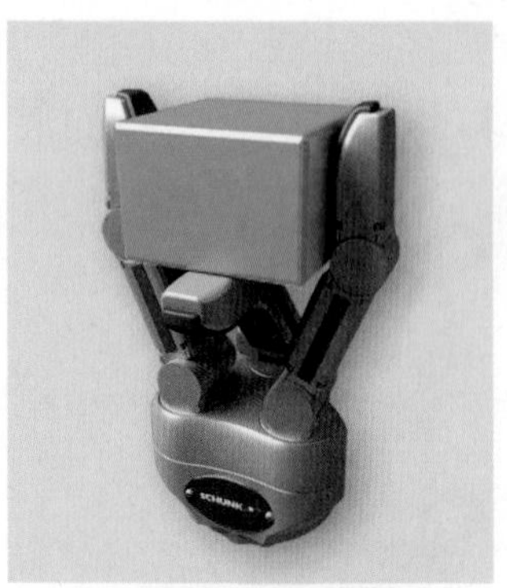
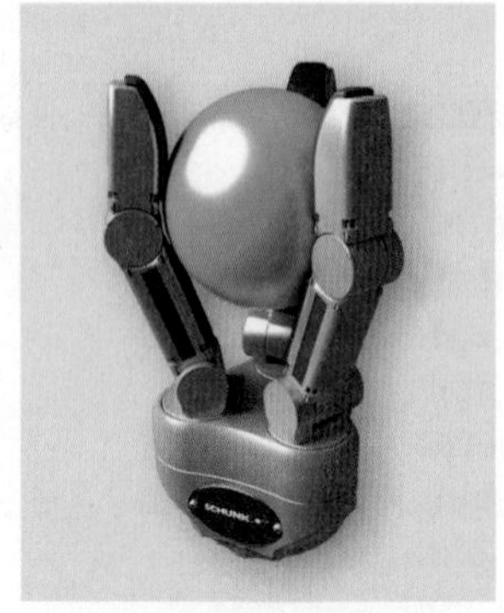

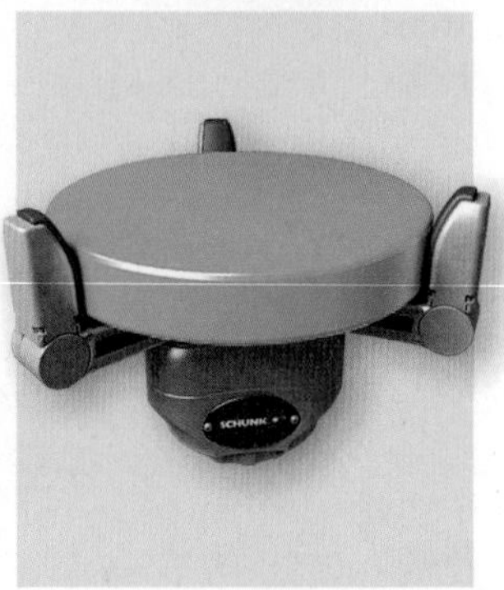

SCHUNK GmbH & Co. KG

Figure 9-8. Three-finger grippers can grasp and handle objects of various shapes.

transfer. They are also used for grinding and deburring operations. Collet grippers are typically controlled by a solenoid and have excellent repeatability. They are available in round, square, or hexagonal shapes.

Vacuum Grippers

Vacuum grippers are end effectors that use suction (vacuum) to pick up an object. They consist of one or more suction cups made of natural or synthetic rubber. They are extremely lightweight and simple in construction. The number, size, and type of suction cups used depends on the weight, size, shape, and type of material being handled. Multi-cup vacuum grippers increase the contact surface area, which allows workpieces of increased size and weight to be handled, **Figure 9-9**. Vacuum grippers can be used on curved and contoured surfaces as well as on flat surfaces.

The flexibility of the suction cups provides the robot with a certain amount of adaptability. Therefore, exact positioning is not as critical as with some other types of grippers. To

Pacific Robotics, Inc.

Figure 9-9. When an object with large surface area must be grasped, a multi-cup vacuum gripper is often used. The gripper pictured also has curved fingers to provide additional support when moving large objects.

allow for unevenness in a part surface, some vacuum cups are spring-loaded or mounted on a ball joint.

Electromechanical Grippers

Electromechanical grippers, also called ***magnetic grippers***, are end effectors that use a magnetic field created by a permanent magnet or an electromagnet to pick up an object. Objects that have flat, smooth, clean surfaces are the easiest to handle.

Grippers that operate using a permanent magnet are well suited for explosive environments because they do not require an electric power source that could spark. Once a part is moved, it is released from the gripper by exerting force to pull it away from the magnetic field.

An electromagnetic gripper is energized by a dc power source. An object is released from the gripper when the power source is interrupted. To speed up release time, the current is not cut off. Instead, the direction of current flow is momentarily reversed.

There are certain disadvantages associated with using electromechanical grippers. These grippers can only be used to handle materials containing iron, which limits possible applications and locations. Metal shavings and other small metal particles are attracted to the magnet when parts are machined. If these metal particles accumulate on the magnet, they can scratch the surface of a part or cause misalignment of parts. The temperature of a workpiece must also be considered. The effectiveness of magnetic force declines when workpieces are heated to several hundred degrees, as may be the case in some processing operations.

Electromechanical grippers have certain advantages over vacuum grippers. An electromechanical gripper has a longer life and can handle hotter and heavier objects. Also, electromechanical grippers immediately grip a part, while vacuum grippers require time to build up the necessary pressure to grip a part. Electromechanical grippers are custom designed for specific applications. Few are available as off-the-shelf items.

9.2.2 Tools

A robot's arm can be equipped with various types of tools to perform specific tasks. Some of the common tools used on robots are spot welding guns, inert gas arc welders, stud welders, gluing guns, spray guns, drills, milling heads, deburrers, polishers, pneumatic screwdrivers, and nut-drivers. The tools used for robotic applications may be categorized as:

- Welding tools.
- Material application tools.
- Machining and assembly tools.

Welding Tools

Welding tools vary according to the type of welding operation performed. Welding robots may have tool changers that provide flexibility and improve productivity. An ***automatic tool changer*** is a device that has more than one end effector and can change end effectors when needed to execute operations, **Figure 9-10**. This automated option reduces the downtime required to change an end effector and decreases the risk of operator injury that can occur when a manual change is performed. One representative type of welding tool is an arc welding torch attached to a robot arm. For this application, a proximity sensor could be used to aid in positioning the end effector.

Spot welding operations can also be performed with robotic systems. Tool changers allow a spot weld device to be attached to a robot arm so multiple functions can be performed within a work cell. In addition, laser and gun welding systems may be used with the proper tool changers.

Material Application Tools

Material application tools are also used in palletizing, packaging, and handling operations. Businesses are very concerned about productivity, so tool changers must be efficient to avoid equipment downtime. Rapid tool changing during material handling automates and improves the process.

Machining and Assembly Tools

Tooling operations performed for machining and assembly use robots to perform multiple tasks that require several different types of tools. These tasks include operations such as drilling and milling machining processes. Rapid tool changing is critical to productivity for these operations, as is the repeatability of every production cycle. Robots can change tooling quickly with tool changers that connect the robot arm to the necessary tool in only a few seconds. Sensors can be attached to the robot arm. For example, a proximity sensor can provide feedback about the distance between the end effector and an object. This feedback helps protect equipment from being damaged due to colliding with objects in the work area.

Schunk-USA

Figure 9-10. Typical automatic tool changer applications. A—An automatic tool changing system permits this robot to execute three operations automatically: cleaning, primer coating, and gluing. B—A large gantry robot is equipped with an automatic tool changing system. Note the extra-long fingers on the gripper that allow the clamping of large rings.

9.3 Changeable End Effectors

Robots used in manufacturing require either multipurpose or changeable end effectors. This type of tooling increases the effectiveness and flexibility of a robotic system. Equipping the manipulator with a multifunction end effector allows a single end effector to perform multiple tasks in a single process or on a single workpiece. Another option is to design an easily changeable end effector, which is also called quick-change tooling.

Quick-change tooling offers several end effector options that can be readily attached and removed from the manipulator as the process requires. With quick-change tooling capabilities, a robot can handle parts of various shapes and perform a wide range of assembly tasks and machining operations. Adding this flexibility in the robotic system increases overall productivity.

When designing a robotic system with changeable end effectors, two considerations must be addressed. The interface adapters must be standardized, and operators must be able to change end effectors with minimum downtime. To make interface adapters standard for use with many different end effectors, they must have compatible mounting and securing systems. All the connectors for electric, hydraulic, and pneumatic components should be provided with the adapters. Since robots come in different sizes, different sizes of standard adapters may need to be provided as well.

9.4 End Effector Design

To determine the type of end effector needed to do a job, a study must be performed to evaluate the operation, the workpiece(s), and the environment. The end effector may be subjected to extreme temperatures or make contact with abrasive or corrosive materials. Special protective materials and shielding devices may be necessary to protect the manipulator.

Objects to be moved may vary in shape, size, and weight. The workpieces may also change in shape, size, or weight during a process. The end effector must be able to adjust to such changes. Other conditions to be considered are the fragility of the workpiece, the surface finish, and the type of material used to construct the workpiece. For example, if an object is made of ferrous material (one that contains iron), a magnetic gripper may be used.

Additional considerations, such as problems involving inertia, center of mass, gripping force, or friction between the part and the gripper, may need to be addressed. Other concerns might involve part orientation, gripper sensing capabilities, or interaction with other equipment.

9.4.1 Desirable Characteristics

End effectors should be designed with the appropriate desirable characteristics. The end effector should have the strength necessary to carry out the required tasks and be able to withstand rigorous use.

End effectors should be equipped with measures to guard against damage from strain. Breakaway devices remove the end effector from the work area using mechanical fuses, detents, or preloaded springs. A breakaway device should be installed to prevent damage to the robot's arm or wrist if the end effector becomes stuck. End effectors that use friction to hold objects are not typically prone to damage caused by excessive strain because the grasped object will slip out of the gripper when an opposing force is applied. However, an opposing force of this type could cause joint slippage and alter the accuracy of the robot's positioning.

Robots may use overload sensors. An ***overload sensor*** is a device that detects obstructions or overload conditions within fractions of a second. The sensor signals the controller to shut down the robot before damage occurs. Unlike breakaway joints, an overload sensor does not have parts to be replaced after it operates and does not require reprogramming.

Another desirable characteristic is compliance. ***Compliance*** is the ability of an end effector to tolerate misalignment of mating parts. For the assembly of close-fitting parts, this characteristic is essential. Compliance prevents the part from jamming, wedging, and wearing. Some end effectors have a certain amount of compliance built into their design. End effectors that lack this capability may use a ***remote center compliance (RCC) device***. This device is installed in the wrist of the robot and helps to compensate for workpiece misalignment or irregularities. Robots equipped with an RCC device can perform precise tasks, such as inserting bearings into a housing with a clearance of only 0.0005″ (0.013 mm).

9.4.2 Custom-Designed End Effectors

In addition to standard grippers and tools, custom end effectors can be designed for a particular application, **Figure 9-11**. Custom-designed end effectors broaden the range of tasks that robotic equipment can perform.

PRI-Precision Robots, Inc.; Schunk-USA

Figure 9-11. Custom-designed end effectors. A—The gripper on this robot transfers large silicon wafers from one container to another. B—This three-fingered, internal gripping end effector is used in a production line for machining precision gears. C—This dual end effector has internal and external gripping capabilities. D—A parallel-motion gripper is used to insert components into a printed circuit board prior to soldering.

Robotics in Society: CyberKnife® System

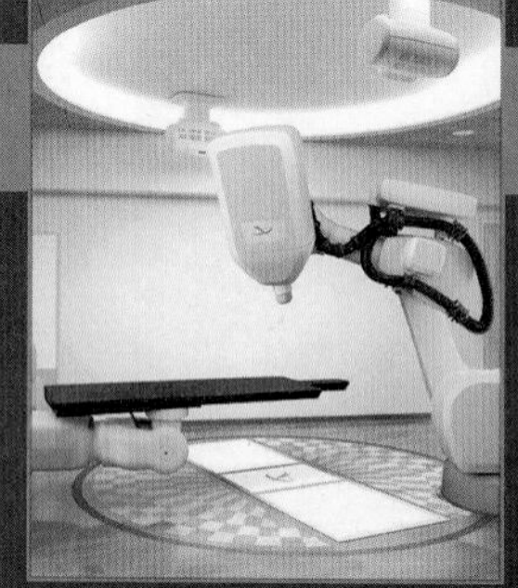

The end effector of the CyberKnife System is a linear accelerator that is used to treat tumors anywhere on the human body.

The CyberKnife System is a robotic radiosurgery system available from Accuray® Incorporated. This system offers a non-invasive treatment option for patients with cancerous tumors. The manipulator of the CyberKnife® System has an extensive range of motion, which allows the end effector (a linear accelerator) to be positioned anywhere on the body. Additionally, high-resolution image detectors continuously track the position and location of the target tissue during the treatment session. This information is communicated to the manipulator, which automatically adjusts the end effector to compensate for movement of the patient or tissue. The continuous tracking capabilities of the CyberKnife® System provide accurate treatment and reduce damage from the radiation beams to surrounding tissues.

Traditional radiosurgery systems are confined to treating tumors in the head and neck. Accuray® made use of new technologies to develop a flexible and precise robotic radiosurgery system that offers a wide range of patients an effective option in the fight against cancer.

Image used with permission from Accuray Incorporated

De-STA-Co

Figure 9-12. This gripper is equipped with a vision guidance system that allows the robot to find objects that are not placed symmetrically. This type of device can also be used in environments with changing conditions, such as lighting levels, and for locating and gripping objects that are placed randomly.

Grippers that are designed to handle fragile objects are an example of custom end effectors. End effectors with soft, flexible cups have been developed for delicate applications, such as handling lightbulbs and eggs. Custom-designed end effectors may combine gripper types or be uniquely shaped to meet the needs of a particular operation, **Figure 9-12**.

Expandable grippers are end effectors that clamp irregularly shaped workpieces using mechanical fingers equipped with hollow rubber envelopes that enlarge when pressurized. The envelopes ensure even distribution of surface pressure. One type of expandable gripper surrounds an object, gripping it from the outside. Another type grips hollow objects from the inside. Expandable grippers are ideal for handling fragile parts or parts that vary a great deal in size.

Summary

- End effectors are devices attached to the wrist of a robot manipulator. They are used to grasp, lift, transport, maneuver, or perform operations on a workpiece.
- End effectors for robots perform many of the same movements as a human hand.
- Prehensile movements require the use of the thumb to grasp objects and include palmar grip, cylindrical grip, spherical grip, lateral grip, and oppositional grip.
- Nonprehensile movements do not require the movement of a thumb and include pushing, poking, punching, hooking, and spread movements.
- End effectors are classified as grippers or tools. Grippers perform prehensile movements, and tools execute nonprehensile movements.
- Types of grippers include mechanical finger grippers (two-, three- or four-finger), collet grippers, vacuum grippers, and electromechanical grippers.
- Tools attached to a robot arm are used for welding, material application (palletizing, packaging, or handling operations), and machining and assembly operations.
- Robots used in manufacturing require either multipurpose or changeable tooling. An automatic tool changer can change end effectors when needed to execute operations.
- End effector design must include study of the type of operation to be performed, the workpieces, and the work environment.
- End effectors should be designed with the appropriate desirable characteristics, such as strength, durability, strain and overload protection, and compliance.
- Custom-designed end effectors can be made for specific operations, such as handling fragile objects or grasping parts that vary greatly in size.

Review Questions

Answer the following questions using the information provided in this chapter.

1. Distinguish between prehensile and nonprehensile movements that robot end effectors can make.
2. Describe five types of gripping movements an end effector can make.
3. Describe the hook movement and the spread movement that can be performed by an end effector.
4. Describe the two major classifications for end effectors.
5. What are four types of grippers?
6. For what tasks are mechanical finger grippers typically used?
7. Explain how a vacuum gripper operates.
8. What is a major disadvantage of an electromechanical gripper?
9. What are three categories of tools used for robotic applications?
10. How does changeable tooling impact the operation of a robotic system?
11. What are the advantages of using automatic tool changers?
12. What are three factors that should be considered when determining the type of end effector needed to do a job?
13. What are some desirable characteristics that should be part of the design for an end effector?

14. Explain how breakaway devices and overload sensors provide protection for robotic systems.
15. What is compliance? How can the compliance of an end effector be increased?

Learning Extensions

1. Pay particular attention to the movements of your hands in the course of a day. Make note of common movements by identifying them as either prehensile or nonprehensile. Name the grip or movement and list the task you performed with the hand movement.
2. Conduct an Internet search to identify end effector manufacturers. Develop a list of the types of end effectors you find and categorize them into common types.
3. Select a specific type of robot operation and describe in detail the design of the end effector used to accomplish the operation.

Advanced Analysis

1. Prepare a description of the operation of moving assembled cardboard boxes from one location to another using a robot. In particular, describe the type of end effector used.
2. Prepare a description of the operation of drilling three different-diameter holes in a metal workpiece using a robot. In particular, describe the type of end effector used.
3. Prepare a description of the operation of moving fragile parts of various sizes from one location to another using a robot. In particular, describe the type of end effector used.

Unit IV Control Systems and Maintenance

The heart of a robotic control system is a microprocessor linked to input/output and monitoring devices. The control system has a series of instructions (a program) stored in its memory. The program supplies the commands that control motors and hydraulic or pneumatic systems to activate the motion control mechanism of the robot. The motion control mechanism is typically an actuator, which is a device that converts power into robot movement.

Chapter 10

Computer Systems and Digital Electronics

Outline

10.1 Computer Systems
10.2 Digital Number Systems
10.3 Binary Logic Circuits
10.4 Computer Programming

Objectives

Upon completion of this chapter, you will be able to:

- Identify the basic components and functions of a computer system.
- Describe the characteristics and functions of digital number systems.
- Discuss the function of logic gates, flip-flops, and digital counters.
- Describe types of programming instructions and the importance of well-designed programs.

Technical Terms

accumulator
address register
analog
AND gate
arithmetic logic unit (ALU)
binary-coded-decimal (BCD) number system
binary counter
binary logic circuit
binary number system
binary point
bistable device
bit
bus network
byte
central processing unit (CPU)
control unit
counter
data register
decade counter
decoder unit
digital
digital electronics
dynamic RAM (DRAM)
electrically erasable programmable read-only memory (EEPROM)
erasable programmable read-only memory (EPROM)
execute
fetch
firmware
flip-flop
hexadecimal number system
input-output (I/O) transfer
interrupt
inverter
logic circuit
logic gate
memory
microprocessor unit (MPU)
MPU cycle
NAND gate
NOR gate
NOT gate
octal number system
OR gate
period
personal computer (PC)
place value
positive logic
program counter
programmable read-only memory (PROM)
random access memory (RAM)
read-only memory (ROM)
register unit
software
static RAM (SRAM)
truth table

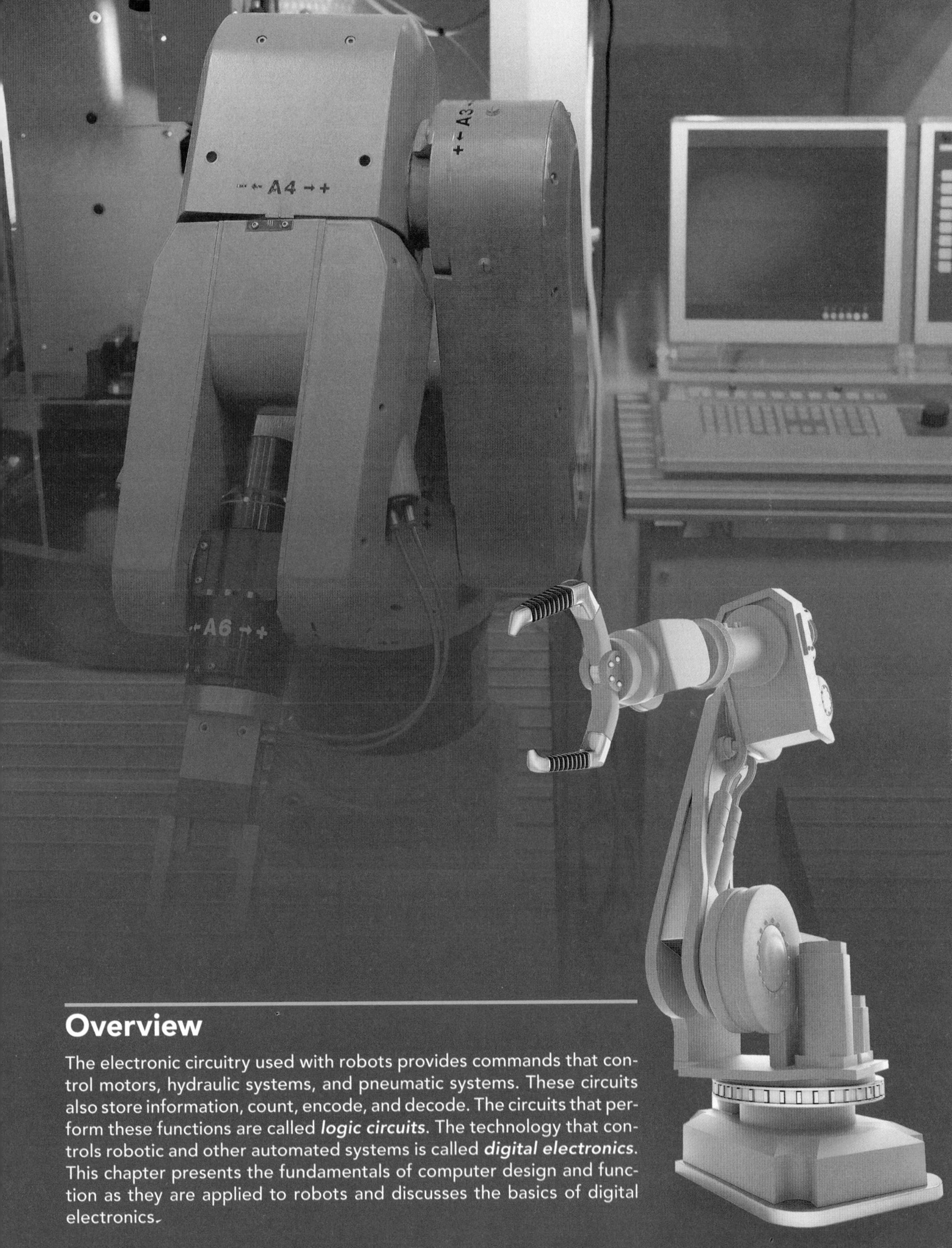

Overview

The electronic circuitry used with robots provides commands that control motors, hydraulic systems, and pneumatic systems. These circuits also store information, count, encode, and decode. The circuits that perform these functions are called ***logic circuits***. The technology that controls robotic and other automated systems is called ***digital electronics***. This chapter presents the fundamentals of computer design and function as they are applied to robots and discusses the basics of digital electronics.

10.1 Computer Systems

The term *computer* is used to cover a number of functions, but primarily it refers to a system that performs automatic computations. Computers range in size and function, from pocket calculators to complex central units that serve an entire organization. Computers process information in two states: analog or digital. ***Analog*** is a term that describes data or signals that vary continuously or devices that use this type of data or signal. An example is temperature measurements, which are in a constant state of change. Also, the mercury indicator of an analog thermometer may read between one degree mark and the next. In contrast, ***digital*** is a term that describes data or signals in a format that occur in separate full units, such as 1 or 0, or devices that use this type of data, such as a computer. With a digital clock, for example, the time is always displayed as one unit (second) or the next, but not between units.

Digital computers use two numbers that represent either the presence or absence of voltage. A voltage pulse is usually represented by a one (1). The absence of voltage is indicated by a zero (0). A single pulse of voltage processed by a computing system is called a ***bit*** or a binary digit. A group of eight bits produces a ***byte***, sometimes called a binary word.

Computer systems have certain basic parts, **Figure 10-1**, which may be arranged in a variety of ways. The organization and design of each circuit differs considerably among manufacturers. The basic components of a computer are input and output devices, arithmetic logic and control circuitry, and a form of data storage.

Coded data in the form of 1s and 0s are written into the computer data storage according to directions provided by a program. This coded data provide operating instructions. The ***central processing unit (CPU)*** is the part of a computer that interprets these instructions, processes the data, and carries out programmed operations. Computers typically have several input and output ports that permit the CPU to communicate with external devices, such as monitors, printers, disc drives, USB flash drives, and modems.

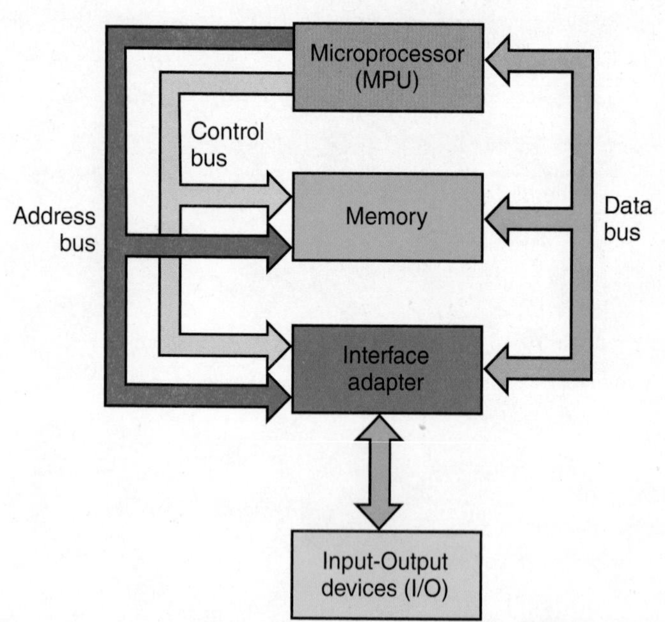

Goodheart-Willcox Publisher

Figure 10-1. This diagram represents the common components of a personal computer.

While the basic parts of a computer are the same, the differences among units is in physical size, amount of ***memory*** (data storage capacity), and processing speed. Mainframes are large computers that have millions of components included in their circuits. Mainframes are used for very large and complex tasks, such as scientific computations and operating industrial plants or governmental departments.

Minicomputer systems are designed for applications that do not require the capacity of a mainframe computer. These computer systems have smaller memory capacity and process data less rapidly than a mainframe system, but they are also less expensive. Minicomputers are used for many business, manufacturing, and educational applications.

Personal computers (PCs) are small desktop or portable units built around a single integrated circuit. PCs were called microcomputers when first produced in the early 1980s. Today, PCs are widely used in applications ranging from average home use to business management tasks, to architectural drafting, to controlling machining cells in factories. While most PCs are standalone units, they may also be connected though a wired or wireless network that allows the computers to share information.

10.1.1 Computer System Components

A personal computer includes several components that receive, store, process, and output data. The functionality of each component depends on the performance of all others in the system.

Microprocessor

A ***microprocessor unit (MPU)*** is a microchip that combines the arithmetic logic unit and control circuitry of the computer, **Figure 10-2**. A microprocessor receives data in the form of one (1) or zero (0). It may store this data for future processing, or it may immediately perform arithmetic and logic operations and deliver the results to an output device. A typical microprocessor contains a number of components that store, decode, and process data and commands.

An ***arithmetic logic unit (ALU)*** is a component of a microprocessor that performs mathematical and logic operations. It works automatically using signals

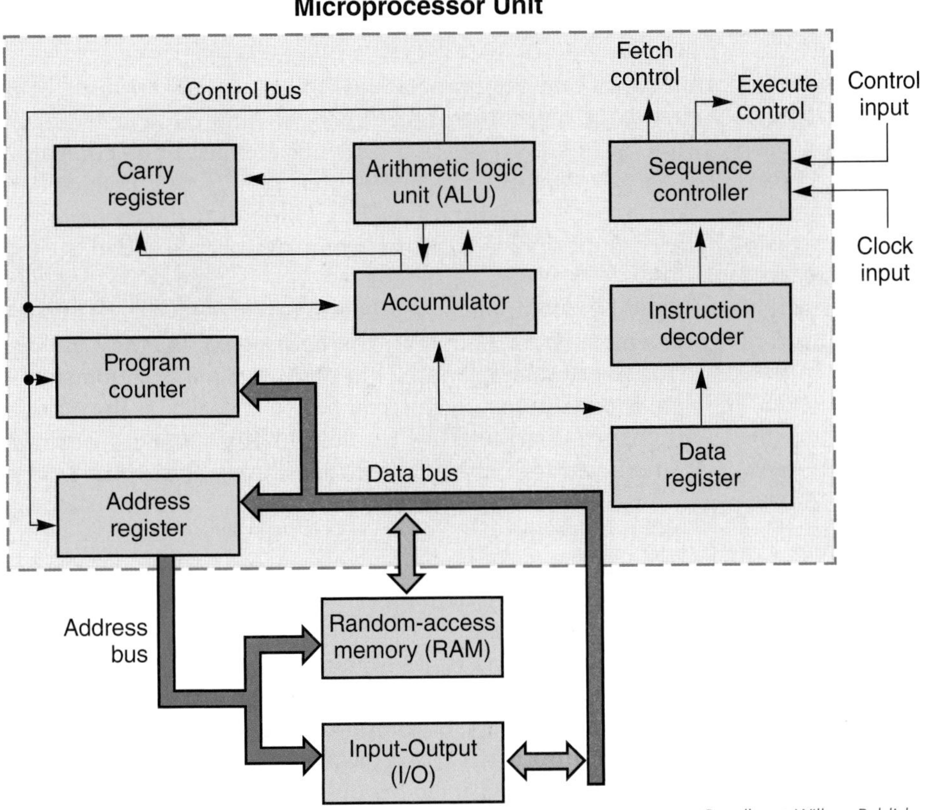

Goodheart-Willcox Publisher

Figure 10-2. This block diagram of a microprocessor has been simplified for clarity.

sent from the instruction decoder. The ***register unit*** is a component of a microprocessor consisting of several compartments that store information. Each compartment in a register unit stores a specific type of data. The ALU combines the input from the data register and the accumulator. A ***data register*** is a compartment of a microprocessor that stores information for ALU input. It may also hold an instruction while that instruction is being decoded, or it may hold data prior to storage in memory. An ***accumulator*** is a type of register that stores values for mathematical and logic operations in process. Primary operations of an ALU are addition, subtraction, and logic comparisons using binary numbers.

A ***decoder unit*** is a component of a microprocessor that examines coded instructions and decides which operation is to be performed by the ALU. After an instruction code is pulled from memory and placed in the data register, it must be decoded. The output of the decoder is sent to the control unit to determine appropriate processing. The ***control unit*** is a component of a microprocessor that receives decoded instructions from the decoder and initiates the proper action. The decoded instructions and data are sent to the register unit for appropriate processing.

The ***address register*** is a compartment that temporarily stores the address of a memory location that is to be accessed. The address register is programmable in some units. This means that new instructions may alter its contents.

The ***program counter*** is a memory device that indicates the location in memory of either the instruction currently being performed or the next instruction to be performed. This unit maintains the programmed sequential order of instructions. The sequence may be modified by using subroutines that interrupt the numeric order of instructions and direct the program counter to another instruction or operation.

Buses

A bus is a set of connections that carries data and control signals within a computer system. The circuits of most microprocessors are connected together by a common bus network. A ***bus network*** is a series of register units connected together, **Figure 10-3**. An advantage of a bus network is the ease with which data can be transferred. Data is packaged in groups of 8, 16, 32, or 64 bits for transfer between registers. The amount of data a bus network can carry is called the bus width, or capacity.

Information processed by a computer one is of two types: instructions or data. A simple addition problem, such as 9 + 2 = 11, can be used to illustrate the distinction. In this problem, the numbers 9 and 2 are data, and the addition sign is an instruction. The data is distributed by the data bus to all parts of the system. Instructions (the program) are distributed by the control bus through a separate path. Data may also be removed from memory and distributed as output.

Memory

The use of a microcomputer depends largely on the amount of memory it has available. Memory refers to the ability of an MPU to store data, so that a single bit or a group of bits can be easily retrieved. Memory can be added by installing auxiliary chips. The two most common types of memory are read-only memory (ROM) and random access memory (RAM).

Read-only memory (ROM) is a type of computer data storage that is permanent and cannot be changed. Information is typically placed in ROM when the chip is manufactured, and ROM often includes operating system and program information. Since ROM data is permanent and cannot be changed, it is not lost

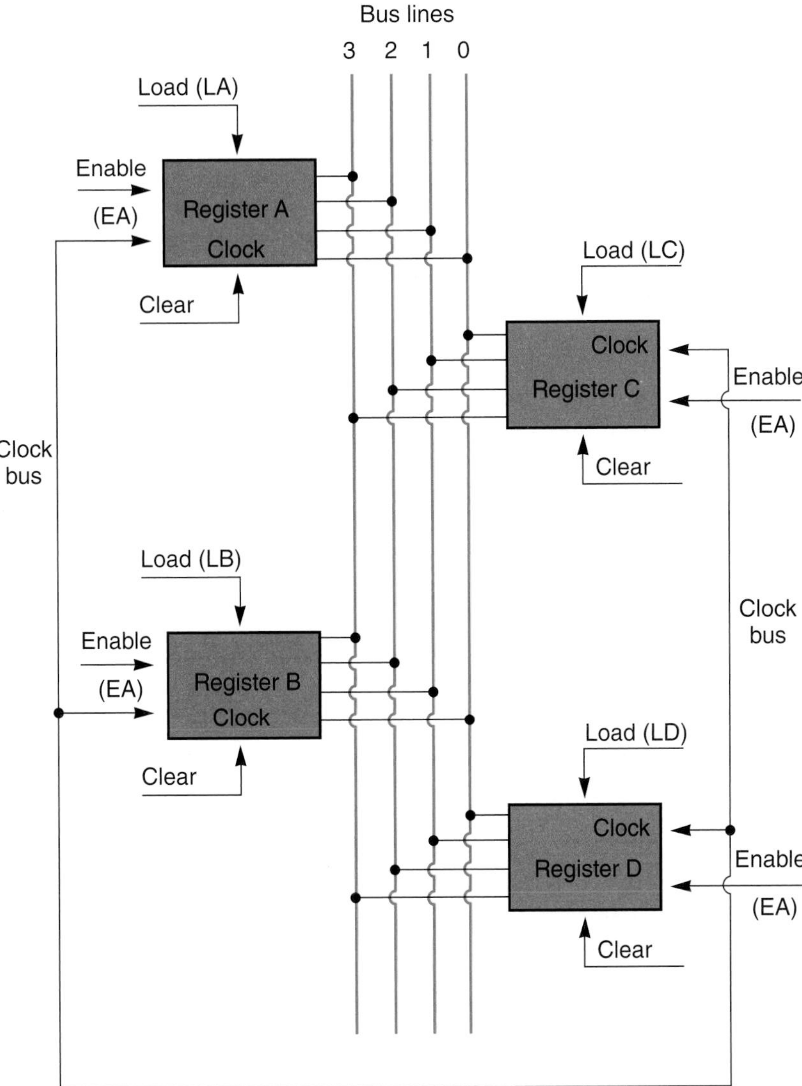

Goodheart-Willcox Publisher

Figure 10-3. In a computer, registers are connected to buses that serve as paths for data transfer.

when the MPU power source is turned off. There are several variations of ROM available:

- A ***programmable read-only memory (PROM)*** chip has data electrically burned onto it and cannot be reprogrammed.
- An ***erasable programmable read-only memory (EPROM)*** chip can be erased and reprogrammed. The chip is erased by exposing it to an ultraviolet light.
- An ***electrically erasable programmable read-only memory (EEPROM)*** chip can be erased one bit at a time and reprogrammed. Instead of an ultraviolet light, a high electrical charge is used to erase portions of the chip.

Random access memory (RAM) is a type of computer data storage that can be altered. Information may be retrieved from and written to RAM. This type of read/write memory is commonly found in microcomputers because it is very fast. The structure of this chip includes a number of separate circuits, which allow data to be accessed in any order. Data can be placed into memory or retrieved at the same rate. There are two basic categories of RAM available:

- ***Dynamic RAM (DRAM)*** is a RAM circuit that stores bits of data using separate capacitors. A capacitor is a component that stores electrical energy. To retain the data, the electrical charge of the capacitors must be constantly refreshed.
- ***Static RAM (SRAM)*** is a RAM circuit that uses semiconductor devices, called flip-flops, to store data. A ***flip-flop*** is a memory device used in digital circuits. It can hold an output state even when the input is completely removed, and it can change the output based on an appropriate input signal. Flip-flops do not have to be constantly recharged, and SRAM transmits data faster than DRAM.

To write data into memory, it is sent along the address bus to a specific address. Data can be read from memory when the microprocessor is turned on. Reading data from memory does not destroy the data. However, a loss of power or turning the unit off will destroy the data. Random access memory is volatile and disappears when power is interrupted.

10.1.2 Basic Functions

Certain functions are basic to most computer systems. Among these are timing, fetch, execute, read memory, write memory, input-output transfer, and interrupt.

Timing

The operation of a computer involves a sequence of instructions. The sequence is controlled by a timing signal. ***Fetch*** is a program operation that retrieves the next instruction from memory. ***Execute*** is an initiate operation that is based on the instruction that is to be performed. The MPU fetches an instruction, executes the required operations, fetches the next instruction, executes it, and so on in a cycling pattern. All the actions performed occur at or during a precise time interval. Such an orderly sequence requires an electronic oscillator or clock. In some systems, the clock may be built into the MPU. In others, separate units send timing signals through a control bus.

System operations are performed within a time frame called the ***MPU cycle***. For example, the fetch operation takes the same amount of time in each instance for

Robotics Innovations: Robonaut

Robonaut is a humanoid robot developed by NASA to work alongside humans in space. Robonaut is currently deployed aboard The International Space Station. Robonaut's dexterous hands allow it to use the same tools and equipment as humans, eliminating the need for specially made tools. The technology being developed and tested on Robonaut will help improve the safety of robots that will work alongside humans. Humanoid robots are stronger than humans, can complete repetitive tasks without getting bored, and do not need to eat or sleep. One day, factories on Earth may employ similar robots to complete tasks currently performed by humans. While it is not the first robot to work in space, Robonaut is the first humanoid robot designed to perform tasks currently being done by astronauts.

each instruction. An execute operation, however, may consist of many events and sequences. Therefore, execute timing varies a great deal. The time interval for a pulse to pass through a complete MPU cycle from beginning to end is called a ***period***.

A microprocessor may take a number of clock periods to perform an operation. For example, a chip that can operate at a clock rate of 33 MHz (33 million cycles per second) has a period of 1/33,000,000 or 0.00000003 second for a single cycle. Periods this small are best expressed in microseconds (millionths of a second, μs) or nanoseconds (billionths of a second, ns). At a clock rate of 33 MHz, the length of a single cycle would be 0.3 μs or 30 ns. So, if this chip requires four clock cycles to execute an instruction change, the change would take 0.12 μs or 120 ns. The operating time of an MPU is a good measure of its effectiveness and power.

Fetch and Execute

After programmed information is placed into memory, related action is directed by a series of fetch and execute operations. The sequence is repeated until the entire program has cycled to its conclusion.

The start signal actuates the control section of the MPU, which automatically starts the sequence of operations. The first instruction is to fetch, or retrieve, the next instruction from memory. The MPU may then issue a read instruction. The contents of the program counter are sent to memory, which returns the next instruction. The first word of the instruction received is placed into the instruction register. If more than one word is included in an instruction, a longer cycling time results. After the complete instruction is in the MPU, the program counter records one count. The instruction is decoded, and the unit is prepared for the next fetch instruction.

The execute, or initiate, operation is based on the instruction that is to be performed. The instruction may be to read memory, write to memory, read the input signal, or transfer to output, among others.

Read Memory

Read memory calls for data to be read from a specific location. The MPU issues a read operation code and sends it to the proper memory address. The read/write memory unit sends the data into the data bus. This number is fed to the MPU, where it is placed in the accumulator after the timing pulse has been initiated.

Write Memory

Write memory calls for data to be stored at a specific location. The MPU issues a write operation code and sends it to a selected read/write memory unit. Data is sent through the data bus and placed into the selected location.

Input-Output Transfer

Input-output (I/O) transfer is a basic computer system function in which an I/O port actuates to receive data from an input or to send data to an output device. Input-output (I/O) transfer operations are similar to read/write operations. The major difference is the code number (opcode) used to call up the operation. This code actuates an I/O port, which either receives data from the input or sends it to the output device. For example, data can move in either direction to the read/write memory and flows from the ROM into the data bus, **Figure 10-4**. The output flows from the data bus to the output device.

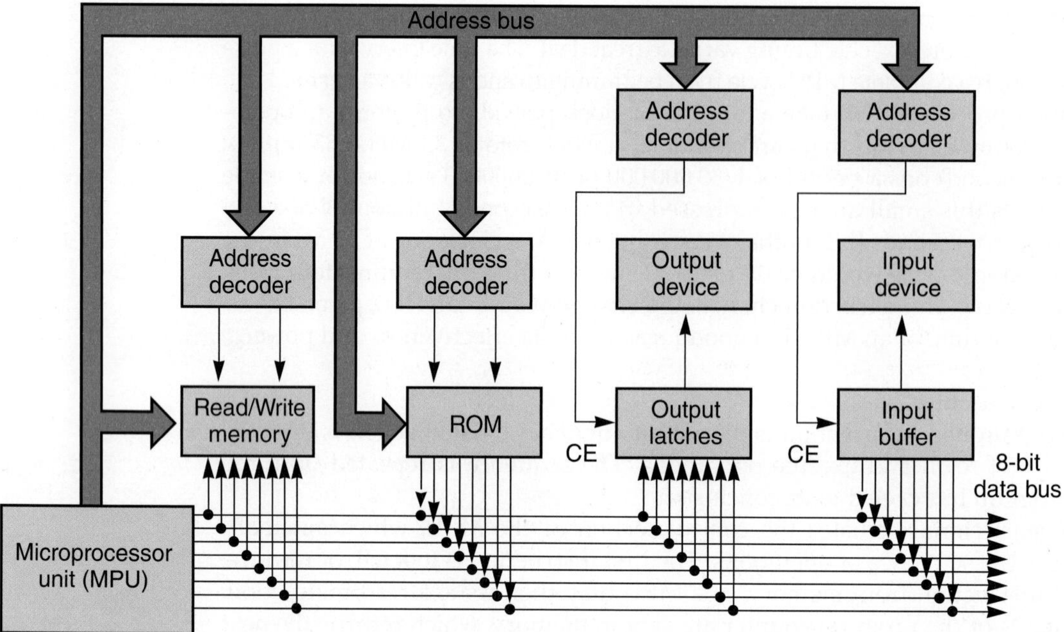

Goodheart-Willcox Publisher

Figure 10-4. This simplified view of an MPU address bus illustrates input-output transfer. Personal computers typically have a 16-bit, 32-bit, or 64-bit bus.

Interrupt

An ***interrupt*** is a signal that originates from peripheral equipment to inform the MPU that the device needs attention. Peripheral equipment that may send an interrupt signal includes items such as keyboards, displays, modems, and printers. For example, a system is designed to process a large volume of data and output to a printer. The MPU can output data much faster than the printer can produce printed pages. This means that the MPU remains idle while waiting for the printer to complete its task. If an interrupt is used, the MPU can output a byte of the data and return to processing functions while the printer operates. When the printer is ready for more data, it sends an interrupt request. The MPU stops, initiates a subroutine to output the next byte, and then continues processing information until it receives another interrupt signal.

10.2 Digital Number Systems

The number system used in everyday life in the United States is the base 10 (decimal) system. In this system, ten digits are used for counting: 0, 1, 2, 3, 4, 5, 6, 7, 8, and 9. The number of digits used in a number system determines its base. Because the decimal system uses 10 digits, it has a base of 10.

Number systems assign digits a ***place value***. This refers to the position of a digit with respect to the decimal point. In the decimal system, the first position to the left of the decimal point is called the units place. Any digit from 0 to 9 can be used in this place. Number values greater than 9 must be expressed using numbers in two or more places. The position to the left of the units place is the 10s place. The number 99 is the largest value that can be expressed using two places in the decimal system. Each place added to the left extends the number value by a power of 10.

A number value in any system can be read by adding the numbers in each place. For example, the decimal value 2583 can be expressed as: (2 × 1000) + (5 × 100) + (8 × 10) + (3 × 1). The values increase by a power of 10 for each place to the left of the decimal point. In the decimal system, this is 10^3, 10^2, 10^1, and 10^0. Mathematically, each place value is the number times a power of the system base, **Figure 10-5**.

10.2.1 Binary Number System

The decimal number system is difficult to use with electronics, so the binary system is typically used instead. The ***binary number system*** is a counting method that uses 2 as its base. The largest value that can be expressed by a specific place is the number 1. As a result, only the numbers 0 or 1 are used. A ***binary point*** is the symbol used to separate the integer part of a number from the fractional part of a number in a base 2 numbering system. The first place to the left of the binary point represents units, or 1s. Places to the left of the binary point are expressed in powers of two, such as $2^0 = 1$, $2^1 = 2$, $2^2 = 4$, $2^3 = 8$, $2^4 = 16$, $2^5 = 32$, $2^6 = 64$, and so on.

When different number systems are used, a subscript number is used to identify the base. The number 100_2 indicates that the binary (base 2) system applies. In the binary system, the number 100 is read as *one-zero-zero* instead of *one hundred*.

When converting a binary number to an equivalent decimal number, write down the binary number first. See **Figure 10-6**. Starting at the binary point, indicate the decimal equivalent for each place location where a 1 is indicated. For each 0 in the binary number, leave a blank space or indicate a 0. Add the place values and record the decimal equivalent.

Converting a decimal number to its binary equivalent involves repeatedly dividing the decimal number by 2, **Figure 10-7**. When the quotient is a whole number with no remainder, the value is 0. When the quotient has a remainder, the value is 1. Division continues until the quotient is zero.

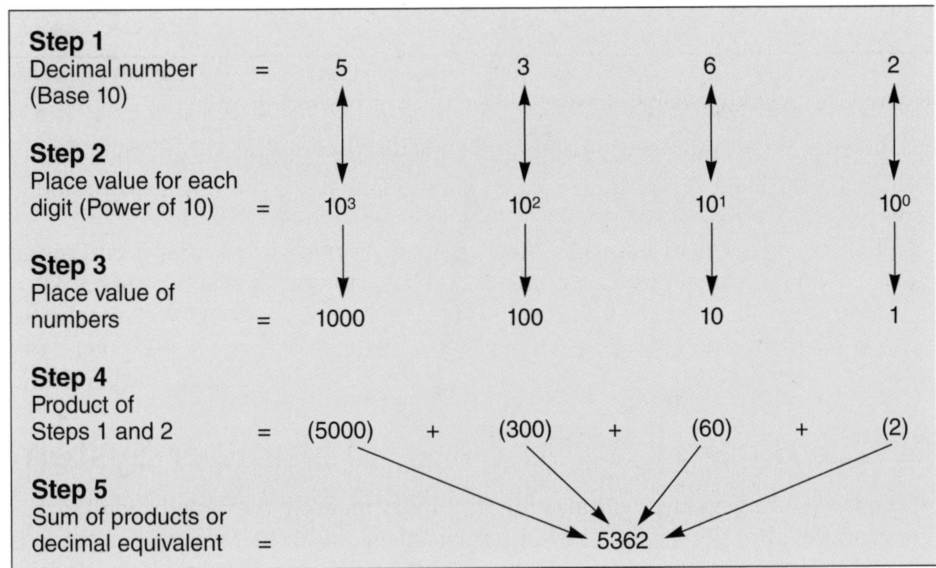

Figure 10-5. Expressing a base 10 (decimal) number.

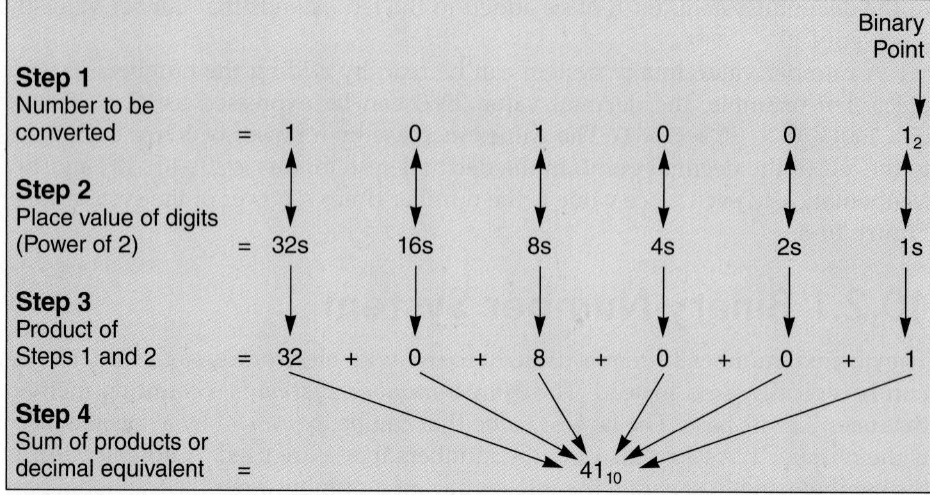

Figure 10-6. Performing a binary-to-decimal conversion.

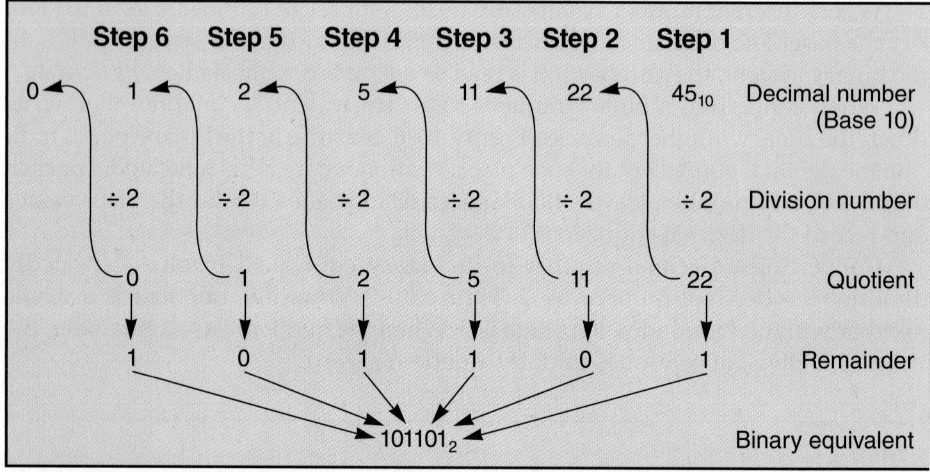

Figure 10-7. A decimal-to-binary conversion is achieved by dividing by 2.

Electronically, the value of 0 indicates low-voltage or no voltage. The number 1 indicates a voltage larger than 0. Binary systems that use these values are said to have ***positive logic***.

The two operational states of a binary system, 1 and 0, are natural circuit conditions. When a circuit is turned off or has no voltage applied, it is in the 0 state. A circuit that has voltage applied is in the 1 state. This makes it possible to change states in less than a microsecond. Therefore, electronic devices, such as computers, can manipulate millions of 0s and 1s per second and process information very quickly.

10.2.2 Binary-Coded-Decimal Number System

When large numbers are written using the binary method, they are difficult to use. For example, $1101001_2 = 105_{10}$. The ***binary-coded-decimal (BCD) number system*** is a method of counting in which four binary digits are used to represent each decimal digit. It was devised to simplify the use of large binary numbers.

The base 10 number is first divided into digits according to place value, **Figure 10-8**. For example, the number 105_{10} is made up of the digits 1, 0, and 5. Each digit is converted to a four-digit binary number. In this example, 105_{10} results in 0001 0000 0101_{BCD}. The largest digit expressed by any group of BCD numbers is 9. Decimal numbers up to 999_{10} may be converted using only three, four-digit binary numbers. It is important to include space between groups of digits when writing the number.

10.2.3 Octal Number System

The ***octal number system*** is a base 8 system used to process large numbers. It uses the same principles as the decimal and binary systems. The digits 0, 1, 2, 3, 4, 5, 6, and 7 are used in the place positions. The place values of digits, moving to the left of the octal point, are powers of eight: 8^0 = units (1s), 8^1 = 8s, 8^2 = 64s, 8^3 = 512s, 8^4 = 4096s, and so on.

The process of converting an octal number to a decimal number is the same as used for binary-to-decimal conversion, but powers of 8 are used. See **Figure 10-9**. Converting an octal number to an equivalent binary number is similar to BCD conversion. The octal number is first divided into digits according to place value.

Given the decimal number		105_{10}	
Step 1 Group the digits	(1)	(0)	(5)
Step 2 Convert each digit to binary group	(0001)	(0000)	(0101)
Step 3 Combine group values	0001 0000 0101 BCD		

Goodheart-Willcox Publisher

Figure 10-8. When performing a BCD conversion, four binary digits represent each decimal digit.

Step 1 Octal number	=	3	7	2_8
Step 2 Place value equivalents (Power of 8)	=	64 (8^2)	7 (8^1)	1 (8^0)
Step 3 Product of Steps 1 and 2	=	192 +	56 +	2
Step 4 Sum of products or decimal equivalent	=		250_{10}	

Goodheart-Willcox Publisher

Figure 10-9. An octal-to-decimal conversion requires the same basic process as a binary-to-decimal conversion.

Each digit is then converted into an equivalent binary number using only three digits, **Figure 10-10**.

Converting a decimal number to an octal number means dividing by the number 8, **Figure 10-11**. After the quotient has been determined, the remainder is brought down as the place value. When the quotient is even, with no remainder, the value is 0.

To change a binary number (110100100_2, for example) to an equivalent octal number, the digits must first be divided into groups of three binary numbers, starting at the binary point, **Figure 10-12**. Each binary group is then converted into an equivalent octal number. These numbers are combined, while remaining in their respective places, to represent the equivalent octal number. Converting a binary number to an octal number is often done for digital circuits. Binary numbers are processed at a very high speed. An output circuit accepts the signal and converts it to an octal number to display on a readout device.

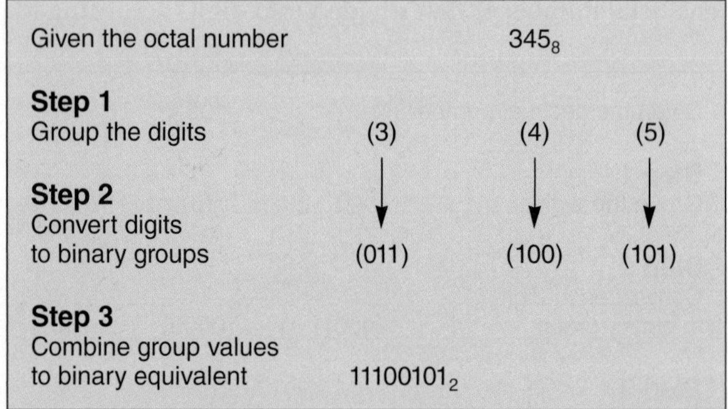

Goodheart-Willcox Publisher

Figure 10-10. In octal-to-binary conversions, a maximum of three binary numbers are used per digit.

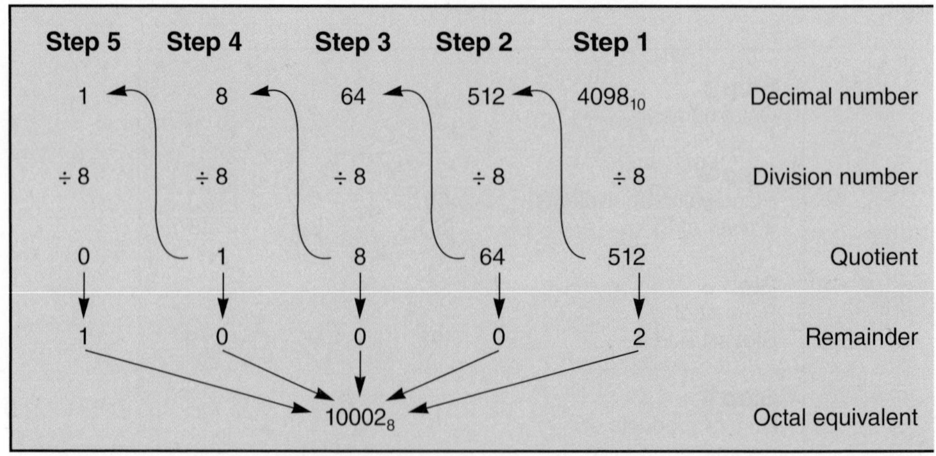

Goodheart-Willcox Publisher

Figure 10-11. A decimal-to-octal conversion involves dividing by 8.

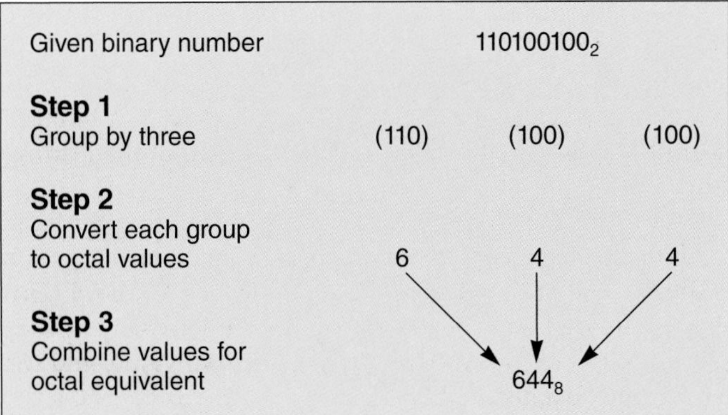

Goodheart-Willcox Publisher

Figure 10-12. Binary-to-octal conversions are often performed for digital circuits.

10.2.4 Hexadecimal Number System

The ***hexadecimal number system*** is a base 16 system that is also used to process large numbers. The largest number used in a place is 15, with digits 0–9 and letters A–F. The letters A–F represent digits 10–15, respectively. The place value of digits to the left of the hexadecimal point are powers of sixteen: $16^0 = 1$, $16^1 = 16$, $16^2 = 256$, $16^3 = 40{,}096$, $16^4 = 65{,}536$, and so on.

The method of changing a hexadecimal number to a decimal number is similar to other conversions, **Figure 10-13**. The hexadecimal number is written in proper digital order and the place values are positioned under each respective digit. The letters are converted to numeric values, and each of the values is multiplied by its corresponding place value. The products are added together, and the resulting sum is the decimal equivalent of the hexadecimal number.

Step 1 Hexadecimal number	=	1		2		C		D_{16}
Step 2 Place values (Power of 16)	=	4096s		256s		16s		1s
Step 3 Convert letters to numbers	=	1		2		12		13
Step 4 Product of Steps 2 and 3	=	4096	+	512	+	192	+	13
Step 5 Sum of products or decimal equivalent	=	4813_{10}						

Goodheart-Willcox Publisher

Figure 10-13. Hexadecimal-to-decimal conversion involves an extra step—converting the letters to numeric values.

The process of changing a hexadecimal number to a binary equivalent is a grouping operation, **Figure 10-14**. The hexadecimal number is separated into digits, and each digit is converted to its equivalent in binary form. For example, *D* in hexadecimal notation equals 13 in decimal form and 1101 in binary. The groups combine to form the equivalent binary number. Note that the leading zeros from the binary number 0010 are dropped.

Decimal-to-hexadecimal conversion is done by dividing by 16, **Figure 10-15**. Remainders can be as large as 15. Converting a binary number to a hexadecimal equivalent is the reverse of the hexadecimal-to-binary process, **Figure 10-16**. The binary number is divided in groups of four digits, starting at the binary point. Each group is converted to the equivalent hexadecimal value, and the groups are combined to form the hexadecimal value.

10.3 Binary Logic Circuits

Binary signals can be processed easily because they represent two stable states of operation: on or off, 1 or 0, up or down, voltage or no voltage, right or left, and so on. There is no in-between state. This makes binary signals far superior to octal, decimal, or hexadecimal signals for use in logic circuits.

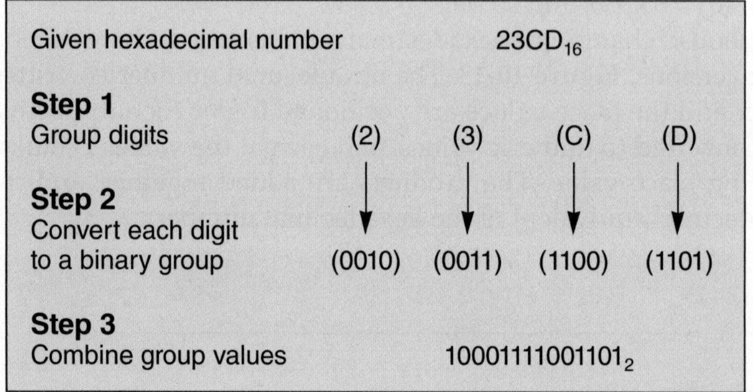

Goodheart-Willcox Publisher

Figure 10-14. A hexadecimal-to-binary conversion requires grouping the binary numbers.

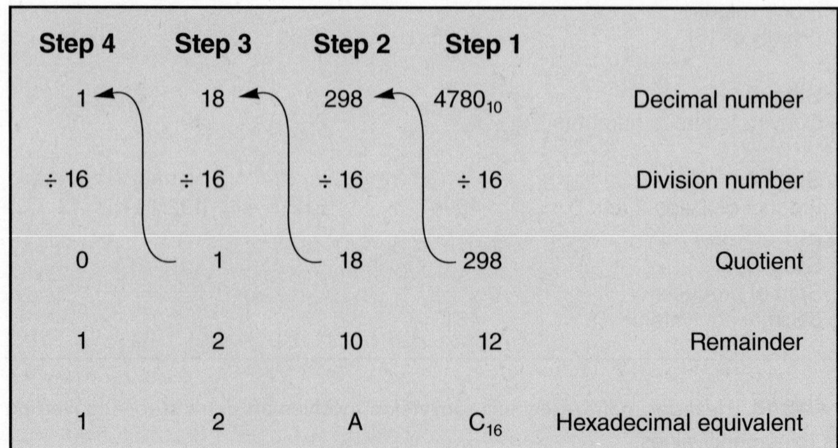

Goodheart-Willcox Publisher

Figure 10-15. A decimal-to-hexadecimal conversion uses a divisor of 16.

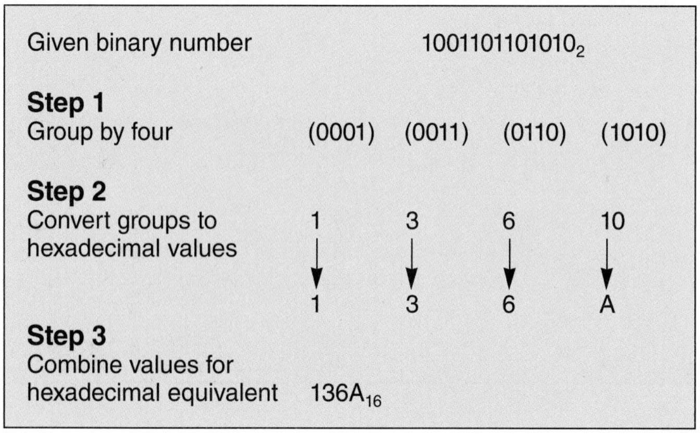

Given binary number	1001101101010_2			
Step 1 Group by four	(0001)	(0011)	(0110)	(1010)
Step 2 Convert groups to hexadecimal values	1 → 1	3 → 3	6 → 6	10 → A
Step 3 Combine values for hexadecimal equivalent	$136A_{16}$			

Goodheart-Willcox Publisher

Figure 10-16. A binary-to-hexadecimal conversion is performed by creating groups of four binary numbers and converting them to hexadecimal values.

The symbols used to represent an operational state are very important. In positive binary logic, *voltage*, *on*, or *true* results are indicated by the 1 operational state. *No voltage*, *off*, or *false* results are indicated by the 0 condition. A circuit can be set to either state, and it remains at that setting until a change occurs.

10.3.1 Logic Gates

Any electronic device that can be set in one of two operational states, either on (1) or off (0), by an outside signal is a ***bistable device***. These devices include relays, lamps, switches, transistors, and diodes. A bistable device can store one binary digit or bit of information. By using many of these devices, it is possible to build a binary logic circuit. A ***binary logic circuit*** is an arrangement of bistable devices that makes decisions based on input signals. The basic binary logic circuits are the AND circuit, OR circuit, and NOT circuit. These circuits are also called ***logic gates***. The term *gate* refers to the circuit's ability to pass or block certain signals. The decision made by each type of circuit is unique. An IF...THEN sentence is often used to describe the basic operation of a logic gate. For example, "IF the inputs applied to a gate are all 1, THEN the output will be 1."

The logic gates discussed in the following sections illustrate basic operation using simple switch and lamp input-outputs. In actual applications, logic gates are typically integrated circuits (ICs).

AND Gates

An ***AND gate*** is a type of logic circuit that has two or more inputs and one output, **Figure 10-17**. When a switch is turned on, the state is 1; when the switch is off, the state is 0. If both inputs are in the 1 state simultaneously, a 1 will be output (the lamp will light). A ***truth table*** is a chart that shows combinations of inputs and the resulting outputs of a logic gate. The operation of the AND gate is simplified by describing the input-output relationships in the truth table. The AND gate produces a 1 output only when switches A and B are both 1.

Each input to an AND gate can create two operational states: 1 and 0. A two-input AND gate has 2^2, or 4, possible combinations that influence the output. A three-input gate has 2^3, or 8, combinations, and a four-input gate has 2^4, or 16, combinations. These combinations are placed in the truth table in binary order.

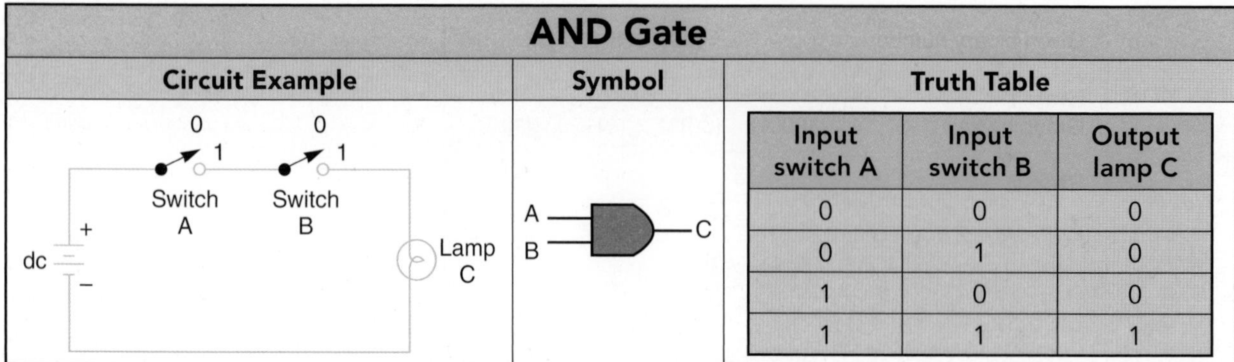

Input switch A	Input switch B	Output lamp C
0	0	0
0	1	0
1	0	0
1	1	1

Goodheart-Willcox Publisher

Figure 10-17. The AND gate circuit.

For a two-input gate, for example, the order is 00, 01, 10, and 11. This represents the binary count of 0, 1, 2, and 3 (in order).

OR Gate

An ***OR gate*** has two or more inputs and one output, **Figure 10-18**. Like the AND gate, each input to the OR gate produces two possible states: 1 and 0. The output is 1 when both switches are 1 *or* when either switch A or B is 1. This gate is used to decide whether or not a 1 appears at either input. The truth table for an OR gate shows that if the state of any input is 1, the output will be 1.

NOT Gate

A ***NOT gate*** has one input and one output, **Figure 10-19**. The output of a NOT gate is opposite to the input. When the switch is on (1 state), it shorts out the lamp. Placing the switch in the off condition (0 state) causes the lamp to be on. A NOT gate is also called an ***inverter***.

Combination Logic Gates

When a NOT gate is combined with an AND gate, it is called a ***NAND gate***. This is an inverted AND gate, **Figure 10-20**. For example, when switches A and B are both on (1 state), lamp C is off (0 state). When either or both switches are off, lamp C is on (1 state).

The combination of a NOT gate and an OR gate results in a ***NOR gate***, **Figure 10-21**. A NOR gate is the opposite of an OR gate. A 1 state is produced as the output only when A is 0 and B is 0.

10.3.2 Flip-Flops

Flip-flops are memory devices used in digital circuits. They are often the basic logic element for counting, temporary memory, and sequential switching operations. Flip-flops can be used to hold an output state even when the input is completely removed, and they can change their output based on an appropriate input signal.

Reset-set (R-S) flip-flops show the different states before an input occurs and how they change afterward, **Figure 10-22**. The logic diagram, symbol, and truth table are more complicated than those for a simple logic gate.

Flip-flops operate in step with a clock pulse. They must often be set and cleared at specific times with respect to other circuits. Both the appropriate R-S inputs and

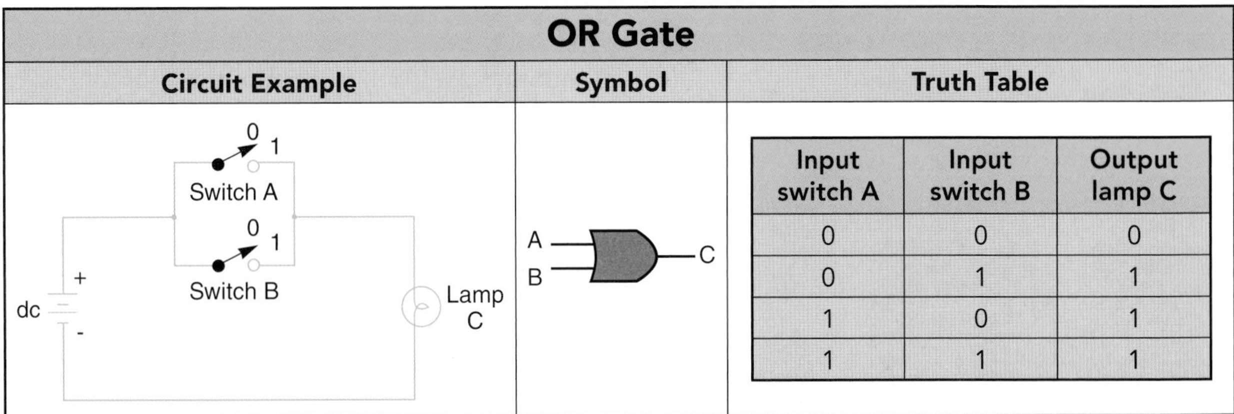

Input switch A	Input switch B	Output lamp C
0	0	0
0	1	1
1	0	1
1	1	1

Goodheart-Willcox Publisher

Figure 10-18. The OR gate circuit.

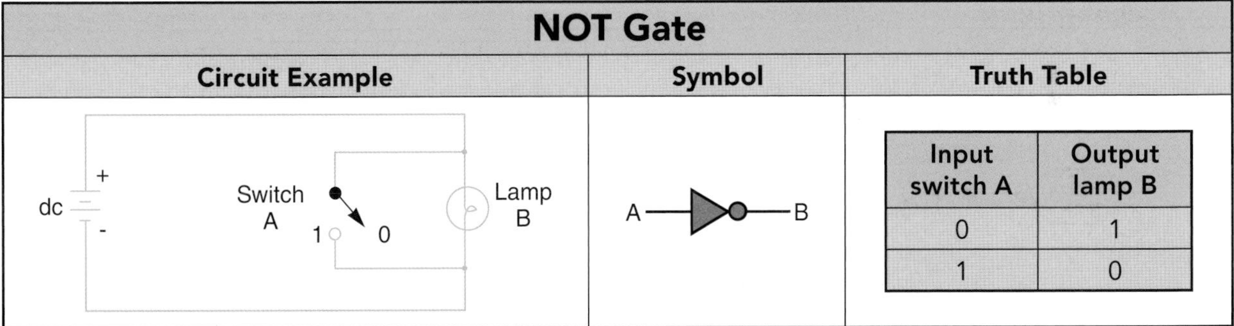

Input switch A	Output lamp B
0	1
1	0

Goodheart-Willcox Publisher

Figure 10-19. The NOT gate circuit.

NAND Gate

Circuit Example | Symbol | Truth Table

dc + - Switch A 1 0 Switch B 1 0 Lamp C

A B C

Input switch A	Input switch B	Output lamp C
0	0	1
0	1	1
1	0	1
1	1	0

Goodheart-Willcox Publisher

Figure 10-20. The NAND gate circuit.

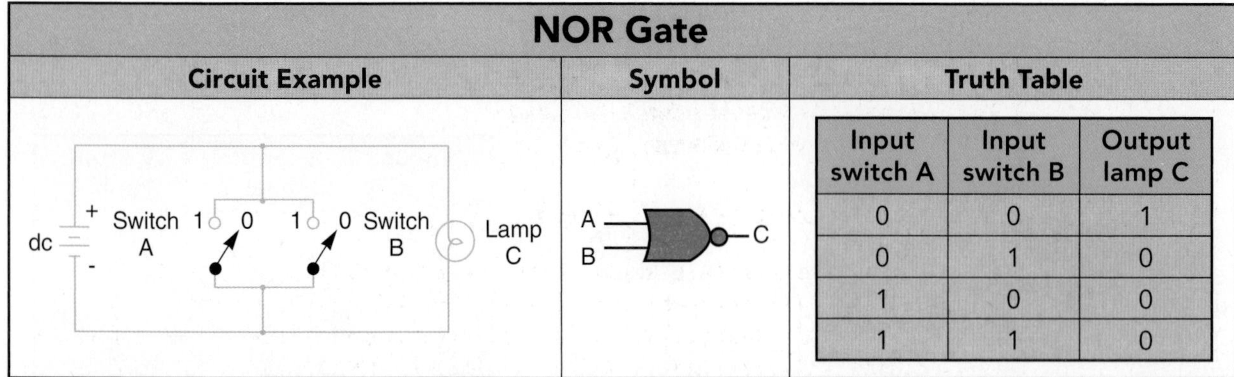

Input switch A	Input switch B	Output lamp C
0	0	1
0	1	0
1	0	0
1	1	0

Goodheart-Willcox Publisher

Figure 10-21. The NOR gate circuit.

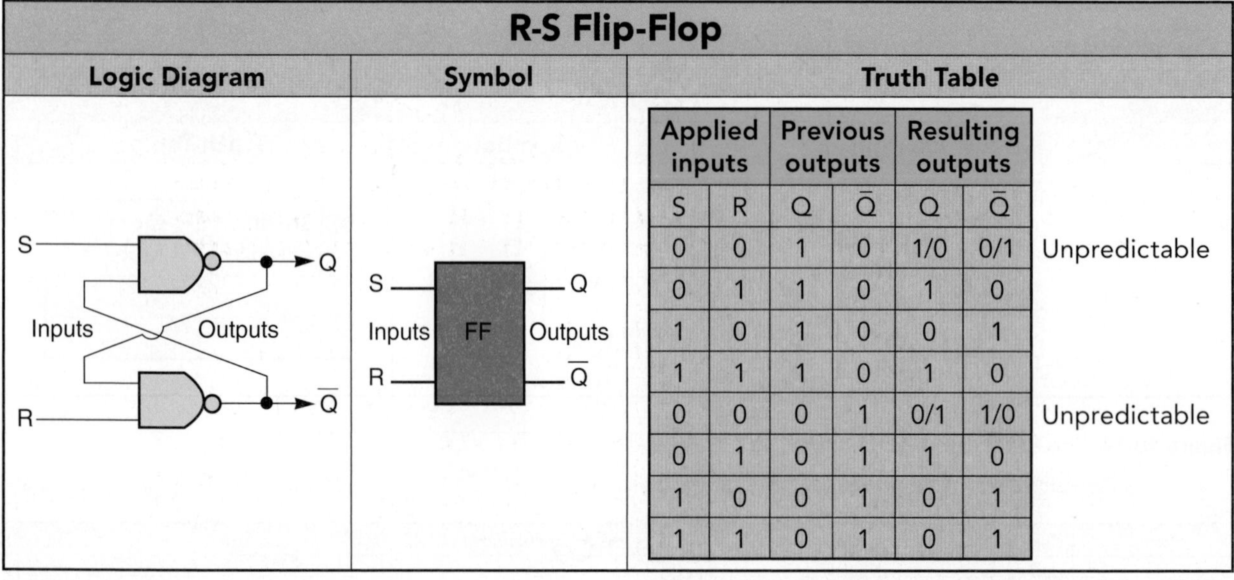

Applied inputs		Previous outputs		Resulting outputs		
S	R	Q	Q̄	Q	Q̄	
0	0	1	0	1/0	0/1	Unpredictable
0	1	1	0	1	0	
1	0	1	0	0	1	
1	1	1	0	1	0	
0	0	0	1	0/1	1/0	Unpredictable
0	1	0	1	1	0	
1	0	0	1	0	1	
1	1	0	1	0	1	

Goodheart-Willcox Publisher

Figure 10-22. The R-S flip-flop.

clock pulse must be present to cause a state change. This device is called an R-S triggered flip-flop, or simply an R-S-T flip-flop, **Figure 10-23**. The truth table for an R-S-T flip-flop is basically the same as that for an R-S flip-flop. A state change occurs only when the clock pulse arrives at the trigger (T) input. A two-input AND gate is added to the set and reset inputs for this purpose.

The JK flip-flop is unique, as it has no unpredictable output states (**Figure 10-24**). It can be set by applying a 1 to the J input and cleared by feeding a 1 to the K input. A 1 signal applied to both J and K inputs simultaneously causes the output to change states, or toggle. A 0 applied simultaneously to both inputs does not initiate a state change. The inputs of a JK flip-flop are controlled directly by clock pulses. Several variations on the basic JK flip-flop are available, including preset and preclear inputs that are used to establish sequential operations at precise times.

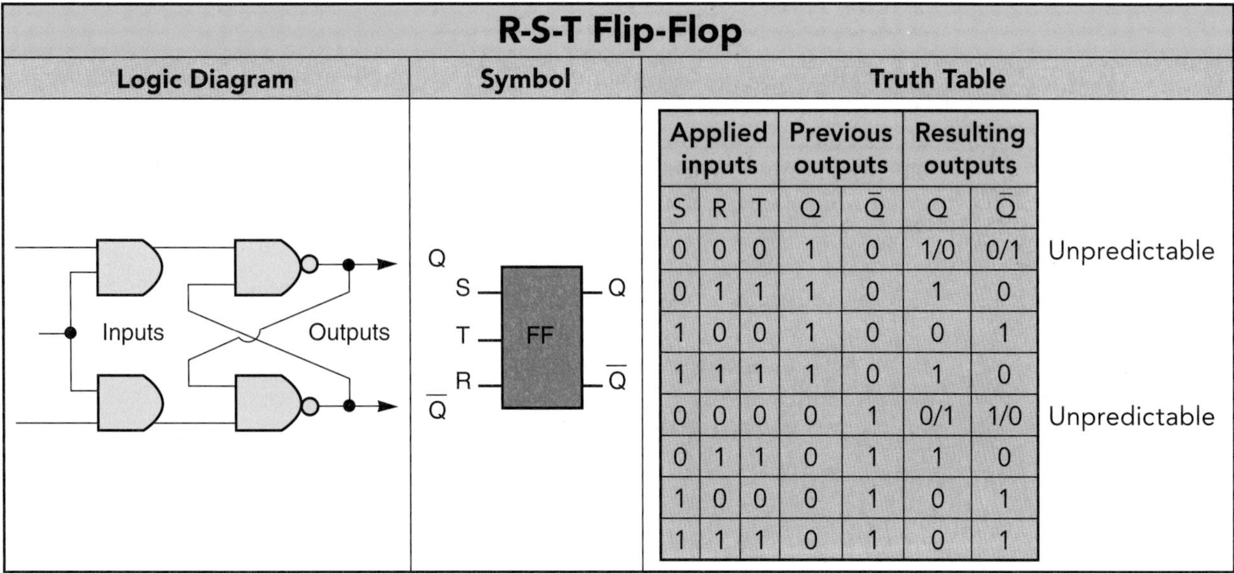

Applied inputs			Previous outputs		Resulting outputs		
S	R	T	Q	$\bar{Q}$	Q	$\bar{Q}$	
0	0	0	1	0	1/0	0/1	Unpredictable
0	1	1	1	0	1	0	
1	0	0	1	0	0	1	
1	1	1	1	0	1	0	
0	0	0	0	1	0/1	1/0	Unpredictable
0	1	1	0	1	1	0	
1	0	0	0	1	0	1	
1	1	1	0	1	0	1	

Goodheart-Willcox Publisher

Figure 10-23. The R-S-T flip-flop.

JK Flip-Flop

Logic Diagram

Clock

Q

$\bar{Q}$

Symbol

J

T

K

FF

Q

$\bar{Q}$

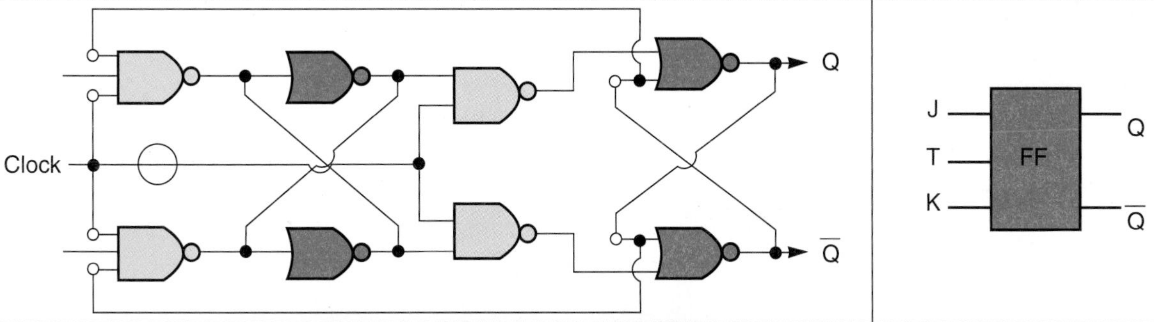

Truth Table

Applied inputs			Previous outputs		Resulting outputs		
J	K	T	Q	$\bar{Q}$	Q	$\bar{Q}$	
0	0	0	0	1	0	1	
0	1	1	0	1	0	1	
1	0	0	0	1	1	0	
1	1	1	0	1	1	0	← Toggle state
0	0	0	1	0	1	0	
0	1	1	1	0	1	0	
1	0	0	1	0	1	0	
1	1	1	1	0	0	1	

Goodheart-Willcox Publisher

Figure 10-24. The JK flip-flop.

10.3.3 Digital Counters

One of the most versatile and important logic devices is the counter. A ***counter*** is a logic device that tallies electronic pulses. Counters are used to count a wide variety of objects in various applications. However, they really count only one thing: electronic pulses. These pulses may be produced mechanically, acoustically, with a clock mechanism, or by a number of other processes. Two common forms of this device are binary counters and decade (BCD) counters.

Binary Counters

A ***binary counter*** is a device used to compute numeric information in binary form. In binary counters, flip-flops are connected so that the output of the first circuit (Q) drives the trigger, or clock input, of the next circuit. See **Figure 10-25.** Each flip-flop, therefore, has a divide-by-two function.

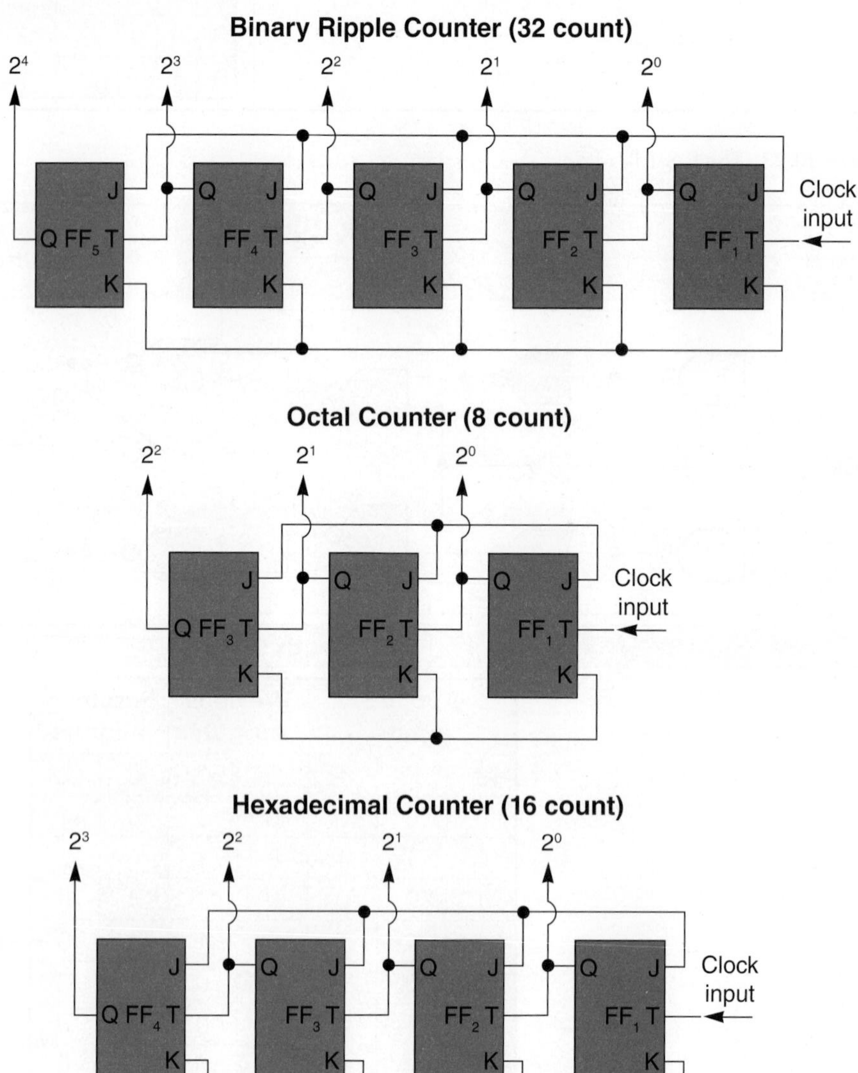

Figure 10-25. Some digital counters that use a JK flip-flop.

In a binary ripple counter, the J and K inputs for each flip-flop are held at a logic 1 level. Each clock pulse applied to the input of the first flip-flop causes a change in state. Since the flip-flops trigger only on the negative-going part of the clock pulse, the output of FF_1 alternates between 1 and 0 with each pulse. Thus, a 1 output appears at Q of FF_1 for every two input pulses. This means that each flip-flop has a divide-by-two function. Five flip-flops connected in this manner produce a 2^5, or 32 count. The largest count in this case is 11111_2 (31_{10}). The next applied pulse clears the counter so that 0 appears at all the Q outputs.

Grouping three flip-flops together creates a binary-coded octal (BCO) counter. For example, the number 111_2 represents the seven count, or seven units, of an octal counter. Two groups of three flip-flops connected in this manner produce a maximum count of 111111_2, which represents 77_8, or 63_{10}.

It is possible to develop the units part of a binary-coded hexadecimal (BCH) counter by placing four flip-flops together in a group. Thus, 1111_2 represents F_{16} or 15_{10}. Two groups of four flip-flops could produce a maximum count of 11111111_2, which represents FF_{16} or 255_{10}. Each succeeding group of four flip-flops raises the count to the next power of 16.

Decade Counters

Since most of the mathematics used today is based on the decimal (base 10) system, it is important to be able to count by this method. The output of a binary counter must be changed into decimal form before it can be used in this way. A ***decade counter*** is a device that is used to change binary signals into a binary-coded decimal (BCD) form. For example, a four-bit binary counter achieves 16 counts using four flip-flops, **Figure 10-26**. To convert this counter into a decade counter, it must skip some of its counts, **Figure 10-27**. In this example, the first seven counts occur naturally. Therefore, FF_D remains at 1. This is applied to the J input of FF_B, permitting it to trigger with each clock pulse. At the seventh count, 1s appearing at the Q outputs of FF_B and FF_C are applied to the AND gate. This action produces a logic 1 and applies it to the J input of FF_D. The next clock pulse triggers FF_A, FF_B, and FF_C into the off state and turns on FF_D. This represents the eighth count.

When FF_D is in the on state, Q is 1 and $\overline{Q}$ is 0. This causes a 0 to be fed to the J input of FF_B, which prevents it from triggering until cleared. The next clock pulse causes FF_A to be set to a 1. This registers a 1001_2, which is the ninth count. The next count clears FF_A and FF_D instantly. Since FF_B and FF_C were previously cleared by

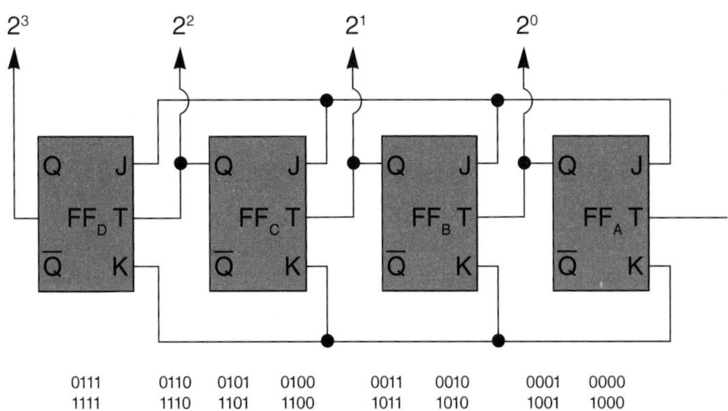

Goodheart-Willcox Publisher

Figure 10-26. A four-bit binary counter uses four flip-flops. The counts are listed below the binary counter, reading right to left.

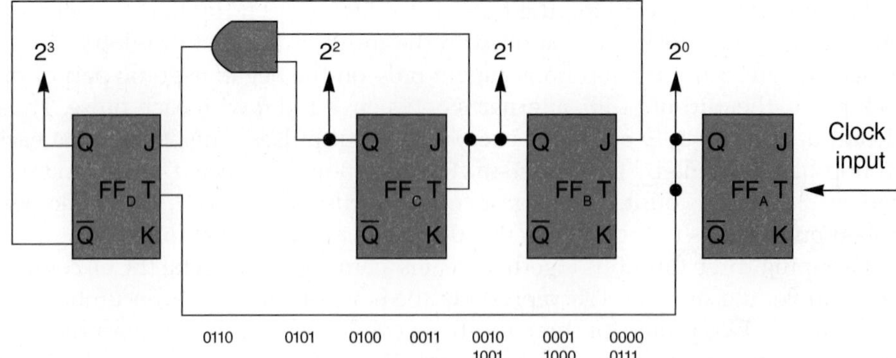

Goodheart-Willcox Publisher

Figure 10-27. Converting a four-bit binary counter into a decade (BCD) counter.

the seventh count, all 0s appear at the outputs. The counter has, therefore, cycled through the ninth count and returned to zero, ready for the next input.

10.4 Computer Programming

Computer programming is accomplished by a set of instructions (characters or symbols) that define specific operations. In microcomputer systems, some programs, called ***firmware***, are held in read-only memory. Other programs are stored in the form of software. ***Software*** is a type of program that is transferred to a computer system by a keyboard, CD-ROM, or download, and the instructions are stored in read/write memory. Firmware is placed on ROM chips, which requires that program changes be made by changing the ROM. Software has more flexibility than firmware because it is transferred to the system by keyboard, CD-ROM, or download.

10.4.1 Instructions

Instructions normally appear as a set of characters or symbols that define a specific operation. These symbols are similar to those on a computer keyboard. They include decimal digits 0–9, letters A–Z, and, in some cases, punctuation marks and special characters. Instructions may also appear in the form of binary numbers, hexadecimal numbers, or mnemonic codes. See **Figure 10-28**. Each computer system has a number of instructions of this one-byte type that contain only an opcode.

Each type of MPU is designed to understand a certain set of instructions. These instructions are in the form of binary data and are normally held in a read-only memory unit. The unit is address-selected and connected to the MPU through the common data bus. Instructional sets are an example of firmware. They are fixed at a specific location and cannot be changed.

Instructions usually consist of one, two, or three bytes of data. This type of data must follow the commands in successive memory locations. These commands are usually called addressing modes.

One-byte instructions are called inherent-mode instructions. They are designed to send data to the accumulator registers of the ALU. No address code is needed because it is an implied machine instruction. For example, the instruction *CLA* is a one-byte opcode that clears the contents of accumulator A. No address or

Mnemonic	Opcode	Meaning
ABA	1B	Add the contents of accumulators A and B. The result is store in accumulator A. The contents of B are not altered.
CLA	4F	Clear accumulator A to all zeros.
CLB	5F	Clear accumulator B.
CBA	11	Compare accumulators: Subtract the contents of ACCB from ACCA. The ALU is involved, but the contents of the accumulators are not altered. The comparison is reflected in the condition register.
COMA	43	Find the ones complement of the data in accumulator A, and replace its contents with its ones complement. (The ones complement is simple inversion of all bits.)
COMB	53	Replace the contents of ACCB with its ones complement.
DAA	19	Adjust the two hexadecimal digits in accumulator A to valid BCD digits. Set the carry bit in the condition register when appropriate. The correction is accomplished by adding 06, 60, or 66 to the contents of ACCA.
DECA	4A	Decrement accumulator A. Subtract 1 from the contents of accumulator A. Store the result in ACCA.
DECB	5A	Decrement accumulator B. Store in accumulator B.
LSRA	44	Logic shift right, accumulator A or B. 0 → b_7 b_6 b_5 b_4 b_3 b_2 b_1 b_0 → C
SBA	10	Subtract the contents of accumulator B from the contents of accumulator A. Store results in accumulator A.
TAB	16	Transfer the contents of ACCA to accumulator B. The contents of register A are unchanged.
TBA	17	Transfer the contents of ACCB to accumulator A. The contents of ACCA are unchanged.
NEGA	40	Replace the contents of ACCA with its twos complements. This operation generates a negative number.
NEGB	50	Replace the contents of ACCB with its twos complements. This operation generates a negative number.
INCA	4C	Increment accumulator A. Add 1 to the contents of ACCA and store in ACCA.
INCB	5C	Increment accumulator B. Store contents in ACCB.
ROLA	49	Rotate left, accumulator A.

Goodheart-Willcox Publisher

Figure 10-28. Example of opcode instructions. Note that the code is in hexadecimal form, and a mnemonic is given for each one.

specific definition of data is needed. Inherent mode instructions differ a great deal among manufacturers.

Immediate Addressing

Immediate addressing is done with a two-byte instruction that contains an opcode and an operand. The opcode appears in the first byte, followed by the eight-byte operand. A common practice is to place intermediate addressing instructions in the first 256 memory locations. Since this is the fastest mode of operation, these instructions can be retrieved very quickly.

Relative Addressing

Relative addressing instructions transfer program control to a location other than the next consecutive memory address. Transfer is often limited to a number of locations in front of or in back of the present location. The two-byte instruction contains an opcode in the first byte and a memory location in the second byte. The second byte points to the location of the next instruction to be executed.

Indexed Addressing

Indexed addressing is similar to relative addressing. The second byte of the two-byte instruction is added to the contents of the index register to form a new, or effective, address. This address is obtained during execution, rather than being held at a predetermined location. It is held in a temporary memory address register to ensure that it does not get altered or destroyed during processing.

Effective use of the Global Hawk requires knowledge of its complex communication and processing systems.

Robotics in Society: RQ-4 Global Hawk

The RQ-4 Global Hawk is an unmanned aerial vehicle (UAV) built by the Northrop Grumman Corporation and used by the United States military forces. This aircraft has a wingspan of about 116 feet, stands about 15 feet tall, and can travel at a maximum altitude of 65,000 feet (about 19 km)—just on the edge of the stratosphere—without a human pilot on board.

The Global Hawk takes off, performs its mission, and lands using the programmed data and monitoring provided by two land-based control systems: the Mission Control Element (MCE) and the Launch and Recovery Element (LRE). Pilots operating the MCE manage the sensors, cameras, and indicators on board as well as the aircraft itself. The LRE team on the ground is responsible for loading the mission plan into the aircraft and monitoring performance during takeoff and landing.

The Global Hawk is equipped with wideband satellite links, line-of-sight links, and a 48″ satellite antenna for communication and data transmission with the land-based control systems. All systems and sensors on board gather and transmit data and images day or night, regardless of weather conditions. Having this real-time information has proven critical in military combat operations. The Global Hawk is improving the efficiency, accuracy, and strategy of military operations, which affects the safety of troops fighting on the ground.

Photo courtesy of Northrop Grumman Corporation

Direct Addressing

In direct addressing, the address is located in the byte of memory following the opcode. This permits addressing the first 256 bytes of memory, from 0000_{16} to $00FF_{16}$.

Extended Addressing

Extended addressing increases the ability of direct addressing to accommodate more data. It is used for memory locations above $00FF_{16}$ and requires three bytes of data for the instruction. The first byte is a standard opcode. The second byte is an address location for the most significant eight bits of data. The third byte holds the address of the least significant eight bits of data being processed.

10.4.2 Program Planning

A computer cannot solve even the simplest problem without the help of a well-planned program. The system follows this program to accomplish a task. This is why programming is an essential part of all computer applications. The programmer must be fully aware of the instructions used by the system.

A programmer should be able to decide what specific instructions are needed to perform a given task. A limited number of operations can usually be developed without the aid of a diagrammed plan of procedure. Complex problems, however, require a specific plan in order to avoid confusion or the loss of an important step. Flowcharts are commonly used to aid in this type of planning, **Figure 10-29**.

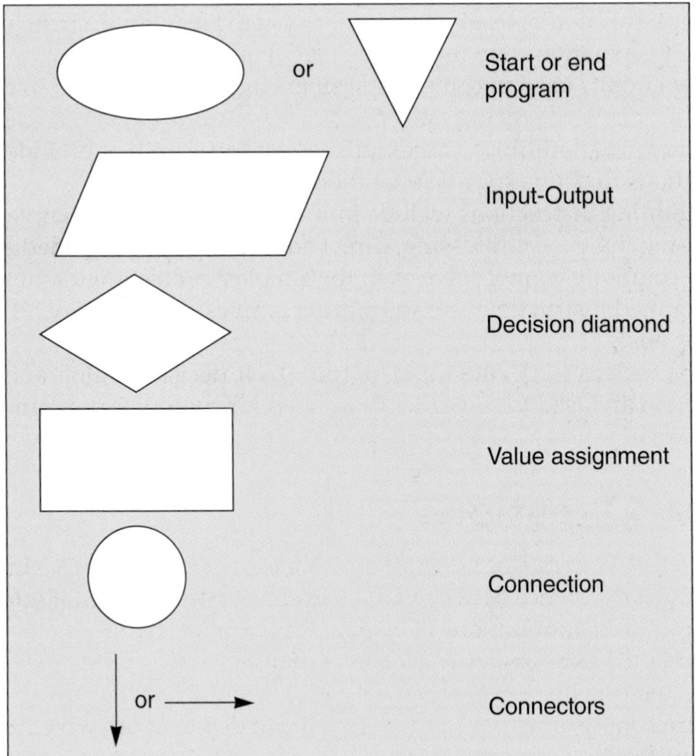

Goodheart-Willcox Publisher

Figure 10-29. These symbols are used by programmers when constructing flowcharts.

Summary

- Electronic circuitry used with robots provides commands to control electric motors, hydraulic systems, and pneumatic systems.
- Computers process either analog information (which varies continuously) or digital information (which is in separate full units, 1s and 0s).
- The basic components of a computer typically include input and output devices, arithmetic logic and control circuitry, and memory.
- A microprocessor unit (MPU) is a microchip that contains the ALU and control circuitry of a computer. The bus network of a microprocessor allows data transfer throughout the system.
- Memory is the ability of a microprocessor to store data that is easily retrieved and includes read-only memory and random access memory.
- Basic functions of a computer include timing, fetch and execute, read memory, write memory, input-output transfer, and interrupt.
- Number systems are important for computer system operation. Commonly used systems include decimal, binary, octal, and hexadecimal.
- Numbers can be converted from one system to another, such as decimal to binary or octal to binary.
- Binary signals can be processed easily because they represent two stable states of operation and are far superior to octal, decimal, or hexadecimal signals for use in logic circuits.
- Logic gates are used in computer systems to pass or block electrical signals and include AND, OR, NOT, NAND, and NOR gates.
- Flip-flops are circuits used in computer systems for counting, memory, and switching operations and include R-S, R-S-T, and JK types.
- Counter circuits used in computer systems include binary and decade (BCD) types.
- Computer programming is accomplished by software that provides a set of instructions to define specific operations.
- Programming instructions include immediate addressing, relative addressing, indexed addressing, direct addressing, and extended addressing.
- Since a computer cannot solve even the simplest problem without the help of well-planned instructions, programming is an essential part of all computer applications.
- Symbols such as start, end, input/output (I/O), decision, value, and connection are used to construct flowcharts for computer programming.

Review Questions

Answer the following questions using the information provided in this chapter.

1. Explain the difference between analog information and digital information.
2. What is a bit of information? How many bits are in a byte?
3. What are the basic components of a computer?
4. What is the function of the CPU?
5. What are some examples of input and output devices with which a CPU may communicate?
6. Define the term *register unit* and list examples of different types of registers.
7. What is the difference between ROM and RAM?
8. List and define the functions that are basic to most computer systems.

9. Describe the number system used in everyday life in the United States.
10. How does use of the binary number system facilitate information processing by electronic devices, such as computers?
11. What is the purpose of the binary-coded-decimal (BCD) number system?
12. What are the base factors for the octal number system and the hexadecimal number system?
13. What is a bistable device?
14. How many inputs and outputs can be found on an OR gate?
15. What is an inverter?
16. What are flip-flops and how are they used in binary counters?
17. What is the purpose of a decade counter?
18. Explain the difference between firmware and software.
19. What are addressing modes and what are five types of addressing modes?
20. Describe the importance of a well-designed computer program.

Learning Extensions

1. Research the history and applications of the binary number system. Prior to computer programming, what were some historical applications of the binary system?
2. Express the following binary (base 2) number equivalents of 14, 23, 61, 72, 128, and 264 (base 10).
3. Change the following binary numbers to decimal equivalents: 1011, 10001, 11110, 1000000, and 1000011.
4. Describe the truth tables of AND, OR, NOT, NAND, and NOR gates. See Figures 10-17 through 10-21.
5. Explain how four flip-flops are used to count to 1111 (base 2). See Figure 10-26.

Advanced Analysis

1. Prepare a written discussion of the procedure to convert decimal (base 10) numbers to binary numbers (base 2).
2. Prepare a simplified summary of the operation of a computer system.

Activity 10-1 Digital Logic Functions

Objectives

Control of a digital system is achieved by a variety of logic gates. Each gate has a particular output response to a combination of input signals. Binary data is represented by ones (1s) and zeros (0s). Multiple-input gates of the AND and OR variety and a single input-output NOT gate represent the three primary logic functions of a digital system. An understanding of these logic functions is essential when analyzing the operation of robotic systems.

In this activity, you will construct each basic logic gate with a single integrated circuit. You will develop truth tables to show the relationship between the input alternatives and the corresponding outputs. Through this activity, you will gain experience in using the basic logic gates and see how the logic functions are achieved electronically.

Equipment and Materials

- SN7408 IC.
- SN7411 IC.
- SN7432 IC.
- SN7404 IC.
- 5V dc power supply.
- 470 Ω, 1/8 W resistors (4).
- Light-emitting diodes (4).
- SPDT toggle switches (3).
- SPST toggle switch.
- IC breadboard construction unit.
- Multimeter.

Procedure

Part A: Gate Testing with the SN7408 and SN7411

1. Connect the logic gate test circuit shown.

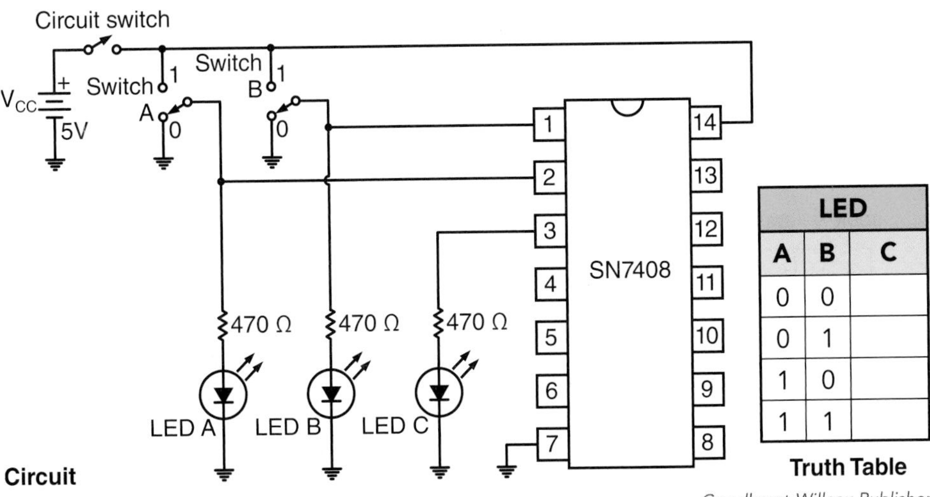

LED		
A	B	C
0	0	
0	1	
1	0	
1	1	

SN7408 IC test circuit and truth table.

2. Before closing the circuit switch, turn on the power supply and adjust it to 5V.
3. Turn on the circuit switch and test the IC gate according to the alternatives listed in the truth table.

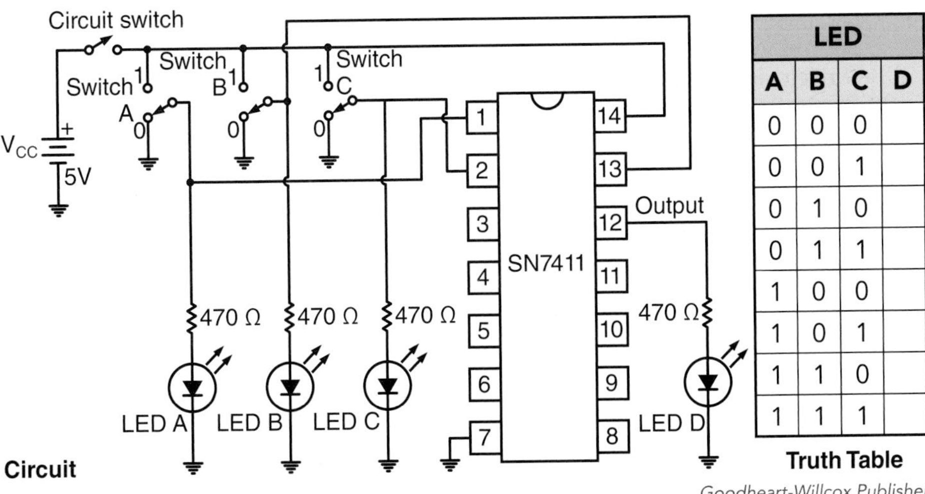

LED			
A	B	C	D
0	0	0	
0	0	1	
0	1	0	
0	1	1	
1	0	0	
1	0	1	
1	1	0	
1	1	1	

SN7411 IC test circuit and truth table.

4. Prepare a multimeter to measure voltage and determine the corresponding input and output voltages corresponding to 1 and 0.
5. There are three other gates included in this IC chip. Pins 4 and 5 are the input and pin 6 is the output of gate 2. Pins 9 and 10 are the input and pin 8 is the output of gate 3. Pins 13 and 12 are the input and pin 11 is the output of gate 4. Test these gates to see if they are working properly.

6. Make a gate symbol drawing on the blank IC dual inline package layout for the SN7408.
7. Open the circuit switch and remove the SN7408. In its place connect an SN7411. Be certain that pin 14 is connected to the +5V V_{CC} source and pin 7 is connected to ground. Connect the remainder of the circuit.
8. Close the circuit switch and test the SN7411. Complete the truth table for the listed input alternatives.
9. Open the circuit switch and disconnect the IC.

Part B: Gate Testing with the SN7432

1. Construct the logic gate test circuit shown.

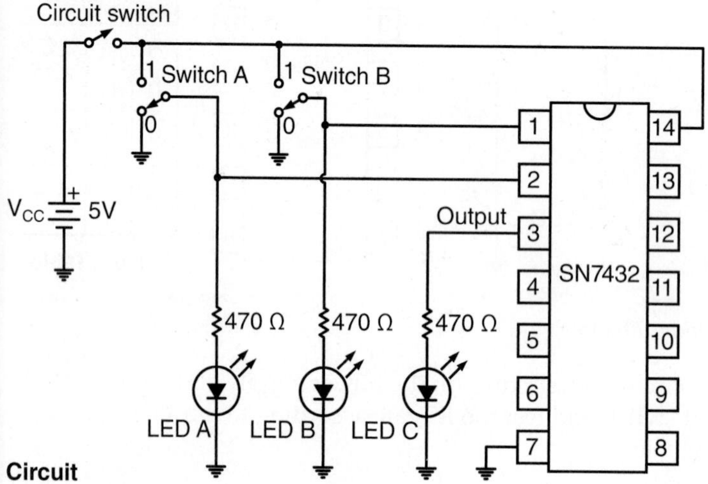

LED		
A	B	C
0	0	
0	1	
1	0	
1	1	

Truth Table

Goodheart-Willcox Publisher

SN7432 IC test circuit and truth table.

2. Before closing the circuit switch, turn on the power supply and adjust it to 5V.
3. Turn on the circuit switch and test the IC gate according to the various alternatives.
4. What logic gate function is achieved by this gate?
5. There are three other gates included in this IC chip. Pins 4 and 5 are the input and pin 6 is the output of gate 2. Pins 9 and 10 are the input and pin 8 is the output of gate 3. Pins 13 and 12 are the input and pin 11 is the output of gate 4. Test these gates to see if they are working properly.
6. Make a gate symbol drawing on the blank IC dual inline package layout showing the input-output connections for the SN7432.

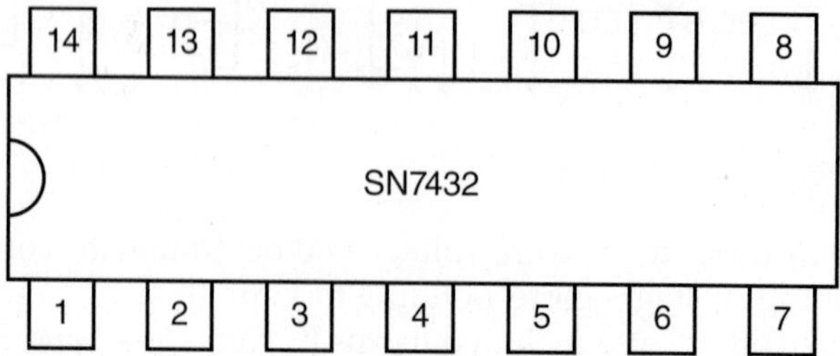

Goodheart-Willcox Publisher

SN7432 IC pin connections.

7. Open the circuit switch and disconnect the circuit.

Part C: Gate Testing with the SN7404

1. Connect the logic gate test circuit shown.

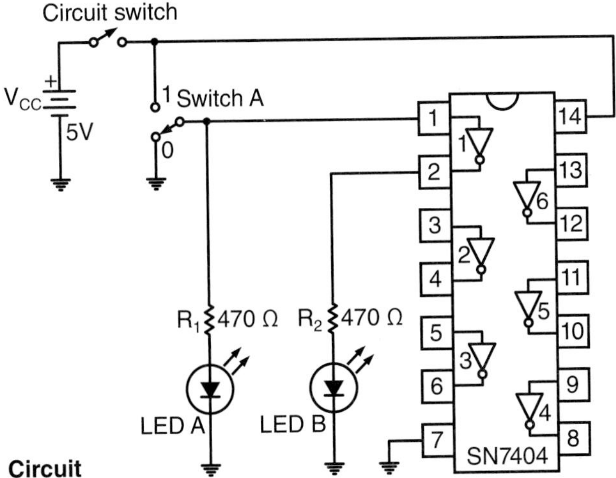

Circuit

Input	Output
LED A	LED B
0	
1	

Truth Table

Goodheart-Willcox Publisher

SN7404 IC test circuit and truth table.

2. Before closing the circuit switch, turn on the power supply and adjust it to 5V dc.
3. Then close the circuit switch and test the IC gate according to the alternate input conditions listed in the truth table. Record the respective output for each of the inputs.
4. What logic function is achieved by this gate?
5. There are three other gates included in this IC chip. Pins 4 and 5 are the input and pin 6 is the output. Pins 9 and 10 are the input and pin 8 is the output. Pins 12 and 13 are the input and pin 11 is the output. Test these gates to see if they are working properly.
6. Make a gate symbol drawing on the blank dual inline package layout showing the input-output connections for each gate of the SN7404.

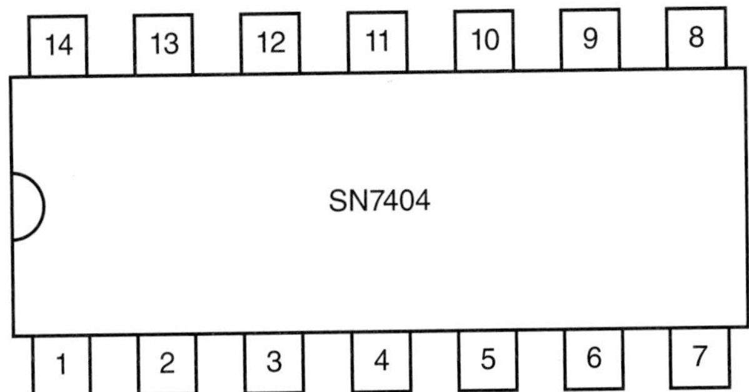

Goodheart-Willcox Publisher

SN7404 IC pin connections.

7. Open the circuit switch and connect the output of gate 1 to the input of gate 2. Connect the output indicating LED to pin 4.
8. Close the circuit switch and test the logic circuit. What function does this indicate?
9. Open the circuit switch and disconnect the circuit. Return all parts to the storage area.

Analysis

1. What logic functions are achieved by the gates studied in part A?
2. What mathematical functions are achieved by the gates studied in part A?
3. What are the symbolic representations of the gates studied in part A?
4. Prepare a statement that describes the operation of the gates studied in part A.
5. What mathematical function is achieved by the gate studied in part B?
6. What are the two logic symbols that are used to represent the circuit constructed in part B of this activity?
7. Prepare a statement that describes the operation of the gate studied in part B.
8. What mathematical function is achieved by the logic gate studied in part C?
9. What is the meaning of the terms *negation* and *double negation*?
10. What would be the resulting output of a 1 applied to three of the gates studied in part C of this activity connected in series?

Activity 10-2
Combination Logic Gates

Objectives

When either an AND gate or an OR gate is connected to a NOT gate, two additional logic functions are achieved. A NOT-AND, or NAND, is one type of combination logic gate. The NOT-OR, or NOR, gate is representative of the second type of combinational logic gate achieved.

NOR, NAND, and NOT gates are often considered to be universal building blocks in digital systems. With these gates, it is possible to build four logic gates in addition to using gates for the original function. A person working with digital systems should be very familiar with the universal building block principle of combinational logic gates.

In this activity, you will investigate the NAND, NOR, and NOT gates in combinational logic gate construction operations. Only gates that have an inverting capability can be used in this building block technique.

Equipment and Materials

- SN7400 IC.
- SN7402 IC.
- SN7404 IC.
- SPST toggle switch.
- SPDT toggle switches (2).
- 470 Ω, 1/8 W resistors (3).
- Light-emitting diodes (LEDs) (3).
- 0–5V dc, 1 A power supply.
- Multimeter.
- IC circuit construction board.

Procedure

Part A: NAND Logic

1. Using the SN7400 quad NAND, build the IC test circuit and check gate 1.

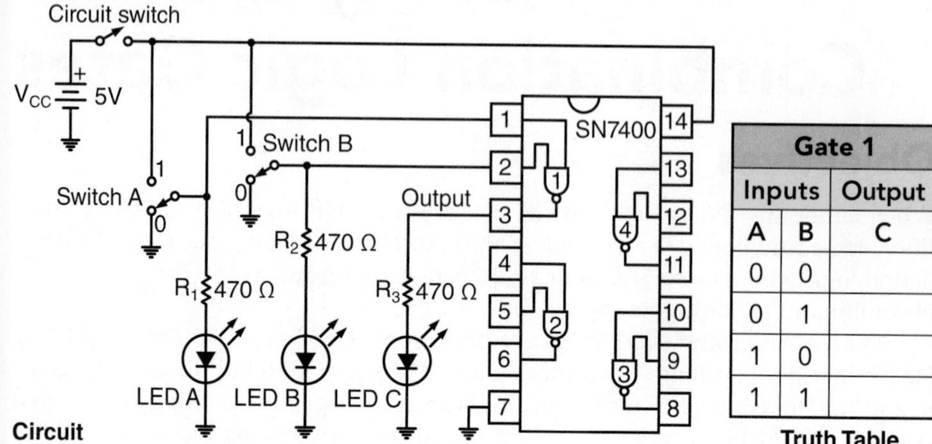

Gate 1		
Inputs		Output
A	B	C
0	0	
0	1	
1	0	
1	1	

Goodheart-Willcox Publisher

SN7400 IC test circuit and truth table.

2. Record the 1s and 0s of the output in a truth table for each of the input alternatives.
3. With a multimeter, measure and record voltage values represented by the 1s and 0s of this gate.
4. Test the other three gates to make certain they are functioning properly.
5. To build a NOT gate, remove the B circuit from pin 2 and connect pins 1 and 2 together. Use switch A as the input, and test the output of the gate.
6. Connect NAND gate 1 as instructed in procedure steps 1 and 3. Connect the output of gate 1 to pins 4 and 5 of NAND 2. Connect the output LED to pin 6. What function does this combination logic gate achieve?
7. Test the circuit to verify your theory. Did it work as you predicted?
8. Connect the combinational logic circuit. Complete a truth table showing the outputs at the indicated points.

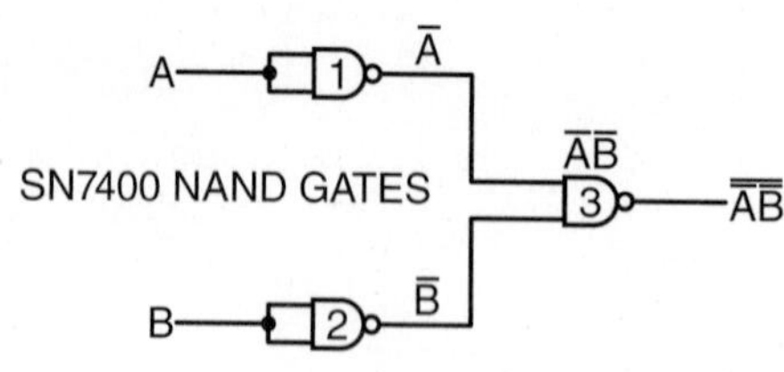

Inputs		Expression			Output
A	B	$\overline{A}$	$\overline{B}$	$\overline{A}\overline{B}$	$\overline{\overline{A}\overline{B}}$
0	0				
0	1				
1	0				
1	1				

Goodheart-Willcox Publisher

Combination NAND gate circuit and truth table.

9. According to the truth table, what gate function is achieved by this combination logic gate? Test the gate to verify your prediction. Did it perform the function you expected?

Part B: NOR Logic

1. Using an SN7402 quad NOR gate, build the IC test circuit, and check gate 1.

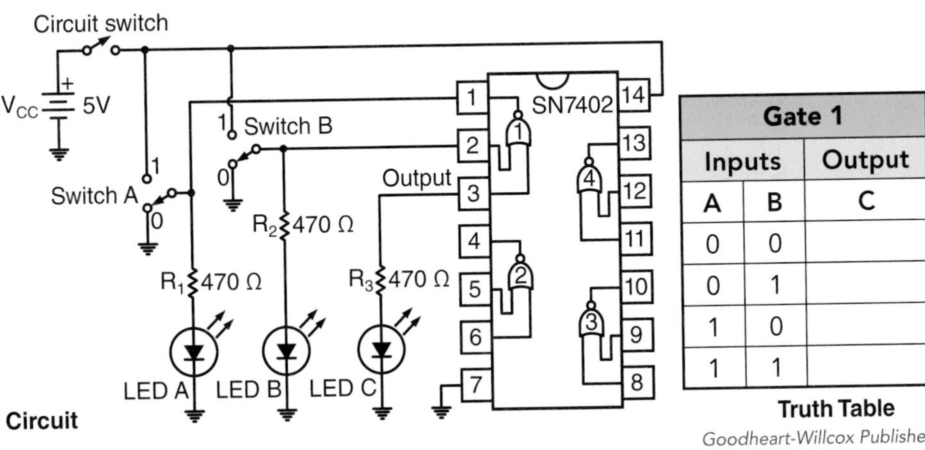

Gate 1		
Inputs		**Output**
A	B	C
0	0	
0	1	
1	0	
1	1	

Truth Table

Goodheart-Willcox Publisher

SN7402 IC test circuit and truth table.

2. Record the 1 and 0 outputs for the input alternatives listed in a truth table.
3. With a multimeter, measure the voltage values represented by the 1s and 0s of this gate.
4. Test the other three gates to make certain they are functioning properly.
5. To build a NOT gate, disconnect the lead to pin 3 and then connect pin 2 to pin 3. Use switch A as the input, and test the output of the gate.
6. Connect NOR gate 1 as instructed in steps 1 and 2. Connect the output of NOR 1 to the input of NOR 2, with NOR 2 connected as a NOT gate. Connect the output LED to pin 4. What combination logic is achieved by this configuration?
7. Test the circuit to verify your theory. Did it work as you predicted?
8. Connect the combinational logic circuit using the SN7402.

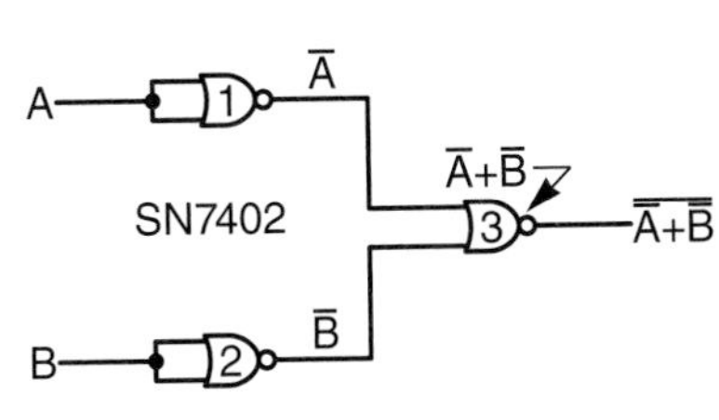

Inputs		**Expression**			**Output**
A	B	$\overline{A}$	$\overline{B}$	$\overline{A}+\overline{B}$	$\overline{\overline{A}+\overline{B}}$
0	0				
0	1				
1	0				
1	1				

Goodheart-Willcox Publisher

Combination NOR gate circuit and truth table.

9. Complete a truth table for outputs at the designated points. Then test the combinational logic circuit. Does it do what you predicted by the truth table?
10. Open the circuit power switch and connect NOR gate 4 as a NOT gate. Attach the output to the input of NOR 4. Connect the LED to the output of NOR 4. Invert the last stage of the truth table and predict the type of gate achieved by your modified circuit.
11. Close the circuit switch and test your prediction. Was it correct?
12. Open the circuit switch and disconnect the circuit.

Part C: NOT Logic

1. Using the SN7404 hex-inverter IC, build the IC test circuit and check gate 1.

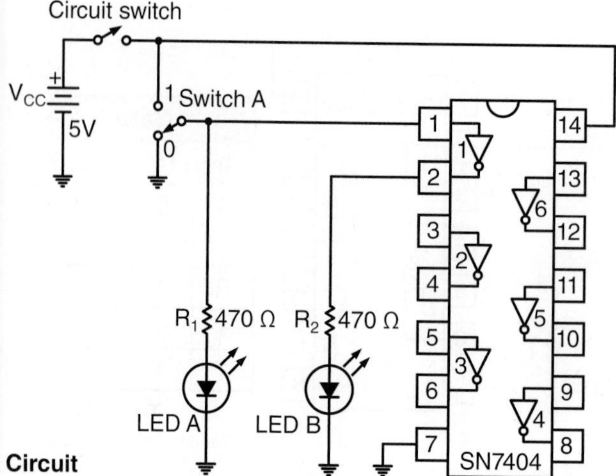

Input	Output
A	B
0	
1	

Truth Table

Goodheart-Willcox Publisher

SN7404 IC test circuit and truth table.

2. Turn on the circuit switch and record the 1s and 0s of the output in a truth table for the input alternatives.
3. With a multimeter, measure the voltage values of a representative 1 and 0 of this gate.
4. Test the other five gates to verify that they are operating properly.
5. Open the circuit switch and combine gates 1 and 2.

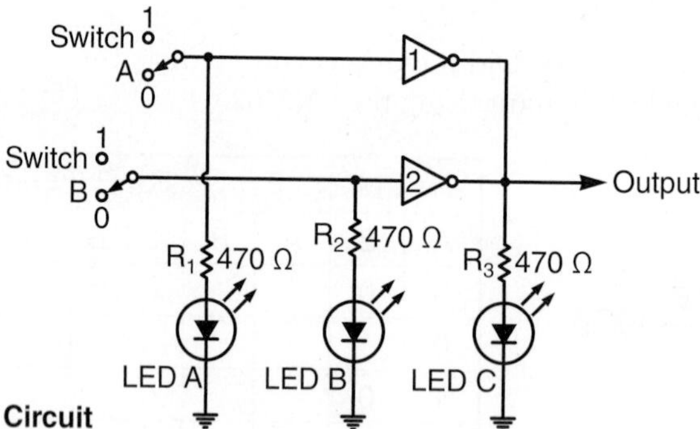

Inputs		Output
A	B	C
0	0	
0	1	
1	0	
1	1	

Truth Table

Goodheart-Willcox Publisher

Combination NOT gate circuit and truth table.

6. Close the circuit switch and test the combination gate. Record the 1 and 0 outputs for each of the input alternatives of the truth table. What logic function does this achieve?
7. Open the circuit switch and connect the output of the circuit to the input of gate 3. Connect the LED resistor to pin 6 and test the circuit. What logic function does it achieve?

8. Open the circuit switch and combine gates of the inverter to form a different circuit.

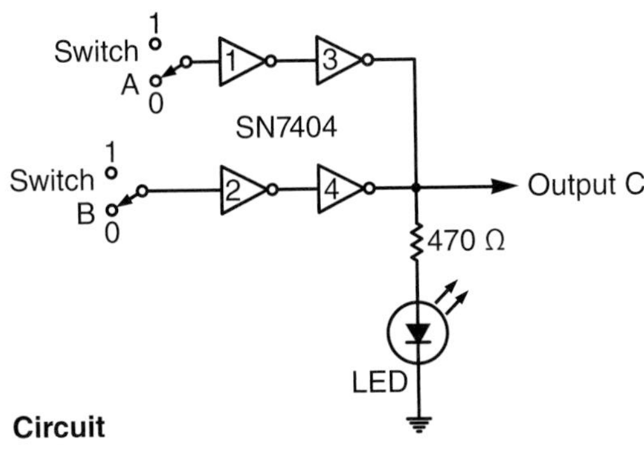

Inputs		Outputs	
A	B	C	D
0	0		
0	1		
1	0		
1	1		

Truth Table

Goodheart-Willcox Publisher

Cascaded double-inverter circuit.

9. Close the circuit switch and test the circuit. Record the 1 and 0 outputs of C in a truth table for the given input alternatives. What gate function does this achieve?
10. Open the circuit switch and add the output to the input of gate 5. Move the LED to the output of gate 5. What gate function do you predict the output will demonstrate?
11. Close the circuit switch and test the circuit. Record your findings at the D output of the truth table. Was your prediction correct?
12. Open the circuit switch and disconnect the components. Return all parts to the storage area.

Analysis

1. In part A, complete a truth table showing the outputs for each logic gate listed.

Inputs		Outputs			
A	B	NAND	AND	OR	NOR
0	0				
0	1				
1	0				
1	1				

Goodheart-Willcox Publisher

General truth table.

2. Make a sketch of the five logic circuit combinations accomplished with the SN7402. Show the actual logic gate combinations used.
3. In any of the combined inverter gate combinations, why is the 0 considered a predominant factor?
4. Make a sketch of the five logic functions achieved by the SN7404.

Chapter 11

Interfacing and Vision Systems

Outline

Objectives

Upon completion of this chapter, you will be able to:

- Explain the purpose of interfacing robotic systems and hardware options for interfacing.
- Explain how machine vision systems are used for robotics.

Technical Terms

digital input port
digital output port
general-purpose interface bus (GPIB)
image acquisition
image analysis
image interpretation
image preprocessing
input/output (I/O) port
interface
machine vision system
parallel port
parallel transmission
pixel
serial port
serial transmission

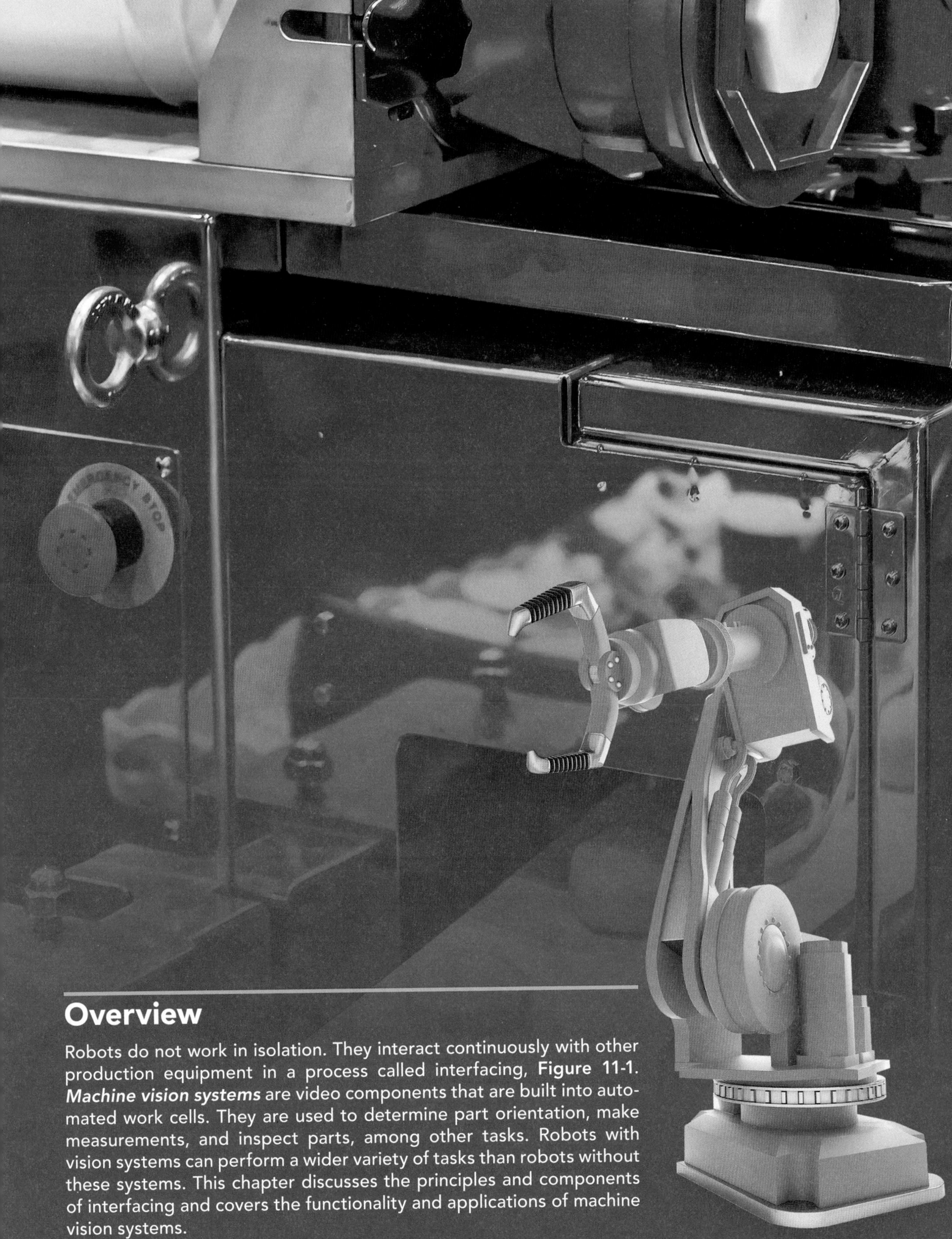

Overview

Robots do not work in isolation. They interact continuously with other production equipment in a process called interfacing, **Figure 11-1**. ***Machine vision systems*** are video components that are built into automated work cells. They are used to determine part orientation, make measurements, and inspect parts, among other tasks. Robots with vision systems can perform a wider variety of tasks than robots without these systems. This chapter discusses the principles and components of interfacing and covers the functionality and applications of machine vision systems.

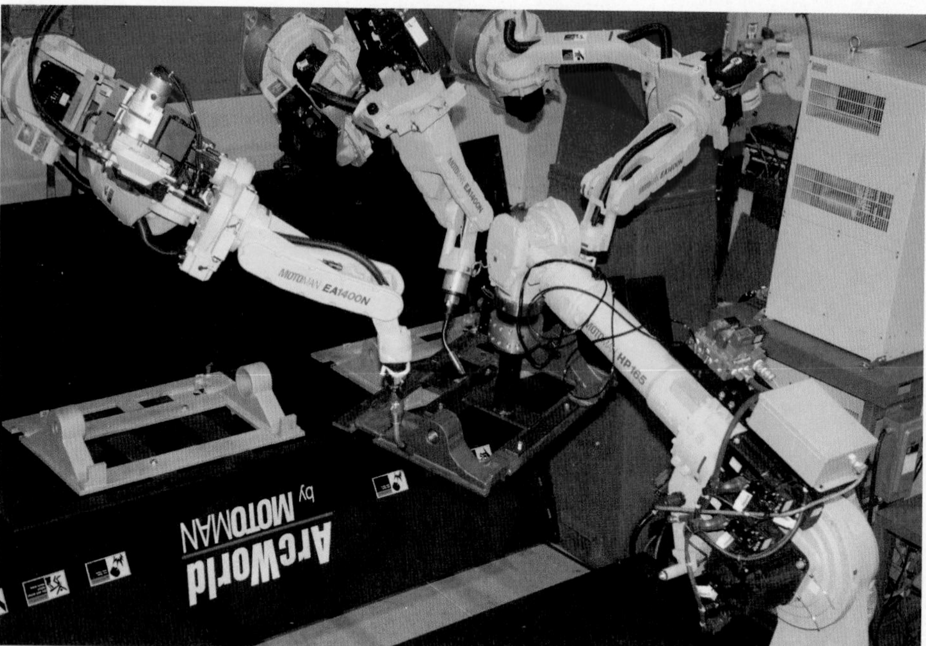

Motoman

Figure 11-1. In a robotic work cell, robots interface with other equipment.

11.1 Interfacing

An ***interface*** is the common point at which two or more systems communicate with each other. External end effector sensors, work cell limit switches, external relays, operator alarms, safety equipment, bar code readers, conveyors, transfer lines, programmable logic controllers, and machine vision systems all interface with robots.

A robot must be able to communicate with related devices. An ***input/output (I/O) port*** is a connection through which digital signals used for communication travel from system to system. Common I/O ports include digital input ports, digital output ports, serial ports, and parallel ports.

11.1.1 Digital Input Ports

Robots receive a signal indicating the state of an external switch (on or off) through an input port. A ***digital input port*** is a connection through which a robot receives information in the form of digital data. Limit switches, **Figure 11-2**, communicate information, such as the presence or absence of a part at a given location. This information is used by the computer program directing the robot to initiate some action, **Figure 11-3**. A common use of digital input ports is with safety devices. Safety devices, such as light curtains, motion detectors, and door switches on safety fences, may be connected to a digital input port. The port sends a signal to the controller to shut down the robot when a person or object enters the work envelope. Many robots are equipped with 16 or more digital input ports.

Motoman

Figure 11-2. Limit switches communicate information through input ports. This limit switch is used to limit a robot's rotary motion.

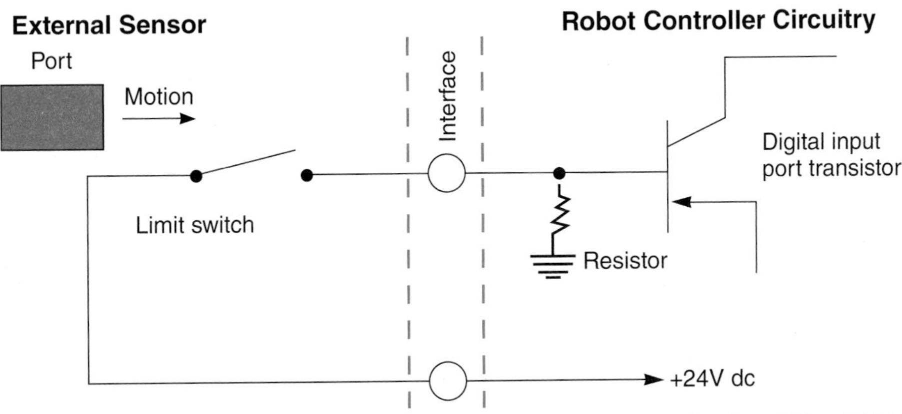

Goodheart-Willcox Publisher

Figure 11-3. This illustration shows a limit switch connected to a typical digital input port. When no object is present, the limit switch is open. When an object, such as an assembly part, is present, the limit switch closes and conducts 24V direct current to the transistor. The controller circuitry detects this change in condition (presence of an object and a change of transistor state) and passes the information to the robot's computer program to determine the appropriate action.

11.1.2 Digital Output Ports

One way that a robot interacts with external equipment, such as a conveyor belt drive motor, solenoid, or another robot, is through output ports. A ***digital output port*** is a connection through which the robot controller sends digital data to peripheral equipment, **Figure 11-4**.

Whenever a control signal from the computer program directing the robot is at a low voltage level, the base (control element) of transistor Q1 does not conduct.

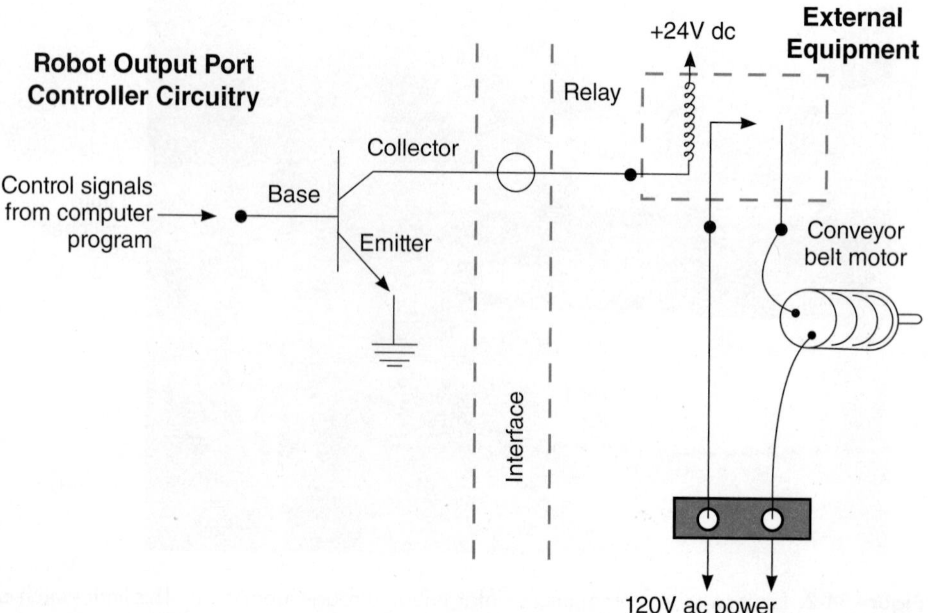

Goodheart-Willcox Publisher

Figure 11-4. When enough voltage is applied to digital output port transistor Q1, the lower end of the control relay is grounded. Grounding the lower end of the control relay coil causes the relay contacts to close. This supplies ac voltage to the conveyor belt motor.

The output current flows from the emitter (current source) to the collector (output element). When the base is at a low voltage level, the output current is zero (0), and the circuit is in a state of high resistance. The motor control relay (CR) is not energized, and the motor is off.

When the computer sends a high-level signal (usually 12V or 24V) to the base control element of transistor Q1, the current flow from the emitter to the collector is at maximum. The resistance between the emitter and collector is now very low. In this condition, the control relay is energized, and the motor is turned on. Robot controllers are equipped with several digital output ports and a matching number of input ports.

11.1.3 Serial Ports

Serial transmission is an interface method in which digital data is sent over long distances at a fairly slow rate. The transmission rate is slow because data must be transmitted one bit at a time. A ***serial port*** is a type of interface connection through which a computer sends or receives digital data using serial transmission. Two computers can communicate with each other over telephone circuits via serial ports connected to modems. Existing transmission lines can be used, which is an advantage.

A common serial interface used in robotics is the standard RS-232C. RS-232C was developed to standardize the interface between data terminal equipment and data communications equipment. This interface is most commonly used when binary data is transmitted over short distances, such as between two computers or between a computer and a piece of peripheral equipment. The circuit performs three functions: data transfer, timing, and control (**Figure 11-5**).

Pin Assignments and Interchange Circuit Functions									
			Ground	Data		Control		Timing	
Interchange Circuit	Pin Number	Description		From DCE	From DCE	From DCE	To DCE	To DCE	To DCE
AA	1	Protective ground	X						
AB	7	Signal ground/common return	X						
BA	2	Transmitted data			X				
BB	3	Received data		X					
CA	4	Request to send					X		
CB	5	Clear to send				X			
CC	6	Data set ready				X			
CD	20	Data terminal ready					X		
CE	22	Ring indicator				X			
CF	8	Received line signal detector				X			
CG	21	Signal quality detector				X			
CH	23	Data signal rate selector (DTE)					X		
CI	23	Data signal rate selector (DCE)				X			
DA	24	Transmitter signal element timing (DTE)							X
DB	15	Transmitter signal element timing (DCE)						X	
DD	17	Receiver signal element timing (DCE)						X	
SBA	14	Secondary transmitted data			X				
SBB	16	Secondary received data		X					
SCA	19	Secondary request to send					X		
SCB	13	Secondary clear to send				X			
SCF	12	Secondary received line signal detector				X			

Goodheart-Willcox Publisher

Figure 11-5. This chart lists pin assignments and interchange circuit functions for the RS-232C serial interface standard.

Applications that do not involve long-distance communications require only a few RS-232C connections. These are the signal ground (AB), transmitted data (BA), and received data (BB), **Figure 11-6**. Long distance communications (those involving modems) require many other connections.

Another common use of the RS-232C serial port is to connect a robot to a personal computer (PC). A PC can be used offline to write and debug the initial program. The program can then be downloaded to the robot using the RS-232C port to provide the robot with the exact moves needed for an application.

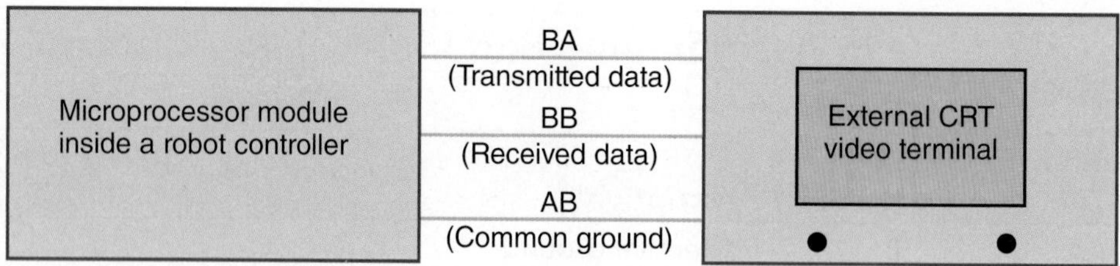

Figure 11-6. The external video terminal is interfaced to a robot controller's microprocessor module via RS-232C connections. This illustration shows how AB, BA, and BB connections communicate between the microprocessor and the video terminal.

When preparing cables to be used with the RS-232C port, do not exceed a total cable capacitance of 2500 picofarads (pF). The capacitance limits the distance between the two pieces of equipment. Another limitation is a maximum 20 kilobit (kb) data transfer rate. The RS-232C standard does not specify the data format, such as 8 bit ASCII.

11.1.4 Parallel Ports

Parallel transmission is an interface method in which multiple bits of data are sent at the same time following side-by-side paths, like the lanes of a highway. A ***parallel port*** is a type of interface connection through which the robot controller sends digital data to peripheral equipment. Although parallel transmission is faster than serial transmission, it is generally considered too costly to use over long distances.

One commonly used parallel port standard is the IEEE 488. It was developed in 1975 by the Institute of Electrical and Electronic Engineers to reduce the amount of time needed to set up test equipment. At that time, test equipment had its own independent interface, and interfacing more than two devices to a computer could be complicated and expensive. The IEEE 488 standard allows computer-controlled test equipment to be set up in a matter of hours, which benefits any industry using automated testing.

The IEEE 488 standard applies when a computer is connected to measuring equipment using a type of cable. The ***general-purpose interface bus (GPIB)*** is a cable that interfaces system controllers with programmable instruments. It can support a maximum of 15 devices, **Figure 11-7**. Total cable length is 20 meters (m)

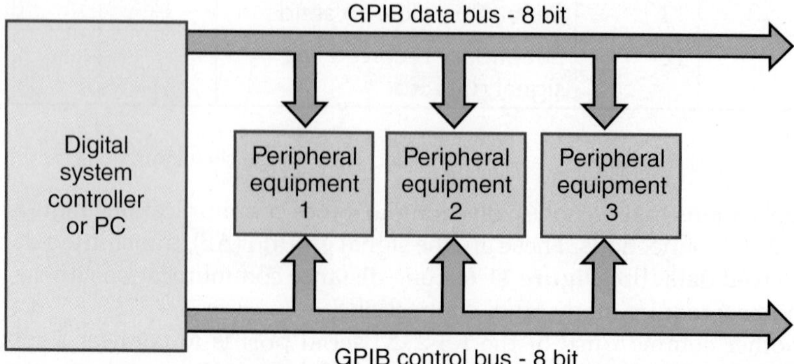

Figure 11-7. The general-purpose interface bus (GPIB) is used here to connect external testing equipment to a digital system controller.

Robotics Innovations: Snake-Legged Robot

This snake-legged robot designed by Carnegie Mellon University is part of a project to develop robots that can navigate difficult natural and manmade terrain. The military has a strong interest in robots that can navigate the complex challenges of both the battlefield and the urban environment. Not all locations have smooth, easy to navigate surfaces. The legs on this robot are designed to walk over obstacles and navigate uneven terrain. While this robot is in the prototype stage, future robotic technologies may use this system, or a modified version, for locomotion. Carnegie Mellon is one of many schools developing cutting edge robotic technology.

Photograph provided courtesy of Carnegie Mellon University

or 2 m times the number of connected devices, whichever is less. The cable between any two devices must not exceed 4 m. The bus carries data in both directions, eight bits at a time, and controls both the data and the direction. The bus can connect various kinds of devices to the controller, such as equipment that can only give input, equipment that can only receive output, and equipment that can both give output and receive input. Many controllers can be attached to the GPIB, but only one controller may have command at any given time.

11.2 Machine Vision

Robots with machine vision systems can perform in a more intelligent manner than robots without these systems. These systems may be used for such applications as guidance, determining part orientation, measuring and inspecting parts, and image identification, **Figure 11-8.** Machine vision has become an essential requirement in automated work cells. Robotic work cells that are not equipped with machine vision require all parts to be pre-oriented before a robot can grasp them and perform a task.

Machine vision eliminates the need for many machine operators by performing many complex tasks once done only by humans. Systems used for assembly and inspection have had a major impact in the automotive and electronics industries. For example, printed circuit boards often contain thousands of individual parts. If even a tiny fraction of these parts is incorrectly placed on the board, the cost of finding and repairing the problem is enormous. When human workers perform circuit board assembly, fatigue can cause as much as a 20% error rate. Using robots with machine vision for circuit board assembly significantly reduces the error rate.

FANUC Robotics

Figure 11-8. The video camera sensors of this system are suspended over the conveyor to scan parts as they travel along the conveyor.

11.2.1 Fundamentals of Machine Vision

Machine vision systems use video cameras and computers to translate light energy into an image. These images are then used to determine part orientation or another designated task. However, even the most effective machine vision systems cannot achieve the level of perception and sophistication of the human eye.

Some of the earliest machine vision systems were introduced in the late 1970s. Early systems were expensive due to the high cost of the computer memory needed to handle the images. They were also inefficient because all computers, except mainframes, contained fairly slow processors. Today, powerful and fast computers with expanded memory capabilities make vision systems for robotics very practical.

11.2.2 Image Processing

The data obtained from machine vision can be used to make decisions on the processing steps or other action taken by either human operators or a robot. Vision systems can also be used in conjunction with material inventory and material flow within a factory. Four functions take place during image processing: acquisition, preprocessing, analysis, and interpretation, **Figure 11-9**.

Image Acquisition

Image acquisition is a function of machine vision in which a workpiece is illuminated and an illustration or picture is digitally created. Fluorescent lamps, incandescent bulbs, strobe lights, or arc lamps usually provide the illumination. The type and amount of light needed depends on the application. Front lighting is used to enhance surface features, such as bar codes or labels. Side lighting is used when three-dimensional (3-D) images are desired. Back lighting provides a silhouette of the object.

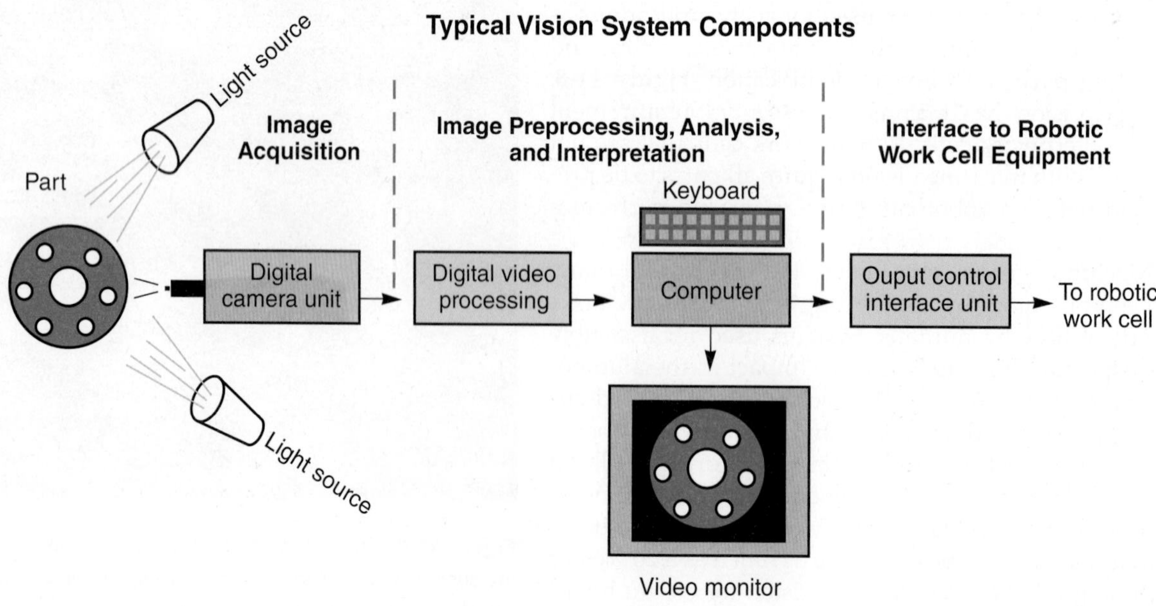

Goodheart-Willcox Publisher

Figure 11-9. These digital electronic subsystems are found in typical machine vision systems.

Creating an image is done with a video camera. The solid-state digital cameras used with vision systems have either a charge-coupled device (CCD) or a charge-injection device (CID). At the heart of the camera is a silicon chip used with an array of photosensitive elements. The light reflected into the camera lens from the workpiece falls onto this photosensitive array and is converted into an analog electrical signal.

Image Preprocessing

During ***image preprocessing***, an analog-to-digital (AD) converter changes the analog signal into an equivalent digital signal, **Figure 11-10**. The image is converted into dots of light called picture elements, or ***pixels***. Pixels are the means used to form the image on a TV or computer screen. A typical digital video camera can produce an image that measures 1024 by 768 pixels. The digital signal represents light intensity values over the entire image. These values are stored in the memory, which allows the digital image to be analyzed and interpreted.

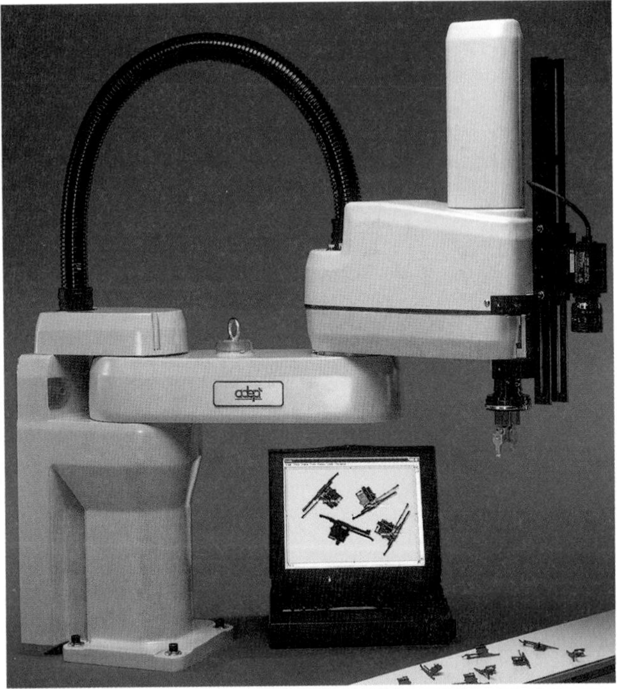

Adept Technology, Inc.

Figure 11-10. This vision system provides 100% inspection of the manufactured pieces. Any defects found in the pieces are automatically reported by type, location, and size on a color display screen.

Image Analysis and Interpretation

Image analysis is a function of machine vision in which information from the illustration or picture is gathered and analyzed by computer software. Using algorithms, the software identifies and measures features of the digital image. After all the features are analyzed, the image information is interpreted. ***Image interpretation*** is a process that involves comparing a scanned image to a model image or to data that relates to a model and deciding whether the scanned image is a match to the model image. The decision is sent to the robot control system and enables a robot to make decisions about the tasks that must be performed.

Image analysis computer programs, **Figure 11-11**, provide a powerful computer-based development environment with flexible programming, debugging, and testing capabilities. Vision guidance and inspection applications may be used to enhance a variety of robotic operations. Robots with vision systems are used to improve the accuracy of machining, material handling, and other industrial operations. Image analysis for inspection is commonly used to collect and record dimensional measurements. This provides a nondestructive method of quality control testing.

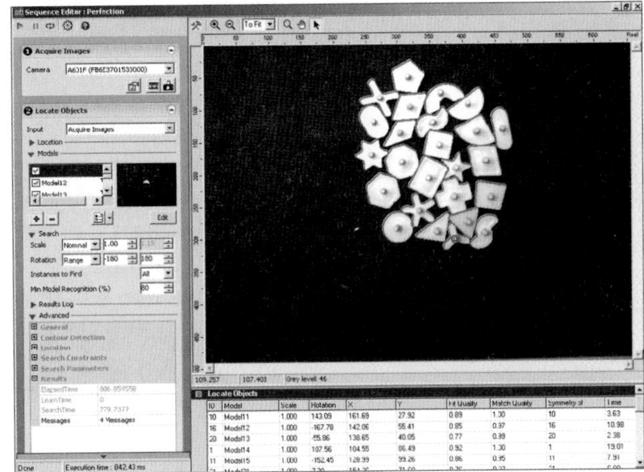

Omron Adept Technologies, Inc.

Figure 11-11. Image analysis is a common method used in quality control processes. Measurements can be specified on a live preview image with tolerances specified.

Summary

- Robots interact continuously with other production equipment in a process called interfacing.
- An interface is a common point at which two or more systems communicate with each other. Operator alarms, safety equipment, and machine vision systems are examples of equipment that interfaces with robots.
- Digital input ports are used to bring information into a robot. Digital output ports allow digital signals to be sent from a robot controller to other equipment.
- Serial transmission delivers digital data over long distances at a slow rate. A common serial port used in robotics is the standard RS-232C.
- Parallel transmission sends multiple bits of data at the same time following side-by-side paths. Parallel transmission is faster than serial transmission. One commonly used parallel port standard is the IEEE 488.
- Robots with machine vision systems can perform in a more intelligent manner than robots without these systems. Example applications include guidance, inspection, and image identification.
- Four functions take place during machine vision image processing: acquisition, preprocessing, analysis, and interpretation.
- During image acquisition, machine vision systems use video cameras and computers to translate light energy into images.
- During image preprocessing, a converter changes an analog signal into its equivalent digital signal. Dots of light, called pixels, are used to form the image on a computer screen or monitor.
- During image analysis, information is gathered and analyzed by computer software.
- Image interpretation involves comparing image data to a model image or data. The resulting information enables a robot to make decisions about the tasks that are being performed.

Review Questions

Answer the following questions using the information provided in this chapter.

1. What is an interface, and how does it apply to robotics?
2. List six external devices that can be connected to digital input and output ports of a robot.
3. What are two advantages and one disadvantage of using serial ports?
4. To send digital data from a robot to external equipment at the highest possible transmission rate, would you use a parallel or a serial port? Why?
5. What is a general-purpose interface bus (GPIB)?
6. Identify three operations for which a robotic vision system could be used.
7. Explain the four functions or phases that take place during image processing.

Learning Extensions

1. Conduct an Internet search to identify vision system manufacturers. Review the system specifications and capabilities of three different vision system products. Create a chart to compare the data gathered for all three systems, and evaluate the qualities that make each system unique.

Advanced Analysis

1. Prepare a written discussion of a specific machine vision application.
2. Select an industrial robot application and prepare a discussion of the interfacing of the robot with other production equipment.

Chapter 12

Maintaining Robotic Systems

Outline

12.1 Troubleshooting Robotic Systems
12.2 General Servicing Techniques
12.3 Preventive Maintenance

Objectives

Upon completion of this chapter, you will be able to:

- Describe successful troubleshooting methods for robotic systems.
- Follow the proper techniques for general servicing of equipment.
- List the steps in developing a preventive maintenance plan.

Technical Terms

preventive maintenance
rework
rosin-core solder
scrap
solder sucker
troubleshooting

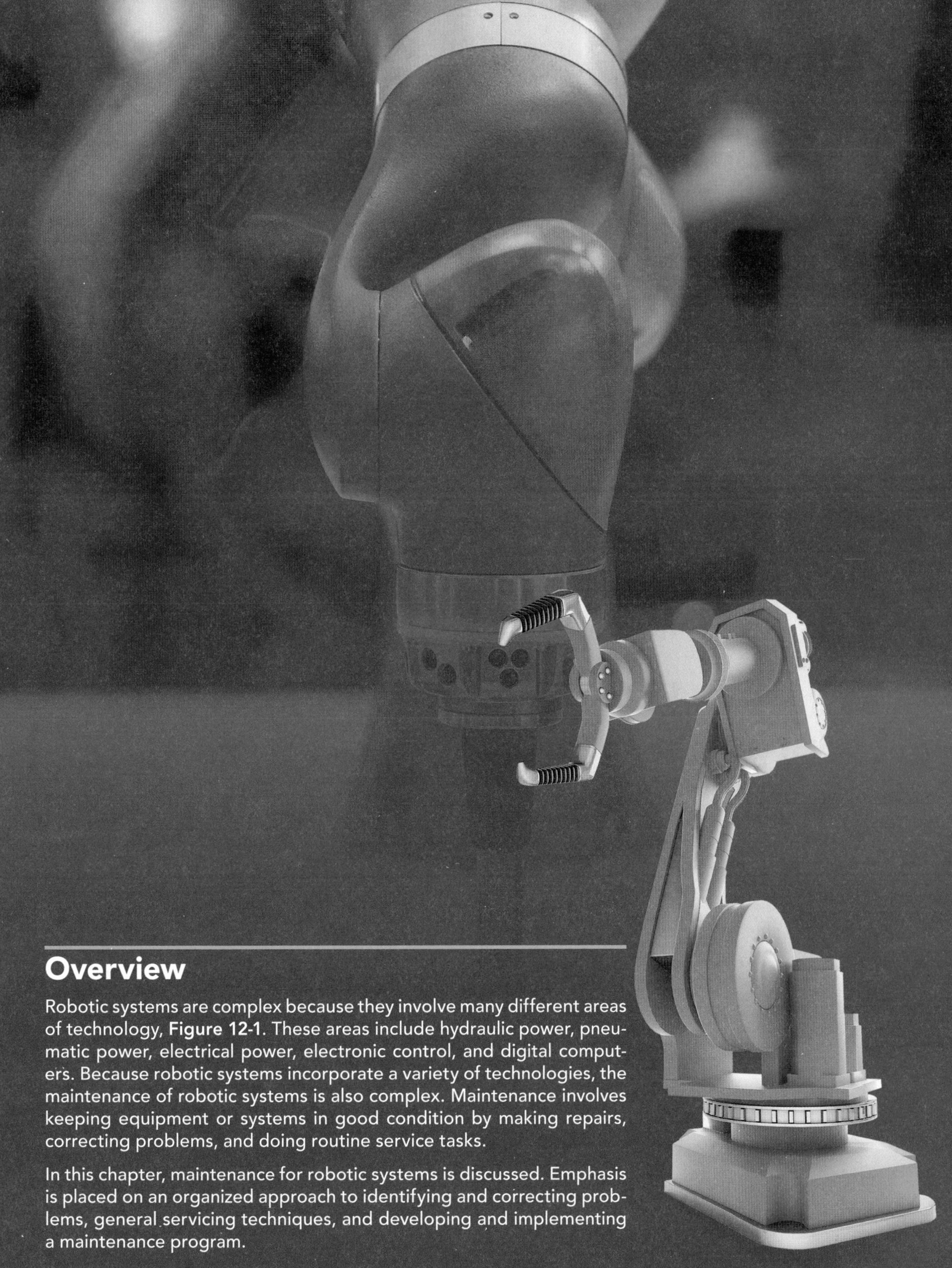

Overview

Robotic systems are complex because they involve many different areas of technology, **Figure 12-1**. These areas include hydraulic power, pneumatic power, electrical power, electronic control, and digital computers. Because robotic systems incorporate a variety of technologies, the maintenance of robotic systems is also complex. Maintenance involves keeping equipment or systems in good condition by making repairs, correcting problems, and doing routine service tasks.

In this chapter, maintenance for robotic systems is discussed. Emphasis is placed on an organized approach to identifying and correcting problems, general servicing techniques, and developing and implementing a maintenance program.

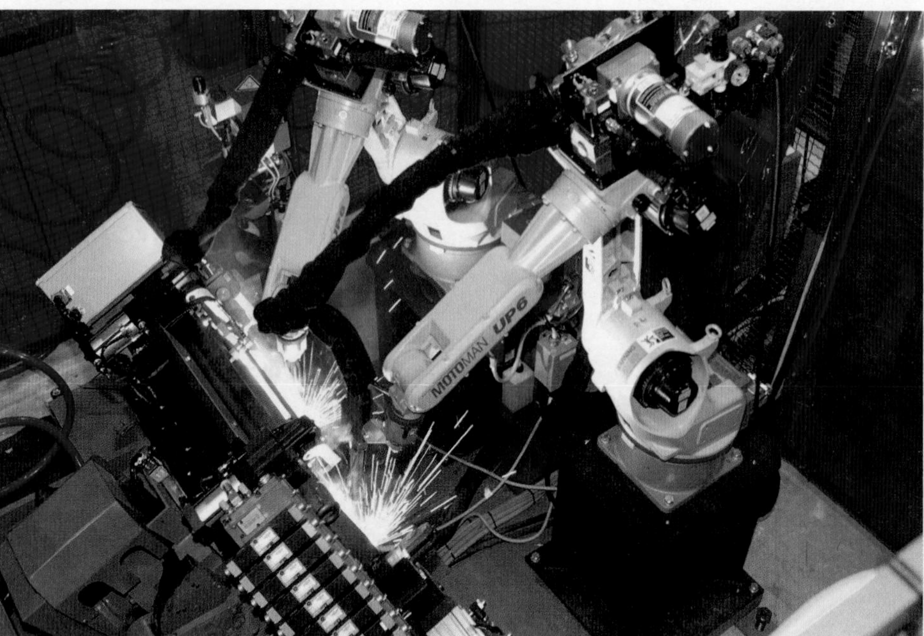

Motoman

Figure 12-1. This operation uses two robotic systems used for welding. All axes of motion are controlled by electric servomotors. Notice that a power supply and associated electronic control equipment are in the area behind the robots. This is a typical industrial application. Other systems use hydraulic or pneumatic power, rather than electrical power.

12.1 Troubleshooting Robotic Systems

Problems or issues that need attention may be found during routine inspections and maintenance tasks. Unexpected problems can cause equipment to malfunction and may require that operations be halted while repairs are made. In either case, troubleshooting may be required. ***Troubleshooting*** is the systematic process of identifying and correcting problems in an operation or system. Some problems are easily solved and require little time to identify or correct. Other problems are more difficult. They occur intermittently or are complex and require many hours of concentration and work. Successful troubleshooting methods include:

- Using common sense.
- Following a logical sequence.
- Applying knowledge of robotic systems and their operation.
- Testing equipment to identify problems.
- Reading and using schematic diagrams effectively.

To begin troubleshooting any type of system or process, organize your thoughts and identify possible courses of action. Without a well-organized approach, troubleshooting can become a time-consuming guessing game. Use a planned approach to save time and resources.

12.1.1 Inspection

The initial inspection is an important step in troubleshooting. When this inspection is performed properly, many problems can be efficiently diagnosed. Look for obvious signs of problems. Use your senses of sight, touch, smell, and hearing. If you suspect that a specific part is causing the problem, turn off the equipment and examine it carefully. A thorough visual inspection should be completed before circuit or system testing is done. Look for these indicators of broken or faulty parts:

- **Burned parts.** Parts may be charred, blistered, or discolored, and may even have holes.
- **Broken parts.** Breaks may appear in the form of cracks, wires pulled out of parts, or parts that have been completely destroyed.
- **Broken wires and poor connections.** These problems can be located visually. Broken wires and poor connections often cause the system to operate inefficiently.
- **Smoke or heat damage.** A part that smokes when the equipment is turned on indicates a damaged part, but it does not necessarily identify the cause of the damage.
- **Oil, air, or water leaks.** These leaks can be located visually and should be corrected quickly.
- **Loose, damaged, or worn parts.** These parts can be located visually or by touch.
- **Noisy parts.** Parts that are creating more noise than is typical for normal operations often indicate defective parts.

Keep in mind that most problems are component failures. Knowing what each component is supposed to do helps in determining the problems it can cause. Properly test each component. To avoid duplicating effort, make a list of circuit or system operations as you test them. Good troubleshooting involves knowing how the system operates and how to use test equipment effectively. Most importantly, be patient and systematic during the troubleshooting process.

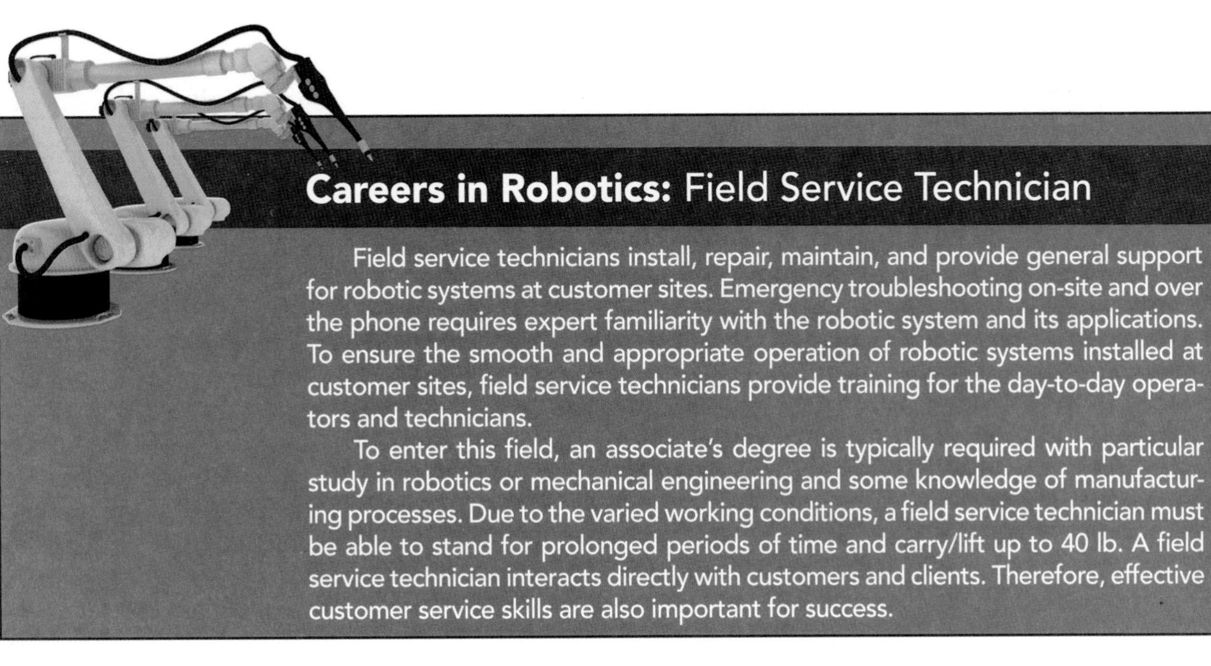

Careers in Robotics: Field Service Technician

Field service technicians install, repair, maintain, and provide general support for robotic systems at customer sites. Emergency troubleshooting on-site and over the phone requires expert familiarity with the robotic system and its applications. To ensure the smooth and appropriate operation of robotic systems installed at customer sites, field service technicians provide training for the day-to-day operators and technicians.

To enter this field, an associate's degree is typically required with particular study in robotics or mechanical engineering and some knowledge of manufacturing processes. Due to the varied working conditions, a field service technician must be able to stand for prolonged periods of time and carry/lift up to 40 lb. A field service technician interacts directly with customers and clients. Therefore, effective customer service skills are also important for success.

12.1.2 Troubleshooting Charts

Most manufacturers of robotic equipment provide troubleshooting manuals with charts and diagrams to aid in repairs. Use this information to help locate the faulty module or part. A typical symptom list itemizes problems associated with a specific robotic system, **Figure 12-2**. Detailed information related to the symptom is provided to help identify the problem, **Figure 12-3**. A flowchart may also be provided to guide you to the problem, **Figure 12-4**.

Understanding how the circuit, device, or system functions and knowing how to use test equipment properly are important factors in successful troubleshooting and testing. This is true for both a simple circuit and the most complex system. Practice is the best way to learn the procedures.

12.2 General Servicing Techniques

The individual parts on robots seldom require servicing. Repairs to complex robotic systems are commonly performed on the subassemblies. A subassembly is a collection of parts or components used together to form a unit in a larger machine or system. Individual parts may not be repaired. Instead, an entire subassembly or module may simply be replaced.

Symptom List	Page
Applications 1 through 5 on at the same time	10.10
Axis does not move	10.11
Axis moves erratically or runs away	10.13
Control panel failures	10.15
Data Error (DE) LED on or blinking	10.17
DI/DO failure (Standard)	10.19
DI/DO (Extended)	10.21
Does not home	10.23
Manipulator power does not work (No other indications)	10.25
MTCB indicator failure	10.26
MTCB switches do not work	10.27
Overtime (OT) LED on	10.28
Overrun (OR) LED on	10.29
Overrun detection failure	10.32
Overrun reset failure	10.33
Power Failure (PF) LED on	10.34
Remote operator panel interface failure	10.35
Repeatability varies	10.36
Servo Error (SE) LED on	10.39
Servo Error (SE) LED blinks and operator panel beeps	10.40
Transmission Error (TE) LED on or communication error	10.42

Goodheart-Willcox Publisher

Figure 12-2. This page from a troubleshooting manual lists symptoms and corresponding pages where detailed service information can be found.

Defective Battery
NOTE: Do not dispose of the battery in a fire. With the system off, check the voltage across the battery. The voltage should be between +3.2V and +4.2V dc. A discharged battery may load down the power supply while charging. To test for this, unplug the battery and reload the applications. If the battery is suspected of only being discharged, leave the controller powered up for eight hours to sufficiently charge the battery. If the voltage is still low after charging, replace the battery.
Low Power Supply Voltages
Check for 24V dc, –12V dc, 12V dc, 5V dc at the power supply. If any of the voltages differ by more than 5%, adjust or replace the power supply. See power supply adjustment (4009). **NOTE:** The power supply may be left powered on with no load attached.
Defective Motor Control Board
If the battery is good and the voltages are good, suspect the Motor Control Board.

A

AXIS DOES NOT MOVE
Defective Servo Pack Fuse
Swap the servo pack fuse for the failing axis. Do not rely on the indicator. Theta 1 is labeled SV1 in the controller. (15 amp) Theta 2 is labeled SV2 in the controller. (10 amp) Z-Axis is labeled SV3 in the controller. (7.5 amp) Roll is labeled SV4 in the controller. (7.5 amp)
Defective Z-Axis Brake
The brake should remain locked until manipulator power is brought up. Then the Z-Axis servo pack takes over by controlling the power to the servo Z-Axis. If the brake remains locked after manipulator power is brought up, suspect the magnetic brake or circuit to the brake (brake is approximately 85 ohms when good). Also, the brake can be tested by disconnecting connector CN16H with manipulator power on, which makes a clicking noise.
Out of Adjustment Zero
Perform the "Servo Pack Zero Adjustment."
Defective Motor Control Board
1. If only the Z-Axis does not move, set the controller power to 0 (off), then ground CN16H-1 (WD16) to release the brake. Set the power switch to I (on). Move the axis to check for binds. If the axis cannot be moved, isolate the bind. Remove the jumper. For any other axis, set the power switch to O (off). Move the axis to check for binds. If the axis cannot be moved, isolate the bind. 2. Set the controller power switch to I (on). Press the Manip Power key on the control panel. **DANGER: The next step requires you to push on the axis with manipulator power on. Do not allow your body to enter the work space.** 3. Attempt to push or turn the axis. If the axis moves easily, then suspect the Servo Pack or wiring from the Servo Pack to the motor (WD07-WD10). >>>Continued on the next page.

B

Goodheart-Willcox Publisher

Figure 12-3. Troubleshooting manuals are an excellent resource for service technicians. A—Detailed troubleshooting information helps to identify the problem. B—Manufacturers usually provide complete directions for diagnosing and solving problems. Note the safety warning included as part of a diagnostic procedure.

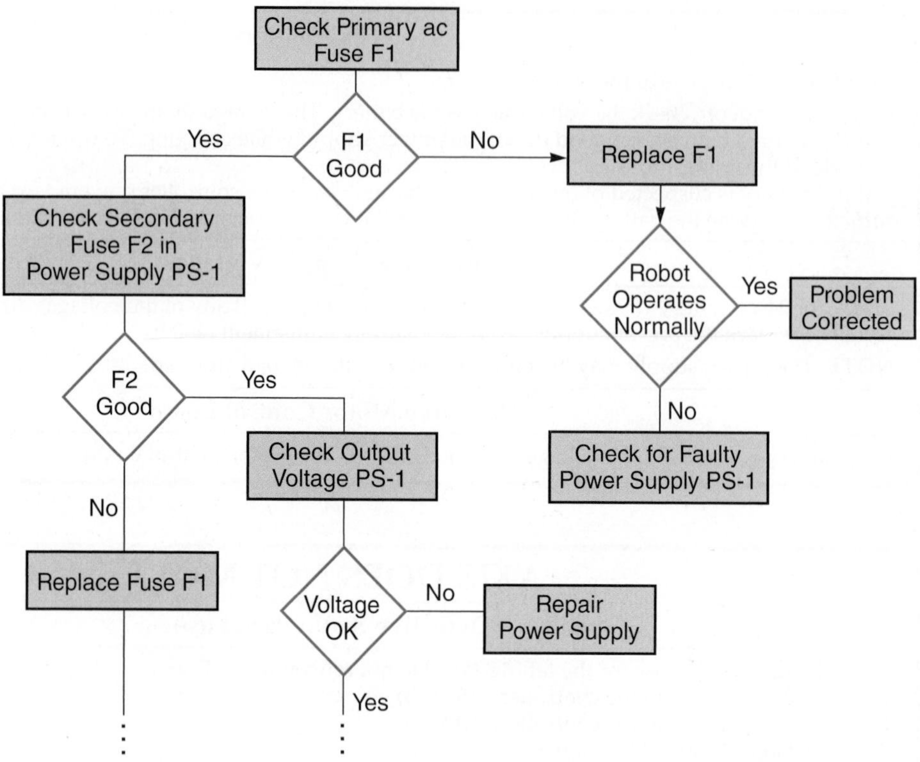

Goodheart-Willcox Publisher

Figure 12-4. A flowchart can provide a step-by-step guide for diagnosing and correcting problems.

12.2.1 Identifying the Problem

To determine the source of a problem, the technician should gather information about the equipment and its operation and evaluate the data to determine a course of action.

Observe and Inspect

Never accept another person's word about the robot's performance. Check the robot or part yourself. Determine if the problem is electrical or mechanical in nature. Inspect the power system. Determine if the proper current (single-phase ac, three-phase ac, or dc) is being used to power the equipment. All magnetic contactors, transformers, switches, circuit breakers, fuses, and other electronic controls should be checked.

Machinery and work areas must be kept clean. This allows the operator to identify any leaks, visible damage, or wear. Observe the physical condition of the equipment. Look for fluid or air leaks, unusual noises, odors, poor electrical connections, or loose wires.

Gather Information

If possible, acquire a schematic, a service or maintenance manual, and an operating manual for the robot. These items are essential in servicing most equipment. If these resources are not available, consult with someone who has experience servicing the robot. An identical robotic system may be used to determine specific values (such as voltage, current, and waveforms). These values can then be compared to those in the robot being serviced.

Analyze the Equipment

After gathering the necessary information, you should have an understanding of how the system functions. Divide the various components into categories according to purpose. Categories may include input power systems, rectifiers, transformers, filters, feedback loops, amplifiers, and interface/control circuits. Then, identify the input and output of each category.

Isolate the Problem

Trace the malfunction to a specific subsystem within the equipment. Use a voltmeter, an oscilloscope, or a multimeter to read measurements or waveforms (**Figure 12-5**). To be successful in finding the problem, you must know how to use the test equipment and how to interpret the measured values. Compare your readings to those indicated on the schematic or in the service manual. Comparing the values of test readings to the recommended values usually isolates the problem.

Verifying a malfunction is sometimes difficult. If the input to a subsystem is correct and the output is not, it may seem logical that the subsystem itself is the problem. This is not true in all cases. The output of one subsystem can become distorted because of another subsystem to which it is connected. When this is suspected, disconnect the subsystem you are inspecting and check the voltage and waveforms again.

12.2.2 Correcting the Problem

Once a problem has been identified and analyzed, the technician should make repairs and keep comprehensive records for reference. To correct the problem, replace the faulty component or device. Any replacement component must be equal to or better than the original part. Important factors to consider are voltage, maximum current, and power rating. Always replace a component with one of identical characteristics. After the repair is complete, operate the system for several minutes to be sure it continues working properly.

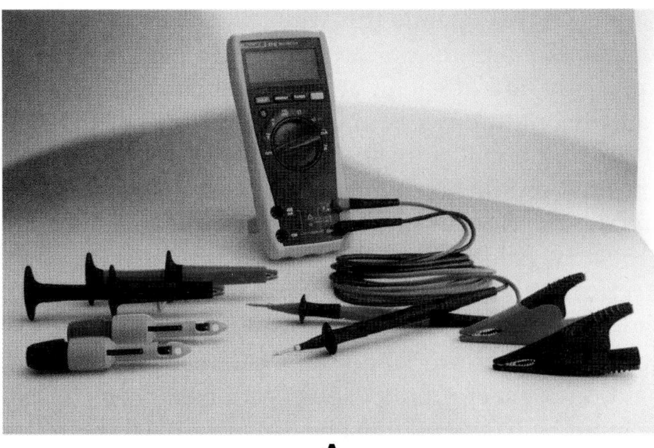

A

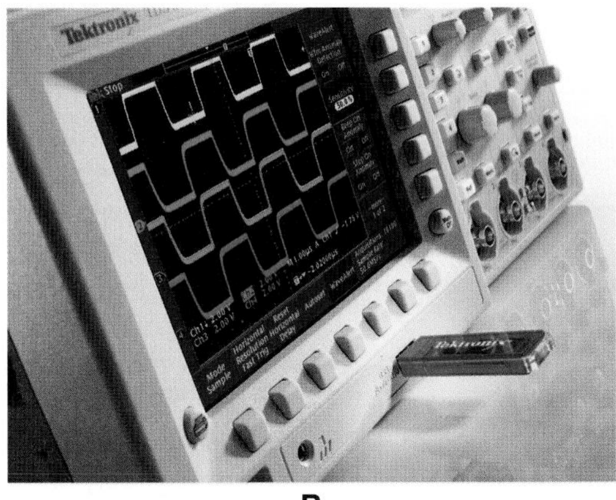

B

Tektronix, Inc.

Figure 12-5. The readings from test equipment assist in isolating equipment problems. A—Voltmeters are portable and have various attachments to accommodate different types of equipment. B—This oscilloscope provides a waveform display of the signal sampled.

Soldering and Desoldering

Soldering and desoldering are important steps in replacing faulty components. Soldering is joining metals by melting them and fusing them together. Solder is an alloy that is used in some soldering operations as a filler in the joint. Desoldering is removing existing solder and possibly components in preparation for making repairs to equipment or devices, such as a circuit board. When making repairs, you may use a ***solder sucker.*** This device uses suction to remove unwanted solder from a printed circuit board safely without damaging the board or other components, **Figure 12-6**. Select the proper iron and soldering tip for the repair and be mindful of nearby components while soldering. Carefully solder the new component into the circuit to avoid damage by overheating. You must also guard against joints that are not connected accurately and do not allow electrical current to flow properly. These joints are sometimes called cold joints. From an electrical standpoint, a cold solder joint causes high resistance in a circuit. Always use ***rosin-core solder*** (*not* acid-core solder) for *all* electronic circuit work.

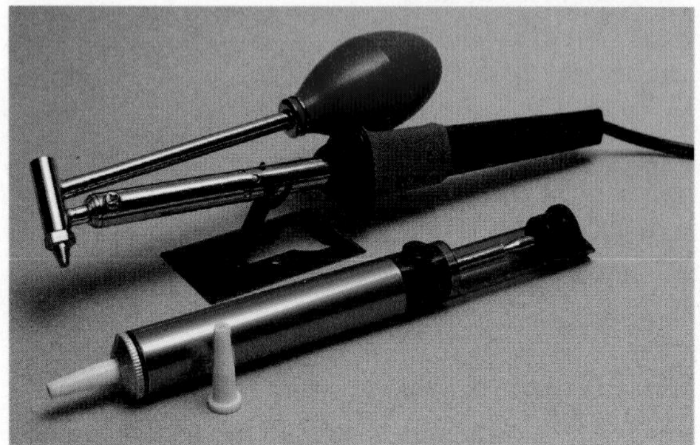

Goodheart-Willcox Publisher

Figure 12-6. Both powered and spring-loaded solder suckers are effective desoldering tools.

Keeping Records

When you are satisfied that the system is functioning properly, document the original problem. State what caused the malfunction and how it was corrected. Keep this information with the service manual or in a separate file, so it is on hand for future reference.

12.3 Preventive Maintenance

Preventive maintenance is performing tasks on a regular schedule that are designed to keep equipment in good working order and avoid breakdowns and loss of production. In some companies, equipment is checked or repaired only when it breaks down. Companies that operate this way leave the production line constantly vulnerable to equipment failure.

Equipment failures are costly and affect all areas in the production process. Production bottlenecks cause work-in-process to back up at work stations. This means the operation must run additional hours to catch up and get back on schedule. This also causes additional production time for the operations that follow. When machines are not in use, idle time results for the operators. Inventory costs increase when materials must be held beyond the normal time it takes to cycle them through the manufacturing process. If an order is not produced on time, the customer may go to another source. If the equipment experienced excessive wear before breaking down, increased rework and scrap can result. ***Rework*** is the process of fixing parts that do not meet product specifications. ***Scrap*** consists of manufactured parts that cannot be fixed and must be discarded. All these factors result in lost revenue for the company.

An effective preventive maintenance program is the best solution to avoid the costly results of equipment failure. Preventive maintenance should be designed to maintain and improve the reliability of equipment by replacing worn components before they fail. Some of these activities include inspecting equipment, changing fluids, lubricating parts, and machine/component refurbishing or replacement at specified periods. Documentation is key to preventive maintenance. Maintenance personnel can observe and record equipment wear to identify the best parts replacement or repair schedule.

12.3.1 Benefits of Preventive Maintenance

Preventive maintenance can play an important role in an organization's management strategy by providing a number of long-term benefits. These benefits may include improved system reliability, decreased cost of replacements, decreased system downtime, and improved inventory management for spare parts. When the process is closely examined in terms of cost and system performance, preventive maintenance is preferred over performing maintenance only when the system fails.

Preventive maintenance also benefits the daily performance of robots and other types of industrial equipment. It can increase flexibility, help maintain production flow, and allow for a continuous analysis of equipment. Additionally, preventive maintenance decreases the need for a large on-call maintenance crew. When used in combination with mathematical analysis, it aids in predicting when equipment needs to be adjusted or replaced to avoid breakdowns. Preventive maintenance can be performed at convenient times, such as when a robot is not actively involved in a process. This reduces overtime costs for skilled operators and balances the workload among shifts. When performed conscientiously, preventive maintenance ensures consistent quality, reduces product costs, and reduces downtime of critical production equipment.

12.3.2 Disadvantages of a Preventive Maintenance Program

While a good preventive maintenance program definitely reduces the frequency of unexpected equipment failure, such programs do have drawbacks. Some disadvantages of using a preventive maintenance program include:

- A greater need for dedicated workers.
- An increased responsibility and paperwork for operators.
- The allotted access time required to work on the equipment.
- The initial cost of implementing a preventive maintenance program.

12.3.3 Developing a Maintenance Program

Developing a new preventive maintenance program involves a number of steps, **Figure 12-7:**

- Establish goals and objectives for the program. Make a list of the desired results of the preventive maintenance plan.
- Develop a strategy to achieve the goals outlined for the program.
- Plan the program in detail, including a schedule for when steps of the program will be implemented. Operators should be involved early in the planning process because they are so important to the success of preventive maintenance schedules and activities.
- Implement the plan. Keep records related to the procedures used and the results.
- Refine the program to add or change procedures that will better achieve the goals of the plan.

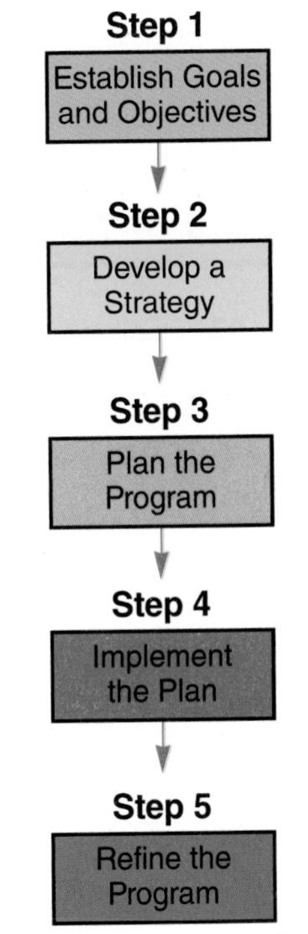

Goodheart-Willcox Publisher

Figure 12-7. Five necessary steps in developing a new preventive maintenance program.

When implementing the plan, assign maintenance responsibilities to each shift. A maintenance schedule can be altered to accommodate the appropriate number of shifts, **Figure 12-8**. However, it must be followed carefully for maximum effectiveness. If one shift fails to follow the schedule, it can cause problems for the entire program. The schedule should be easy to maintain and should be posted directly on the machines.

Use assignment sheets that describe tasks that must be performed on a regular basis, **Figure 12-9**. Each assignment sheet identifies the unit to be maintained, the maintenance intervals, and the type of action required. The steps may be carried out in any order, but each should be completed thoroughly. Assignment sheets should also be posted directly on the machines.

Daily/Weekly Preventive Maintenance Schedule

Machine Type: IBM-7547 **Serial Number:** 257698ZMW

Department: 179 **Building Number:** 5

Shift	Week	Mon.	Tues.	Wed.	Thurs.	Fri.	Weekly	Operator	Comments
Shift 1	1	X	X	X	X	X	X	John Wasson	Air leak
Shift 2	1	X	X	X	X	X	X	Stephen Fardo	Bad relay
Shift 3	1	X	X	X	X	X	X	Robert Towers	No problems
Shift 1	2	X	X	X	X	X	X	Tim Ross	No problems
Shift 2	2	X	X	X	X	X	X	Stephen Fardo	Air leak
Shift 3	2	X	X	X	X	X	X	John Wasson	Bolt loose
Shift 1	3	X	X	X	X	X	X	Robert Towers	Motor noise
Shift 2	3	X	X	X	X	X	X	Tim Ross	No problems
Shift 3	3	X	X	X	X	X	X	John Wasson	Bad I/O port
Shift 1	4	X	X	X	X	X	X	Stephen Fardo	Loose clamp
Shift 2	4	X	X	X	X	X	X	Robert Towers	Air leak
Shift 3	4	X	X	X	X	X	X	Tim Ross	No problems

Goodheart-Willcox Publisher

Figure 12-8. Following a preventive maintenance schedule helps identify and eliminate small problems before they become large ones.

IBM-7547 Preventive Maintenance Assignment Sheet

Unit	Action	Interval	Lubrication
Manipulator	Clean Check Oil Levels Check for Air Leaks	Daily Daily Daily	— — —
Controller	Check Air Filters	Monthly	—
Roll Axis Belts	Check Tension	Monthly	—
Roll Axis Gears	Lubricate	Monthly	Molykote G
Theta 1 Axis	Change Oil	Semi-Annually	Number 10 Oil
Theta 2 Axis	Change Oil	Semi-Annually	Number 10 Oil
Bearings	Pack Bearings	Every Replacement	Number 23 Grease

Goodheart-Willcox Publisher

Figure 12-9. Assignment sheets should be posted on the machines to identify maintenance tasks and the intervals at which they should be performed.

Summary

- Robotic systems are complex because they involve different areas of technology, such as hydraulic, pneumatic, and electrical power; electronic control; and computer systems. This means that maintenance for the systems is also complex.
- Troubleshooting is the systematic process of identifying and correcting problems. It involves knowing how a system operates and how to use test equipment effectively.
- An initial inspection, involving the senses of sight, touch, smell, and hearing, is an important step in troubleshooting.
- Troubleshooting charts are provided by manufacturers to help technicians locate and isolate faulty modules or parts of a robotic system.
- Repairs to robotic systems are commonly performed on subassemblies rather than individual components.
- To determine the source of a problem, the technician should observe and inspect the equipment, gather information about the equipment and its operation, and evaluate the data to determine a course of action.
- Once a problem has been identified and analyzed, the technician should make repairs and keep comprehensive records for reference.
- Preventive maintenance is performing tasks on a regular schedule that are designed to keep equipment in good working order and avoid equipment breakdowns or loss of production.
- Preventive maintenance can play an important role in an organization's management strategy by providing a number of benefits, such as improved system reliability, decreased cost of replacements, and decreased system downtime.
- Preventive maintenance programs also have disadvantages, such as increased responsibility and paperwork for operators and the allotted access time required to work on the equipment.
- Developing a new preventive maintenance program involves establishing goals and objectives, developing a strategy, planning the program, implementing the plan, and refining the program.

Review Questions

Answer the following questions using the information provided in this chapter.

1. Identify several areas of technology used by robotic systems.
2. What does the term *troubleshooting* mean?
3. What are some methods that technicians find helpful in troubleshooting?
4. What are some indicators of broken or faulty parts to look for during equipment inspection?
5. Why are equipment troubleshooting manuals important?
6. What are four steps or activities involved in identifying a problem with a robotic system?
7. During the initial observation of equipment to be serviced, what should you look for?
8. How does a technician proceed in gathering information about equipment?
9. What are some types of measurements that can be taken to check equipment?

10. How do you narrow down the cause of an equipment problem?
11. What factors should be considered when replacing a component?
12. What is preventive maintenance?
13. What do the terms *scrap* and *rework* mean?
14. What are some benefits of using a preventive maintenance plan?
15. What are some disadvantages of using a preventive maintenance plan?
16. List the steps to follow in developing a preventive maintenance program.
17. What is a preventive maintenance assignment sheet?

Learning Extension

1. Perform an Internet search to find information on preventive maintenance programs and other robotic maintenance tasks. Prepare a brief presentation on one of the programs, schedules, or methods you found to be interesting.

Advanced Analysis

1. Perform an Internet search and locate a troubleshooting chart. Prepare a brief discussion of the use of this chart.
2. Prepare a written discussion of the process of developing a preventive maintenance program for an industry.

Activity 12-1
Soldering and Terminal Connections

Objectives

In this activity, you will demonstrate your skill in three important exercises: soldering, desoldering, and terminal connection. An understanding of the proper soldering and terminal connection techniques is needed by technicians who design, service, or repair industrial equipment, including robots.

Equipment and Materials

- Soldering iron—25 W–50 W.
- Desoldering tool.
- Solder.
- No. 22 insulated wire—3″ length.
- Printed circuit board.
- Component mounting strip.
- Resistor (any value).
- Wire stripper.
- Side-cutting pliers.
- Needle-nose pliers.
- Solderless connector and wire.
- Crimping tool.

Safety

Exercise caution while soldering. Use protective eyewear and appropriate clothing during soldering operations.

Procedure

Section A. Soldering Iron Preparation and Component Mounting

1. Allow the soldering iron to heat to its operating temperature.
2. Apply a small amount of solder to coat the tip of the iron. (This is called *tinning*.)
3. Obtain a printed circuit (pc) board and a 3″ length of No. 22 insulated wire or a resistor.
4. Place the wire (or resistor) near the holes in the pc board where you intend to solder it to determine the length required.
5. Strip 3/8″ to 1/2″ of insulation from one end of the wire (if used) with the wire stripper.

6. Cut off any excess wire and strip the insulation off the other end of the wire. The wire should lie flat on the pc board when soldered. If a resistor is used, it should lie flat also.
7. Using the proper soldering techniques, solder both ends of the wire (or resistor) to the pc board terminals.

Section B. Printed Circuit Board Desoldering

1. Obtain a desoldering tool.
2. Allow the desoldering tool to heat to its operating temperature.
3. Using the proper desoldering techniques, remove the piece of wire from the pc board without damaging the conductive strips on the pc board.

Section C. Terminal Connection Installation

1. Obtain a solderless terminal connector crimping tool and a short length of insulated wire from your instructor.
2. Strip enough insulation from the wire to allow the terminal connector to fit properly.
3. Use the crimping tool to fasten the terminal connector to the wire.

Activity 12-2 Troubleshooting and Testing

The activities included in this book provide an opportunity to practice troubleshooting and testing electronic circuits, devices, and systems. Troubleshooting and testing procedures may be accomplished while performing the activities. Emphasis is placed on understanding circuit operation, safety, and the proper use of test equipment. If the technician understands how the circuit, device, or system functions and knows how to use test equipment, troubleshooting and testing are relatively easy to accomplish. This is true for the simplest type of circuit or for more complex systems.

Objectives

Upon completion of this activity, you should be able to outline basic troubleshooting procedures, test devices and circuits using the correct procedure, and find the defective part or circuit by using a common sense approach.

Troubleshooting is a systematic process by which malfunctions are traced and repaired. A technician should be able to locate systematically and repair many malfunctions. For effective troubleshooting, you should have a practical understanding of the type of system that is being repaired. With this basic knowledge of operation and some test equipment, you should be able to perform many repair jobs.

Electrical motors are commonly used with robotic systems. Some of the following instruments may be used to make various tests on the motor:

- Tachometer—speed test.
- Megger—insulation test.
- Multimeter—resistance, current, and voltage test.
- Growler—wound armature test.
- Motor dynamometer analysis unit—speed vs. torque test.

You should check with the instructor on the use of some of this equipment as your troubleshooting progresses. After you complete this activity, you will understand how to test systematically and repair a motor of any type. Many parts of the procedure listed may not be applicable to your particular troubleshooting job. However, try to follow most of the listed steps to complete the analysis although it may not be feasible to complete the entire activity.

Equipment and Materials

- Motor that does not function properly.
- Multimeter.
- Growler (optional).
- Megohmmeter (optional).

- Dynamometer (optional).
- Tachometer (optional).

Procedure

1. Examine the external features of the motor you are to troubleshoot.
2. Find the motor nameplate and record the information listed below. If the information is not available, indicate so.
   - Manufacturing company.
   - Motor type.
   - Identification number.
   - Model number.
   - Frame type.
   - Number of phases (ac).
   - Horsepower.
   - Cycles (ac).
   - Speed (rpm).
   - Voltage rating.
   - Current rating (amperes).
   - Thermal protection.
   - Temperature rating.
   - Time rating.
   - Other information.
3. Determine the type of motor you are working with, such as ac induction, dc stepping, shaded-pole, or ac servo. Consult with your instructor if you cannot determine the specific type. What is the specific motor type?
4. Draw a schematic diagram of the motor. This will provide a general reference for later tests.
5. After the proper type of power source and operating voltage has been determined, apply the correct voltage to the motor for a very brief period of time while attempting to detect the specific malfunction.
6. Describe any malfunctions that seem to exist in your motor.
7. If available, consult a symptom chart in a motor operation book, manufacturer's data, or other reference. List some possible causes of the malfunction.
8. Determine and list some possible corrections for the malfunction.
9. If the motor must be disassembled, remove the end plates carefully. Be particularly careful with the connecting wires and bearings. Remove the rotor assembly at this time. Check the bearings.
10. If connecting wires must be removed, use tape or some other means to label the proper terminal connections.
11. Inspect all wiring connections, splices, and other mechanical connections for possible opens or shorts.
12. Inspect the stator portion of the motor for excessive dirt, oil, or other damage.
13. Check the rotor for visible damage.
14. If the rotor is the squirrel-cage type, it may need cleaning with fine sandpaper.
15. If a wound rotor is used, check the commutator for high mica insulation, roughness, irregular wear, or other damage. You may wish to dress the commutator on a metal lathe, clean it with sandpaper, or remove high mica with an undercutting tool.

16. If the motor has brushes, check for excessive wear, oil, grease, chipping, or improper contact.
17. You may need to perform some resistance tests with the multifunction meter to determine the condition of the stator windings. (Example: Ac single-phase induction motor; start winding—20 ohms resistance; run winding—5 ohms resistance.) Of course, those resistances will vary. Record your resistance readings.
18. What can you conclude from the resistance test regarding the condition of your motor?
19. You may test the rotor windings for damaged insulation with a megohmmeter. You may test the rotor for shorts, grounds, or open circuits with a test instrument called a growler, if available.
20. Perform any other tests you feel may be needed to locate the trouble. Consult with your instructor.
21. Perform the necessary procedures to repair the motor.
22. Consult with your instructor when you think you have completed the troubleshooting procedure and have repaired the motor malfunction.
23. This should complete your troubleshooting job. Return all materials to the storage area.

Analysis

1. Write a detailed description of the troubleshooting process you have used in this experiment. Explain how you determined the trouble for this particular type of motor. Also, explain how you corrected (or would correct) the malfunction.

Chapter 13

Robots in Modern Manufacturing

Outline

Objectives

Upon completion of this chapter, you will be able to:

- Identify working environments that are candidates for robot implementation.
- Evaluate potential uses for robots.
- Prepare a robot implementation plan.

Technical Terms

avoidance costs
capital investment
cost savings
engineering economics
payback period

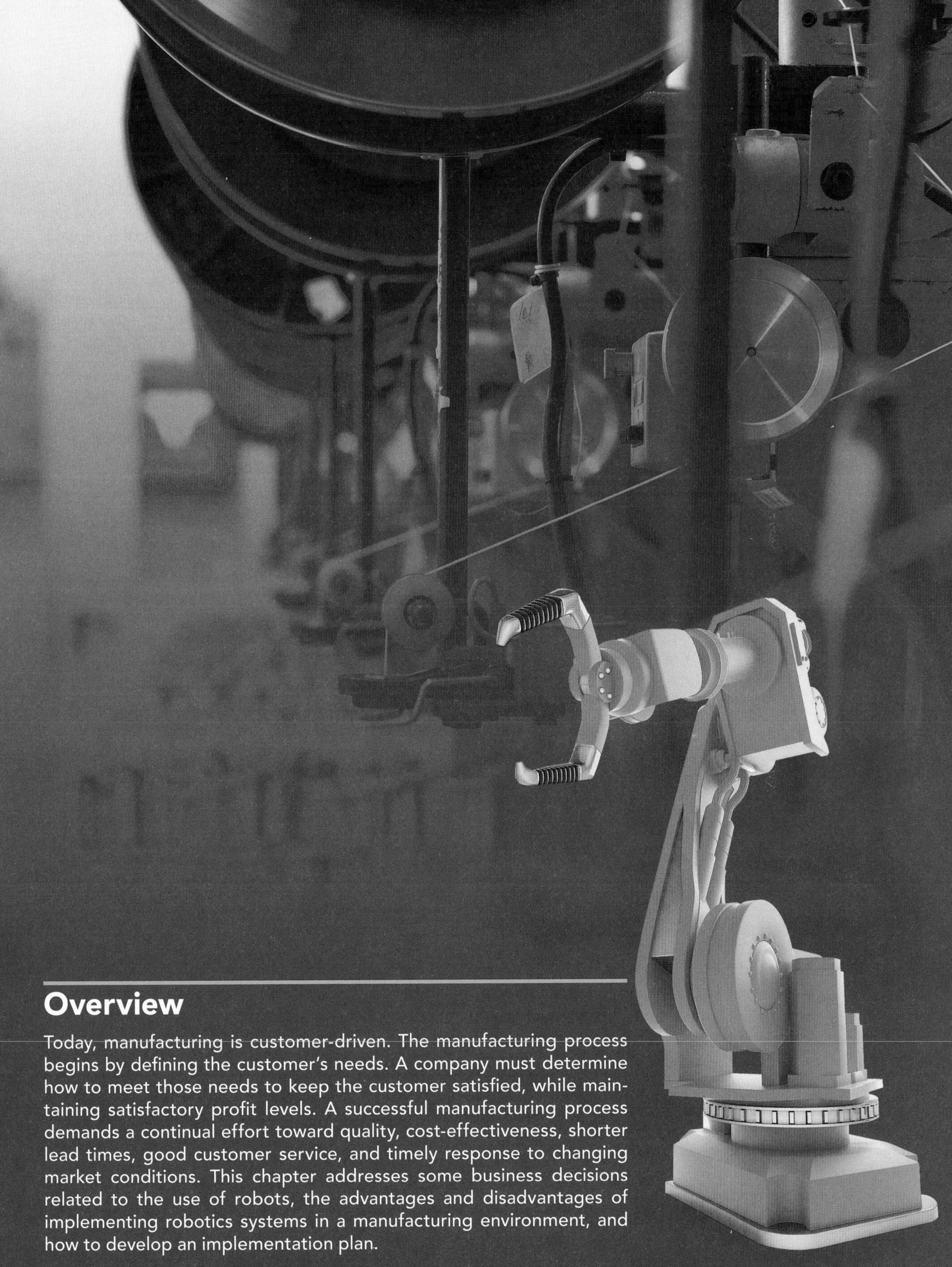

Overview

Today, manufacturing is customer-driven. The manufacturing process begins by defining the customer's needs. A company must determine how to meet those needs to keep the customer satisfied, while maintaining satisfactory profit levels. A successful manufacturing process demands a continual effort toward quality, cost-effectiveness, shorter lead times, good customer service, and timely response to changing market conditions. This chapter addresses some business decisions related to the use of robots, the advantages and disadvantages of implementing robotics systems in a manufacturing environment, and how to develop an implementation plan.

13.1 Using Robots in Manufacturing

During the mid-80s, manufacturers believed that robotic technology was a key element in the factory of the future. Today, many major companies that initially invested millions in robotic systems tell stories of failure as well as of success. Robots and other complex automation systems did not solve every manufacturing problem. In fact, when faced with the prospect of re-tooling an entire line, many organizations scaled back the level of automation and took a more simplified approach.

For some organizations, the following incorrect assumptions produced costly mistakes:

- Incorporating robots automatically increases productivity.
- Intelligent machines can do the same job as trained workers and save labor costs by replacing workers.
- A large investment in robotic automation pays off in time.

13.1.1 Productivity

Does robotic technology automatically increase productivity? The answer is no. No machine has been invented that is more flexible or more efficient than a motivated human worker equipped with the right tools. In certain applications, robots can be more productive than humans over a long period of time because robots do not suffer from fatigue and do not require downtime for food or rest. Productivity does increase when the product being manufactured is specifically designed for automated assembly. However, manual assembly processes benefit from specific design considerations as well.

Goodheart-Willcox Publisher

Figure 13-1. Maintenance technicians and programmers are needed to ensure that the robot correctly performs the tasks required and operates efficiently.

13.1.2 Replacing Employees

Many companies assumed that robots would replace a significant number of employees and the resulting savings would offset the initial cost for the robotic equipment. When robotic systems were implemented to replace production workers, computer programmers and other highly trained technicians were hired to program and maintain the robotic systems, **Figure 13-1**. As a result, companies were left with fewer workers who understood the manufacturing processes.

In general, direct labor costs are not a major portion of the total cost in producing a product. Typically, a robot replaces fewer than two workers in an industrial process or application and results in little direct labor savings, **Figure 13-2**. In fact, adding robots and other high-tech equipment can increase direct labor cost due to the additional skilled individuals required to support the technology.

13.1.3 The Payoff

Do large investments in robotic systems pay off over time? Not necessarily. After spending millions of dollars on such systems, a company may be forced into long production runs to pay for the equipment. A simple downturn in the economy can destroy such a plan overnight. Additionally, manufacturing methods have changed. Industry trends, such as fast product turnaround, can add significant costs to a production process that relies on robotic systems. Re-tooling automated manufacturing systems is expensive. If robots or other automated systems cannot be changed or updated economically, a large investment in these systems is not worthwhile.

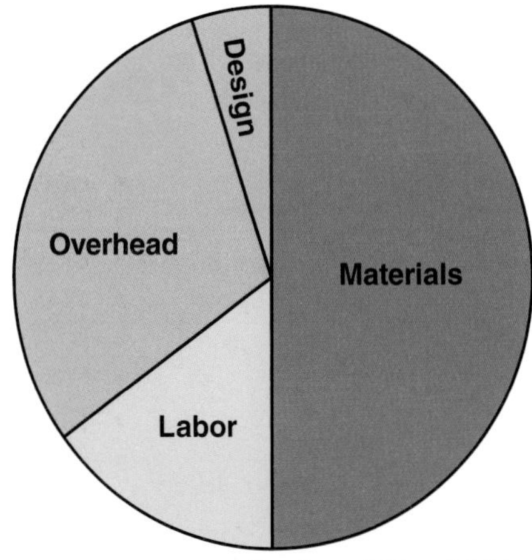

Goodheart-Willcox Publisher

Figure 13-2. Direct labor costs comprise about 15% of the total cost in producing a product.

13.1.4 Successful Robotic Implementations

Robotic systems are being used successfully in numerous situations. The automotive industry and electrical/electronic industries are continuing to make large investments in robotics, **Figure 13-3**. Global marketing, worldwide competition, and attempting to meet a wide range of customer preferences have resulted in shorter product development time and less changeover lead time. To accommodate the robotic systems, task complexity was reduced by altering product and fixture designs. Production line flexibility has been enhanced by the use of programmable robots.

Robotic installations are most easily justified in undesirable environments. Whenever working conditions are unpleasant or unsafe, worker productivity is affected. The use of robots in such environments has proven to be an effective

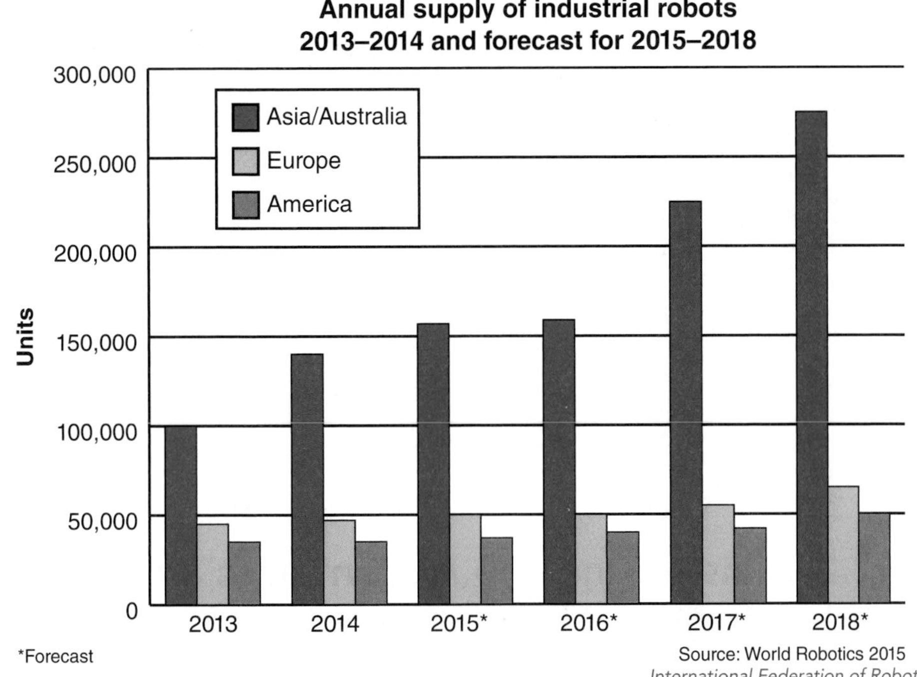

International Federation of Robotics

Figure 13-3. A comparison of the use of robots in various parts of the world.

solution. If working conditions cannot be improved, the operation should be automated. Undesirable conditions suitable for robot implementation include the following:

- Tasks that involve hazardous materials.
- Environments that lack clean, breathable air.
- Operation of hazardous machines.
- Environments with extreme temperatures.
- Tasks that include handling heavy materials.
- Areas with poor lighting.
- Environments containing hazardously high noise levels.
- Tasks that cause mental strain or fatigue.
- Excessively dusty or dirty environments.

13.2 Evaluating Potential Uses for Robots

Most robotic applications cannot be simply defined. A potential robotic application can be evaluated by answering these five questions:

- Does the technology match the need?
- Have all the advantages and disadvantages been considered?
- What are the economic justifications?
- What are the non-economic justifications?
- Is the organization ready for robots?

13.2.1 Matching the Technology to the Need

When evaluating a robotic system for a process, consider the production characteristics of the process. A robotic system may be beneficial if production volume ranges from medium to high, production cycles are short- to medium-length, lot sizes are small, and several workpieces are produced per process. The need for material and product handling should also be reviewed. Robotic systems can be an effective addition when the process requires handling heavy loads, precise positioning of materials and products, and general (not specific) orientation of parts.

A process is not a candidate for implementing robotic systems when one or more of the following conditions exist:

- Critical aspects of the workpiece or the process cannot be measured.
- Specific workpiece and process standards do not exist.
- Identification of the workpiece is difficult.
- Positioning the workpiece is complicated.
- Evaluating the condition of the workpiece is intricate.
- The range of movement needed for the robot in the process cannot be readily identified.
- The robot's movement cannot be contained within a limited space.

13.2.2 Considering the Advantages and Disadvantages

When determining whether a robotic implementation is justified, carefully consider and evaluate the advantages and disadvantages involved in making the change. The balance of advantages and disadvantages shifts with each unique environment and situation.

Advantages

Robots can offer increased productivity under certain conditions. While robots are no faster than humans at completing many tasks, the increased productivity results from the constant work pace of the robot. During an eight-hour shift, the robot usually outperforms a human operator. This is the case especially if the task is repetitive, boring, heavy, or performed under poor working conditions. Further benefits to productivity can be accomplished by allowing the robot to operate for more than one shift.

Robots can improve product quality because of their accuracy and repeatability. For example, a robot used in a spot welding application may be able to place a weld within 0.050″ (0.127 cm), time after time. This accuracy is hard for a human to duplicate, since the spot-welding gun weighs approximately 100 lb (45 kg). Even though the gun is counterbalanced, fatigue takes its toll on a human worker after several hours, **Figure 13-4**.

Automated equipment requires consistently uniform components to function efficiently. This requirement helps improve product quality, **Figure 13-5**. A human operator may be able to adjust the process and assemble components that are a bit out of spec. Automated equipment, however, cannot compensate for irregularities in the product. Parts must be within programmed specifications and are held to closer tolerances. To take full advantage of automated production, manufacturers demand 100% quality parts from vendors.

The use of robots in some situations reduces personal injuries and increases employee safety. The passage of the Occupational Safety and Health Act (OSHA) provided an incentive for manufacturers to introduce robots in many operations that were considered dangerous. Using robots in these areas reduces the need

A

B

Goodheart-Willcox Publisher

Figure 13-4. A robot's performance is not affected by fatigue. A—During a typical work shift, a human operator tires from handling and operating heavy welding equipment. B—A robot does not experience any strain while spot welding an automotive body assembly during a work shift.

Goodheart-Willcox Publisher

Figure 13-5. To function efficiently, assembly robots must work with components of consistently high quality. The result is a final product of high quality. This robot is assembling electronic printed circuit boards.

for complex safety procedures and equipment. Automating unpleasant or dangerous tasks also increases worker morale. If workers are moved from dirty, hostile, and hazardous environments, their attitude toward working improves; and they can be placed in more challenging positions with greater responsibility.

Robots can contribute to holding down or reducing production costs. Some manufacturers boast of productivity gains of up to 400%. Others have reduced scrap material and time spent reworking products. For example, General Motors once reported that four-fifths of rejected products in spray painting came from manual operations. Reducing the amount of material used also helps reduce production costs. Using robots for spray painting can result in a 50% reduction in the amount of paint used. Additional savings may result from the reduction of energy required to light, heat, cool, or ventilate work areas where robots operate.

Most automated equipment is dedicated, or designed to perform one function. Trying to adapt dedicated equipment to new operations often proves costly or difficult to accomplish. Robots offer the advantage of flexibility; they can be reprogrammed to perform new tasks. For example, a robot can perform operations on different automobile models on the same assembly line. Manufacturing operations are typically performed using groups of 50 or fewer parts. Robots are well-suited for operations of this type.

Disadvantages

The cost of implementing robots in a manufacturing environment is a primary disadvantage. The money that a business spends to purchase fixed assets, such as buildings and equipment, is known as a ***capital investment***. The addition of robots, like other pieces of equipment, must be economically justified. However, other factors, such as worker health and safety, should be considered when deciding whether to use robots.

Using robots directly impacts the production line. For example, parts must be properly positioned and oriented for use by the robot. The manufacturing workflow changes, and additional work space is required. After a robot is acquired, all the tooling and equipment required to interface the robot with the manufacturing line is necessary. This tooling and peripheral equipment is very expensive, and, depending on the application, the cost of the equipment can exceed the cost of the robot itself.

Another disadvantage is employee opposition. Some workers feel threatened by robots and need to be reassured of their value to the company. If management and workers do not develop a positive attitude regarding the use of robots, installation and implementation will most likely fail.

Robotics Innovations: Segway Robot

The Segway robot, built on the popular gyro-balanced personal transportation device, is a fun and unique robot. The Segway robot is strong enough to carry a rider, and then transition back into robot mode and follow along behind a person, much like a dog following its owner. The digital display also contains cameras that allow the robot to interact with and navigate its environment.

Photo courtesy of Segway, Inc.

13.2.3 Economic Justifications

Many business decisions are based in part on economic factors. Business managers typically want to decrease costs and increase profits. The decision to invest in robots may be based solely on economic factors. ***Engineering economics*** involves analyzing the monetary impact and technical aspects of equipment or other projects. All the factors involved in purchasing, implementing, and maintaining robotic equipment for a manufacturing business are analyzed in this process.

There are two situations where economic analysis can assist in making an investment decision. The first is investing in equipment for a new application that will prevent the need to spend additional money in the future. The money spent for this type of investment is referred to as ***avoidance costs***. The second is replacing an existing production or assembly method to reduce the money needed for current operations. The money spent for this type of investment is known as ***cost savings***.

Business managers want money spent on a robotic system to provide a return equal to or greater than the required investment. Most companies have established benchmark criteria for return on investment and payback period. Return on investment is the amount received or gained from an investment compared to the cost of the investment. It is usually expressed as a percentage. A ***payback period*** is the time required to recover the amount spent for new equipment through increased profits or savings in labor or materials. For example, a company may require a 25% return on investment and a payback period of less than two years. In addition to return on investment and payback period, other financial factors are considered.

Looking at the long-term impact of using robots may justify the expense of implementing robots. Over a period of years, will the use of robotic technology:

- Produce better products of more consistent quality?
- Reduce new product lead time?
- Increase flexibility of the production line?
- Improve customer service and product satisfaction?
- Improve market response time?

If the answer to several of these questions is yes, an analysis should be done to determine if the capital investment is justified. However, other alternatives should also be explored. One alternative may be to modify or rebuild present equipment. The same goals may also be accomplished through retraining and motivating the workforce.

In a highly competitive market, new technology may help maintain or improve a company's competitive position. See **Figure 13-6**. This, too, may justify investing in robotic equipment. Robots are often used in research and development. The expertise gained in this area can be applied later to their general use. Economic returns occur because the time needed to get robots up and working is reduced.

Impact of Robotic Equipment on Auto Manufacturing and Assembly Operations		
Production Year	Models	Body Types
1997	8	5
2005	12	9
2010	14	10
2017	19	11

Goodheart-Willcox Publisher

Figure 13-6. The robotic manufacturing and assembly equipment used by automotive companies allows them to change, update, and expand vehicle model and body options in order to meet consumer demand and to stay competitive in the automotive industry.

Sometimes it takes competitive pressure to initiate change. If one competitor starts to use robots and gains a price advantage, robots may become more attractive to other companies in that market. However, buying a robot should be based on firm business factors. Robots should not be implemented simply to follow trends.

13.2.4 Non-Economic Justifications

Non-economic considerations sometimes outweigh economic ones when deciding whether to use robots. The use of robots in hostile environments is an example. Some jobs require handling toxic materials or operating hazardous machines. Other jobs must be completed in very noisy or dirty environments. Extreme temperatures or very heavy materials are needed to complete some tasks. These conditions or job requirements can be unpleasant or dangerous for people. If poor working conditions are ignored by management, productivity can decrease, and product quality may suffer. Workers may become ill or may be absent from work to avoid unpleasant conditions. Using robots to complete dangerous or unpleasant tasks is often a good solution.

13.2.5 Is the Organization Ready for Robots?

A critical question for many companies is, "Is the organization ready for robots?" This is not a simple question, since many variables are involved in making a judicious decision. Robots can do the some of the same work as humans. They can also accomplish some work more efficiently. Robots never get sick or need rest, and they can work around the clock. Robots perform precise operations on repetitive work at a rapid rate.

On the other hand, robots do not have the capacity to be innovative or collaborative, and they cannot think independently. Robots do not learn from their mistakes. They do not have the ability to make major changes and independently adapt without help from a human operator. Once the technology, workflow impact, and financial aspects have been researched and discussed, a decision can be made about the implementation of robots. However, this decision can only be made if the organization is ready for robots.

13.3 Preparing an Implementation Plan

In many cases, company managers will decide that using robot equipment in a manufacturing process will help the company reach its production and business goals. The following steps can be used to develop a successful implementation plan:

1. Identify potential applications.
2. Analyze potential applications.

3. Construct a matrix for comparing equipment.
4. Review the available equipment.
5. Match the robot to the application.
6. Develop the proposal.
7. Develop and refine the application.
8. Begin implementation.
9. Provide training.
10. Provide maintenance programs.

13.3.1 Identify Potential Applications

One way to identify possible applications for robotic equipment is through a plant survey. Certain job, product, or process characteristics can be used to identify potential uses for robots. Some general areas that can be used to begin the survey include material handling, component insertion, inspection, and testing. Potential applications can also be identified by reviewing personnel and safety records. For example, if a job is monotonous, dirty, hot, noisy, or a possible health or safety hazard, it is a candidate for a robot. Other characteristics of manufacturing jobs that may be performed by robots include short product life, frequent design changes, using families of parts, using a minimum of tools and parts, and operating for multiple shifts.

13.3.2 Analyze Potential Applications

Once possible robotic applications are identified, select the most promising applications and construct profiles of each. This typically involves several trips to the production floor and talking to people connected with the operations. The importance of drawing on the expertise of operators, supervisors, engineers, and other manufacturing personnel cannot be overemphasized.

Installing a robot can have a ripple effect on the whole production line. For example, additional material handling and feeding devices may be required. Blueprints, specification sheets, production control records, quality records, and safety records must all be explored to create a complete picture of the production process. Compile data on the number and sequence of operations, parts flow, cycle times, product volume, personnel requirements, product life, results of quality checks, and the effects of environmental factors. This information is useful in determining if a robot is really capable of handling a job.

13.3.3 Construct a Matrix for Comparing Equipment

Matching a robot to a particular manufacturing application is not an easy task. To evaluate various robots, construct a matrix to simplify the comparison of key elements, **Figure 13-7**. The matrix should include categories for the model number, price, number of axes, type of controller, load capacity, work envelope, type of power supply, speed, accuracy, repeatability, and methods of programming.

13.3.4 Review the Available Equipment

Research all the available products that meet the requirements of the selected applications. Review information about the capabilities and limitations of various

Equipment Comparison Matrix			
	System #1	**System #2**	**System #3**
Manufacturer	Motoman	FANUC	ABB Robotics
Model Number	UP200	M 420iA	IRB 6600
Horizontal Reach	2951 mm	1855 mm	2550 mm
Number of Axes	5	4	6
Contoller Type	Teach Pendant	Teach Pendant	Teach Pendant
Load Capacity	200 kg	40 kg	225 kg
Mass	1350 kg	620 kg	820 kg
Structure	Vertical Jointed Arm	Articulated	Vertical Jointed Arm
Repeatability	+/– 0.2 mm	+/– 0.5 mm	+/– 0.1 mm
Programming Method	Teach Pendant	Teach Pendant	Teach Pendant
Other Features	Material Handling	Material Handling	Material Handling

Goodheart-Willcox Publisher

Figure 13-7. This sample matrix presents the features and capabilities of three robotic systems.

models. Information can be acquired from robot manufacturers as well as from the Internet. In researching various products, you can become familiar with common elements as well as model-specific features and packages. Compile the data from this research into the matrix for review.

13.3.5 Match the Robot to the Application

At this point, you must determine if the job can be handled by a robot. If yes, you must develop a method for using the robot. In addition to the robot itself, tooling and peripheral equipment must be assessed. The robot's cycle time should also be established to further match the robot with the application. If laboratory facilities are available, construct a prototype to establish cycle times and evaluate robot performance and the process sequence.

13.3.6 Develop the Proposal

A proposal is a document that states the problem and proposes one or more solutions supported by a rationale. The rationale should include both economic and non-economic factors. The proposal addresses personnel requirements, required resources, the schedule, and the budget. Additionally, a financial analysis is prepared that reviews the investment required, expense items, savings, return on investment, payback periods, and other economic factors for each proposed alternative. Proposals should always contain economic justification.

13.3.7 Develop and Refine the Application

After the proposal is approved, the details of the plan are itemized. The necessary tooling and safety devices are designed and ordered. The specifics of the robot application are refined and debugged. Much of the time spent on the project is devoted to this step. This is the initial proving ground for the application.

13.3.8 Begin Implementation

During implementation, the site is prepared for production. The floor space is prepared, and the various service drops and safety devices are installed, **Figure 13-8**. While a service drop usually consists of an electrical line run to a piece of equipment, it can also include air, water, or any other utility necessary to the equipment operation. Simply plugging a robot into a production line is not possible. All the analysis, planning, and preparation up to this point should ensure a successful implementation process.

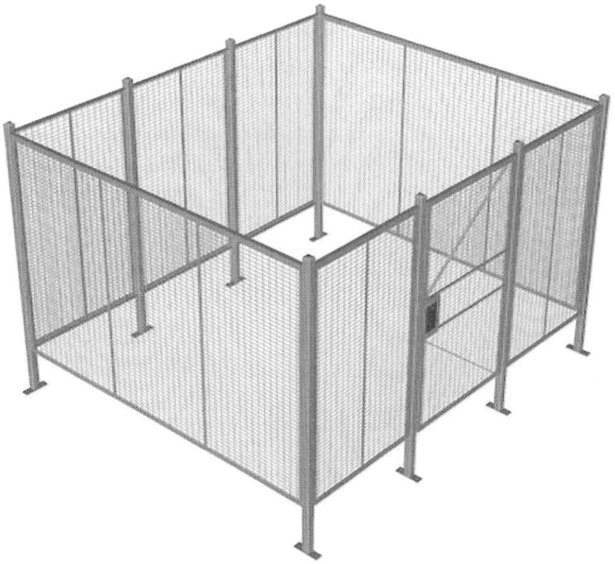

Cisco-Eagle

Figure 13-8. Safety fences must be installed around the robotic workstation during the implementation phase of the project.

13.3.9 Provide Training

During implementation, everyone involved in the operation should be trained prior to the actual installation. Maintenance workers may already possess some of the technical skills needed to work on robots. However, they need to become familiar with the manufacturing process and the basics of programming. Robot manufacturers usually provide training for maintenance and other personnel, either in their own facilities or at the installation site. In-service and retraining programs are also a vital part of any manufacturing operation. These programs are needed not only for technical personnel but for all employees. Identify the various groups involved in the manufacturing process, so training can be tailored to meet the needs of each group.

13.3.10 Provide a Maintenance Program

Establishing a preventive maintenance program is important to success. The company should hire an adequate number of maintenance personnel. Consider that workers become sick, go on vacation, are transferred, retire, or accept positions with other companies. The company should have enough maintenance personnel to keep the equipment running when these personnel problems or changes occur.

A maintenance program also includes stocking an ample quantity of spare parts. Having spare parts on hand allows repairs to be made quickly, which avoids loss of production time. Robot manufacturers supply a list of spare parts to stock. As experience is gained with the robot, compile a list of parts that applies to its specific operation.

Summary

- Manufacturing companies must meet the needs of customers while maintaining satisfactory profit levels.
- In the past, some organizations have made incorrect assumptions related to robots. For example, some assumed that incorporating robots automatically increases productivity.
- Robotic technology does not automatically increase productivity. In certain applications, however, robots can be more productive than humans.
- Adding robots and other high-tech equipment can increase direct labor cost due to the additional skilled individuals required to support the technology.
- Companies should consider whether using robots will pay off over a period of time.
- Robot use is particularly effective in undesirable environments. For example, tasks that involve hazardous materials or extreme temperatures may be appropriate for robots.
- Potential use of robots in manufacturing involves matching the technology to the need or process.
- Advantages of using robots in manufacturing include increased productivity, improved safety, and improved product quality.
- Disadvantages of using robots in manufacturing include capital investment costs, production line changes, and employee opposition.
- Business managers typically want to decrease costs and increase profits. The decision to invest in robots may be directly related to these economic goals.
- Non-economic considerations sometimes outweigh economic ones when deciding whether to use robots. The use of robots in hostile environments is an example.
- Company managers must determine whether the organization is ready for the implementation of robots. Many variables are involved in making a judicious decision.
- The use of robots in a manufacturing company should involve the development of an implementation plan.
- When planning robotic use for specific applications, companies should compare robots from different manufacturers carefully.
- During implementation of robotic systems, everyone involved in the operation should receive proper training.
- A well-planned maintenance program is very important for the success of robotic operations.

Review Questions

Answer the following questions using the information provided in this chapter.

1. Identify three incorrect assumptions company managers make about implementing robots in manufacturing and discuss why each is incorrect.
2. List several examples of undesirable working environments where robots are an effective solution.
3. When evaluating an application for potential robot installation, what are the five questions that should be answered?
4. What are some examples of a process that is not a candidate for implementing robotic systems?

5. Why are robots sometimes more productive than humans, even if they do not complete a task faster than a human?
6. How does the use of automated equipment affect the quality of a product?
7. What are three disadvantages of implementing robots in a manufacturing environment?
8. What is engineering economics as it applies to robotic systems?
9. What are five questions to consider when deciding whether the long-term impact of using robots will justify the expense of implementing them?
10. What is an example of a non-economic factor to consider when deciding whether to use robots?
11. What are the ten steps involved in preparing an implementation plan?
12. What types of information are included in a proposal?
13. Why is it important to stock spare parts for robotic equipment as part of a maintenance program?

Learning Extensions

1. Identify a process for which you would consider the use of a robot.
   - Identify the advantages and disadvantages of implementing robots into the process.
   - Provide economic justification for purchasing the robot.
   - Prepare an implementation plan for the robot that addresses each of the ten steps.

Advanced Analysis

1. Select a product that is produced by a manufacturing company. Do an Internet search to determine more specific information about the product. Describe your concept of the operations involved in manufacturing this product, and develop a robot implementation plan.
2. Select another product and describe your concept of operations for manufacturing and develop a robot implementation plan.

Chapter 14

Robotic Applications and Innovations

Outline

Objectives

Upon completion of this chapter, you will be able to:

- Describe fully automated factories.
- Discuss uses of robots outside of factories.
- Explain the impact of artificial intelligence on the field of robotics.
- Describe the impact computer-controlled machinery has on the workforce.
- Describe the recent evolution of robotic applications and innovations.
- Describe types of training and educational opportunities for robotics.

Technical Terms

expert system
microbot
serbot
telerobotics
humanoid robot

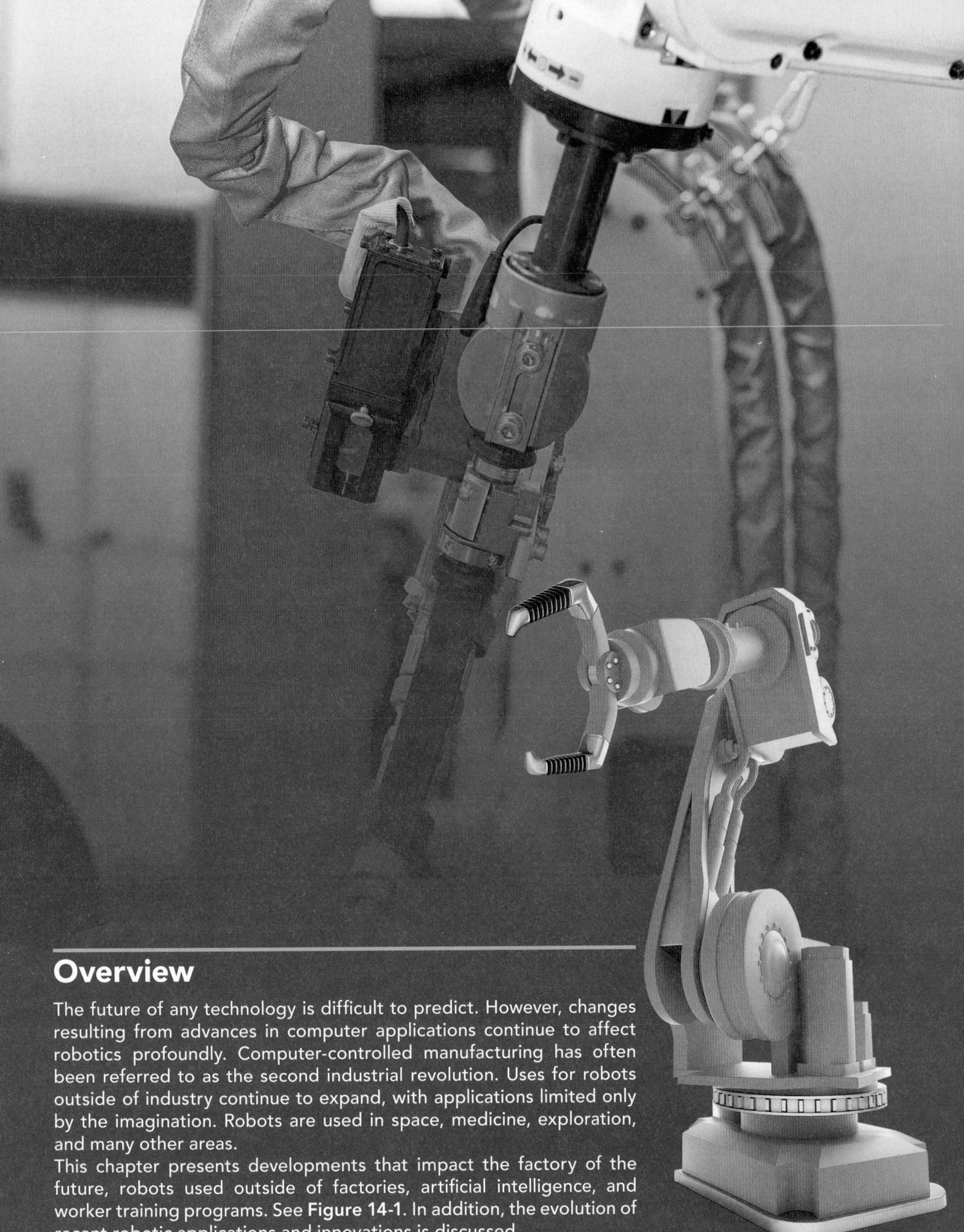

Overview

The future of any technology is difficult to predict. However, changes resulting from advances in computer applications continue to affect robotics profoundly. Computer-controlled manufacturing has often been referred to as the second industrial revolution. Uses for robots outside of industry continue to expand, with applications limited only by the imagination. Robots are used in space, medicine, exploration, and many other areas.

This chapter presents developments that impact the factory of the future, robots used outside of factories, artificial intelligence, and worker training programs. See **Figure 14-1**. In addition, the evolution of recent robotic applications and innovations is discussed.

Microstock-Thailand/Shutterstock.com

Figure 14-1. Robots are commonly used outside of factories. This robot can handle tough floor debris and maneuvers over many different surface types.

14.1 Fully Automated Factories

For decades, scientists, engineers, and plant managers have dreamed of a fully automated factory without human workers. In such factories, lasers are used to inspect parts and check for wear on machine tools, and assembly components are shaped by automated equipment. Driverless, automated vehicles move through the factory picking up and delivering parts, and robot arms assemble components the way a human would. This type of operation eliminates complicated fixtures and parts feeders, and it has only a few humans working among the multitude of robots. The entire automated system can be re-tooled simply by switching the software program that guides the robots. This fully automated factory model was described in a *Business Week* magazine article in the late 1980s. Today, many of the robotic systems described have been realized and are in use.

The technology for fully automated, workerless factories exists today. See **Figure 14-2**. Implementing such technology is not easy, however. Automotive companies have found that simply increasing the number of robots on an assembly line does not necessarily translate into increased sales and profits. The time and cost invested in getting multiple robots programmed and running error-free has been prohibitive for many companies. The hesitation of companies to automate production lines fully has been due, in great part, to a misunderstanding of how to use current technology. For example, applying robots to poorly understood production problems can make a bad situation even worse.

Praphan Jampala/Shutterstock.com

Figure 14-2. While the technology for fully automated factories exists, implementing the plan is not always easy.

For years, American manufacturers lagged behind their Japanese and European counterparts in the knowledge and expertise needed to apply robotics technology successfully. George Devol, who patented the first industrial robot in 1954, criticized the "lack of interest in our manufacturing industry and automated systems." He noted that "European and Japanese companies embrace this concept." High degrees of automation are possible with developments and advancements in robot technology, including vision systems, smart sensing systems, and artificial intelligence.

14.2 Robots Outside the Factory

In the 1980s, the robotic industry was in high gear. Hundreds of companies were involved in manufacturing and installing industrial robotic systems. By the early 1990s, however, only a very small number of companies remained actively involved. Major growth outside the industrial sector began in the 1990s with service robotics. ***Serbot*** (service robot) applications range from mobile robots on building security patrol to automated robotic vehicles that explore other planets. Diverse serbot applications have developed, including:

- Serbots used to clean floors.
- Agricultural applications, such as a robot that works on a dairy farm. The robot can adjust a cow's legs, wash the udders, and attach a milking machine.
- Space-exploration serbots designed to examine the surface of Mars.
- Robots are used for education and entertainment by museums and zoos. For example, a robot can be used to operate replicas of prehistoric animals.

The field of ***telerobotics*** is a branch of engineering that has produced robot arms and mobile robots that are operated by remote control (using a radio transmitter and receiver). These robots have been used for both deep-sea and outer-space applications. An example is the robot rovers used on the NASA Pathfinder Mission in the late 1990s. The remote-controlled rover used in the Pathfinder Mission was placed on Mars to transmit data about the surface of Mars back to Earth. Such highly visible and publicized events have increased interest in remotely controlled telerobotic applications. The use of robotics in the home is a very popular and expanding area of robotic applications. Consumer remote-control robotic systems can accomplish several domestic tasks, such as cleaning floors and mowing lawns, **Figure 14-3**.

Microbots, the tiniest robots, can perform operations inside the human body. Doctors can guide their work by

A

B

iRobot Corporation

Figure 14-3. Telerobotic products are available to consumers and can make life more enjoyable by handling home maintenance tasks. A—The Looj™ travels through gutters and clears debris and clogs. B—Robots may clean in-ground pools in a little as 60 minutes and circulate 80 gallons of water per minute.

iRobot Corporation

Figure 14-4. The SUGV (small unmanned ground vehicle) is a mobile robot that gathers data on its environment and surrounding activities to keep military and security personnel out of harm's way.

means of computer control. Imagine a robot submarine small enough to cruise through your veins and arteries. Some people predict that by the year 2029, insect-sized robots will construct, alter, and clean everything from ranch houses to silk suits.

Several manufacturers have developed robots that are used primarily for educational experimentation. These robots are smaller versions of those used in industry, and they have almost identical operational characteristics. Students in the field of robotics or related manufacturing fields can use these educational robots to learn basic operation principles.

Military operations have made use of robotic technology for several years, **Figure 14-4**. Remote-control capabilities provide a safety measure when performing dangerous tasks, such as minefield detection and removal of improvised explosive devices (IEDs). Applications for government and security robots have received much emphasis in recent years, **Figure 14-5**. Such applications include the handling and disposal of hazardous materials using robotic systems rather than exposing humans to the risk of exposure and injury.

A

B

iRobot Corporation

Figure 14-5. Robots used for security tasks must be mobile and versatile. A—The PackBot® can climb stairs and navigate uneven terrain. It is light enough to be transported and positioned by hand. With numerous accessory kits available, the PackBot can be configured to meet a variety of security and inspection needs. B—In addition to moving payloads of more than 150 lb and surveillance functions, the Warrior™ robot is also capable of handling bomb-disposal operations.

14.3 Artificial Intelligence (AI) and Expert Systems

Machines may be developed that are genuinely intelligent and seem humanlike in their problem-solving abilities. Machines that appear to think have long been the goal of scientists and researchers. However, most researchers agree that science has quite a way to go in developing a truly humanlike thinking machine. Some people think that such a machine can never exist. These people ask the age-old question: "Can any mechanism really think?" They believe that a human being is more complex and sophisticated than a machine could ever be. However, this does not mean that machines cannot be made more intelligent than they are currently.

Marvin Minsky is one of the fathers of the artificial intelligence (AI) field and defines it as "the science of making machines do things that would require intelligence if done by man." AI has been called the force behind the second computer revolution. AI researchers come from many scientific disciplines, including computer science, psychology, cognitive science, philosophy, biology, and engineering.

For many years, researchers have conducted developmental work on software using AI techniques and related hardware. The goal is to address one of industry's most costly problems—the lengthy setup and process planning time required to get machine tools ready for a new operation. With the help of AI, a new generation of intelligent machine tools is now able to run with little human help. These machines choose the correct tool, tool speed, cutting fluids, and optimum cutting strategies. The tools can then assemble and inspect the parts, maintaining correct tolerances.

AI research can be broadly classified into three major interrelated areas:

- **Robotics.** AI relates to robotics in the areas of machine vision, movement, and tactile sensing. The goal is for the robot to perform these operations in an intelligent, humanlike manner.
- **Natural language processing.** AI is used to enhance the user interface that allows humans and machines to communicate. The goal is to develop machines that understand natural human speech. See **Figure 14-6**.
- **Expert systems.** AI is used to provide knowledge and reasoning capabilities. An ***expert system*** is a software program that contains a knowledge base acquired from people who are specialists or experts in a certain field. The goal is to help nonexperts diagnose problems and make decisions.

Natural language processing is also called *voice recognition*. Voice recognition is often used for robotic applications because it allows robots to identify words spoken by a human and convert these words to specific operations. A robot may be controlled through voice commands to complete specific tasks.

Artificial intelligence for voice recognition requires study of human thought processes and application of these to a machine, such as a robot. This process is made possible through computer software that allows the robot to understand voice input and convert it to specific actions. Human words are matched with those stored

polkadot_photo/Shutterstock.com

Figure 14-6. Natural language processing AI allows humans and machines to communicate in everyday life.

internally in the operating system of the computer, allowing communication with the machine (robot). Voice recognition is an important type of AI for robotic applications.

The fastest-growing branch of AI is expert systems. An expert system involves a knowledge base, an inference engine, and a user interface. Engineers gather expert information for a specific field and place it into the knowledge base of the system. The information is retrieved as data by users of the system, or it is used to carry out logic operations.

Expert systems differ from regular computer programs in a number of ways. Expert systems can reason, handle uncertainties, and learn from previous experiences. Additionally, this type of system can explain why it needs further information and how it reached a specific conclusion. Expert systems have come out of the laboratory. They have various applications in industry and in the service and public sectors to solve a range of problems.

One of the first expert systems developed was known as MYCIN. It was developed in the mid-1970s as a medical expert system. It was designed to diagnose the possible cause and treatment of bacterial infections. Even though it was never put to practical use, the MYCIN system became a model for later reasoning, rule-based expert systems.

Many other applications for expert systems have been put into practice:

- Automotive expert systems can detect motor problems before a breakdown occurs.
- System design applications allow salespeople, who are not engineers, to design computer networks for their customers.
- Inventory systems help sales personnel keep track of the thousands of parts required by some complex systems.
- A filler-metal selection system uses AI to recommend optimal filler materials for the welding of aluminum.

AI is an important frontier in the advancement of robotic and computer technology, and a growing number of practical applications are becoming commercially available.

14.4 Impacts on Society

There is little doubt that robots and other forms of computer-controlled machinery will continue to increase in number and sophistication. A major area in which the use of automation impacts workers is in the types of jobs available and skills needed by workers. To date, workers in a few types of jobs have been replaced by this technology. However, the employment opportunities for semiskilled and unskilled workers will decrease, **Figure 14-7**. Studies on the impact of industrial robots on job displacement and job creation reached the conclusion that a remarkable skill difference is needed when the jobs eliminated are compared to the jobs created. The jobs eliminated are semiskilled or unskilled, while jobs that are created require a significant technical background and skills. Retraining workers is an important activity for every company that uses technology if it is to remain efficient and competitive. Currently, computer literacy is an essential employment prerequisite for workers in many job fields.

Jobs that are highly labor intensive or that pose risks of injury for human workers are often considered candidates for automation. Robots are used in a

Columbia/Okura LLC

A

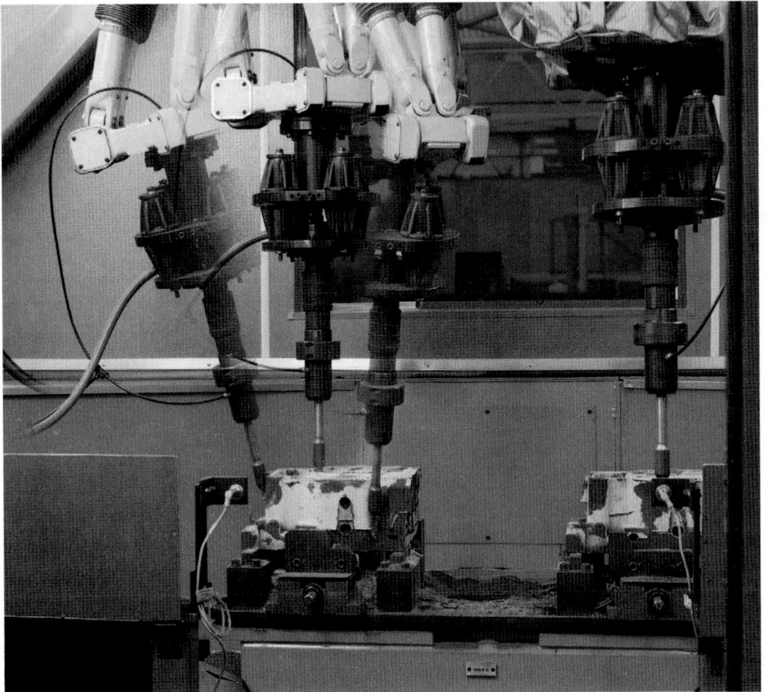

FANUC Robotics

B

Figure 14-7. Robots are currently used to perform many manufacturing tasks. A—This robotic system can perform palletizing operations at a rate of up to 20 bags per minute. B—Dual-grinding operations are performed by this robotic system.

wide variety of industries for a broad range of applications, including automotive, foundry, heavy material handling, packaging, and palletizing. With rising costs, increased competition, and a shortage of skilled workers, companies search for ways to achieve new levels of manufacturing efficiency. This includes using robots for applications that were not previously possible. Vision-guided robotic systems use digital imaging and intelligent software. Vision-guided robots can accurately perform inspection, handling, and assembly tasks. This type of robot relies on camera images and can perform multiple operations on the same line without the need to re-tool. This increases productivity and reduces ergonomic risks.

The future of robotics may be highly dependent on combining computer capabilities and peripheral systems, such as vision-guided technologies and remote control, to develop machines capable of assisting in homes, offices, and public areas. For example, a humanoid butler could assist disabled people at home, and a humanoid porter could carry heavy bags for passengers in airports or train stations. The application of such robotic systems is limited only by the imagination of the designer. Many people, including students, have expressed the desire to use robots to improve the world.

ASIMO (Advanced Step in Innovative Mobility) integrates advancements in robot mobility and intelligence.

Robotics in Society: Humanoid Robots

Developing robots with the physical characteristics and cognitive capabilities of humans has been a challenge and fascination since the time of Leonardo da Vinci. ***Humanoid robots*** are typically created with the human features of two arms, two legs, a torso, and a head. More importantly, a true humanoid robot is also capable of receiving, processing, and responding to information from its environment. From musical performances and verbal communication to walking and maneuvering stairways, advancements in robot technology have led to robots with some amazing humanoid abilities.

Honda's ASIMO robot is a well-known example of a humanoid robot. It has made appearances on television talk shows and nationally televised parades. It has also been featured at amusement parks. ASIMO can walk, run, climb stairs, and pick up objects. Its cognitive capabilities include reacting to its environment, understanding and responding to simple commands, and recognizing a familiar face.

As humanoid robot technology advances, the robot will likely take on a more human form and develop more complex reasoning and communication abilities. Future applications for humanoid robots may include assisting the elderly or disabled, performing hazardous tasks, or extended exploration missions.

Photo courtesy of American Honda Motor Co., Inc.

14.5 Evolution of Robotic Applications and Innovations

In the last few years, there have been many new robotic applications and innovations. These changes have taken place due to research conducted by companies, groups, and individuals who work with robots and other systems that require controlled movement.

Figure 14-8 summarizes (by year) several of the applications and innovations that have taken place. Keep in mind that an innovation a few years ago may be a common application today. Some of these innovations are discussed, with illustrations, throughout the book.

14.6 Your Future in Robotics

Robotics and automated manufacturing offer careers that can be both exciting and challenging. A variety of educational opportunities are available to students and workers from diverse backgrounds. Specific training is needed for those who operate, maintain, or supervise robots. Training is also needed for workers displaced by automation. Most companies attempt to retrain displaced workers for more satisfying jobs. In some cases, workers may be retrained to program, install, or service robots. Operation and servicing is a major industry concern. Although many workers have not been in an educational environment for a long time, well-organized training programs offer a career boost for most workers.

Many of the applications and functions mentioned are provided by suppliers of robotic systems. Training provided by suppliers is usually short and focuses on operation and programming of the systems being installed. Coverage of system maintenance and some special system applications that encourage the sale of products may also be provided.

Colleges and technical schools offer programs in robotics. Some focus on operator training, while others prepare workers to be supervisors for computer-integrated manufacturing. Essential subjects include robotic fundamentals, programming, electricity, basic electronics, hydraulics, pneumatics, digital electronics, microprocessors, programmable logic controllers, and machine vision systems. Basic courses in math, science, and communications are also recommended.

2011–2012	
8/3/11	Humanoid robot resides at the International Space Station
9/7/11	Industrial robots are used for high-speed machining
9/9/11	Robot has the ability to climb buildings and deploy a paraglider
11/10/11	Humanoid robot equipped with behavior control technology
12/5/11	Miniature industrial robots used for welding and metal fabrication
3/21/12	Medical robots used to aid surgeons and assist quadriplegics
3/21/12	Robots assist the military in making three-dimensional maps
5/2/12	Military robots used to keep soldiers out of life-threatening situations
8/8/12	Researchers testing the use of robots to dive into water for exploration
8/9/12	Tactical military robot used to negotiate rough terrain to collect and relay data
9/6/12	Person walks with assistance of brain-controlled robotic legs
9/6/12	Monorail-riding robot developed that can continually rearrange solar panels to follow the sun's angle
9/10/12	Robot used for underwater environmental monitoring
9/27/12	Toyota Company has developed a robot to assist elderly and handicapped with limited movement
10/11/12	Carnegie-Mellon University launches a major in robotics
12/19/12	Robotic medical assistants are helping with surgery, diagnostics, and remote presence
2013–2014	
1/10/13	Robots used for dangerous underground mining tasks
1/14/13	Robotics courses offered at public high schools
2/1/13	Robots used to sort and sterilize surgical tools in hospitals
3/1/13	Researchers have developed a robot that can be used for cold, hazardous, Antarctic research
3/13/13	Researchers have developed a dexterous robot capable of movement on rough terrain
3/25/13	ASEE sponsors 15th Annual Robotics Competition in June
3/27/13	Wearable robot applies torque to hip, knee, and ankle to help people move
3/29/13	Mitsubishi Company has developed a dexterous vertical-axis robot to perform complex tasks formerly operated by humans
4/1/13	Robots have been developed to be used in post-disaster rescue missions
4/11/13	US robot safety regulations are changing to accommodate human-robot operations
4/22/13	Next-generation robot can use visual servo control for guidance and motion control
5/3/13	Robots used with self-guided vehicles to collaborate with humans
5/29/13	Military robots extend human hands using tank-like unmanned ground vehicles (UGVs) moving on treads or wheels
6/13/13	Aerial robots perform surveillance, reconnaissance, and search-and-rescue missions for the military
7/1/13	A firefighting robot with stereo cameras and other sensors can be used to search burning buildings
7/3/13	Robots used to inspect cables of bridges, elevators, and cable cars that are exposed to high levels of stress
7/15/13	Robots are being designed to revolutionize farming and ease labor woes
8/9/13	Medical robot uses steerable needles to treat brain clots
9/20/13	A bio-robotics lab is being developed to design sensing and control networks for robots that work cooperatively with people

Goodheart-Willcox Publisher

Figure 14-8. This time line shows some of the more recent and notable changes in robotics applications from 2011 on.

2013–2014 (continued)	
1/24/14	A robot has been programmed to work at a supermarket checkout lane
2/4/14	ABB Robotics and Ford Motor Company have developed robots with real-time tactile feedback for delicate assembly and machining operations
2/28/14	An unmanned vehicle, the Guardbot, is able to travel over rough terrain, including snow, sand, mud, and water, for a variety of security applications
3/18/14	Researchers at Harvard University have designed a soft robotic cardiac simulator that can move with 3-D motion similar to an actual heart valve
4/22/14	The demand for service robots is rising, particularly for military and agriculture applications
5/2/14	A robotic wheelchair controlled by your voice has been developed
6/6/14	Medical robot use in the pharmaceutical industry is projected to increase significantly
6/11/14	Military robots now include unmanned ground vehicles (UGVs), large autonomous vehicles, and both large and small winged robots, one carrying its own fuel cell
7/2/14	Researchers are developing shoulder- and hip-mounted robotic arms to help workers in aircraft manufacturing perform difficult assembly tasks that would normally require two people
11/7/14	New industrial robots have been designed to work safely around humans, with no need to use safety barriers
11/10/14	A robot gripper has been developed that uses static electricity to pick up virtually anything
11/25/14	Researchers are developing a humanoid robotic pilot that could potentially replace humans in the cockpits of planes
12/16/14	NASA has installed a giant robot to help its research on composite aerospace materials
2015–Present	
1/14/15	Robotic grippers designed for use in hazardous locations
2/10/15	Industrial robots replacing humans at a faster pace
3/10/15	Agricultural robots are saving dairy farms
3/11/15	Robotic welding should grow sharply by 2020
3/30/15	Robot *surgeons* are being engineered for medical applications
4/13/15	National Robotics Week has adapted a broader definition of robotics that includes autonomous vehicles
4/13/15	Students compete in team robot battles
4/16/15	An in-store robot makes your glasses while you wait
4/18/15	Job and salary prospects for robotics engineers are available
6/13/15	DARPA now holds a robotics competition
7/21/15	Researchers have developed the first 3-D printed robot
7/31/15	A Japanese hotel is staffed by robots
8/17/15	A tiny robotic *wrist* has the skill to perform brain surgery
9/16/15	A giant robot will build huge composite parts for Mars and other space exploration
10/12/15	The world's first robot-operated farm is being researched
11/3/15	Self-driving delivery robots are being designed
11/20/15	NASA has launched a humanoid robot design contest
12/1/15	ABB Robotics North America has introduced the largest industrial robot ever developed
1/12/16	Industrial robots can be controlled with human gestures

Goodheart-Willcox Publisher

Figure 14-8. *(Continued)*

Summary

- Advances in computer technology directly affect robotic applications, and the use of robots outside manufacturing continues to expand.
- The technology for fully automated, workerless factories exists today. However, implementing such technology is not easy.
- Major growth of robot use outside the industrial sector began in the 1990s with service robotics, which can do a wide variety of tasks.
- The field of telerobotics has produced robot arms and mobile robots that are operated by remote control.
- Tiny microbots can now perform applications such as medical, military, and hazardous operations.
- Artificial intelligence (AI) is the science of making machines do things that would require intelligence if done by a person.
- AI research can be broadly classified into three major interrelated areas: robotics, natural language processing, and expert systems.
- Robots and other forms of computer-controlled machinery will continue to impact society.
- The jobs eliminated due to automation are semiskilled or unskilled, while jobs that are created require a significant technical background and skills.
- The future of robotics may be highly dependent on combining computer capabilities and peripheral systems to develop machines capable of assisting in homes, offices, and public areas.
- Many robotic applications and innovations have been developed due to extensive and continuous research by individuals and organizations.
- A variety of educational opportunities are available to students and workers for training in robotics. Training may be provided by suppliers of robotic systems. Colleges and technical schools offer programs in robotics.

Review Questions

Answer the following questions using the information provided in this chapter.

1. Describe some characteristics of a fully automated factory.
2. What are some developments and advancements in robot technology that have made high degrees of automation possible?
3. What are four applications for service robotics outside the factory?
4. Define the term *telerobotics.*
5. What is an example of how military operations make use of robots?
6. Define the term *artificial intelligence (AI)* and give an example of its use in industry.
7. What are three interrelated areas of artificial intelligence?
8. What is an expert system? How does an expert system differ from regular computer programs?
9. What are two examples of applications for expert systems?
10. What is one major area in which the use of automation impacts workers?
11. What level of skill is typically needed in jobs eliminated due to automation, and what level of skill is required in jobs created due to automation?
12. What has resulted in the many new robotic applications and innovations that have emerged in the last few years?

13. Describe two developments or innovations in robotics that occurred in 2015 that you think are significant.
14. What are some sources for training and education in robotics?
15. What are some of the essential areas of study for a career in robotics?

Learning Extensions

1. Visit the websites of the robotics laboratories of Stanford University and Carnegie Mellon University. Review the current research projects at each of these universities. How many of the research projects relate to manufacturing and industrial automation? Briefly describe one of the projects.
2. Discuss one significant robotic application or innovation that occurred in each of the following time periods:
   - 2011 to 2012.
   - 2013 to 2014.
   - 2015 to the present.

Advanced Analysis

1. Research robotics technology as shown in movies. Produce a poster comparing and contrasting fictional robotics technology with historical advances in technology.
2. Perform further research on the societal impacts of robots, including potential future impacts. Write a paper discussing how future issues may be resolved.

Chapter 15

Designing and Constructing a Robot

Outline

Objectives

Upon completion of this chapter, you will be able to:

- Describe factors to consider in planning the design for a robot.
- Select appropriate materials to construct a model robot.
- Select and implement a motion control system for a model robot.
- Integrate a microcontroller system into a robotic system.
- Integrate sensors into a robotic system.
- Troubleshoot and repair faults within a robotic system.

Technical Terms

bootloader
breadboard
machine language
microcontroller
motor controller
pseudocode
pulse width modulation (PWM)

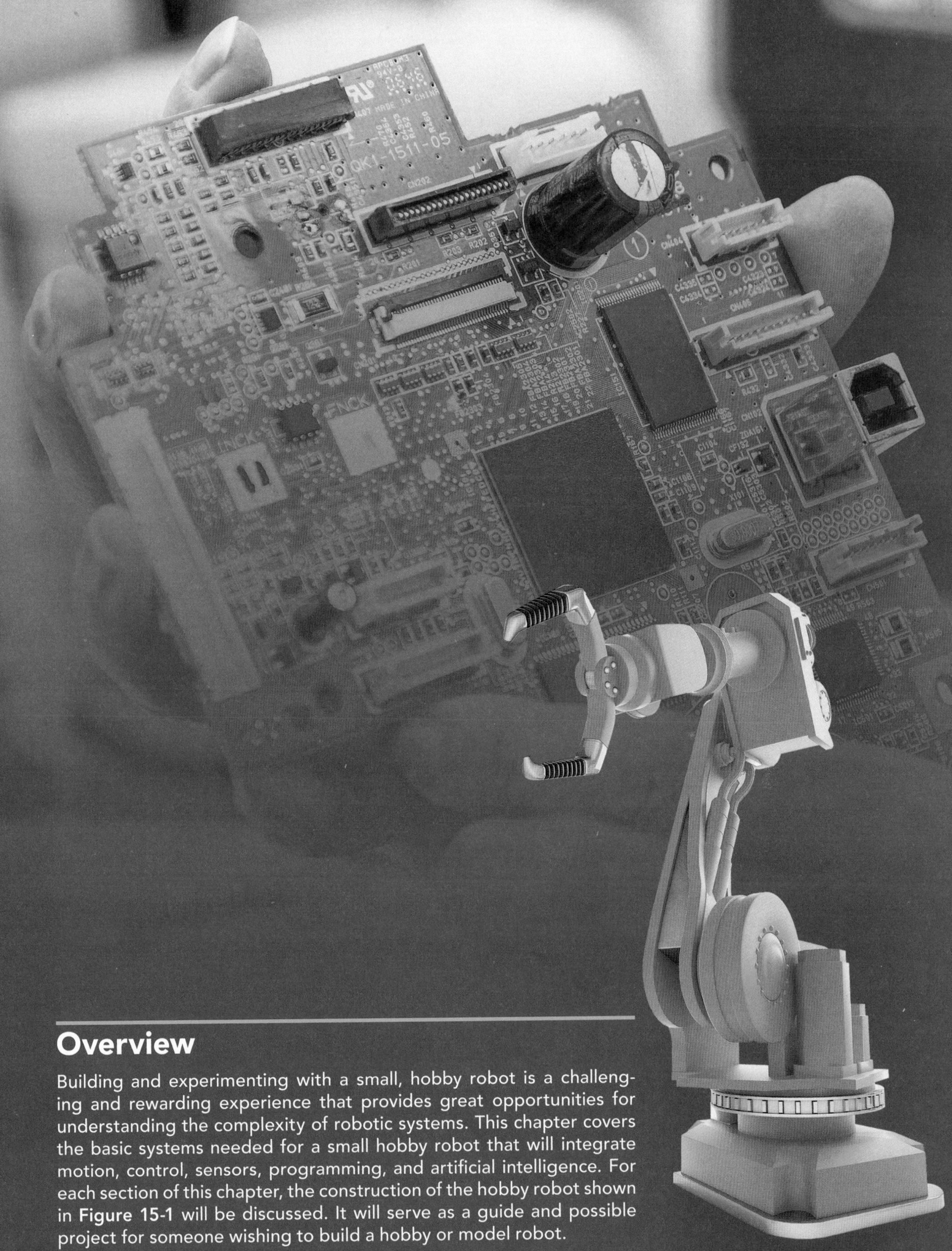

Overview

Building and experimenting with a small, hobby robot is a challenging and rewarding experience that provides great opportunities for understanding the complexity of robotic systems. This chapter covers the basic systems needed for a small hobby robot that will integrate motion, control, sensors, programming, and artificial intelligence. For each section of this chapter, the construction of the hobby robot shown in **Figure 15-1** will be discussed. It will serve as a guide and possible project for someone wishing to build a hobby or model robot.

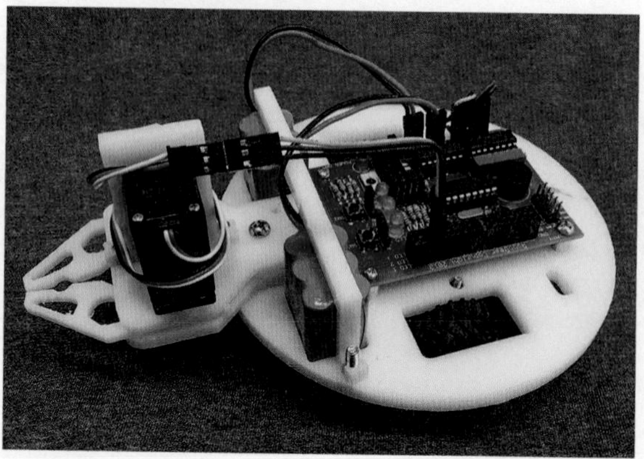

Goodheart-Willcox Publisher

Figure 15-1. A homebuilt, hobby robot made from high-density polyethylene (HDPE) plastic. A combination of homemade components and off-the-shelf parts have been incorporated into this design.

15.1 Robot Design

Before a robot can be built, a design for the robot must be planned. The first step in designing a robot is to determine the purpose, or mission, for the robot. For an industrial robot, the purpose might be to assemble components on an assembly line. A commercial robot might be used for search and rescue missions. A hobby or model robot might be created for the learning experience or used for friendly competitions. The purpose of the robot should drive the design and construction.

There will likely be other attributes of a robot that need to be considered in the design phase. Mobility is one such issue. Will the robot need to move to various locations or will it stay in one place? The cost of the robot and the funds available will also be factors. Components can be purchased off the shelf and ready to use, or components can be built. Accuracy, speed, and safety are other considerations. What type of accuracy is expected from this robot? How fast does the robot need to move? What types of safety issues should be considered? Will the robot be used in close proximity to people? Will the robot pose a risk to persons or property? Research on available technology and the costs of such technology should be carried out before any design work begins. A practical list of the requirements for building the robot should be created.

15.2 Materials Selection

When building a hobby robot, the materials selected are determined by the skill set of the builder, access to tools and machinery, and the purpose of the robot. There are many robotic kits that can be purchased that decrease the amount of work to be done. LEGO® has been producing robotic systems for many years, and the

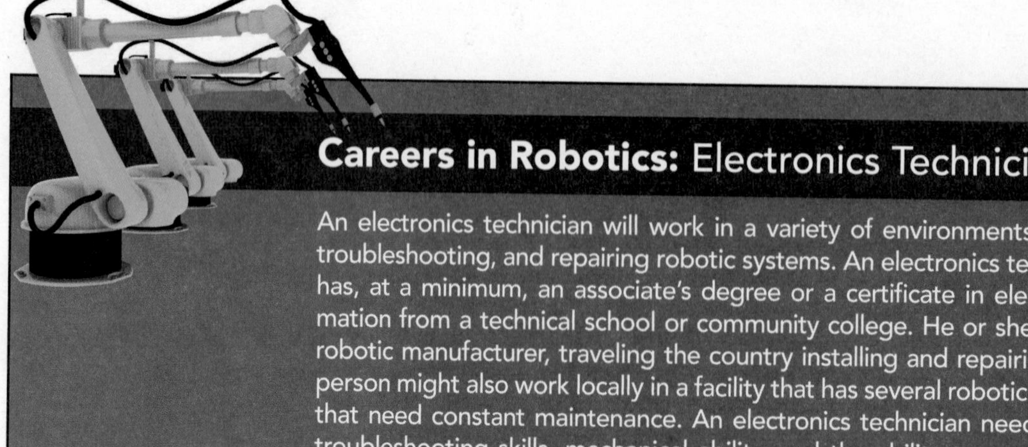

Careers in Robotics: Electronics Technician

An electronics technician will work in a variety of environments while installing, troubleshooting, and repairing robotic systems. An electronics technician typically has, at a minimum, an associate's degree or a certificate in electronics or automation from a technical school or community college. He or she may work for a robotic manufacturer, traveling the country installing and repairing systems. This person might also work locally in a facility that has several robotic systems in place that need constant maintenance. An electronics technician needs to have good troubleshooting skills, mechanical ability, and the ability to work in a variety of environmental conditions.

company has a strong presence in the elementary and middle schools. Carnegie Mellon University has developed a robotic system called VEX®, **Figure 15-2** and **Figure 15-3**, that has metal building components, motors, and sensors. They also have extensive educational support materials to go with their kits. Pololu produces several robotic platforms. Some, such as the 3pi, **Figure 15-4**, are already built and ready to program. Many sensors are built into the robot. The builder needs to decide how much fabrication he or she wishes to do. An honest assessment of the builder's current skills and access to resources is important before any construction begins.

15.2.1 Materials for the Base

Building a robot from scratch can sometimes be more affordable than other options, and it certainly offers a much greater challenge to the builder. The most common materials used for building robots are cardboard, wood, plastic, and metal.

Cardboard is a building material that can be easy to use. It can be cut easily, can be held together with glue or tape, and is cheap enough to throw away if mistakes are made. Scissors or a small hobby knife can be used to cut and trim cardboard.

VEX Robotics, Inc.

Figure 15-2. Shown are components of the VEX Robotics kit.

VEX Robotics, Inc.

Figure 15-3. This robot was created using the VEX Robotics kit robot.

Pololu Robotics

Figure 15-4. The 3pi robot is built and ready to program.

Wood is probably the next most likely material choice for use in building a hobby robot. Wood is easy to cut with basic hand or power tools. Plywood or pine are relatively inexpensive, lightweight, and strong materials. Wood can be drilled easily, and components can be mounted directly to it. Wood does not conduct electricity, so there is no concern of short-circuiting electronics that are mounted to it. Wood can be glued, screwed, or nailed together, offering many options for construction.

Plastic makes a good building material, although it is more expensive than wood. Plastic is typically just as easy to cut as wood. There are many types of plastic. High-density polyethylene (HDPE) works well for robotics. HDPE is strong and impact resistant. It can be cut with woodworking tools and can be drilled and tapped for threads just like metal. Plastic (such as HDPE) typically is not locally available and must be ordered. Some hardware and building supply stores carry acrylic. However, acrylic does not process as well as HDPE.

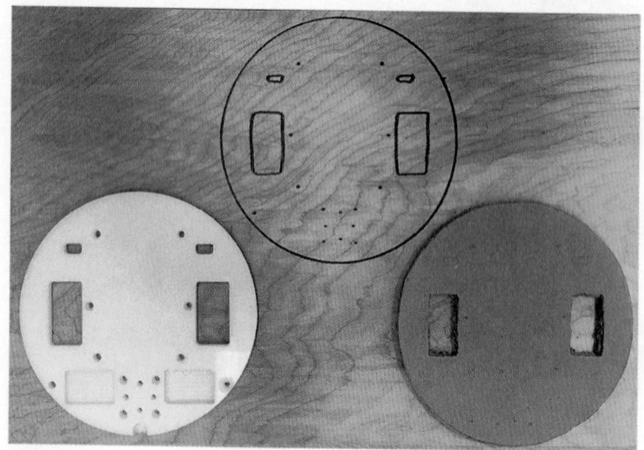

Goodheart-Willcox Publisher

Figure 15-5. The same robot base made from three different building materials: wood, cardboard, and plastic. The wood base is shown laid out and ready to cut.

Metal is the strongest option for use in building robots, but it is the most costly and the most difficult to work with. Using metal requires special metal cutting tools. Metal can be welded, brazed, riveted, or bolted together. There are many metal building systems available, such as Erector or Tetrix, that work well for robotic systems as well as the many systems already mentioned above.

Fiberglass is another option for material used to build a robot. However, it can release toxic fumes, is messy, and typically requires a mold to form the cloth around. Fiberglass might work well for finished products. Typically, however, it does not work as well as a prototype material.

Similar 5" circular robot bases made from cardboard, plastic, and wood are shown in **Figure 15-5**. The plastic base shown was cut using a CNC router. The cardboard robot base was cut using scissors, and the wooden base shown (laid out ready to cut) can be cut with a band saw, scroll saw, or small hand saw with a thin blade. Screw holes are drilled with a hand drill or drill press.

15.2.2 Base Design

The robot design presented in **Figure 15-6** uses a 1/4″ HDPE plastic base that is 5" in diameter. (A base made of 1/4″ plywood can be substituted as well.) Precut plastic bases are available from several vendors for minimal cost. This small, round, roving robot will be used to explain the building process and required systems presented in this chapter. This design may be modified to suit the builder's requirements, budget, skill level, and available resources.

Goodheart-Willcox Publisher

Figure 15-6. Two samples of robot bases cut from HDPE plastic. Note the rounded corners on the square base. These bases were cut using a CNC router, another type of robotics system.

A round base prevents the robot from catching a square corner and getting stuck. If a square base is used, rounding the corners helps avoid some hang-ups on objects, **Figure 15-6**. This robot uses two powered wheels and one caster for balance. This commonly used tank steering design is simple and effective. Tank steering allows the robot to turn by controlling the speed and direction of the rotation of the motors, rather than using a more complex rack-and-pinion steering system, such as an automobile might have. With a tank steering system, forward motion is achieved by both motors rotating forward. Turns are accomplished by one motor rotating forward and one rotating in reverse. Reverse travel is achieved by rotating both motors in reverse. Tank steering simplifies control to three basic motor states: off, forward, and reverse. Motor selection and motor control are discussed next.

15.3 Motion and Motion Control

A robot needs a way to move. The builder must select a type of motor and wheels that are appropriate for the robot. A motor control unit and power are also required.

15.3.1 Motors

Electric motors are one of the most common choices for providing motion for a hobby robot. Small toy dc motors, **Figure 15-7**, are cheap and easy to work with, but they do present some difficulties. They create a lot of electromagnetic interference, which can cause problems with microprocessors that share the same power supply. These interference issues can be addressed, but the builder should be aware of the limitations. Remote control (RC) servo motors, **Figure 15-8**, are another inexpensive option. RC servos typically only have 180° of rotation. However, they can be modified for continuous rotation. RC servos also use dc motors, but they have the added benefit of a built-in gearbox. Stepper motors, **Figure 15-9**,

Goodheart-Willcox Publisher

Figure 15-7. A simple dc toy motor. Electrical connections are soldered onto the small metal tabs located on the right side of the image.

Pololu Robotics

Figure 15-8. Remote control (RC) servo.

Pololu Robotics

Figure 15-9. A stepper motor is more complex than a simple dc motor.

work well; however, they require a more advanced control system than a simple dc motor. Servo motors with encoders, **Figure 15-10**, are another option. However, they are more costly and more difficult to control than a stepper or RC servo motor. The example used in this chapter uses two small 6V dc toy motors and a plastic off-the-shelf gearbox, **Figure 15-11**.

Pololu Robotics

Figure 15-10. Servo motors with encoders are more complex than dc motors or stepper motors.

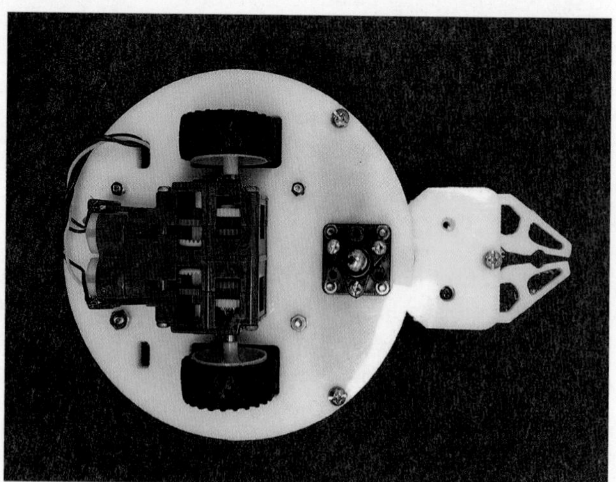

Goodheart-Willcox Publisher

Figure 15-11. In this figure, a two-motor gearbox with tank steering is shown. A ball caster is used for balance. Note that the wheels are located on the centerline of the robot. This facilitates tank steering and allows the robot to turn within its own diameter. Gearboxes are required in order to covert speed into torque.

15.3.2 Wheels

Wheels are an important consideration in any robot design. The type of motor to be used, as well as the type of terrain to be covered, determines the best choice of wheels. A robot that is used indoors typically uses hard plastic, soft foam, or rubber wheels. A robot that is used outdoors may be better served with inflatable rubber wheels or tank treads. Simple plastic wheels are often good, inexpensive solutions for the novice builder, **Figure 15-12**. The sample robot uses two small, hard plastic wheels that press fit directly onto the shafts on the plastic gearbox.

Wheels cannot be connected directly to a motor shaft because motors typically output a lot of speed, but have very little torque (twisting force). Having a gearbox between the motor and the wheels is required on most robots. The example shown in **Figure 15-11** uses two small toy dc motors and a plastic gearbox. Although the examples shown have a plastic robot base, cardboard or wood can be substituted.

Goodheart-Willcox Publisher

Figure 15-12. Many off-the-shelf wheels will work well for the hobby robot. Pictured on the left side in white is a homemade plastic wheel made from HDPE plastic machined on a CNC router. Model remote control cars and airplane wheels that are available from hobby stores also work well.

15.3.3 Motor Control

Goodheart-Willcox Publisher

Figure 15-13. SN754410 IC chip.

A basic motor control circuit can be used to control small dc motors, making use of a SN754410 motor controller integrated circuit (IC) chip, **Figure 15-13**. A ***motor controller*** is a small IC chip that allows control of the direction of rotation of two motors independently, as well as enabling and disabling each motor's rotation. A ***microcontroller*** is a small computer on a single integrated circuit (IC) chip that is the brain of this robotic system. It only needs to send an enable/disable signal and a direction signal to the motor controller. Starting, stopping, turning left, turning right, forward, or reverse travel is accomplished with these four simple inputs (motor one direction and enable and motor two direction and enable). The wire diagram, **Figure 15-14,** shows that there are 16 pins on the SN754410. Pins 4, 5, 12, and 13 all get connected to ground. Pins 3 and 6 connect to motor one, and pins 11 and 14 connect to motor two. Pins 1 and 2 are the enable and direction (respectively) inputs for motor one. Pins 9 and 10 are the enable and direction (respectively) inputs for motor two. Pin 8 is motor voltage, and pin 16 is 5V logic power. A 7404 hex inverter is connected between pins 2 and 7, and another is connected between pins 10 and 15. Pin 7 is the ground, and 14 is the power on the 7404 hex inverter. The inverter input is pin 1, and the output is pin 2. The second inverter input is pin 3, and its output is pin 4, as shown in the schematic, **Figure 15-14**. That completes the motor control circuit.

A sample of this circuit on a solderless breadboard can be found in **Figure 15-15.** A ***breadboard*** is a construction base used for building samples or models of electronics. Building and testing this circuit should be the first electronics task completed when building a robot. Once complete, the circuit can be attached to the motor and gearbox assembly. A prototype robot may have all its circuits constructed (plugged in) on solderless breadboards, **Figure 15-16**. If the builder wishes, circuits can be soldered to a permanent soldered breadboard, **Figure 15-17**. There are also circuit board design software programs, such as EAGLE PCB, that

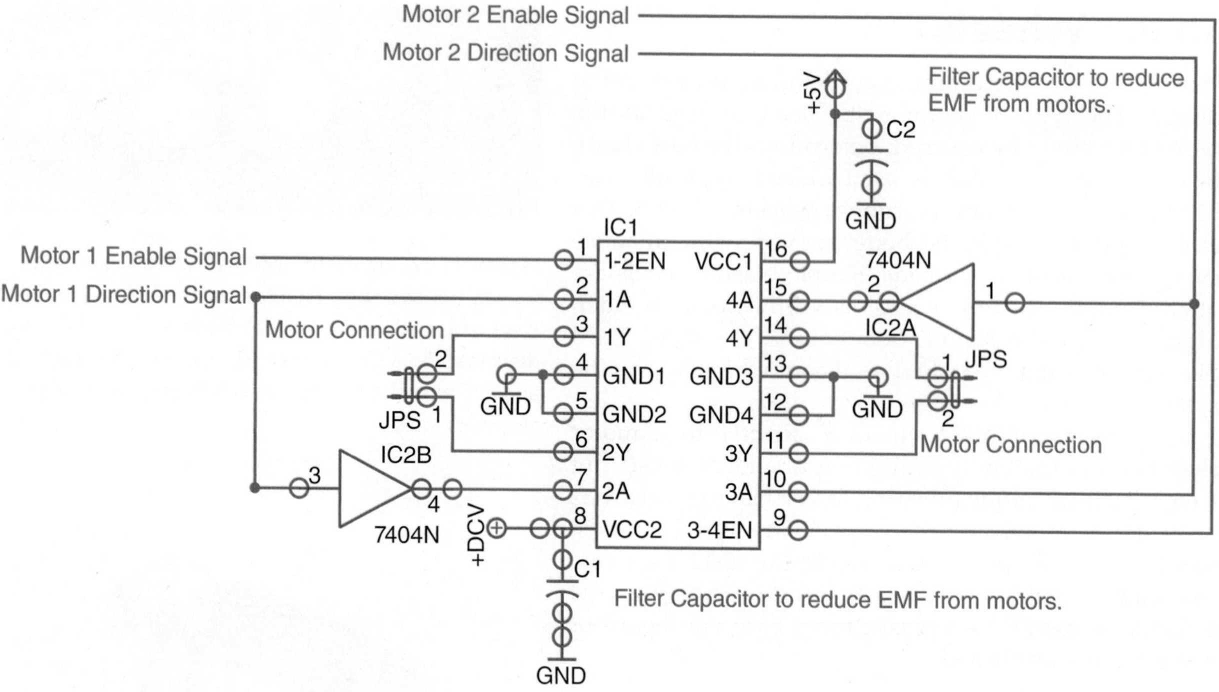

Goodheart-Willcox Publisher

Figure 15-14. SN754410 motor driver IC wiring diagram.

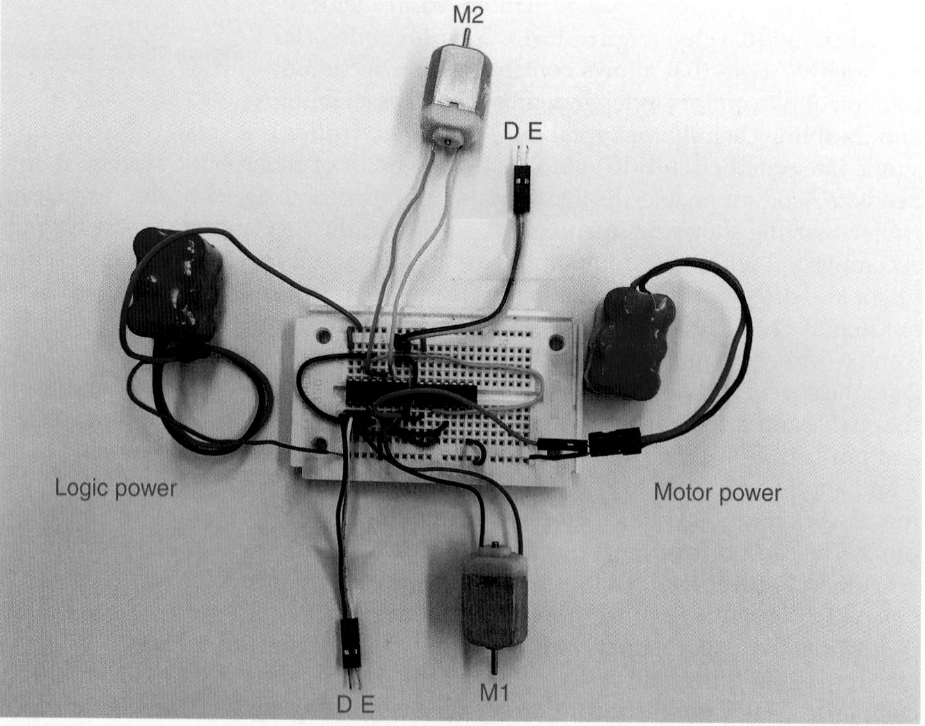

Goodheart-Willcox Publisher

Figure 15-15. A motor control circuit shown on a solderless breadboard. Prototyping new designs on a breadboard is a quick and easy way to test new ideas before assembling using permanent methods, such as soldering parts to a printed circuit board.

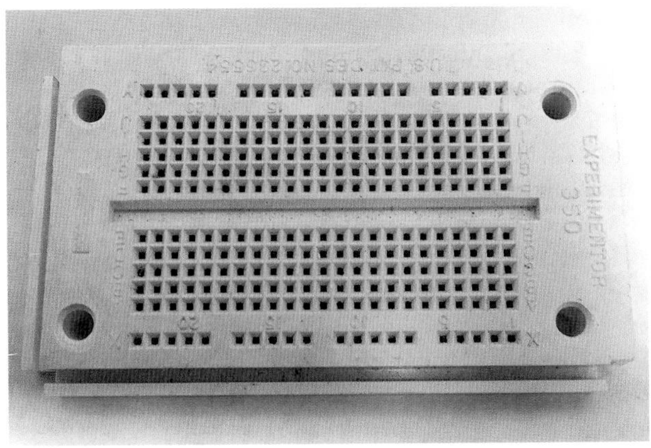

Goodheart-Willcox Publisher

Figure 15-16. Solderless breadboard.

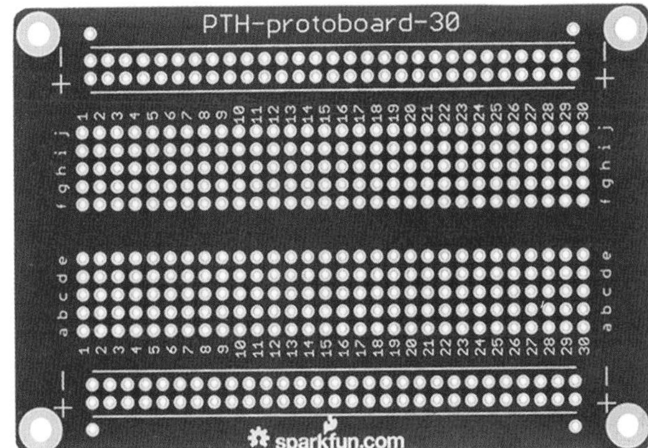

Sparkfun Electronics, Inc.

Figure 15-17. Soldered breadboard.

can be used to design and build professional-looking circuit boards, **Figure 15-18**. CAD files from a program such as EAGLE can be uploaded to a PCB printing house that will fabricate the design and mail a completed board to you for a minimal cost. Components soldered onto a professional manufactured PCB look nice and typically work better than breadboard circuits. Time and expense, however, limit their use for the hobbyist.

Motor power should be kept separate from logic power because the interference from the motors can cause issues with the digital components on the robot, especially the microcontroller. The term *digital* describes data or signals in a format that occurs in separate full units, such as 1 or 0, or devices that use this type of data, such as a computer. The best power solution is to have two separate battery packs. One battery pack powers the motors; the other battery powers the digital electronics.

Goodheart-Willcox Publisher

Figure 15-18. Homemade circuit board.

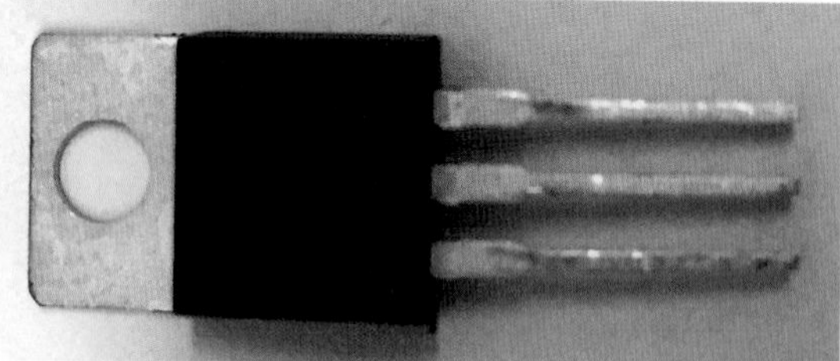

Goodheart-Willcox Publisher

Figure 15-19. A 5V regulator chip.

The two power supplies can share a common ground, and in most cases, a common ground is necessary in order for the separate systems to function properly. Digital circuitry will typically require 5V dc power, although some circuits require 3.3V dc. A 5V regulator chip, **Figure 15-19**, will convert a voltage higher than 5V into a steady 5V. Most batteries output 1.5V. A battery pack that holds four AA batteries, for example, will output 6V. This 6V output fed into a 5V regulator will provide digital logic power. Another battery pack of similar size can then power the motors. The negative side of both battery packs should be connected together, creating a common ground. This separation of motor power and logic power eliminates most interference issues. In **Figure 15-1**, the two separate green battery packs are visible on the left side of the image under the white battery clamp. Red and black wires come from each battery pack and connect to the robot control board. The ground (negative terminal) of each battery pack is electrically connected on the circuit board. The positive power terminals are kept separate.

Remote control (RC) servo motors are useful for part manipulators, such as grippers or claws. A claw controlled by an RC servo can be seen in **Figure 15-1**. This servo has a unique method of control. RC servos typically are controlled with a transmitter and receiver. The servo plugs into a receiver, and a transmitter sends signals to the receiver. The receiver interprets those signals and drives the servo. On a robot, however, the transmitter and receiver circuit are typically omitted if the robot is going to be autonomously controlled (controlled by a computer, not a person). The computer, or microcontroller, therefore needs to generate the same signals a transmitter and receivers would have generated. A servo has three wires: one for power (typically red), one for ground (typically black), and one for a signal (typically white but may be any color). On most RC servos, the middle wire is power, but this may vary from manufacturer to manufacturer. Check the data sheet with the servo used to be sure of the wiring. A servo expects to see a pulsating dc signal on its signal wire. The signal pulses on and off with a period of 20 milliseconds (0.020 seconds), **Figure 15-20**. During the 20 ms (millisecond) period, the signal is off for most of that time and on for 1–2 ms. A 1 ms on (5V) pulse moves the servo to one limit (either clockwise or counterclockwise), and a 2 ms on pulse moves it to its opposite limit. A 1.5 ms pulse centers the servo. These pulses occur 50 times per second (50 Hertz). To demonstrate what this might look like from a programming perspective, pseudocode will be used. ***Pseudocode*** is simple language used to describe instructions and actions that will be part of a computer program; it is not a programming language. These instructions are written with the assumption that a servo's signal wire is connected to pin number 1 of a microcontroller and the servo's power and ground wires are connected to 5V power and ground. Everything after the // characters is a comment and not part of the code.

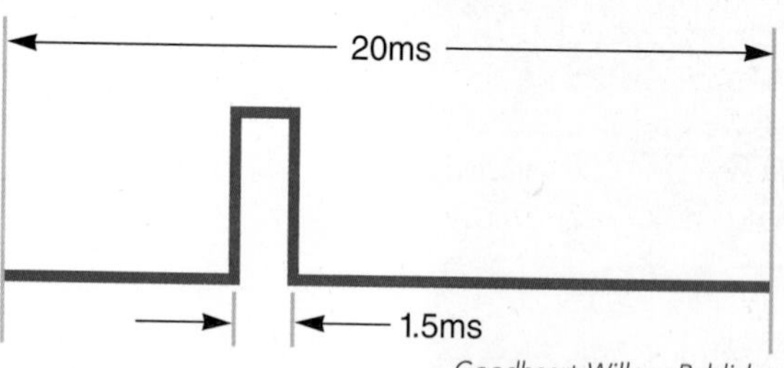

Goodheart-Willcox Publisher

Figure 15-20. A 50 Hz RC servo position signal showing a 1.5 ms on pulse and a 20 ms period.

Line 1: Pin 1 high // Send a 5V signal out of pin 1.
Line 2: Wait 1 ms // Do nothing for 1 ms. This time delay controls the pulse width.
Line 3: Pin 1 low // Send a 0V (ground) signal out of pin 1.
Line 4: Wait 19 ms // Do nothing for 19 ms.
Line 5: Loop // Repeat this code forever.

The electrical output of pin 1 of this microcontroller might look similar to **Figure 15-21**. The actual syntax of the code will be different from one programming language to the next, but the concepts are the same. Varying the time delays will change the width of the pulse. A longer time delay on line 2 will increase the pulse width, and a shorter time delay will decrease the pulse width. If line 2 is changed, line 4 will need to be adjusted so that the total period of the signal adds up to 20 ms. For example, a delay of 2 ms on line two will require a delay of 18 ms on line 4. A delay of 1.5 ms on line two will require a delay of 18.5 ms on line 4.

As the last example shows, it is difficult to discuss motion control without understanding the brain of the system, a microcontroller. The next section discusses microcontrollers as a means of controlling robotic systems.

15.4 Intelligence

The brain of the robot will typically come from a computer in the form of a microcontroller. A microcontroller is a microprocessor with some added electronics that allows for easy input/output of signals and data, all located on a single integrated circuit (IC) chip. Most microcontrollers will have several digital input/output lines that can be used as either inputs or outputs, and several analog inputs. The term *analog* refers to data or signals that vary continuously or devices that use this type of data or signal.

15.4.1 Pulse Width Modulation

Pulse width modulation (PWM) is a method of digitally encoding analog signal levels. If the microcontroller has PWM, then the PWM outputs can be used to control the speed of a motor. Pulse width modulation is an ingenious way to control motor speed without wasting power. An easy (but less effective) way to control the speed of a motor is to use a variable resistor (similar to a dimmer switch on a light) to limit the power to the motor, thus controlling the speed. The problem, however, is that the resistor will waste some of the power that would have gone to the motor in an effort to regulate speed. The wasted energy is lost as heat, limiting the life of the batteries used for the robot. PWM circuitry, on the other hand, generates a pulsating square waveform. Basically, it turns the motor off and on. The amount of time the motor is off compared to the amount of time it is on (the pulse width) controls the speed. No energy is wasted, and speed is controlled. The turning on and off of the motor happens so fast it is not noticeable. The motor is coasting during

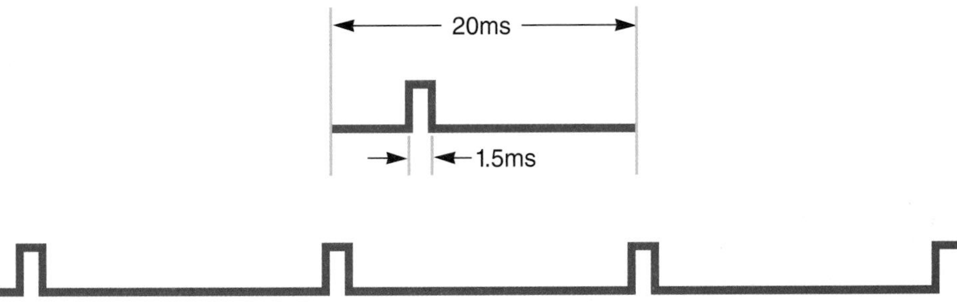

Goodheart-Willcox Publisher

Figure 15-21. A stream of 50 Hz RC servo position signals.

the off cycles, so it never stops rotating. This same process is used with RC servos as described earlier. Servos, however, use PWM to control the position of a motor shaft. Here, PWM is being used as a form of speed control.

15.4.2 Machine Language

A microcontroller must be programmed in order to perform specific tasks. Program code is written on a computer and then compiled and downloaded to the microcontroller. A high-level programming language, such as BASIC or C, is used. Then a compiler converts the code into machine language. ***Machine language*** is a code consisting of binary (1s and 0s) instructions that a microprocessor can run or respond to directly. At one point in time, all code was written directly in machine language. The process of coding in machine language is slow, and mistakes are easily made using this method. High-level programming languages increase the speed of programming and use code that is closer to spoken language.

15.4.3 Microcontroller Boards

There are many microcontrollers to choose from, such the PIC® and ATMEL® lines of microcontrollers. Several companies have used these or similar microcontrollers to produce microcontroller boards. These boards hold the microcontroller chip and other components that work with the microcontroller chip. This makes the microcontroller easier to program and interface with motors, sensors, or other input/output devices. The advantage of the microcontroller board is that it typically only requires a power source and a programming cable or programmer, **Figure 15-22**. Many microcontroller boards only need a USB cable in order to download code. With some microcontroller boards, the microcontroller chip must be removed from the board for programming. Removing the chip every time it needs to be programmed will quickly become an annoyance.

Many manufacturers have developed microcontroller boards that combine easy input/output connections, battery connection points, power switches, LEDs, and switches for testing and debugging program software. Some even include small solderless breadboards, LCDs (liquid crystal displays), and motor drivers. It is highly recommend that a microcontroller board with built-in circuity that allows for direct connection to a computer be used. The Arduino®, Pololu, or Parallax line of microcontroller boards are commonly used in robot design. All these have built-in support for programming, most requiring only a USB cable.

Microcontroller boards, such as the Arduino, have a bootloader installed on the microcontroller board. A ***bootloader*** is a small program that helps transfer program code from a computer to the microcontroller without the need for removing the microcontroller chip from the board. The Arduino boards often have the hardware already installed on them to allow connection between the computer and the microcontroller board via a USB cable. Program code is written on a computer in the Arduino programming environment. Then it is uploaded to the microcontroller via a USB port on the board, **Figure 15-23**, directly from the Arduino environment.

Goodheart-Willcox Publisher

Figure 15-22. A microcontroller board.

There are too many different microcontrollers and programming environments to discuss them all here. The web-based support for many of these environments is very good. A review of what is available should be made, and then the best option should be selected.

Each microcontroller board has advantages and disadvantages, and several will be discussed below. One of the most popular microcontroller boards is the Arduino, **Figure 15-24**. Arduino was invented by a group of electrical and software engineers in Italy using an ATMEL microcontroller. The developers of Arduino made their product open source, meaning the code and the hardware designs are free for anyone to use. While a ready-to-use board from Arduino can be purchased, many Arduino clones are available, or the hobbyist can make one at home. (A clone is a board made by someone other than Arduino, that is built using the same components, and contains the Arduino bootloader.) The open-source environment encourages builders to innovate and be creative. Source code and schematics for the Arduino are available online for download. An Internet search will reveal a wide range of resources. A few shortcomings of the Arduino are that it cannot control a motor directly (unless it is a very low current motor); it must be programmed in C, which can be a difficult language to learn for the first-time programmer; and a design error on the original board (an unusual pin spacing) has been replicated on the Arduino line and third-party boards that connect to the Arduino. While not a major design error, builders using the Arduino may find the pin spacing awkward at times.

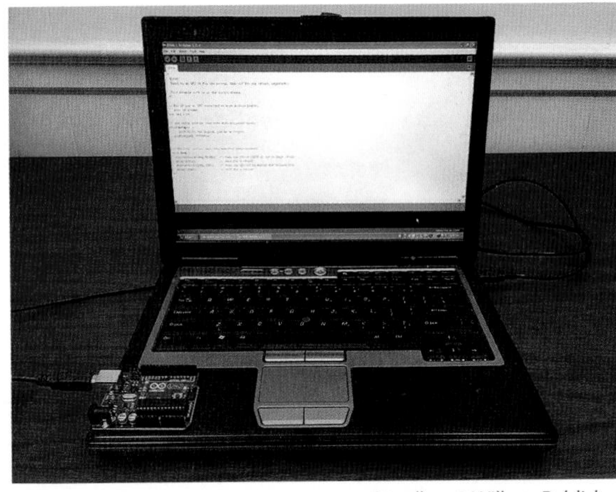

Goodheart-Willcox Publisher

Figure 15-23. An Arduino microcontroller board hooked up to a laptop ready to be programmed. The Arduino programming environment is visible on the computer screen. It does not take a powerful computer to program microcontrollers, as is evident by the older laptop used here.

Parallax, **Figure 15-25**, has been selling microcontroller boards for a long time, and the company has a strong presence in the education market. Their platform, built around the BASIC Stamp®, is a small microcontroller board that can be used for a variety of purposes including robotic control. Parallax offers a whole line of products and continues to innovate and adapt to new technologies. Several complete kits that include books and experiments are available from Parallax and make a complete learning system. The biggest drawback is that the kits can be a

Goodheart-Willcox Publisher

Figure 15-24. Arduino microcontroller board.

Goodheart-Willcox Publisher

Figure 15-25. Parallax BASIC Stamp® HomeWork board.

little expensive. There is less support for them now that Arduino has taken over so much of the robotics market.

Pololu, a robotics company in the United States, sells a variety of microcontroller boards, sensors, and other robotic products. The Orangutan line offers an all-in-one solution with built-in motor drivers, LCD display, and test switches, and LEDs built into the microcontroller board. The downside of the Orangutan line is that the programmer needs to have a programming environment to write code and download to the controller. Unlike the Arduino, Pololu does not have a programming environment. The learning curve is a little steeper. The user must find, download, and install a program environment that will be compatible. In response to this, their latest model, the A-Star line, **Figure 15-26**, is Arduino compatible. This means it comes with the Arduino bootloader preloaded on the microcontroller, and it can communicate with the Arduino PC programmer software. The items in the A-star line from Pololu are all Arduino compatible, and their high-end controllers will connect to a Raspberry Pi.

The Raspberry Pi, **Figure 15-27**, was developed in Wales by the Raspberry Pi Foundation as a means of supplying a low-cost computer for home and educational use. The Raspberry Pi board typically costs around $30 and can perform word processing, video graphics, and many other basic computer functions. The Raspberry Pi combined with a microcontroller board, such as the A-Star, makes a very powerful computing and robotic platform. The ability to do physical computing (that involves hardware and software) is one of the key features of the Raspberry Pi.

Most microcontrollers cannot control a motor directly. A microcontroller output is typically limited to 5V dc at very low current. In order for a microcontroller to be used to control motors, it needs to have a motor control circuit added to it. Some microcontroller boards, such as the Pololu line, **Figure 15-22**, have built-in motor controllers. For the microcontroller boards that do not have motor drivers, they can be added. These can be built on a small circuit board or purchased. One option is the SN754410, **Figure 15-13**, described earlier. Other off-the-shelf motor controllers can also be used.

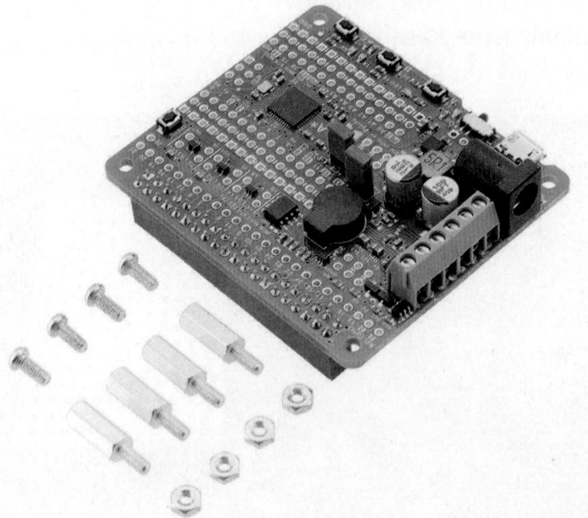

Pololu Robotics

Figure 15-26. Pololu add-on board for the Raspberry Pi. Add-on parts allow the Raspberry Pi to interface with and control motors and sensors.

Pololu Robotics

Figure 15-27. Raspberry Pi mini-computer board.

The robot shown in **Figure 15-1** uses a homemade microcontroller board. This board was designed using the EAGLE PCB design software and created by a company that does printed circuit board fabrication. Three of these boards were printed for about $30 US ($10 each). The components to fill the board total about $10 for each board. Considerable time went into prototyping and testing the design, drawing, and having the circuit board produced. Buying an off-the-shelf control board is considerably easier and will save a lot of time. The learning experience, however, can be of significant value for someone trying to learn about robotics and electronics.

15.4.4 Testing

Once a microcontroller, motor controller, and motors are connected to a robotic base, test programs can be run to bring the robot to life. It is best to start with simple control programs that will turn on an LED (light emitting diode) or will turn motors on and off at constant intervals. These simple tests will confirm that all components are functioning, that software is loaded and running correctly, and that there are no wiring errors. Switches can be used to test sensor inputs. LEDs are good for checking digital outputs. Sensors will be discussed further later in this chapter.

15.4.5 Programming

Figure 15-28 shows a hobby robot that is ready to be programmed for the first time. The microcontroller board has a connector that allows it to be programmed without removing the microcontroller chip from the board. The small black plastic box, **Figure 15-29**, is called a programmer. (Not to be confused with a person who writes computer code, also called a programmer.) The programmer shown here moves the text file containing program code on the computer to the memory on the microcontroller. Microcontrollers use a type of memory called flash memory. Flash memory is erasable, but it will store information even if power is lost. Most flash memory can store information for about 10 years or longer, although

Goodheart-Willcox Publisher

Figure 15-28. A homebuilt robot ready for programming. Notice the programmer and cable between the computer and the robot.

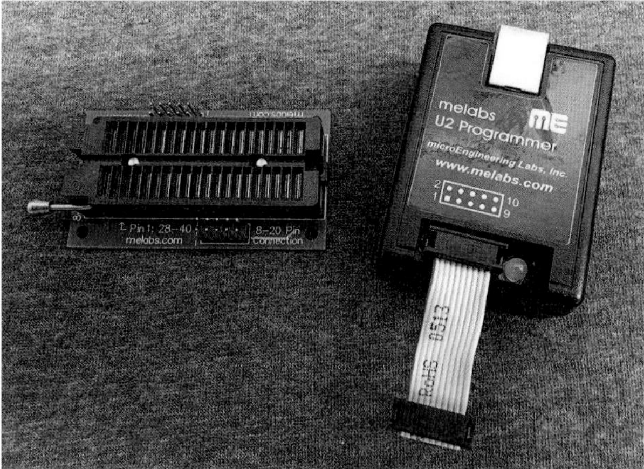

ME Labs, Inc.

Figure 15-29. A Micro Engineering Labs programmer. The green and black attachment is for programming chips outside of a circuit board. The chip holder can accommodate microcontroller chips of many different sizes.

this technology is constantly changing. The amount of times data is written and erased, as well as the environment the electronics are stored in, can affect the length of time data will last.

Technically, program code on a computer is called software, and program code on an integrated circuit (IC) chip or microcontroller is called firmware. Software can be edited easily, but firmware is somewhat more permanent. Firmware can be replaced, edited, or deleted but not as easily as software on a personal computer. Hardware consists of the physical parts of a computer.

A microcontroller with a bootloader program does not need a programmer because it can connect to the computer with a USB cable via the microcontroller board. The bootloader program helps move the files from the computer to the microcontroller memory. The downside of using a bootloader is that the bootloader takes up program memory space and the bootloader program has to run for a few moments (sometimes as much as a few seconds) before the robot's program code can begin to execute. The bootloader program waits and monitors the USB port to see if there is any new code to download. If it does not detect anything, it begins executing the robot's program code. The Arduino in **Figure 15-24** is an example of a microcontroller with a bootloader program installed. The Arduino programming environment is shown on the laptop screen, and the Arduino is pictured connected to the laptop via a USB programming cable. The Arduino does not need a programmer to install code onto its microcontroller. Be cautious when using a programmer. If a programmer is used to program a microcontroller, all code in the microcontroller's memory will be erased, including any bootloader program that might be installed.

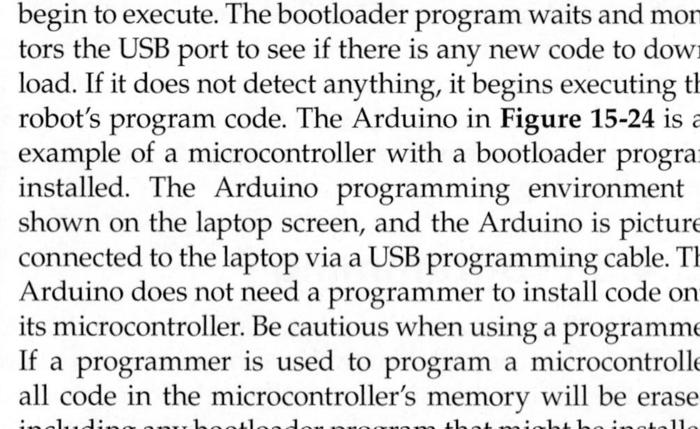

Goodheart-Willcox Publisher

Figure 15-30. A homebuilt robot ready to navigate a maze. This robot uses time delays and direction commands to navigate. Time delays are fine for beginner experiments; however, using sensors will make a much more versatile robot.

Once the microcontroller and motor control circuit of the robot are tested and are working, more complex experiments can be conducted. Making the robot navigate a simple maze (**Figure 15-30**) using time delays, and giving directional control inputs is a good way to start to learn the programming aspect of robotic control. This is also a good way to see the limitations of time delays as a means of motor control.

15.5 Sensors

The next phase of building a hobby robot is to incorporate sensors. Sensors allow the robot to interact with the world around it. On a basic hobby robot, typical sensors include light sensors (in the form of photoresistors, phototransistors, or infrared sensors), touch sensors, and sound sensors (in the form of microphones). These sensors can be built with very inexpensive items, and they can make a robot very versatile. There are sensors to detect smoke, gas (**Figure 15-31**) heat, pressure, temperature, radiation, salinity, humidity (**Figure 15-32**) and many other values in the natural and manmade world. Regardless of the sensor, they all produce the same basic data, either a digital

Pololu Robotics

Figure 15-31. A gas sensor.

logic high/low or an analog value (a range of numbers). Once an understanding of digital and analog inputs is achieved, any type of sensor can be added and integrated. A simple robot can do many things with just switches for touch sensing, phototransistors or photoresistors for light sensing, and microphones for sound sensing.

Sparkfun Electronics, Inc.

Figure 15-32. A humidity sensor.

15.5.1 Touch Sensors

The simplest and most commonly used sensor is the touch sensor. Touch sensors, also called switches, help a robot interact with the environment. When the switch bumps into an object, the switch closes a set of electrical contacts. Those contacts send current to an input on the microcontroller. The microcontroller senses the change in logic state (high to low or low to high) and executes some code based on the program. There are many types of touch sensors; a few examples are shown in **Figure 15-33**.

A sensor will output a digital value. Most robotic controllers will use either 5V dc or 3.3V dc for a digital logic high and 0V dc (ground) for a digital logic low. Sensors that output only these two values are considered digital sensors. Some more advanced sensors may be able to encode data into a stream of digital values and transmit this data to the microcontroller. In these examples, however, a single binary digit (bit) will be used. The pseudocode below demonstrates how a switch might be used. This code assumes that a switch is connected to pin 1. Pin 2 and pin 3 output direction signals to a motor controller. Pin 2 controls the direction of motor 1, and pin 3 controls the direction of motor 2.

```
Line 1: Start:
Line 2: Pin 2 high // Motor 1 forward. (Move the robot forward.)
Line 3: Pin 3 high // Motor 2 forward. (Move the robot forward.)
Line 4: If "Pin 1 high" then // The switch was pressed, do something.
Line 5: Pin 2 high // Motor 1 forward. (This line is redundant. This motor is
        already moving forward, but the note was placed here for clarity.)
Line 6: Pin 3 low // Motor 2 reverse. (This makes the robot turn.)
Line 7: Wait 500 ms // Keep the robot in this turn for half a second.
Line 8: End If // This is the end of this section of code.
Line 9: GOTO Start // Return to the start (line 1) and do it again.
```

Pololu Robotics

Figure 15-33. This figure shows several touch sensors.

With the code sample on the previous page loaded onto the sample robot, the robot will move forward until the switch is pressed. Then it will turn for 500 ms and move forward again. Every time the robot bumps into something, it will then try to turn. It will then go forward again. This robot will drive around looking a little bit like a bumper car, bouncing off objects it hits. (Assuming the switch is located in front of the robot so that it gets pressed when the robot crashes into a solid object.) If the switch has not been pressed when the microcontroller gets to line 4, the microcontroller will skip to line 9 with no changes in motor direction.

Switches allow sensing of position, or let a robot know when it has hit a barrier. The negative aspect of using limit switches is that the switch physically has to touch an object to detect it. In some applications, touching an object might not be ideal. Imagine a robot that detects land mines, explosives, or fire. It would be better to know the location of such dangers before contact was made.

15.5.2 Proximity Sensors

Some sensors will output a range of values. These sensors are called analog sensors. Sensors that measure light or sound are good examples of analog sensors. Typically, an analog signal will range from 0V to 5V. The microcontroller will have a circuit called a DAC (digital to analog converter). The digital equivalent of the analog value will become a binary number between 0 and 255. (Depending on the microcontroller, this number may be larger or smaller.) This number will be stored in a register inside the microcontroller. A program developed by the robot builder can compare this value to a constant value or variable in the software and make decisions. A proximity sensor is an example of an analog sensor.

Proximity sensors, **Figure 15-34**, can detect objects from several inches away using high-frequency sound. By measuring the time it takes sound to travel from the sensor to an object, bounce off and return to the sensor, a distance can be calculated. These sensors usually supply either an analog value or a digital value back to the microcontroller. The value can then be interpreted by the program code and a decision made as to how the robot should respond.

Pololu Robotics

Figure 15-34. A proximity sensor.

15.5.3 Light Sensors

Light sensing can be accomplished with photoresistors or phototransistors. A phototransistor, **Figure 15-35**, will act much like a switch. The phototransistor will output a high or low value, and that value will change when the light hitting the transistor passes a threshold. A photoresistor, **Figure 15-36**, will provide an analog value. A photoresistor will change resistance as the amount of light striking it changes. A simple circuit using a photoresistor is shown in **Figure 15-37**. The signal (voltage) between the photoresistor and the fixed resistor will change as the light striking the resistor changes. This signal can be fed into the analog input on the microcontroller, and a threshold value can be set in the software. Two

Goodheart-Willcox Publisher

Figure 15-35. A phototransistor.

Goodheart-Willcox Publisher

Figure 15-36. A photoresistor.

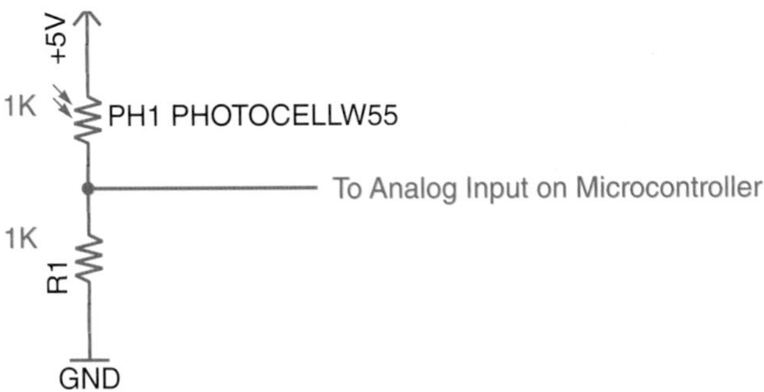

Goodheart-Willcox Publisher

Figure 15-37. This schematic shows a voltage divider made from two resistors. The lower resistor is a fixed 1/4 W carbon resistor. The upper resistor is a variable light sensitive resistor. As light strikes the photoresistor, its resistance changes (either increases or decreases, depending on the type) and changes the voltage at the junction between the two resistors. This change in voltage is read by the microcontroller.

such sensors provide basic vision. Each sensor can control a separate motor and, in effect, drive the robot. A simple application of this circuit is to make the motor direction change when light hits the sensor. A robot wired to chase light (an *angry* robot) or a robot that runs from light (a *shy* robot) can be built using only light sensors.

Figure 15-38 shows two photoresistors wired as light sensors and connected to a robot. The yellow wires are the signal wires that carry voltage back to the microcontroller. The voltage on the yellow wires will vary (from 0V to 5V) as the amount of light striking the photoresistors changes in intensity. The red and green wires are for 5V power and ground to power the circuit. Software on the microcontroller will compare the voltage present on the yellow wires and compare it to a fixed (threshold) value. Once the voltage crosses that threshold, the microcontroller will send a control signal to change the direction of one or both of the motors. Pseudocode is used below to demonstrate how this might look in the software. A light sensor is connected to pin 1. Motor direction outputs are pins 2 and 3, just as they were in the switch example. The only difference here is that an analog light sensor is used instead of a digital switch.

Goodheart-Willcox Publisher

Figure 15-38. This robot has a light sensor circuit built on a solderless breadboard. The output signals are being fed into the microcontroller inputs.

```
Dim X as word // Create a variable named "x" that can hold an analog value.
Line 1: Start:
Line 2: Pin 2 high // Motor 1 forward. (Move the robot forward.)
Line 3: Pin 3 high // Motor 2 forward. (Move the robot forward.)
Line 4: x=ADC_read(1) // Read the analog to digital conversion (ADC) value
        from pin 1 into variable "x".
Line 4: If "x>500" then // The sensor is higher than our threshold value.
Line 5: Pin 2 high // Motor 1 forward. (This line is redundant. This motor is
        already moving forward, but the note was placed here for clarity.)
Line 6: Pin 3 low // Motor 2 reverse. (This makes the robot turn.)
Line 7: Wait 500 ms // Keep the robot in this turn for half a second.
Line 8: End If // This is the end of this section of code.
Line 9: GOTO Start // Return to the start (line 1) and do it again.
```

The above code loaded on the sample robot would allow the robot to be controlled by a light source, such as a flashlight. To make the robot turn, the user would only need to shine a light on the light sensor and the robot would turn.

Infrared sensors work well for line following applications. An infrared sensor is typically an infrared LED and an infrared transistor pair. The infrared LED shines light down on a white or black surface. The white surface reflects the infrared light; the black surface does not. When the infrared light strikes the infrared transistor, the transistor turns on just like a switch. This signal can then be used to drive the robot. An example of this application is a maze drawn with black tape on a white surface. The robot is placed over the line so that one sensor is on the left side of the robot and on the left side on the line. The other sensor is on the right side of the robot and on the right side of the line. The robot will drive both wheels forward if neither sensor detects a line. If the left sensor detects a line, the robot will continue to move forward. However, it will also turn slightly left until it no longer detects the line with its left sensor. If the right sensor detects the line, the robot will continue forward and turn slightly right until the right sensor no longer detects the line. The robot will continue this way until it is told to do something else.

Line following competitions are popular with middle, high school, and college robotic teams. In a line following competition, students build and program robots that can follow a maze. Sometimes, the mazes have breaks in the solid lines or other distractions that the programmers will have to deal with. Typically, the fastest robot to navigate the maze will win. Some robots are programed to drive the maze once and then travel it a second time using the *map* it built of the maze the first time. An example line following program in pseudocode is written below. In this example, pins 1 and 2 are connected to infrared transistor inputs. Pins 3 and 4 are motor direction outputs.

```
Line 1:  Start:
Line 2:  If "Pin 1 high" then // Line detected, turn left.
Line 3:  Pin 3 low // Motor 1 reverse. (Left turn.)
Line 4:  Pin 4 high // Motor 2 forward. (Left turn.)
Line 5:  Else // No line detected, move straight.
Line 6:  Pin 3 high // Motor 1 forward. (Move the robot forward.)
Line 7:  Pin 4 high // Motor 2 forward. (Move the robot forward.)
Line 8:  End If
Line 9:  If "Pin 2 high" then // Line detected, turn right.
Line 10: Pin 3 high // Motor 1 forward. (Turn right.)
Line 11: Pin 4 high // Motor 2 reverse. (Turn right.)
Line 12: Else
Line 13: Pin 3 high // Motor 1 forward. (Move the robot forward.)
Line 14: Pin 4 high // Motor 2 forward. (Move the robot forward.)
Line 15: End If
Line 16: GOTO Start // Return to the start (line 1) and do it again.
```

The code above scans the sensor inputs constantly to determine the motor direction. As the robot drifts into the line, the sensors detect the line and reverse the correct motor to get the robot back on course. If no line is detected, the robot continues to move forward.

15.5.4 Sound Sensors

Sound sensing can be accomplished with a small microphone and amplifier circuit. Similar to the light sensor, the sound sensor will feed an analog signal into the microcontroller. The microcontroller will sample the sound level and compare it to a threshold value. The electrical signal from a small microphone is too weak, so a signal amplifier is required, **Figure 15-39**. A simple program that uses this sound-sensing circuit might make the robot stop, rotate 90°, then continue moving forward each time the microphone picks up a loud noise, such as a hand clap or a shout. Such a program will allow a robot to be controlled by sound. The same code used for the light sensor can be used for the sound sensor. The only line of code that might need adjustment is the threshold value on line 4.

Another application for a sound sensor is to use the microphone sensor to control a gripper/claw on a robot. The robot used as an example throughout this chapter has a small gripper attached to the front of its base. The same signal described above to control the direction of the robot could be used instead to open and close

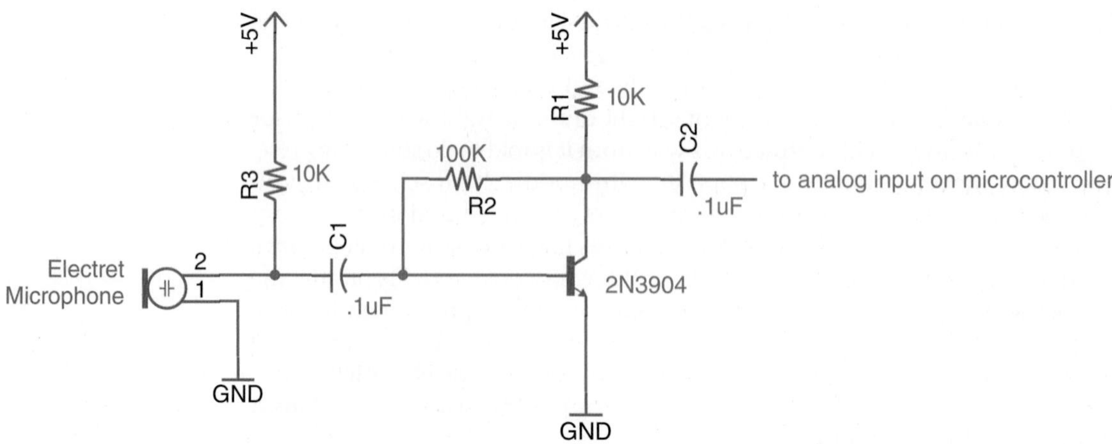

Goodheart-Willcox Publisher

Figure 15-39. An audio amplifier circuit is required to amplify the weak signal from a small microphone into a strong analog signal to be read by a microcontroller.

the gripper. The example below demonstrates how a sound sensor might control a servo motor. Refer to section 15.3 on motion control for an explanation of servo motion control. This code assumes that a sound sensor is connected to pin 1. A servo control wire is connected to pin 2. The servo power and ground wires are connected to power and ground.

Dim X as word // Create a variable named "x" that can hold an analog value.
Line 1: Start:
Line 2: x=ADC_read(1) // Read the analog to digital conversion (ADC) value from pin 1 into variable "x".
Line 3: If "x>500" then // The sensor is higher than the threshold value. // Rotate servo clockwise.
Line 4: Pin 2 high // Send a 5V signal out of pin 2.
Line 5: Wait 2 ms // Do nothing for 1 ms. This time delay controls the pulse width.
Line 6: Pin 2 low // Send a 0V (ground) signal out of pin 2.
Line 7: Wait 18 ms // Do nothing for 19 ms.
Line 8: Else // Rotate servo counterclockwise.
Line 9: Pin 2 high // Send a 5V signal out of pin 2.
Line 10: Wait 1 ms // Do nothing for 1 ms. This time delay controls the pulse width.
Line 11: Pin 2 low // Send a 0V (ground) signal out of pin 2.
Line 12: Wait 19 ms // Do nothing for 19 ms.
Line 13: End If // This is the end of this section of code.
Line 14: GOTO Start // Return to the start (line 1) and do it again.

While the code above may look complex, basically, all that it instructs the robot to do is listen to the analog input on pin 1 from a microphone. If the input is above a certain threshold value, in this case 500, it moves the servo clockwise. If the value is less than the threshold value of 500, it moves the servo counterclockwise. If this servo is connected to the claw on the robot, the claw would now be sound controlled. A loud noise would open the claw, and quiet or little noise would allow the claw to close. The value *500* used here is just an arbitrary number. The analog to digital converter in the microcontroller takes the voltage present on the input pin (which will be a value between 0V and 5V) and converts it into a number between 0 and 1000 (roughly). The number 500 does not represent decibels, voltage, or light levels. It is just an arbitrary number. It is up to the programmer to determine the range of values that will be read and what threshold to use. In the experiments with the sample robot used in this chapter, trial and error was used to determine a good threshold value. The maximum value stored in the microcontroller will depend on the size in bits of the register that holds the value. In this example, the register is 10 bits in size, so the maximum value it can hold is 1024 (210). 1024 is 2 to the tenth power. A 10-bit register is able to hold a binary value from 0000000000 to 1111111111. 1111111111 is a binary (base 2) number. That binary number (1111111111) in decimal (base 10) is 1,024. To determine the largest number a register can hold, the formula is 2 to the "X" power. The "X" power is the number of bits in the register, which is 10 in this case.

15.6 Troubleshooting

Troubleshooting is a process used to locate and eliminate problems in a system. A robot is a collection of many systems working together. The better you understand how each small system functions, the better you will be able to troubleshoot and repair systems when they malfunction. Troubleshooting your hobby robot system can be very frustrating if you do not have a sequential system for doing so. A robot has electrical systems, mechanical systems, and software. Any one or all of these systems can have faults, and determining which system is at fault can be difficult.

When troubleshooting, systems should be isolated whenever possible so that the fault can be limited to one system or another. Also, recognizing when a fault occurred can help locate the fault. If the robot was working before but not after a software change was made, the software is a good place to start looking for the cause of the problem. A logical first step would be to load the old, working software back on the robot and see if the problem goes away. If the problem is in software, it should be possible to bring the problem back, then remove it again. If the problem is intermittent, it is not likely to be a software issue.

When prototyping circuits on a breadboard, neatness can be very helpful. A sloppy circuit with an untidy nest of wires, **Figure 15-40**, represents poor design and construction methods. Trying to troubleshoot such a system can be very difficult and frustrating.

15.6.1 Power Systems

A common place for problems to occur is in the power system. Even if the batteries have some charge, but are not fully charged, the robot may not function. A microcontroller needs an even 5V (or in some cases 3.3V) power source, and it will not tolerate much variation from those values. Just because an LED power indicator is lit does not mean the system has adequate power. An LED only requires 0.7V and minimal current to work. The rest of the hardware in the system will require much more power. When in doubt, replace or recharge the batteries.

Troubleshooting should have a logical progression from one end of a system to the other. Starting at the power supply and tracing power to each device in the system is one approach. Make sure that power of the correct voltage is available at all the locations where it should be present. A digital multimeter is an important tool and should be part of any electronic tool kit. Also, an LED can be used as a logic probe to check digital signals in a circuit. To make a simple logic probe, connect one end of an LED through a resistor (100 ohm to 1,000 ohm) to ground. Make

Goodheart-Willcox Publisher

Figure 15-40. This "rat's nest" of wires makes a troubleshooting session very difficult when something is not working. Keep wires arranged neatly and try to color code them when possible.

sure the negative lead is closest to ground. Connect a jumper wire to the positive side of the LED. Touch the other end of the jumper wire to power and the LED should light. If it does not, try flipping the LED connections. This logic probe can be used to follow digital signals around the digital components, and it can be used to check for power on all integrated circuit chips.

15.6.2 Interference

Motors can create a lot of interference. Try to keep the motor wires away from sensor wires. Sometimes the interference is enough to cause the microcontroller to reboot. A rebooting microcontroller is hard to detect as it has no monitor screen and makes no noise. If motors are suspected as being the problem, replace them with LEDs. Connect LEDs to the outputs on the microcontroller that are driving the motors and observe what happens when the program runs. The LEDs can simulate the motors, but without causing the interference. If the problem goes away with the LEDs but returns with the motors, interference is likely the cause.

Limiting the interference from dc motors can be difficult, but not impossible to solve. The easiest and most effective way is to have a separate power source for the motors and the digital electronics. Any electrical system that uses a motor, coil, relay, solenoid, etc., will create a lot of interference. The interference comes from two sources. When a coil of wire has current flowing through it, a magnetic field is created. When the power to that coil is turned off, the magnetic field collapses. The energy stored in the magnetic fields is turned back into electrical current. When a dc motor spins it also generates electricity. The electricity generated flows in the opposite direction to the current being supplied to the motor. All this interference is not tolerated well by a microprocessor. All devices that have a coil (motors, relays, solenoids, etc.) should have their own power source. All the digital logic devices, such as the microcontroller, should have their own power source. The two power sources can share a common ground, however, as this will be necessary for many systems to function.

15.6.3 Capacitors

Capacitors can also be placed on or very near the power connections to the microcontroller. A small capacitance capacitor will help to short small transient voltage spikes to ground and protect the microcontroller. A 0.1μF or smaller capacitor typically works well. One side of the capacitor should connect to power and the other side to ground. The same size capacitor can also be soldered across the electrical contacts on the motors. Check the documentation for any microcontroller used to see if a specified capacitor is recommend.

15.6.4 Software

Software can be a bit more challenging to troubleshoot; however, a similar logical process should be used. It is good to have a simple program that can be used as a test program. An example of a test program is one that makes an LED blink on and off. This program can be loaded into the robot. Its proper operation will signify that the software, bootloaders, power system, and output lines of the microcontroller are all functioning. If a more complicated program is causing issues, this sample program can be loaded to eliminate hardware as the issue and ensure that software debugging is the cause of the problem. When writing new software, test it in stages. Testing a large amount of new code at once is fine if it all works. However, when it does not work, locating the specific error can be very difficult. If possible, remove suspect sections of code to see if the problem can be made to go away. Once a section of code is identified as a problem, then that section of code can be further broken down until the bad parts are eliminated.

Summary

- The first step in designing a robot is to determine the purpose for the robot. The cost of the robot, the funds available, mobility, accuracy, speed, and safety are also design considerations.
- When building a robot, the materials selected will be determined by the skill set of the builder, access to tools and machinery, and the mission of the robot. Cardboard, wood, plastic, and metal can be used to build hobby robots.
- Electric motors are one of the most common choices for providing motion for a small, hobby robot. Small toy dc motors and remote controlled servo motors are often used.
- The type of motor to be used as well as the type of terrain to be covered will determine the best choice of wheels. A robot may use hard plastic, soft foam, or rubber wheels or tank threads.
- Motion control circuits are necessary for direction and speed control of motors.
- A microcontroller is a small computer on a single integrated circuit that can serve as the brain of a robotic system.
- Motor power should be kept separate from logic power because the interference from the motors can cause issues with the digital components of the robot.
- Most microcontrollers will have several digital input/output lines that can be used as either inputs or outputs and several analog inputs.
- A microcontroller must be programmed to perform specific tasks. Program code is written on a computer and then compiled and downloaded to the microcontroller.
- Many manufacturers have developed microcontroller boards. Each microcontroller board has advantages and disadvantages.
- Sensors allow the robot to interact with the world around it. Proximity sensors can detect objects from several inches away.
- Light sensing can be accomplished with photoresistors or phototransistors. Sound sensing can be accomplished with a small microphone and amplifier circuit.
- A robot has electrical systems, mechanical systems, and software that may malfunction or have faults. Troubleshooting is a specific and deliberate process used to locate and eliminate problems in a system.

Review Questions

Answer the following questions using the information provided in this chapter.

1. What are some factors to consider when planning the design of a robot?
2. What factors should be considered when selecting materials to use for building a robot?
3. What are some materials that are good choices for constructing the base for a robot?
4. What are some advantages of using wood to construct the base for a robot?
5. What are two disadvantages of using metal to construct the base for a robot?
6. What are some types of motors that are common choices for providing motion for a hobby robot?
7. What factors should be considered in determining the type of wheels to use for a robot? What are some types of wheels that may be used?
8. Why is having a gearbox between the motor and the wheels required on most hobby robots?

9. What is the function of a motor controller?
10. Why should motor power be kept separate from logic power when constructing a robot?
11. What is a microcontroller?
12. What is pulse width modulation (PWM), and what are two ways PWM may be used as controls for a robot?
13. What types of languages are used to program a robot to do tasks?
14. What is an advantage of using a microcontroller board?
15. Explain the purpose of a bootloader.
16. Why is it important to be cautious when using a programmer?
17. What are some types of sensors that can be used for robots?
18. What does a proximity sensor detect, and what method is used?
19. What type of sensor works well for line following applications? Describe how the sensor works.
20. What are the two basic components used for sensing sound?
21. What is troubleshooting?
22. What are two sources of interference that may cause malfunctions in a hobby robot?
23. Why should new software code be tested in small sections?

Learning Extensions

1. Do an Internet search for microcontrollers. Locate the data sheets to several microcontrollers and compare their features.
2. Brainstorm some robot design ideas. Make several sketches of possible designs. Sketch at least three versions of each design in order to develop unique solutions to problems.
3. Compare the cost and features of several of the microcontroller boards presented in this chapter, such as the Arduino®, Raspberry Pi, Pololu, Parallax, LEGO®, and VEX®.
4. Create a cost sheet for constructing a hobby robot based on the materials you would most likely use. Experiment with different materials and determine what the total estimated cost will be. Do not forget to include shipping, applicable taxes, and other charges.
5. Review several programming environments for microcontrollers. If free trials are available, download and sample different systems. Make note of the features you like, and rate the ease of operation of each system. Factor cost into your decision to determine the best solution for your application.
6. Search the Internet for robotic design competitions. Watch some online videos of competitions. Make notes about designs and challenges that are of interest to you.

Advanced Analysis

1. Complete a search of media reports related to the use of amateur-built robotic systems, such as drones or unmanned aerial vehicles (UAVs). Write a short paper discussing some of the legal and ethical issues related to such systems.
2. Draft a set of rules for a robotics competition that would include challenges to be completed by a robot similar to those described in this chapter. Be sure to consider design constraints, such as size, weight, cost, and related safety factors.

Math Skills Handbook

Table of Contents

Getting Started

Math skills are needed in everyday life. You will need to be able to estimate your purchases at a grocery store, calculate sales tax, or divide a recipe in half. This section is designed to help develop your math proficiency for better understanding of the concepts presented in the textbook. Using the information presented in the Math Skills Handbook will help you understand basic math concepts and their application to the real world.

Using a Calculator

There are many different types of calculators. Some are simple and only perform basic math operations. Become familiar with the keys and operating instructions of your calculator so calculations can be made quickly and correctly.

Shown below is a scientific calculator that comes standard with the Windows 8 operating system. To display this version, select the **View** pull-down menu and click **Scientific** in the menu.

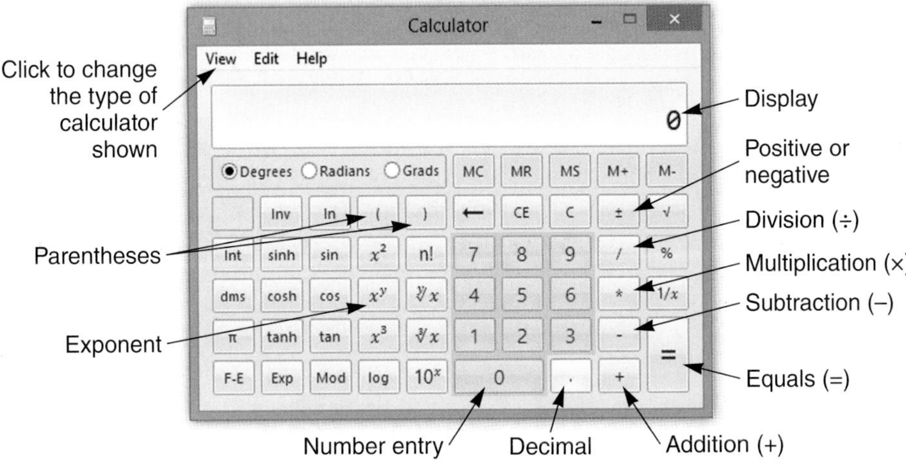

Solving Word Problems

Word problems are exercises in which the problem is set up in text, rather than presented in mathematical notation. Many word problems tell a story. You must identify the elements of the math problem and solve it.

There are many strategies for solving word problems. Some common strategies include making a list or table; working backward; guessing, checking, and revising; and substituting simpler numbers to solve the problem.

Strategy	How to Apply
List or table	Identify information in the problem and organize it into a table to identify patterns.
Work backward	When an end result is provided, work backward from that to find the requested information.
Guess, check, revise	Start with a reasonable guess at the answer, check to see if it is correct, and revise the guess as needed until the solution is found.
Sustitute simpler information	Use different numbers to simply the problem and solve it, then solve the problem using the provided numbers.

Number Sense

Number sense is an ability to use and understand numbers to make judgments and solve problems. Someone with good number sense also understands when his or her computations are reasonable in the context of a problem.

Example

Suppose you want to add three basketball scores: 35, 21, and 18.

- First, add 30 + 20 + 10 = 60.
- Then, add 5 + 1 + 8 = 14.
- Finally, combine these two sums to find the answer: 60 + 14 = 74.

Example

Suppose your brother is 72 inches tall and you want to convert this measurement from inches to feet. Suppose you use a calculator to divide 72 by 12 (number of inches in a foot) and the answer is displayed as 864. You recognize immediately that your brother cannot be 864 feet tall and realize you must have miscalculated. In this case, you incorrectly entered a multiplication operation instead of a division operation. The correct answer is 6.

Numbers and Quantity

Numbers are more than just items in a series. Each number has a distinct value relative to all other numbers. They are used to perform mathematical operations from the simplest addition to finding square roots. There are whole numbers, fractions, decimals, exponents, and square roots.

Whole Numbers

A whole number, or integer, is any positive number or zero that has no fractional part. It can be a single digit from 0 to 9, or may contain multiple digits, such as 38.

Place Value

A digit's position in a number determines its *place value.* The digit, or numeral, in the place farthest to the right before the decimal point is in the *ones position.* The next digit to the left is in the *tens position,* followed by next digit in the *hundreds position.* As you continue to move left, the place values increase to thousands, ten thousands, and so forth.

Example

Suppose you win the lottery and receive a check for $23,152,679. Your total prize would be *twenty-three million, one hundred fifty-two thousand, six hundred seventy-nine dollars.*

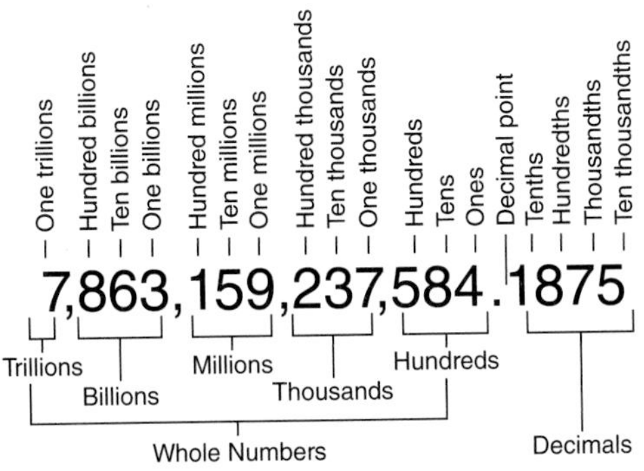

Addition

Addition is the process of combining two or more numbers. The result is called the *sum*.

> **Example**
> A plumber installs six faucets on his first job and three faucets on his second job. How many faucets does he install in total?
> $$6 + 3 = 9$$

Subtraction

Subtraction is the process of finding the *difference* between two numbers.

> **Example**
> A plumber installs six faucets on her first job and three faucets on her second job. How many more faucets did she install on the first job than the second? Subtract 3 from 6 to find the answer.
> $$6 - 3 = 3$$

Multiplication

Multiplication is a method of adding a number to itself a given number of times. The multiplied numbers are called *factors*, and the result is called the *product*.

> **Example**
> Suppose you are installing computers and need to purchase four adapters. If the adapters are $6 each, what is the total cost of the adapters? The answer can be found by adding $6 four times:
> $$\$6 + \$6 + \$6 + \$6 = \$24$$
> However, the same answer is found more quickly by multiplying $6 times 4.
> $$\$6 \times 4 = \$24$$

Division

Division is the process of determining how many times one number, called the *divisor*, goes into another number, called the *dividend*. The result is called the *quotient*.

Example

Suppose you are installing computers and buy a box of adapters for $24. There are four adapters in the box. What is the cost of each adapter? The answer is found by dividing $24 by 4:

$$\$24 \div 4 = \$6$$

Decimals

A decimal is a kind of fraction with a denominator that is either ten, one hundred, one thousand, or some power of ten. Every decimal has three parts: a whole number (sometimes zero), followed by a decimal point, and one or more whole numbers.

Place Value

The numbers to the right of the decimal point indicate the amount of the fraction. The first place to the right of a decimal point is the tenths place. The second place to the right of the decimal point is the hundredths place. As you continue to the right, the place values move to the thousandths place, the ten-thousandths place, and so on.

Example

A machinist is required to produce an airplane part to a very precise measurement of 36.876 inches. This measurement is *thirty-six and eight hundred seventy-six thousandths* inches.

36.876

Addition

To add decimals, place each number in a vertical list and align the decimal points. Then add the numbers in each column starting with the column on the right and working to the left. The decimal point in the answer drops down into the same location.

Example

A landscaper spreads 4.3 pounds of fertilizer in the front yard of a house and 1.2 pounds in the backyard. How many pounds of fertilizer did the landscaper spread in total?

$$\begin{array}{r} 4.3 \\ +\ 1.2 \\ \hline 5.5 \end{array}$$

Subtraction

To subtract decimals, place each number in a vertical list and align the decimal points. Then subtract the numbers in each column, starting with the column on the right and working to the left. The decimal point in the answer drops down into the same location.

Example

A landscaper spreads 4.3 pounds of fertilizer in the front yard of a house and 1.2 pounds in the backyard. How many more pounds were spread in the front yard than in the backyard?

$$\begin{array}{r} 4.3 \\ -\ 1.2 \\ \hline 3.1 \end{array}$$

Multiplication

To multiply decimals, place the numbers in a vertical list. Then multiply each digit of the top number by the right-hand bottom number. Multiply each digit of the top number by the bottom number in the tens position. Place the result on a second line and add a zero to the end of the number. Add the total number of decimal places in both numbers you are multiplying. This will be the number of decimal places in your answer.

Example

An artist orders 13 brushes priced at $3.20 each. What is the total cost of the order? The answer can be found by multiplying $3.20 by 13.

$$\begin{array}{r} \$3.20 \\ \times \quad 13 \\ \hline 960 \\ +\ 3200 \\ \hline \$41.60 \end{array}$$

Division

To divide decimals, the dividend is placed under the division symbol, the divisor is placed to the left of the division symbol, and the quotient is placed above the division symbol. Start from the *left* of the dividend and determine how many times the divisor goes into the first number. Continue this until the quotient is found. Add the dollar sign to the final answer.

$$\begin{array}{rl} 3.20 & \\ 3\overline{)9.60} & \\ -9 & \text{Product of } 3 \times 3 \\ \hline 0\,6 & \text{Bring down the 6} \\ 6 & \text{Product of } 2 \times 3 \\ \hline 00 & \text{No remainder} \end{array}$$

Example

An artist buys a package of three brushes for $9.60. What is the cost of each brush? The quotient is found by dividing $9.60 by 3.

$$\begin{array}{r} 3.20 \\ 3\overline{)9.60} \\ -9 \\ \hline 0\,6 \\ \hline 00 \end{array}$$

Rounding

When a number is rounded, some of the digits are changed, removed, or changed to zero so the number is easier to work with. Rounding is often used when precise calculations or measurements are not needed. For example, if you are calculating millions of dollars, it might not be important to know the amount down to the dollar or cent. Instead, you might *round* the amount to the nearest ten thousand or even hundred thousand dollars. Also, when working with decimals, the final answer might have several more decimal places than needed.

To round a number, follow these steps. First, underline the digit in the place to which you are rounding. Second, if the digit to the *right* of this place is 5 or greater, add 1 to the underlined digit. If the digit to the right is less than 5, do not change the underlined digit. Third, change all the digits to right of the underlined digit to zero. In the case of decimals, the digits to the right of the underlined digit are removed.

Example

A company's utility expense last year was $32,678.53. The owner of the company is preparing a budget for next year and wants to round this amount to the nearest 1,000.

Step 1: Underline the digit in the 1,000 place.

$3**2**,678

Step 2: The digit to the right of 2 is greater than 5, so add 1.

2 + 1 = 3

Step 3: Change the digits to the right of the underlined digit to zero.

$33,000

Fractions

A fraction is a part of a whole. It is made up by a numerator that is divided by a denominator.

$$\frac{\text{numerator}}{\text{denominator}}$$

The *numerator* specifies the number of these equal parts that are in the fraction. The *denominator* shows how many equal parts make up the whole.

Proper

In a *proper fraction,* the numerator is less than the denominator.

Example

A lumber yard worker cuts a sheet of plywood into four equal pieces and sells three of them to a carpenter. The carpenter now has 3/4 of the original sheet. The lumber yard has 1/4 of the sheet remaining.

Improper

An *improper fraction* is a fraction where the numerator is equal to or greater than the denominator.

Example
A chef uses a chili recipe which calls for 1/2 cup of chili sauce. However, the chef makes an extra-large batch that will serve three times as many people and uses three of the 1/2 cup measures. The improper fraction in this example is 3/2 cups of chili sauce.

Mixed

A mixed number contains a whole number and a fraction. It is another way of writing an improper fraction.

Example
A chef uses a chili recipe that calls for 1/2 cup of chili sauce. However, the chef makes an extra-large batch that will serve three times as many people and uses three of the 1/2 cup measures. The improper fraction in this example is 3/2 cups of chili sauce. This can be converted to a mixed number by dividing the numerator by the denominator:

$$\begin{array}{r} 1 \\ 2\overline{)3} \\ \underline{-2} \\ 1 \end{array}$$

The remainder is 1, which is 1 over 2. So, the mixed number is 1 1/2 cups.

Reducing

Fractions are reduced to make them easier to work with. Reducing a fraction means writing it with smaller numbers, in *lowest terms.* Reducing a fraction does not change its value.

To find the lowest terms, determine the largest number that *evenly* divides both the numerator and denominator so there is no remainder. Then use this number to divide both the numerator and denominator.

Example
The owner of hair salon asks ten customers if they were satisfied with the service they recently received. Eight customers said they were satisfied, so the fraction of satisfied customers is 8/10. The largest number that evenly divides both the numerator and denominator is 2. The fraction is reduced to its lowest terms as follows.

$$\frac{8}{10} = \frac{8 \div 2}{10 \div 2} = \frac{4}{5}$$

Addition

To add fractions, the numerators are combined and the denominator stays the same. However, fractions can only be added when they have a *common denominator.* The *least common denominator* is the smallest number to which each denominator can be converted.

Example

A snack food company makes a bag of trail mix by combining 3/8 pound of nuts with 1/8 pound of dried fruit. What is the total weight of each bag? The fractions have common denominators, so the total weight is determined by adding the fractions.

$$\frac{3}{8}+\frac{1}{8}=\frac{4}{8}$$

This answer can be reduced from 4/8 to 1/2.

Example

Suppose the company combines 1/4 pound of nuts with 1/8 cup of dried fruit. To add these fractions, the denominators must be made equal. In this case, the least common denominator is 8 because 4 × 2 = 8. Convert 1/4 to its equivalent of 2/8 by multiplying both numerator and denominator by 2. Then the fractions can be added as follows.

$$\frac{2}{8}+\frac{1}{8}=\frac{3}{8}$$

This answer cannot be reduced because 3 and 8 have no common factors.

Subtraction

To subtract fractions, the second numerator is subtracted from the first numerator. The denominators stay the same. However, fractions can only be subtracted when they have a *common denominator.*

Example

A snack food company makes a bag of trail mix by combining 3/8 pound of nuts with 1/8 pound of dried fruit. How much more do the nuts weigh than the dried fruit? The fractions have common denominators, so the difference can be determined by subtracting the fractions.

$$\frac{3}{8}-\frac{1}{8}=\frac{2}{8}$$

This answer can be reduced from 2/8 to 1/4.

Example

Suppose the company combines 1/4 pound of nuts with 1/8 cup of dried fruit. How much more do the nuts weigh than the dried fruit? To subtract these fractions, the denominators must be made equal. The least common denominator is 8, so convert 1/4 to its equivalent of 2/8. Then the fractions can be subtracted as follows.

$$\frac{2}{8}-\frac{1}{8}=\frac{1}{8}$$

This answer cannot be reduced.

Multiplication

Common denominators are not necessary to multiply fractions. Multiply all of the numerators and multiply all of the denominators. Reduce the resulting fraction as needed.

Example

A lab technician makes a saline solution by mixing 3/4 cup of salt with one gallon of water. How much salt should the technician mix if only 1/2 gallon of water is used? Multiply 3/4 by 1/2:

$$\frac{3}{4} \times \frac{1}{2} = \frac{3}{8}$$

Division

To divide one fraction by a second fraction, multiply the first fraction by the reciprocal of the second fraction. The *reciprocal* of a fraction is created by switching the numerator and denominator.

Example

A cabinet maker has 3/4 gallon of wood stain. Each cabinet requires 1/8 gallon of stain to finish. How many cabinets can be finished? To find the answer, divide 3/4 by 1/8, which means multiplying 3/4 by the reciprocal of 1/8.

$$\frac{3}{4} \div \frac{1}{8} = \frac{3}{4} \times \frac{8}{1} = \frac{24}{4} = 6$$

Negative Numbers

Negative numbers are those less than zero. They are written with a minus sign in front of the number.

Example

The number –34,687,295 is read as *negative thirty-four million, six hundred eighty-seven thousand, two hundred ninety-five.*

Addition

Adding a negative number is the same as subtracting a positive number.

Example

A football player gains nine yards on his first running play (+9) and loses four yards (–4) on his second play. The two plays combined result in a five-yard gain.

$$9 + (-4) = 9 - 4 = 5$$

Suppose this player loses five yards on his first running play (–5) and loses four yards (–4) on his second play. The two plays combined result in a nine-yard loss.

$$-5 + (-4) = -5 - 4 = -9$$

Subtraction

Subtracting a negative number is the same as adding a positive number.

Example

Suppose you receive a $100 traffic ticket. This will result in a –$100 change to your cash balance. However, you explain the circumstance to a traffic court judge, and she reduces the fine by $60. The effect is to subtract –$60 from –$100 change to your cash balance. The final result is a –$40 change.

$$-\$100 - (-\$60) = -\$100 + \$60 = -\$40$$

Multiplication

Multiplication of an odd number of negative numbers results in a *negative* product. Multiplication of an even number of negative numbers results in a *positive* product.

Example

If you lose two pounds per week, this will result in a –2 pound weekly change in your weight. After five weeks, there will be a –10 pound change to your weight.

$$5 \times (-2) = -10$$

Suppose you have been losing two pounds per week. Five weeks ago (–5) your weight was 10 pounds higher.

$$(-5) \times (-2) = 10$$

Division

Division of an odd number of negative numbers results in a *negative* quotient. Division of an even number of negative numbers results in a *positive* quotient.

Example

Suppose you lost 10 pounds, which is a –10 pound change in your weight. How many pounds on average did you lose each week if it took five weeks to lose the weight? Divide –10 by 5 to find the answer.

$$-10 \div 5 = -2$$

Suppose you lost 10 pounds. How many weeks did this take if you lost two pounds each week? Divide –10 by –2 to find the answer.

$$-10 \div -2 = 5$$

Percentages

A percentage (%) means a part of 100. It is the same as a fraction or decimal.

Representing Percentages as Decimals

To change a percentage to a decimal, move the decimal point two places to the left. For example, 1% is the same as 1/100 or 0.01; 10% is the same as 10/100 or 0.10; and 100% is the same as 100/100 or 1.0.

Example

A high school cafeteria estimates that 30% of the students prefer sesame seeds on hamburger buns. To convert this percentage to a decimal, move the decimal point two places to the left.

$$30\% = 0.30$$

Representing Fractions as Percentages

To change a fraction to a percentage, first convert the fraction to a decimal by dividing the numerator by the denominator. Then convert the decimal to a percentage by the moving the decimal point two places to the right.

Example

A high school cafeteria conducts a survey and finds that three of every ten students prefer sesame seeds on hamburger buns. To change this fraction to a percentage, divide 3 by 10, and move the decimal two places to the right.

$$3 \div 10 = 0.30 = 30\%$$

Calculating a Percentage

To calculate the percentage of a number, change the percentage to a decimal and multiply by the number.

Example

A car dealer sold ten cars last week, of which 70% were sold to women. How many cars did women buy? Change 70% to a decimal by dividing 70 by 100, which equals 0.70. Then multiply by the total number (10).

$$0.70 \times 10 = 7$$

To determine what percentage one number is of another, divide the first number by the second. Then convert the quotient to a percentage by moving the decimal point two places to the right.

Example

A car dealer sold 10 cars last week, of which seven were sold to women. What percentage of the cars were purchased by women? Divide 7 by 10 and then convert to a percentage.

$$7 \div 10 = 0.70$$

$$0.70 = 70\%$$

Ratio

A ratio compares two numbers through division. Ratios are often expressed as a fraction, but can also be written with a colon (:) or the word *to*.

Example

A drugstore's cost for a bottle of vitamins is $2.00, which it sells for $3.00. The ratio of the selling price to the cost can be expressed as follows:

$$\frac{\$3.00}{\$2.00} = \frac{3}{2}$$

$$\$3.00{:}\$2.00 = 3{:}2$$

$$\$3.00 \text{ to } \$2.00 = 3 \text{ to } 2$$

Measurement

The official system of measurement in the United States for length, volume, and weight is the US Customary system of measurement. The metric system of measurement is used by most other countries.

US Customary Measurement

The following are the most commonly used units of length in the US Customary system of measurement.

- 1 inch
- 1 foot = 12 inches
- 1 yard = 3 feet
- 1 mile = 5,280 feet

Example

An interior designer measurers the length and width of a room when ordering new floor tiles. The length is measured at 12 feet 4 inches (12′ 4″). The width is measured at 8 feet 7 inches (8′ 7″).

Example

Taxi cab fares are usually determined by measuring distance in miles. A recent cab rate in Chicago was $3.25 for the first 1/9 mile or less, and $0.20 for each additional 1/9 mile.

Metric Conversion

The metric system of measurement is convenient to use because units can be converted by multiplying or dividing by multiples of 10. The following are the commonly used units of length in the metric system of measurement.

- 1 millimeter
- 1 centimeter = 10 millimeters
- 1 meter = 100 centimeters
- 1 kilometer = 1,000 meters

The following are conversions from the US Customary system to the metric system.

- 1 inch = 25.4 millimeters = 2.54 centimeters
- 1 foot = 30.48 centimeters = 0.3048 meters
- 1 yard = 0.9144 meters
- 1 mile = 1.6093 kilometers

Example

A salesperson from the United States is traveling abroad and needs to drive 100 kilometers to meet a customer. How many miles is this trip? Divide 100 kilometers by 1.6093 and round to the hundredth place.

```
                  62.138
1.6093.)100.0000.000
        -96558
          34420
         -32186
          22340
         -16093
          62470
         -48279
          141910
         -128744
           13169
```

Estimating

Estimating is finding an *approximate* answer and often involves using rounded numbers. It is often quicker to add rounded numbers, for example, than it is to add the precise numbers.

Example

Estimate the total miles a delivery truck will travel along the following three segments of a route.

- Detroit to Chicago: 278 miles
- Chicago to St. Louis: 297 miles
- St. Louis to Wichita: 436 miles

The mileage can be estimated by rounding each segment to the nearest 100 miles.

- Detroit to Chicago: 300 miles
- Chicago to St. Louis: 300 miles
- St. Louis to Wichita: 400 miles

Add the rounded segments to estimate the total miles.

$$300 + 300 + 400 = 1{,}000 \text{ miles}$$

Accuracy and Precision

Accuracy and precision mean slightly different things. *Accuracy* is the closeness of a measured value to its actual or true value. *Precision* is how close measured values are to each other.

Example

A machine is designed to fill jars with 16 ounces of peanut butter. The machine is considered accurate if the actual amount of peanut butter in a jar is within 0.05 ounces of the target, which is a range of 15.95 to 16.05 ounces. A machine operator tests a jar and measures the weight to be 16.01 ounces. The machine is accurate.

Suppose a machine operator tests 10 jars of peanut butter and finds the weight of each jar to be 15.4 ounces. The machine is considered precise because it fills every jar with exactly the same amount. However, it is not accurate because the amount differs too much from the target.

Algebra

An *equation* is a mathematical statement that has an equal sign (=). An *algebraic* equation is an equation that includes at least one variable. A *variable* is an unknown quantity.

Solving Equations with Variables

Solving an algebraic equation means finding the value of the variable that will make the equation a true statement. To solve a simple equation, perform inverse operations on both sides and isolate the variable.

Example

A computer consultant has sales of $1,000. After deducting $600 in expenses, her profit equals $400. This is expressed with the following equation.

sales – expenses = profit

$1,000 – $600 = $400

Example

A computer consultant has expenses of $600 and $400 in profit. What are her sales? An equation can be written in which sales are the unknown quantity, or variable.

sales – expenses = profit

sales – $600 = $400

Example

To find the value for sales, perform inverse operations on both sides and isolate the variable.

sales	– $600	=	$400
	+ $600	+	600
sales		=	$1,000

Order of Operations

The order of operations is a set of rules stating which operations in an equation are performed first. The order of operations is often stated using the acronym *PEMDAS*. PEMDAS stands for parentheses, exponents, multiplication and division, and addition and subtraction. This means anything inside parentheses is computed first. Exponents are computed next. Then, any multiplication and division operations are computed. Finally, any addition and subtraction operations are computed to find the final answer to the problem. The equation is solved from left to right by applying PEMDAS.

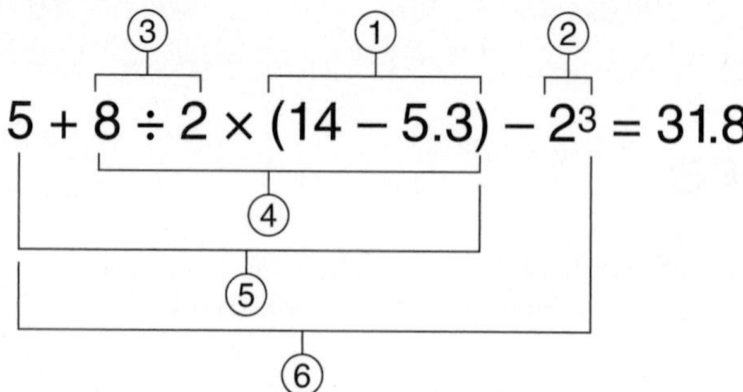

Recursive Formulas

A *recursive formula* is used to determine the next term of a sequence, using one or more of the preceding terms. The terms of a sequence are often expressed with a variable and subscript. For example, a sequence might be written as a_1, a_2, a_3, a_4, a_5, and so on. The subscript is essentially the place in line for each term. A recursive formula has two parts. The first is a starting point or seed value (a_1). The second is an equation for another number in the sequence (a_n). The second part of the formula is a function of the prior term (a_{n-1}).

Example

Suppose you buy a car for $10,000. Assume the car declines in value 10% each year. In the second year, the car will be worth 90% of $10,000, which is $9,000. The following year it will be worth 90% of $9,000, which is $8,100. What will the car be worth in the fifth year? Use the following recursive equation to find the answer.

$$a_n = a_{n-1} \times 0.90$$

$$\text{where } a_1 = \$10{,}000$$

$$a_n = \text{value of car in the } n^{th} \text{ year}$$

Year	Value of Car
n = 1	$a_1 = \$10{,}000$
n = 2	$a_2 = a_{2-1} \times 0.90 = a_1 \times 0.90 = \$10{,}000 \times 0.90 = \$9{,}000$
n = 3	$a_3 = a_{3-1} \times 0.90 = a_2 \times 0.90 = \$9{,}000 \times 0.90 = \$8{,}100$
n = 4	$a_4 = a_{4-1} \times 0.90 = a_3 \times 0.90 = \$8{,}100 \times 0.90 = \$7{,}290$
n = 5	$a_5 = a_{5-1} \times 0.90 = a_4 \times 0.90 = \$7{,}290 \times 0.90 = \$6{,}561$

Geometry

Geometry is a field of mathematics that deals with shapes, such as circles and polygons. A *polygon* is any shape whose sides are straight. Every polygon has three or more sides.

Parallelograms

A *parallelogram* is a four-sided figure with two pairs of parallel sides. A *rectangle* is a type of parallelogram with four right angles. A *square* is a special type of parallelogram with four right angles (90 degrees) and four equal sides.

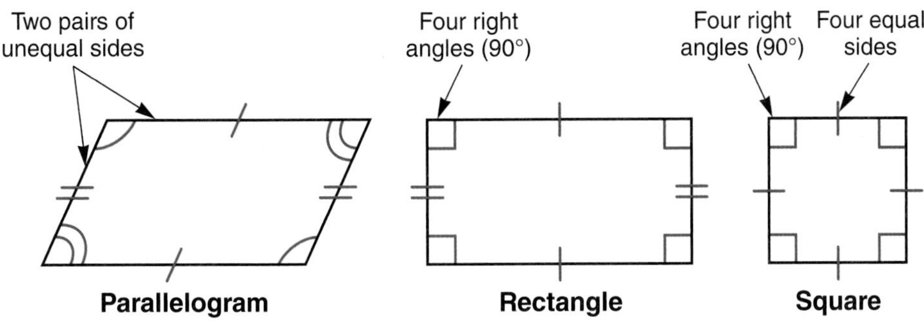

Example
Real-life examples of squares include ceramic floor and wall tiles, and each side of a die. Real-life examples of a rectangle include a football field, pool table, and most doors.

Triangles

A three-sided polygon is called a *triangle.* The following are four types of triangles, which are classified according to their sides and angles.

- *Equilateral:* Three equal sides and three equal angles.
- *Isosceles:* Two equal sides and two equal angles.
- *Scalene:* Three unequal sides and three unequal angles.
- *Right:* One right angle; may be isosceles or scalene.

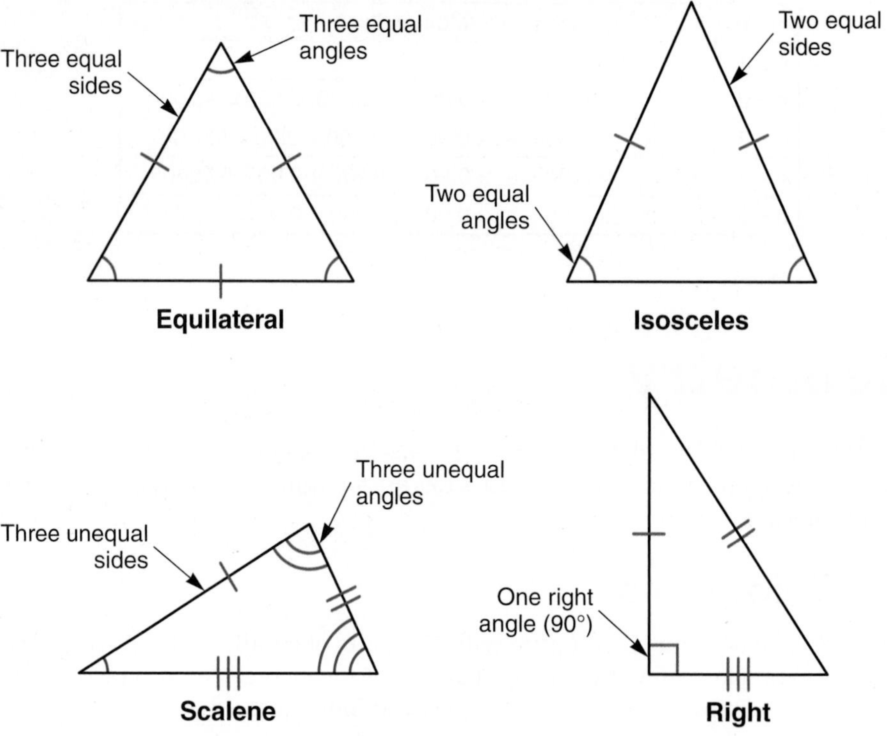

Example
Real-life examples of equilateral triangles are the sides of a classical Egyptian pyramid.

Circles and Half Circles

A *circle* is a figure in which every point is the same distance from the center. The distance from the center to a point on the circle is called the *radius.* The distance across the circle through the center is the *diameter.* A half circle is formed by dividing a whole circle along the diameter.

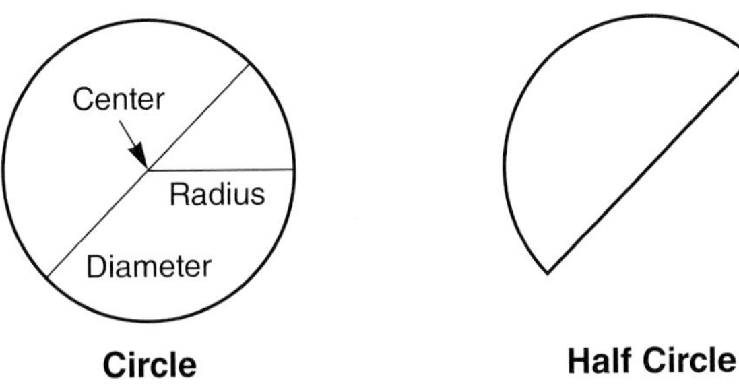

Example
Real-life examples of circles include wheels of all sizes.

Perimeter

A *perimeter* is a measure of length around a figure. Add the length of each side to measure the perimeter of any figure whose sides are all line segments, such as a parallelogram or triangle. The perimeter of a circle is called the *circumference.* To measure the perimeter, multiply the diameter by pi (π). Pi is approximately equal to 3.14. The following formulas can be used to calculate the perimeters of various figures:

Figure	Perimeter
parallelogram	2 × width + 2 × length
square	4 × side
rectangle	2 × width + 2 × length
triangle	side + side + side
circle	π × diameter

Example
A professional basketball court is a rectangle 94 feet long and 50 feet wide. The perimeter of the court is calculated as follows:

2 × 94 feet + 2 × 50 feet = 288 feet

Example
A tractor tire has a 43-inch diameter. The circumference of the tire is calculated as follows:

43 inches × 3.14 = 135 inches

Area

Area is a measure of the amount of surface within the perimeter of a flat figure. Area is measured in square units, such as square inches, square feet, or square miles. The areas of the following figures are calculated using the corresponding formulas:

Figure	Area
parallelogram	base × height
square	side × side
rectangle	length × width
triangle	1/2 × base × height
circle	π × radius² = π × radius × radius

Example

An interior designer needs to order decorative tiles to fill the following spaces. Measure the area of each space in square feet.

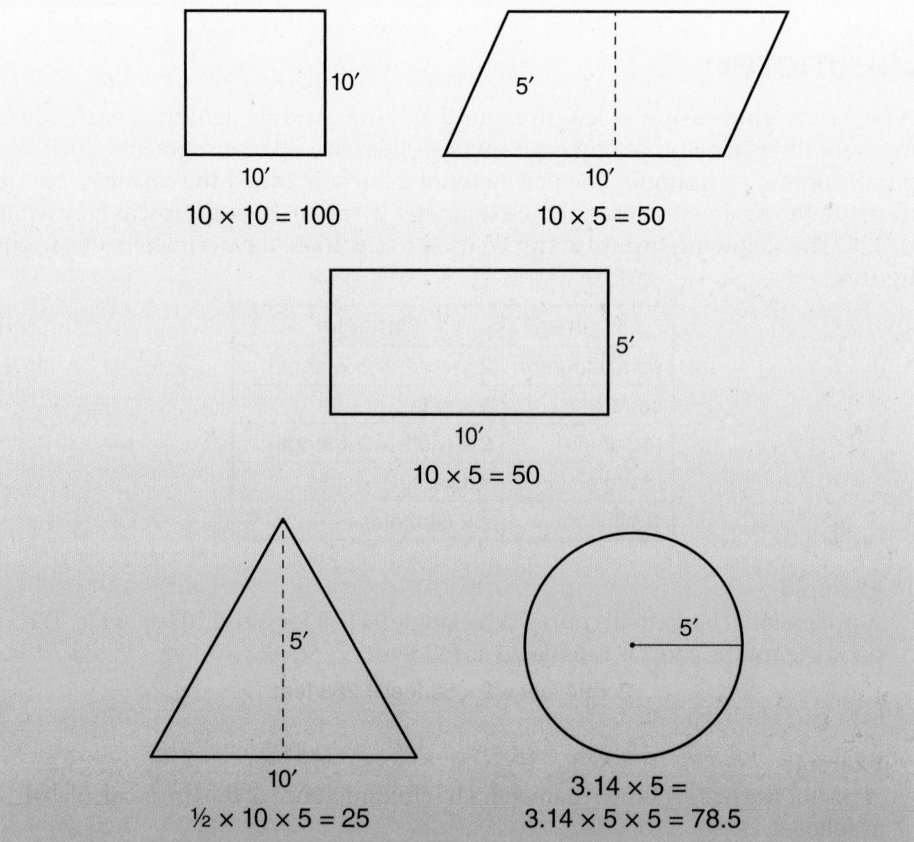

Surface Area

Surface area is the total area of the surface of a figure occupying three-dimensional space, such as a cube or prism. A *cube* is a solid figure that has six identical square faces. A *prism* has bases or ends which have the same size and shape and are parallel to each other, and each of whose sides is a parallelogram. The following are the formulas to find the surface area of a cube and a prism.

Object	Surface Area
cube	6 × side × side
prism	2 × [(length × width) + (width × height) + (length × height)]

Example

A manufacturer of cardboard boxes wants to determine how much cardboard is needed to make the following size boxes. Calculate the surface area of each in square inches.

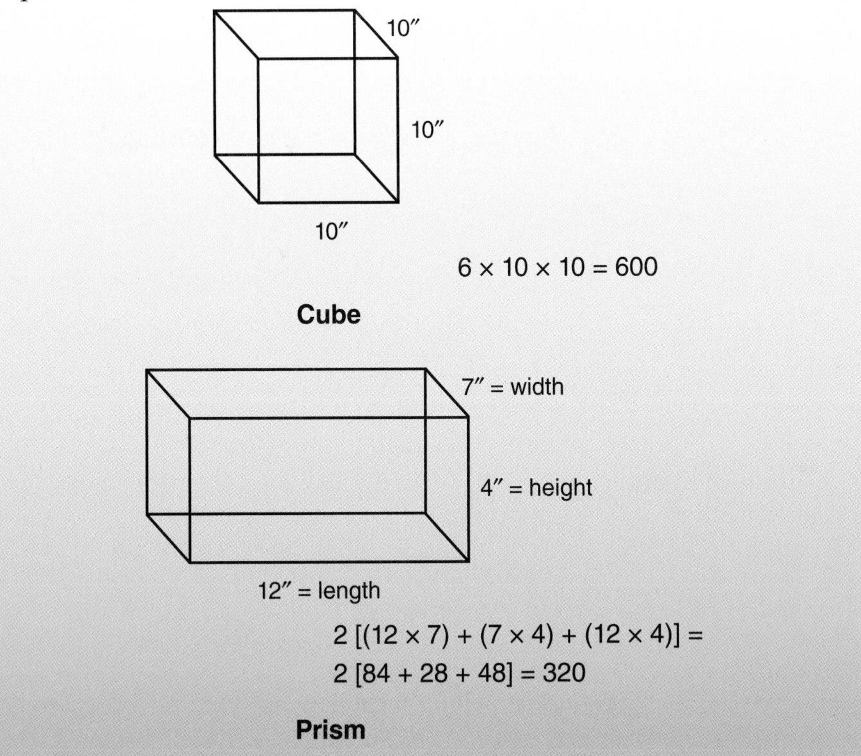

Volume

Volume is the three-dimensional space occupied by a figure and is measured in cubic units, such as cubic inches or cubic feet. The volumes of the following figures are calculated using the corresponding formulas.

Solid Figure	Volume
cube	$side^3$ = side × side × side
prism	length × width × height
cylinder	$\pi \times radius^2 \times height = \pi \times radius \times radius \times height$
sphere	$4/3 \times \pi \times radius^3 = 4/3 \times \pi \times radius \times radius \times radius$

Example

Find the volume of packing material needed to fill the following boxes. Measure the volume of each in cubic inches.

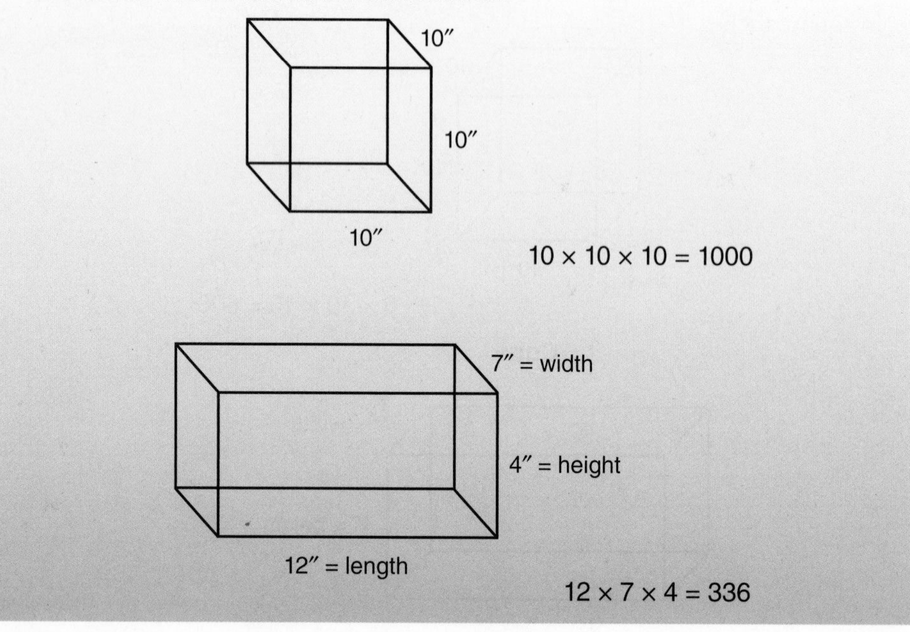

Example

Find the volume of grain that will fill the following cylindrical silo. Measure the volume in cubic feet.

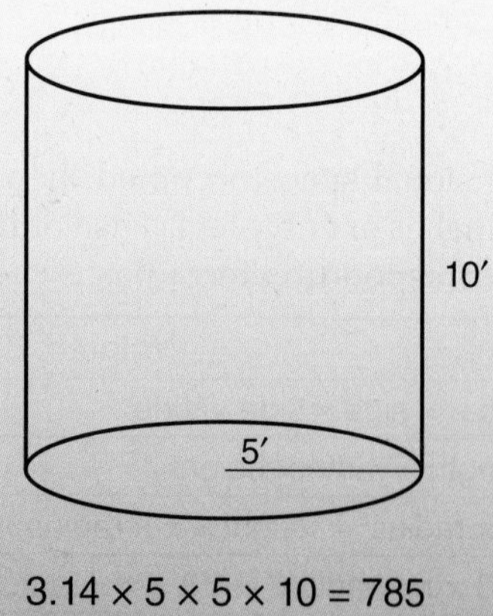

3.14 × 5 × 5 × 10 = 785

Example

A manufacturer of pool toys wants to stuff soft material into a ball with a 3-inch radius. Find the cubic inches of material that will fit into the ball.

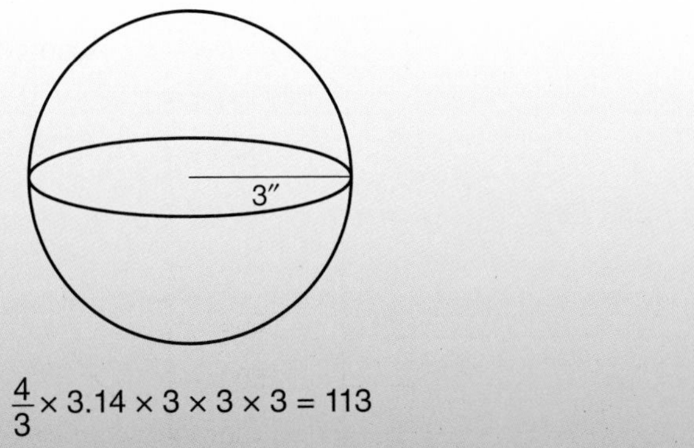

$$\frac{4}{3} \times 3.14 \times 3 \times 3 \times 3 = 113$$

Data Analysis and Statistics

Graphs are used to illustrate data in a picture-like format. It is often easier to understand data when they are shown in a graphical form instead of a numerical form in a table. Common types of graphs are bar graphs, line graphs, and circle graphs.

A *bar graph* organizes information along a vertical axis and horizontal axis. The vertical axis runs up and down one side; the horizontal axis runs along the bottom.

A *line graph* also organizes information on vertical and horizontal axes; however, data are graphed as a continuous line rather than a set of bars. Line graphs are often used to show trends over a period of time.

A *circle graph* looks like a divided circle and shows how a whole object is cut up into parts. Circle graphs are also called *pie charts* and are often used to illustrate percentages.

Example

A business shows the following balances in its cash account for the months of March through July. These data are illustrated below in bar and line graphs.

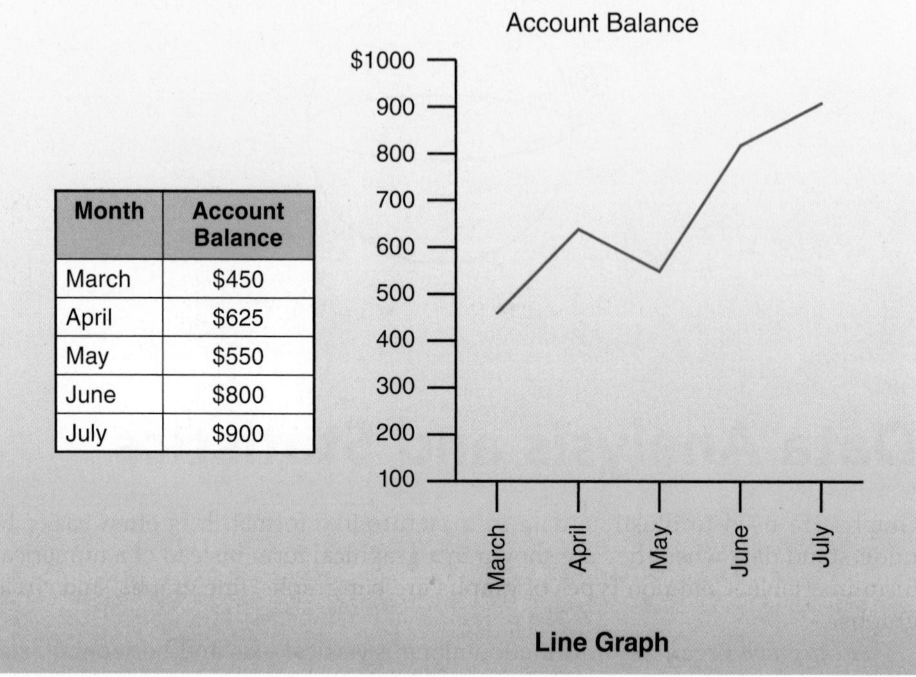

Month	Account Balance
March	$450
April	$625
May	$550
June	$800
July	$900

Line Graph

Example

A business lists the percentage of its expenses in the following categories. These data are displayed in the following circle graph.

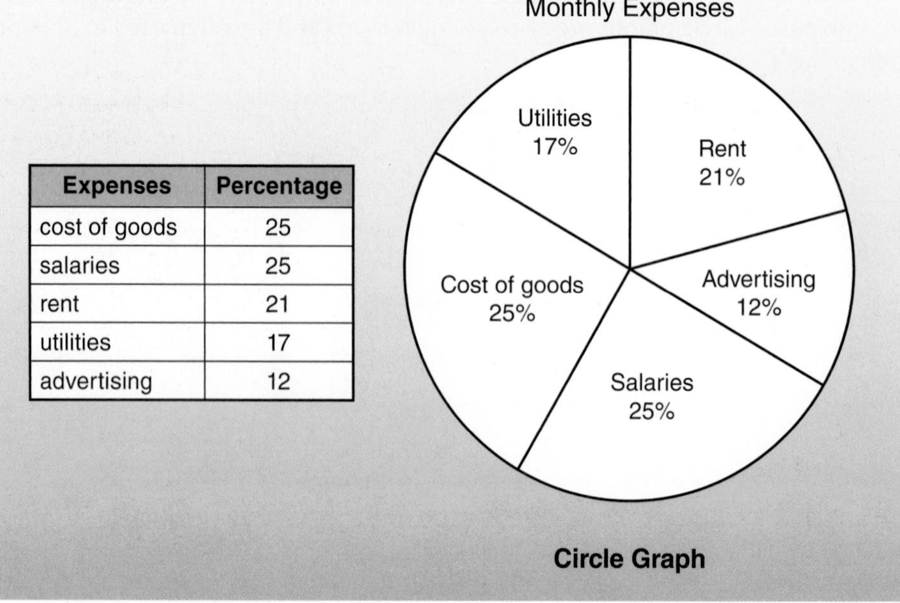

Expenses	Percentage
cost of goods	25
salaries	25
rent	21
utilities	17
advertising	12

Circle Graph

Math Models for Business and Retail

Math skills used in business and retail are the same math skills required in everyday life. The ability to add, subtract, multiply, and divide different types of numbers is very important. However, this type of math is often focused on prices, taxes, profits, and losses.

Markup

Markup is a retailing term for the amount by which price exceeds the cost. One way to express markup is in dollars. Another way to express markup is percentage. The *markup percentage* is the amount of the markup as a percentage of the cost.

Example

A retailer pays $4 for a pair of athletic socks and prices them for sale at $7. The dollar markup is $3.

selling price – cost = dollar markup

$7 – $4 = $3

Example

A pair of athletic socks, which costs $4, is priced at $7. The dollar markup is $3. To find the markup percentage, divide $3 by $4. The markup percentage is 75%.

markup dollars ÷ cost = markup percentage

$3 ÷ $4 = 0.75 = 75%

Percentage Markup to Determine Selling Price

The selling price of an item can be determined if you know the markup percentage and the cost. First, convert the markup percentage to a decimal. Next, multiply the cost by the decimal. Then, add the markup dollars to the cost to determine the selling price. Another way to find the selling price is to convert the markup percentage to a decimal and add 1.0. Then multiply this amount by the cost.

Example

A pair of athletic socks costs $4, which the retailer marks up by 75%. Find the selling price.

1. Convert the markup percentage to a decimal.

   75% = 0.75

2. Multiply the cost by the markup.

   cost × markup = dollar markup

   $4 × 0.75 = $3

3. Add the $3 markup to the $4 cost to find the selling price. The selling price is $7.

   $4 + $3 = $7

Example

A pair of athletic socks costs $4, which the retailer marks up by 75%. Find the selling price.

1. Convert the 75% markup percentage to 0.75 and add 1.0.

   0.75 + 1.0 = 1.75

2. Multiply 1.75 by the $4 cost to find the selling price.

   $4 × 1.75 = $7

Markdown

A *markdown* is the amount by which the selling price of an item is reduced. Sometimes a markdown is also called a *discount*. To find the amount of a markdown, subtract the new or discounted price from the original price. A markdown can also be expressed as a percentage of the original price. Sometimes this is called a *percentage discount*.

Example

A package of meat at a supermarket is originally priced at $10. However, the meat has not sold and is nearing its expiration date. The supermarket wants to sell it quickly, so it reduces the price to $6. This is a markdown of $4.

selling price – discounted price = dollar markdown

$10 – $6 = $4

Example

A package of meat at a supermarket is originally priced at $10. However, the meat has not sold and is nearing its expiration date. The supermarket wants to sell it quickly, so it marks down the price by $4. The markdown percentage is determined by dividing the $4 markdown by the original $10 price.

markdown ÷ selling price = markdown percentage

$4 ÷ $10 = 40%

Gross Profit

Gross profit is a company's net sales minus the cost of goods sold. *Gross margin* is often expressed as a percentage of revenue.

Example

A wristband manufacturer generated net sales of $100,000 last year. The cost of goods sold for the wristbands was $30,000. The net sales of $100,000 minus the $30,000 cost of goods sold leaves a gross profit of $70,000.

net sales – cost of goods sold = gross profit

$100,000 – $30,000 = $70,000

Example

The gross profit of $70,000 divided by the net sales of $100,000 is 0.70, or 70%.

gross profit ÷ net sales = gross margin percentage

$70,000 ÷ $100,000 = 0.70 = 70%

Net Income or Loss

Net income or loss is a company's revenue after total expenses are deducted from gross profit. Total expenses include marketing, administration, interest, and taxes. A company earns a *net income* when gross profit exceeds expenses. A *net loss* is incurred when expenses exceed gross profit.

Example

A wristband manufacturer had a gross profit of $70,000. In addition, expenses for marketing, administration, interest, and taxes were $50,000. Net income is calculated by subtracting the total expenses of $50,000 from the gross profit of $70,000. The net income was $20,000.

gross profit on sales – total expenses = net income or loss

$70,000 – $50,000 = $20,000

Break-Even Point

A *break-even point* is the number of units a company must sell to cover its costs and expenses and earn a zero profit. Use the following formula to find a company's break-even point:

total costs ÷ selling price = break-even point

Sales Tax

Sales tax is a tax collected on the selling price of a good or service. The sales tax rate is usually expressed as a percentage of the selling price. Sales tax is calculated by multiplying the sale price by the tax rate.

Example

Suppose you buy a T-shirt for $10.00. How much is the sales tax if the tax rate is 5%? Convert 5% to a decimal (.05) and multiply it by the sale price.

sale price × sales tax rate percentage = sales tax

$10 × 0.05 = $0.50

Return on Investment (ROI)

Return on investment (ROI) is a calculation of a company's net profit as a percentage of the owner's investment. One way to determine ROI is to divide net profit by the owner's investment.

Example

Suppose you start a dry-cleaning business with a $100,000 investment, and you earn a $20,000 net profit during the first year. Divide $20,000 by $100,000, which equals a 20% return on your investment.

net income ÷ owner's investment = return on investment (ROI)

$20,000 ÷ $100,000 = 0.20 = 20%

Glossary

A

accumulator: A type of register that stores values for mathematical and logic operations in process. (10)

accuracy: A measurement that expresses how precisely the end effector of a robot is programmed to reach a predetermined point. (5)

acoustical proximity sensor: A device that reacts to sound. Standing sound waves are generated within a cylindrical, open-ended cavity inside the sensor. The presence of a nearby object interferes with these sound waves, which alters the sensor output. (8)

actuator: A motor or valve that converts power into robot movement. (2)

address register: A register compartment that temporarily stores the address of a memory location that is to be accessed. It is programmable in some units, which means that new instructions may alter its contents. (10)

alternating current (ac): A source of electrical power in which electrons flow in one direction and then change course to flow in the opposite direction. (6)

analog information: A term that describes data or signals that vary continuously or devices that use this type of data or signal. (10)

AND gate: A type of logic circuit that has two or more inputs and one output. (10)

angstrom (Å): A unit of light measurement, equal to one-tenth of a nanometer. (8)

anthropomorphic: An object that is humanlike in form or has human attributes. (1)

arithmetic logic unit (ALU): A component of a microprocessor that performs mathematical and logic operations. It works automatically using signals sent from the instruction decoder. (10)

armature: The electromagnet in an electric motor that rotates within a magnetic field and creates torque. (6)

artificial intelligence (AI): The ability of a computer program to make decisions based on known information. (1)

automated guided vehicle (AGV): A computer-controlled, battery-operated transportation device that operates using one of several navigation options, such as a buried wire guidepath, a magnetic tape guidepath (inertial guidance), laser target guidance, or a global positioning system (GPS). (5)

automatic tool changer: A device that has more than one end effector and can change end effectors when needed to execute operations. (9)

automaton: An object made by a person that moves automatically. This term was originally used for what we now consider to be a robot. (1)

avoidance cost: An investment made in equipment or other business/production assets that is projected to prevent the need to spend additional money in the future. (13)

Note: The number in parentheses following each definition indicates the chapter in which the term can be found.

B

bifilar construction: A method of winding stator coils in which two separate wires are wound into the coil slots at the same time. The two wires are small, permitting twice as many turns as with a larger wire. (6)

binary-coded-decimal (BCD) number system: A method of counting in which four binary digits are used to represent each decimal digit; devised to simplify the use of large binary numbers. (10)

binary counter: A device used to compute or tally numeric information in binary form. (10)

binary logic circuit: An arrangement of bistable devices that makes logical decisions based on input signals. (10)

binary number system: A counting method that uses 2 as the base. The largest value that can be expressed by a specific place is the number 1; therefore, only the numbers 0 or 1 are used. (10)

binary point: The symbol used to separate the integer part of a number from the fractional part of a number in a base 2 numbering system. (10)

bistable: An electronic device that can be set in one of two operational states by an outside signal and can store one binary digit or bit of information. (10)

bit: A single pulse of voltage processed by a computing system. Also called a *binary digit*. (10)

bootloader: A small program that helps transfer program code from a computer to a microcontroller without the need for removing the microcontroller from the microcontroller board. (15)

breadboard: A construction base used for building samples or models of electronics. (15)

brushes: Carbon devices that rub against the commutator in an electric motor. When power passes through the brushes and commutator on its way to the armature, magnetic fields are created. (6)

bus network: A series of registers connected together. (10)

byte: A group of eight bits. Also called a *binary word*. (10)

C

capacitance: The ratio of charge on a conductor to the potential difference between conductors or the ability of a material to hold an electrical charge. (8)

capacitive transducer: A device that measures a change in capacitance, or the ability of a material to hold an electrical charge. (8)

capital investment: The money that a business spends to purchase fixed assets, such as buildings or equipment. (13)

Cartesian configuration: A type of layout that allows the movement of a robot along three intersecting, perpendicular, straight lines that are referred to as the X, Y, and Z axes. (2)

central processing unit (CPU): The part of a computer that reads program instructions, processes data, and carries out programmed operations. (10)

centrifugal pump: A non-positive displacement pump that moves a varying amount of fluid with each rotation using an impeller blade. (7)

circuit breaker: A component of an electrical system that provides overload protection by interrupting the flow of the current when a short circuit occurs or a power surge reaches it. (3)

closed-loop system: A type of control method that allows feedback to affect the output of the robot. (2)

collet gripper: An end effector that delivers 360° of clamping contact and is typically used to pick and place cylindrical parts that are uniform in size. (9)

command resolution: The closest distance between the movements of a robot. It is calculated by dividing the travel distance of each joint by the number of control increments. (5)

commutator: Part of a dc motor that switches current flow. (6)

comparator: A control system element that compares the feedback signal from the controlled element to a reference signal or standard. It develops a correction signal, if needed, which is sent to the control unit. (6)

compiler: A computer program that translates high-level programming languages into machine code that computers can understand. (4)

compliance: The ability of an end effector to tolerate the misalignment of mating parts. (9)

compound-wound dc motor: A type of dc motor that has two sets of field windings. (6)

computer vision sensor: A device that detects spatial relationships and provides depth information using stadimetry and triangulation. (8)

conditioning: The process of removing contaminants, such as dirt and moisture, from hydraulic fluid and air. (7)

continuous-path (CP) motion: The smooth and continuous movement of a robot end effector through many points in space (up to several thousand) to a desired location. (4)

control: Some type of operational change in a system, such as changes in electric current, hydraulic pressure, light intensity, or airflow. (6)

controller: A microprocessor linked to input/output and monitoring devices that coordinates all movements of the mechanical system and receives input from the immediate environment through various sensors. (2)

control unit: A component of a microprocessor that receives decoded instructions from the decoder and initiates the proper action. (10)

cost savings: An investment made to replace existing equipment or assets in order to reduce the money required for current operations. (13)

counter: A logic device that tallies (counts) electronic pulses and is used in a variety of applications. (10)

counter electromotive force (cemf): Voltage generated as the armature of a dc motor rotates. This voltage opposes the voltage applied to the motor. (6)

cycle: Each repeated pattern of change in the direction of the flow of electrons in alternating current power. (6)

cycle timing system: Components that turn a device on or off at specified intervals or in time with an operational sequence. (6)

cylindrical configurations: A type of work envelope layout that consists of two slides placed at a 90° angle and mounted on a rotary axis. (2)

cylindrical grip: A prehensile movement in which the fingers and thumb form a C shape to grasp a cylindrical object. (9)

D

data register: A compartment of a microprocessor that stores information for input to the arithmetic logic unit. It may also hold an instruction while that instruction is being decoded, or it may hold data prior to storage in memory. (10)

dc stepping motor: A type of dc motor in which the rotor has a permanent magnet. These motors are primarily used to change electrical pulses into rotary motion. (6)

decade counter: A device used to change binary signals into a binary-coded-decimal (BCD) form. (10)

decoder unit: A component of a microprocessor that examines coded instructions and decides which operation is to be performed by the ALU. (10)

dedicated equipment: An automated device that is designed to perform only one function. (5)

degrees of freedom: A value that describes a range of motion (as for a robotic arm and end effector) in three-dimensional space—specifically, the ability to move forward and backward, up and down, and to the left and to the right. (2)

delay timing system: Components that provide a lapse in time before the load device becomes energized. (6)

desiccant: A very dry substance designed to attract moisture, which is used in filters. (7)

design for manufacturability: A product-planning process that takes into account the robots that will assemble the products. (5)

detector: A component of a sensing system that responds to energy from the source and outputs a signal that is used to control the load device. (6)

detents: Two or more elements that are held in position by spring-loaded mechanisms. When placed under excessive stress, they shift from their original position and power is removed from the machine to prevent an overload. (3)

digital: A term that describes data or signals in a format that occur in separate full units, such as 1 or 0, or devices that use this type of data, such as a computer. (10)

digital electronics: The technology that controls robotic and other automated systems. (10)

digital input port: The connection through which a robot receives information in the form of digital data. This data is used by the computer program to initiate some action by the robot. (11)

digital output port: The connection through which the robot controller sends digital data to peripheral equipment. (11)

digital system: Components that work together to process numeric information. (6)

direct current (dc): A source of electrical power in which electrons flow in only one direction. (6)

direct-drive motor: A high-torque engine that drives the robot arm directly without the need for reducer gears. (2)

direction control device: An element used to start, stop, or reverse fluid flow without causing a significant change in pressure or flow rate. (7)

dynamic performance: A value that represents how quickly the robot can accelerate, decelerate, and come to a stop at a given point. Also called *operational speed*. (5)

dynamic RAM (DRAM): A random access memory circuit that stores bits of data using separate capacitors. To retain the data, the electrical charge of the capacitors must be constantly refreshed. (10)

E

eddy current proximity sensor: A device that produces a magnetic field in a detector unit. The magnetic field induces eddy currents into any conductive material that is near the probe. A pick-up coil senses a change in magnetic field intensity when an object enters the field. (8)

electric motor: A device that converts electrical energy to mechanical energy in the form of rotary motion. (6)

electrical shock: An injury that occurs when a person comes into contact with electric current. (3)

electromagnetic spectrum: The range of visible and invisible light, which includes frequencies for radio, television, radar, infrared radiation, visible light, ultraviolet light, X-rays, and gamma rays. (8)

electromechanical gripper: An end effector that uses a magnetic field to pick up an object. Also called *magnetic gripper.* (9)

electromechanical system: Components that transfer power from one point to another through mechanical motion that is used to do work. (6)

electronically erasable programmable read-only memory (EEPROM): A type of data storage in a computer chip that can be erased one bit at a time (using a high electrical charge) and reprogrammed. (10)

end effector: A device that is attached to the end of the manipulator (robot arm) that operates to grasp and manipulate objects. Also called the *end-of-arm tooling.* (2)

end stop: Devices placed on an axis to physically block movement past a certain point along the axis. (4)

engineering economics: The analysis of the monetary impact and technical aspects of equipment or other projects. (13)

erasable programmable read-only memory (EPROM): A type of data storage in a computer chip that can be completely erased (by exposing it to an ultraviolet light) and reprogrammed. (10)

error detector: A component of servo systems that receives data from both the input source and the output device and compares that data to determine if a correction is needed. (6)

error signal: A message generated when the programmed actions of a robot and the actual actions do not match. (2)

execute: An initiate operation that is based on the instruction that is to be performed. (10)

expandable gripper: An end effector that clamps irregularly shaped workpieces using mechanical fingers equipped with hollow rubber envelopes that enlarge when pressurized. (9)

expert system: A program that contains a knowledge base and uses artificial intelligence techniques to offer advice or alternatives for making decisions. (14)

F

feedback: A signal or data that provide information about the interaction between the control unit and the controlled element. (6)

fetch: A program operation that retrieves the next instruction from memory. (10)

field winding: The coil wrapped around the electromagnet inside a dc motor. (6)

filter: A conditioning device that removes very small pieces of debris and is typically made of some porous medium, such as paper, felt, or very fine wire mesh. (7)

fire extinguisher: A safety device, typically a pressurized container, that contains a substance that can be used to put out flames. (3)

firmware: Programs that are held in a read-only memory. (10)

first aid kit: Supplies and equipment used to treat a person in the event of an accident or injury. (3)

fixed-sequence robot: A manipulator that repetitively performs successive steps of a given operation according to a predetermined order, condition, and position. Its instructions cannot be easily changed. (1)

fixturing: The ability of a robot to provide precise points in the process of accurately locating parts. (5)

flexible automation: Machinery capable of performing a variety of tasks. (1)

flip-flop: A memory device used in digital circuits that can hold an output state even when the input is completely removed and can change the output based on an appropriate input signal. (10)

flow control device: A element that alters the volume or flow rate of a fluid. (7)

flow indicator: A device used to test flow rates from pumps and at the inlet and outlet ports of actuators. (7)

fluid motor: A device that converts the force of a moving fluid into rotary motion through the use of vanes, gears, or pistons. (7)

fluid power system: A power system that uses air or liquid, or a combination of both, to transfer power. (7)

flux: Strength of the electromagnetic field within a dc motor; directly affected by changes in the field current. (6)

force: Any factor that tends to produce or modify the motion of an object; normally expressed in units of weight. (7)

FRL unit: A conditioning device used in many pneumatic systems that combines the air filter, regulator, and lubricator components. (7)

fuse: A device that provides overload protection in an electrical system. When a power surge reaches a fuse, an element in the fuse melts, interrupting the flow of current. (3)

G

general-purpose interface bus (GPIB): A cable that interfaces system controllers with programmable instruments and can support a maximum of 15 devices. (11)

gripper: An end effector that performs prehensile movements by grasping objects and moving them. (9)

grounding: Providing a path that allows current to flow from an electrical system or device to the ground in the event of a short circuit or power surge to protect against electrical shocks. (3)

H

hand tool: A device or implement used to perform work that is held by the hands and does not use electricity or some other type of power. (3)

hard automation: Machinery specifically designed and built to perform particular tasks within an assembly line. (1)

heat exchanger: A device that cools the fluid in a hydraulic system to maintain a constant fluid temperature. (7)

hexadecimal number system: A base 16 number system that is used to process large numbers. The largest number used in a place is 15, with digits 0–9 and letters A–F. (10)

hierarchical control: A system that organizes commands and instructions from the software to the parts and devices of the robot into levels or layers. Each level is dependent on the level above for instructions. (2)

hierarchical control programming: A method of creating instructions for a robot in which the commands are structured in levels. Each program level accepts commands from the level above and responds by generating simplified commands for the level below. (4)

high-level language: A programming language that more closely resembles standard English than traditional programming code. (4)

hook movement: A nonprehensile hand movement that involves curling the tips of the fingers to pull or lift objects. (9)

humanoid robot: A robot with the human features of two arms, two legs, a torso, and a head that is capable of receiving, processing, and responding to information from the environment. (14)

hydraulic drive: A power system that uses fluid and consists of a pump connected to a reservoir tank, control valves, and an actuator. (2)

I

image acquisition: A function of machine vision in which a workpiece is illuminated and an illustration or picture is digitally created. (11)

image analysis: A function of machine vision in which information from an illustration or picture is gathered and analyzed by computer software. Using algorithms, the software identifies and measures features of the digital picture. (11)

image interpretation: A process that involves comparing a scanned image to a model image or to data that relates to a model and deciding whether the scanned image is a match to the model image. The robot makes decisions about tasks that must be performed based on the decision. (11)

image preprocessing: A function of machine vision in which an analog-to-digital converter changes the analog signal of the image into an equivalent digital signal. The digital signal represents light intensity values over the entire image. (11)

indicator: A subsystem that displays information about operating conditions at various points throughout the system. (6)

inductance: A property of electrical circuits caused by the magnetic field that surrounds a coil when current is flowing. In ac circuits, it opposes changes in current and increases as frequency increases. (8)

inductive transducer: A device with a stationary coil and a movable core that measures movement and creates signals that affect current flow. (8)

industrial robot: A multifunction manipulator programmed to perform various tasks. (1)

inertia: The resistance to change of an object to be moved. (7)

infrared sensor: A device that responds to radiation in the infrared region of the electromagnetic spectrum. (8)

input/output (I/O) port: A connection through which digital signals used for communication travel from system to system. (11)

input-output (I/O) transfer: A basic computer system function in which an I/O port actuates to either receive data from an input or send data to an output device. (10)

intelligent robot: A robot that can, by itself, detect changes in the work environment by means of sensors (visual and/or tactile). Using decision-making capabilities, the robot can then proceed with the appropriate operations. (1)

interface: The common point at which two or more systems communicate with each other. (11)

interlocks: Safety devices that are designed to prevent unauthorized access to hazardous areas by requiring a key for entry. (5)

interrupt: A signal that originates from peripheral equipment to inform the microprocessor unit that the device needs attention. (9)

interval timing system: Components that are used after a load is energized and operate using specified time periods. (6)

inverter: A type of logic gate that has one input and one output. The output is opposite to the input. Also called a *NOT gate*. (10)

L

laser: A device that produces a highly focused or intense beam of light. Also known as *light amplification by stimulated emission of radiation*. (8)

laser interferometric gauge: A range-sensing system that is sensitive to humidity, temperature, and vibration. (8)

lateral grip: A prehensile movement in which the fingers and thumb grasp flat objects from the sides, rather than around the object. (9)

light curtain: A safety device that consists of a series of photoelectric sensors. If the beam of light between two sensors is interrupted, the power to a nearby machine, such as a robot, is stopped. (3)

light-emitting diode (LED): A small, lightweight opto-electronic device (solid-state lamp) that produces different colors of light. (8)

light pipe: A fiber-optic rod used in a transmission path to direct light energy around corners or from its source to a distant location. (6)

limit switch: An electrical device with an actuator that is mechanically linked to a set of contacts. When activated, contacts that are normally open are closed, and contacts that are normally closed are opened. This causes a change of state to an electrical circuit. (8)

linear actuator: A type of motor or valve that provides motion along a straight line. (2)

load: The part or parts of an automated system designed to produce work. (6)

lockout/tagout procedures: Practices used to disable machinery and prevent injury to workers from hazardous energy releases. (3)

logic circuit: The electronic circuitry used with robots that provides commands to control motors, hydraulic systems, and pneumatic systems. These circuits also store information, count, encode, and decode. (10)

logic gate: A basic binary circuit that passes or blocks certain signals (AND circuit, OR circuit, and NOT circuit). (10)

lubricator: A conditioning device used in pneumatic power systems that adds a small quantity of oil to the air after it leaves the regulator. (7)

M

machine language: A code consisting of binary instructions (1s and 0s) that a microprocessor can run or respond to directly. (15)

machine vision system: Video systems that are built into automated work cells and are used for applications, such as guidance, part orientation, measuring and inspecting parts, and image identification. (11)

magnetic field sensor: A device that identifies a change in an existing magnetic field without making physical contact with objects in the environment. (8)

magnetic gripper: An end effector that uses a magnetic field to pick up an object. Also called *electromechanical gripper.* (9)

manipulator: The arm of the robot that moves materials, parts, tools, or special devices through various motions to provide useful work. (1)

manual manipulator: A manipulator worked by a human operator. (1)

manual programming: A type of machine setup in which an operator adjusts the necessary end stops, switches, cams, electric wires, or hoses to set up the movement sequence for a robot. (4)

manual rate control box: A device that consists of a knob and some switches used to move each axis of a robotic arm in the process of programming a path of motion. (4)

mechanical finger gripper: An end effector used for grasping parts within a confined space, reaching into channels, or picking and placing any object that has a simple shape. (9)

mechanical fuse: An overload protection device in the form of a pin or tube that breaks or buckles under extreme stress. (3)

memory: The data storage capacity of a computer that allows information to be saved and retrieved. (10)

metering: Controlling the rate of fluid flow. (7)

microbot: A tiny robot that can be used in applications such as medical surgery. (14)

microcontroller: A small computer on a single integrated circuit chip. (15)

microprocessor unit (MPU): A microchip that combines the arithmetic logic unit and control circuitry of the computer. (10)

microswitch: An electrical device that is turned on or off with a very small amount of force. Microswitches are small, durable, and easy to activate. (8)

motor controller: A small integrated circuit chip that will allow control of the direction of rotation of two motors independently as well as enable and disable rotation in each motor. (15)

MPU cycle: The time frame in which system operations are performed. (10)

N

NAND gate: A NOT gate combined with an AND gate. (10)

nanometer (nm): A unit of light measurement that is one billionth of a meter. (8)

National Electrical Code (NEC): A set of rules that covers safety related to electrical installation in residential and commercial structures. (3)

non-positive displacement pump: A type of fluid pump that moves fluid with each rotation of the impeller blades. The amount of fluid that passes through the pump with each rotation varies. (7)

nonprehensile movement: An action (of the hand or an end effector) that does not require finger dexterity or use of the thumb. Examples include pushing, poking, punching, and hooking. (9)

non-servo robot: The simplest type of robot, which uses an open-loop system that does not allow for system feedback. Also called *limited sequence, pick-and-place*, or *fixed-stop robots.* (2)

NOR gate: A NOT gate combined with an OR gate. (10)

NOT gate: A type of logic circuit that has one input and one output. The output is opposite to the input. Also called an *inverter.* (10)

numerically controlled (NC) robot: A manipulator that can perform the sequence of movement, conditions, and positions of a given task, which are communicated by means of numerical data. (1)

O

Occupational Safety and Health Administration (OSHA): An agency of the US federal government whose purpose is to ensure safe and healthful conditions for workers by setting and enforcing standards related to safety. (3)

octal number system: A base 8 number system used to process large numbers. The place values of digits, moving to the left of the octal point, are powers of eight. (10)

off-line programming: Creating the instructions for a robot on a computer that is not connected to the robot. (4)

on-line programming: Creating the instructions for a robot on a computer that is connected to the robot console. (4)

open-loop system: A type of control method that does not include a feedback mechanism to compare programmed positions to actual positions. (2)

operational speed: A value that represents how quickly the robot can accelerate, decelerate, and come to a stop at a given point. Also called *dynamic performance.* (5)

oppositional grip: A prehensile movement in which the tip of the index finger and thumb hold an object. (9)

optical fibers: Strands of material made of glass or plastic that can transmit light from one point to another regardless of how the material is bent or shaped. (8)

optical proximity sensor: A device that measures the amount of light reflected from an object and can respond to either visible or infrared light. (8)

opto-electronic: Using a combination of optical and electronic components. (8)

OR gate: A type of logic circuit that has two or more inputs and one output. (10)

overload sensor: A device used on an end effector that detects obstructions or overload conditions within fractions of a second and signals the controller to shut down the robot before damage occurs. (9)

P

palmar grip: A prehensile movement in which the fingers and thumb wrap around an object to grasp it. (9)

parallel port: A type of interface connection through which the robot controller sends digital data to peripheral equipment using parallel transmission. (11)

parallel transmission: An interface method in which multiple bits of data are sent at the same time, following side-by-side paths, like the lanes of a highway. (11)

Pascal's law: A principle of fluid properties that states when pressure is applied to a confined fluid, the pressure is transmitted undiminished throughout the fluid. Additionally, this pressure acts on all surfaces of the container at right angles to those surfaces. Discovered by Blaise Pascal (1653). (7)

payback period: The time required to recover the money spent for new equipment through increased profits or savings in labor or materials. (13)

payload: The maximum weight or mass of material a robot is capable of handling on a continuous basis. (5)

period: The time interval for a pulse to pass through a complete MPU cycle from beginning to end. (10)

permanent-magnet dc motor: A type of dc motor in which the dc power supply is connected directly to the conductors of the rotor through the brush-commutator assembly. This type of motor is used when a low amount of torque is needed. (6)

personal computer (PC): A small desktop or portable computing unit built around a single integrated circuit. (10)

personal protective equipment (PPE): Clothing or devices worn to shield the worker's body and help prevent injuries, such as boots with steel toes, safety glasses, earplugs, gloves, respirators, and hard hats. (3)

photoconductive device: An opto-electronic element that varies in conductivity according to fluctuations in light. Electrical resistance decreases when light is more intense and increases when light intensity decreases. (8)

photoemissive device: An opto-electronic element that emits electrons in the presence of light. (8)

photovoltaic device: An opto-electronic element that converts light energy into electrical energy. Also called a *solar cell.* (8)

pick-and-place motion: Movement of the end effector of a robot to the correct position or through a fixed pattern of movement. (4)

piezoelectric effect: The characteristic of certain crystals to develop an electrical potential when subjected to mechanical stress. Rochelle salt and quartz are examples of these crystals. (8)

pitch: The up-and-down movement of an object, such as a robot's wrist. (2)

pixel: Abbreviation for picture elements, which are dots of light that form an image. (11)

place value: The position of a digit in a number system with respect to the decimal point. (10)

playback robot: A manipulator that can reproduce operations originally executed under human control and stored in memory. (1)

pneumatic drive: A power system that uses air-driven actuators. (2)

point-to-point (PTP) motion. The movement of a robot end effector through a number of points in space to a desired location. (4)

positive displacement pump: A type of fluid pump that moves a definite amount of fluid through the pump during each revolution. (7)

positive logic: A binary system in which the digits 1 and 0 represent two operational conditions. The value of 0 indicates low voltage or no voltage. The number 1 indicates a voltage larger than 0. (10)

power: A measurement that considers the amount of work accomplished in relation to the amount of time taken to perform the work. (7)

power supply: A mechanism that provides energy to drive the robot's controller and actuators. (2)

power tool: A device or implement that is used to perform work and that uses electricity or some other type of energy. (3)

prehensile movement: An action (of the hand or an end effector) that requires the use of the thumb and the fingers to grasp objects. (9)

preloaded spring: An overload protection device that responds to excess stress by releasing a spring that causes the end effector on a robot to break away from the work area. (3)

pressure: The amount of force applied to a specific area; often expressed in pounds per square inch (lb/in^2 or psi) or as kilopascals (kPa) in the metric system. (7)

pressure indicator: A device that monitors the fluid pressure within a hydraulic system. (7)

pressure regulator valve: A device that allows the air pressure in a pneumatic system to be adjusted to a specific level. (7)

pressure relief valve: A type of valve that controls the pressure in a system. In a hydraulic system, the pressure relief valve allows fluid to flow back into the reservoir when the pressure rises to a dangerous level. In a pneumatic system, pressure relief valves release excess air in the system into the atmosphere. (7)

pressure sensitive safety mat: A floor covering that contains sensors that detect force applied to it. If a person steps on this, a signal is sent to power down a nearby machine, such as a robot. (3)

preventive maintenance: Performing tasks on a regular schedule that are designed to keep equipment in good working order and avoid breakdowns and loss of production. (12)

prime mover: The component of a power system that provides the initial power for movement in the system. (7)

program: A series of instructions stored in the controller's memory that direct the robot's actions. (2)

program counter: A memory device that indicates the location in memory of either the instruction currently being performed or the next instruction to be executed. (10)

programmable read-only memory (PROM): A type of read-only computer chip that has data electrically burned onto it and cannot be reprogrammed. (10)

proximity sensor: A device that detects either the absence of an object or the presence of an object within a certain distance. (8)

pseudocode: A simple language used to describe instructions and actions that will be part of a computer program; not a programming language. (15)

pulse width modulation (PWM): A method of digitally encoding analog signal levels. (15)

R

radial traverse: One of the three degrees of freedom in the robot arm, which involves the extension and retraction of the arm and creates in-and-out motion relative to the base. (2)

random access memory (RAM): A type of computer system data storage that can be altered. Information may be retrieved from and written to random access memory. (10)

range sensor: A device that can determine the precise distance from the sensor to an object. (8)

read-only memory (ROM): A type of computer system data storage that is permanent and cannot be changed. (10)

reciprocating pump: A type of positive displacement pump that uses the reciprocating action (backward and forward motion) of a moving piston to move fluid into and out of a chamber. (7)

rectification: The process of converting alternating current (ac) to direct current (dc) power. (6)

reed switch: A simple type of magnetic field sensor that either makes or breaks contact in response to changes in a magnetic field. (8)

register unit: A storage component of a microprocessor consisting of several compartments that store information. Each compartment stores a specific type of data. (10)

remote center compliance (RCC) device: An element that is installed in the wrist of the robot to help compensate for workpiece misalignment or irregularities. (9)

repeatability: A value that expresses the variation in movements that occurs when an action or task is performed several times. (5)

reprogrammable: Having the ability to be given new instructions, as in a robot that can meet changed requirements to perform new tasks. (1)

resistance: Friction that forms in a fluid system and causes a decrease of power from the input to the output. (7)

resistive transducer: A device that converts a variation in resistance into electrical variations. (8)

resolution: The smallest incremental movement a robot can make, which is determined by the control system for the robot. (5)

revolute configuration: A type of layout that is irregularly shaped and requires a jointed-arm robot. Also called the *jointed-arm configuration.* (2)

rework: The process of fixing manufactured parts that do not meet product specifications. (12)

robot: A machine or device that automatically performs tasks or activities that are typically completed by a person using skill and intelligence. (1)

robotics: The study, engineering, and use of robots. (1)

roll: The swivel or rotating movement of an object, such as a robot's wrist. (2)

rosin-core solder: A type of alloy used for joining fusible metals that is noncorrosive and conductive and that is used on electronic circuits. (12)

rotary actuator: A type of motor or valve that provides rotation, moving a load in an arc or circle. (2)

rotary gear pump: A type of positive displacement pump that uses rotary motion to produce pumping action. (7)

rotary vane pump: A type of positive displacement pump that uses a series of sliding vanes to move fluids. (7)

rotational traverse: One of the three degrees of freedom in the robot arm, which involves movement on a vertical axis and allows the side-to-side movement of the robot arm on the base. (2)

rotor: The rotating component of a motor that includes the armature, shaft, and associated parts. (6)

S

SCARA configuration: A robot arm configuration that is horizontally articulated and generally has one vertical (linear) and two revolute joints. Also known as *selective compliance assembly robot arm.* (2)

scrap: A manufactured part that cannot be fixed and must be discarded. (12)

sensing system: Components that signal a response to a particular form of energy, such as light. (6)

sensory feedback: Information about the environment that is received through various input devices (sensors). (4)

serbot: A service robot, or a mobile robot that can move to a work area to perform the necessary tasks. (14)

serial port: A type of interface connection through which a computer sends or receives digital data using serial transmission. (11)

serial transmission: An interface method in which data is sent one bit at a time over long distances. (11)

series-wound dc motor: A type of dc motor in which the armature and field circuits are connected in a series arrangement. This is the only type of dc motor that also can be operated using ac power. Also called a *universal motor.* (6)

service robot: A mobile robot that can move to a work area to perform the necessary tasks. (5)

servo amplifier: A device that translates feedback signals from the controller into motor voltage and current signals. (2)

servomotor: A component of a servo system that produces controlled shaft displacements to achieve a precise degree of rotary motion. (6)

servo robot: A type of robot that uses a closed-loop system, which allows feedback signals to affect the performance of the robot. (2)

servo system: Machines that change the position or speed of a mechanical object in response to system feedback or error signals. (6)

shunt-wound dc motor: A type of dc motor in which the field coils are connected in parallel with the armature and have relatively high resistance. (6)

single-phase ac motor: A type of ac motor that operates using a single-phase ac power source. (6)

single-phase induction motor: A type of ac motor that has a solid rotor and must be set into motion by some auxiliary starting method. (6)

slip: The difference between the synchronous speed and the rotor speed of a motor. (6)

software: A program that is transferred to a computer system by a keyboard, CD-ROM, or download, and the instructions are stored in read/write memory. (10)

solder sucker: A device that creates suction to remove unwanted solder from a printed circuit board without damaging the board. (12)

sound sensor: A device that relies on the piezoelectric effect to convert sound to electrical energy. (8)

spatial resolution: A value that expresses the accuracy of movement of the tool tip of a robot. Spatial resolution takes into account command resolution and mechanical inaccuracy. (5)

speed sensing: Measuring the rotary motion of shafts, gears, pulleys, and other rotating components of industrial equipment. (8)

spherical configuration: A type of work envelope layout that resembles a sphere, relative to the base. Also called *polar configuration*. (2)

spherical grip: A prehensile movement in which the fingers and thumb are used to hold round objects. (9)

spread movement: A nonprehensile movement of the hand in which the fingers and thumb are extended outward until they make contact with the interior walls of a hollow object. The force of the fingers against the walls of the object allows it to be picked up and carried. (9)

squirrel cage rotor: The solid rotor in a single-phase induction motor. (6)

stadimetry: A method used to determine depth that considers the distance to an object based on the apparent size of the camera image. (8)

static RAM (SRAM): A random access memory circuit that uses semiconductor devices, called flip-flops, to store data. (10)

stator: The stationary portion of a motor that includes the permanent magnets (or electromagnetic windings), a frame, and other stationary components. (6)

strainer: An in-line device that captures large particles of foreign matter within hydraulic systems. (7)

strain gauge: A sensor made of fine-gage resistance wire mounted on a strip of insulation that measures mechanical movement. (8)

subroutine: A set of instructions within programming code that has an independent beginning and end. (4)

subsystem: Various components that work together to form a larger unit, such as inputs, controls, and specialized machinery for robotic systems. (6)

synchronous motor: A type of servomotor that is comprised of a rotor and a stator assembly. It contains no brushes, commutators, or slip rings. (6)

synchronous speed: Speed of the rotating magnetic field of an ac motor. (6)

synthesized system: The combination of subsystems that create a cohesive, operational process or unit. (6)

system: A combination of components that work together to form a unit. (6)

T

tachometer: A device used to measure the speed of rotation of an object, such as the acceleration and deceleration of a robot's manipulator (arm) movements. (2)

tactile sensor: A device that uses touch to determine whether an object is present. (8)

task-level programming: A method of creating instructions for a robot in which the goals of each task, rather than the motions or points, are entered using simple English-like terms. (4)

teach pendant: A device used to demonstrate the movements required for a robot to perform a useful task. Also called a *teach box* or *handheld programmer.* (2)

teach pendant programming: A method of creating instructions for a robot in which the operator moves the robot through the various positions involved in an operation. The points are recorded and used to generate a point-to-point path that the robot will follow during operation. (2)

telerobotics: A field of engineering that develops and produces robot arms and mobile robots that are operated by remote control (radio transmitter and receiver). (14)

thermistor: A resistor commonly used for heat sensing. The resistance of a thermistor decreases as temperature increases (and vice versa). (8)

thermocouple: A device that converts heat energy into electrical energy. (8)

thermoelectric sensor: A device that produces a change in electrical output in response to fluctuations in temperature. (8)

three-phase ac motor: A type of motor that uses a three-phase ac power source. (6)

three-phase induction motor: A type of motor that uses a three-phase ac power source and has a squirrel cage rotor. (6)

three-phase synchronous motor: A type of motor in which direct current is applied to the wound rotor to produce an electromagnetic field and three-phase ac power is applied to the stator. (6)

timing system: Components that turn a device on or off at a specific time or in step with an operating sequence. (6)

tool: An end effector that executes nonprehensile movements to perform specific tasks, such as welding or painting. (9)

torque: A twisting force that causes rotation, such as the rotary motion, or turning force, that is produced by the armature in a motor. (6)

touch-sensitive proximity sensor: A device that operates on capacitance developed by a large conductive object (such as the human body). This capacitance changes the frequency of an electronic circuit. A conductive plate or rod may be used to sense contact. (8)

trajectory: Path of movement. (2)

transducer: A part of a sensor that converts light, heat, or mechanical energy into electrical energy. (8)

transmission path: A subsystem that provides a channel for the transfer of energy. (6)

triangulation: The process of measuring angles and the baseline of a triangle to determine the position of an object. (8)

troubleshooting: The systematic process of identifying and correcting problems in an operation or system. (12)

truth table: In computer circuitry, a chart that displays combinations of inputs and resulting outputs of a logic gate. (10)

turbulence: Irregular motion or agitation of air or fluids. (7)

U

ultraviolet sensor: A light-sensing device that responds to electromagnetic radiation in the ultraviolet range. (8)

Unimate: The first robot produced by Unimation Inc. in 1961. (1)

universal motor: A motor that can be powered by either an ac power source or a dc power source and is built like a series-wound dc motor. (6)

V

vacuum gripper: An end effector that uses suction to pick up an object. The effector has one or more suction cups that attach to the surface of a workpiece to grip. (9)

variable-sequence robot: A manipulator that repetitively performs successive steps of a given operation according to a predetermined order, condition, and position. The order of movement it follows can be easily changed. (1)

vertical traverse: One of the three degrees of freedom in the robot arm, which allows the up-and-down motion of the robot arm. (2)

voice recognition: An input system in which recognizable words or phrases are used as a form of audio data entry for a computer. Also called *speech recognition*. (4)

W

walk-through programming: A method of recording instructions for continuous-path robots in which an operator physically moves the end effector through the desired motions and points (locations) are recorded into memory. (4)

WAVE: The first programming language for robots, which was developed at the Stanford Artificial Intelligence Laboratory in 1973. (4)

weight: The gravitational force exerted on a body by the earth; expressed in pounds (lb) in the US customary measurement system and in newtons (N) in the metric system. (7)

work: An action that occurs when energy is transformed into mechanical motion, heat, light, chemical action, or sound. (6)

work envelope: The area within reach of a robot's end effector or end-of-arm tooling. (2)

X

X-rays: An invisible band of radiation within the electromagnetic spectrum between ultraviolet and gamma rays. (8)

Y

yaw: The side-to-side movement of an object, such as a robot's wrist. (2)

Index

D

J

L

M

N

O

P

R